Smart Nanomaterials Technology

Nanotechnology is a rapidly growing scientific field and has attracted a great interest over the last few years because of its abundant applications in different fields like biology, physics and chemistry. This science deals with the production of minute particles called nanomaterials having dimensions between 1 and 100 nm which may serve as building blocks for various physical and biological systems. On the other hand, there is the class of smart materials where the material that can stimuli by external factors and results a new kind of functional properties. The combination of these two classes forms a new class of smart nanomaterials, which produces unique functional material properties and a great opportunity to larger span of application. Smart nanomaterials have been employed by researchers to use it effectively in agricultural production, soil improvement, disease management, energy and environment, medical science, pharmaceuticals, engineering, food, animal husbandry and forestry sectors.

This book series in Smart Nanomaterials Technology (SNT) publishes cutting edge research monographs and edited volumes by top researchers and professionals across the globe. This book series aims to comprehensively cover topics in the fabrication, synthesis and application of these materials for applications in the following fields:

- Energy Systems—Renewable energy, energy storage (supercapacitors and electrochemical cells), hydrogen storage, photocatalytic water splitting for hydrogen production
- Biomedical—controlled release of drugs, treatment of various diseases, biosensors
- Agricultural—agricultural production, soil improvement, disease management, animal feed, egg, milk and meat production/processing
- Forestry—wood preservation, protection, disease management
- Environment—wastewater treatment, separation of hazardous contaminants from wastewater, indoor air filters

Review Process

The proposal for each volume is reviewed by the main editor and/or the advisory board. The chapters in each volume are individually reviewed single blind by expert reviewers (at least two reviews per chapter) and the main editor.

Ethics Statement for this series can be found in the Springer standard guidelines here - https://www.springer.com/us/authors-editors/journal-author/journal-author-helpdesk/before-you-start/before-you-start/1330#c14214

Rakesh Kumar Bachheti
Editor

Nanotechnology in the Pulp and Paper Industry

Springer

Editor
Rakesh Kumar Bachheti
Department of Research Analytics
Saveetha Dental College & Hospitals
Saveetha Institute of Medical and Technical
Sciences, Saveetha University
Chennai, Tamil Nadu, India

ISSN 3004-8273 ISSN 3004-8281 (electronic)
Smart Nanomaterials Technology
ISBN 978-981-95-3472-2 ISBN 978-981-95-3473-9 (eBook)
https://doi.org/10.1007/978-981-95-3473-9

This Springer imprint is published by the registered company Springer Nature Singapore Pte Ltd.
The registered company address is: 152 Beach Road, #21-01/04 Gateway East, Singapore 189721, Singapore

Preface

The pulp and paper industry, historically recognized as a cornerstone of global manufacturing and knowledge dissemination, is undergoing a major transformation driven by nanotechnology. The increasing demand for sustainable, high-performance, and multifunctional paper products has catalyzed the adoption of nanoscale innovations that are redefining the capabilities, efficiency, and environmental footprint of paper-based materials. This book, *Nanotechnology in the Pulp and Paper Industry*, provides an in-depth examination of how nanoscience and nanomaterials are revolutionizing traditional pulp and paper processes, leading to advanced applications and sustainable solutions.

The book begins with a broad overview of nanotechnology applications within the pulp and paper sector, establishing the foundation for understanding its rapid integration into industrial practices. It explores the unique properties, synthesis, characterization, and diverse applications of nanocellulose, one of the most promising renewable nanomaterials for strengthening paper products and enhancing their mechanical durability, flexibility, and functional performance. Further discussions highlight nanocoatings that improve surface attributes, such as barrier protection, water resistance, and printing quality, as well as the role of nanoparticles in imparting advanced functionalities like antimicrobial resistance, flame retardancy, and electrical conductivity. Innovations in paper coating and printing, recycling and deinking, and packaging solutions designed to meet modern sustainability goals are addressed comprehensively. Paper-based nanocomposites and their challenges are discussed in detail, alongside recent advancements at the intersection of nanotechnology and biotechnology in pulp bleaching.

The book also emphasizes cutting-edge nanosensors for real-time production monitoring, critical issues in nanocellulose utilization, and life cycle assessment studies that evaluate environmental impacts. It explores specialty paper development, advancements in wood pulping and lignin extraction, and nanotechnology-driven sustainable forest and paper management practices. The application of nanotechnology in wastewater treatment for the pulp and paper sector is reviewed, concluding with a forward-looking discussion on future trends and emerging opportunities in the industry. This volume is designed to serve as a valuable reference for researchers, industry professionals, academicians, and policymakers interested in leveraging nanotechnology for sustainable growth and innovation in the pulp and paper sector. We extend our heartfelt gratitude to all the contributors for their expertise and dedication in preparing the chapters, and to the peer reviewers whose constructive insights and evaluations have greatly enhanced the clarity and quality of the content.

Chennai, India Rakesh Kumar Bachheti

Contents

About the Editor

Dr. Rakesh Kumar Bachheti obtained his M.Sc. degree in Organic Chemistry from Hemwati Nandan Bahuguna (HNB) University, Garhwal, in 1998. Later, in 2001, he completed a one-year Post Graduate Diploma in Pulp and Paper Technology from the Forest Research Institute, Dehradun, India. He earned his Ph.D. in Organic Chemistry from Kumaun University, Nainital, in 2007. Dr. Bachheti has over 22 years of national and international research and teaching experience and serves as an adjunct faculty member at various national and international universities. His research interests include natural products for industrial applications, green synthesis of nanoparticles, and pulp and paper technology. Dr. Bachheti has a strong research profile, having presented his work at several national and international conferences, including in Malaysia, Thailand, and India. He has an h-index of 29(Scopus), over 181 research publications, and two published patents. He also serves as an external examiner for Ph.D. thesis evaluations. As a dedicated academic mentor, he has successfully guided 40 M.Sc. and eight Ph.D. students to completion. Dr. Bachheti has served as an Editor and Editorial Board Member for several international journals. Dr. Bachheti is also recognized for his editorial work, having edited 14 books. He has also contributed 41 book chapters published by renowned publishers such as Elsevier, Springer, CRC Press, Taylor & Francis, among others. He has received several accolades, including being recognized among the World's Top 2% Scientists by Stanford University and Elsevier for the year 2025 and 2024

Nanotechnology in the Pulp and Paper Industry: An Overview

Limenew Abate Worku, Sheetal Tyagi, D. P. Pandey, Rakesh Kumar Bachheti, and Archana Bachheti

Abstract Nanomaterials made of cellulose and lignocellulose have special qualities such a huge surface area, great mechanical strength, and biodegradability. They may find applications in various industries, including environmental remediation, composites, biomedical materials, and packaging. Using the processes developed for other nanomaterials as platforms to create new wood-based materials and applications will require new methods to liberate these components, such as nanodimensional cellulose fibrils, macromolecules, and nanomaterials. These materials can be made multifunctional and self-assembled into distinct architectures. Numerous facets of papermaking have shown encouraging outcomes from the application of nanotechnology; studies have demonstrated that incorporating cellulose nanofibers can enhance the mechanical properties of paper by up to 50%. Cellulose nanofibers have been prepared by several research groups using techniques such as electrospinning, chemical treatment followed by mechanical procedures, or mechanical separation methods. In specific processes, yields of cellulose nanofibers have reached as high as 90%. These developments demonstrate the great potential of cellulose and lignocellulose as environmentally friendly nanomaterials for a range of industrial uses. This chapter explores the use of nanotechnology in the pulp and paper

L. A. Worku
College of Natural and Computational Science, Department of Chemistry, Debre Tabor University, Debre Tabor, Ethiopia

S. Tyagi
Department of Chemistry, School of Basic and Applied Sciences, Shri Guru Ram Rai University, Dehradun, Uttarakhand, India

D. P. Pandey
Department of Chemistry, Government Degree College Dehradun Shahar, Suddhowala, Dehradun, India

R. K. Bachheti
Department of Research Analytics, Saveetha Dental College & Hospitals, Saveetha Institute of Medical and Technical Sciences, Saveetha University, Chennai, Tamil Nadu, India

A. Bachheti (✉)
Department of Environment Science, Graphic Era (Deemed to be University), Dehradun, Uttarakhand, India
e-mail: bachheti.archana@gmail.com

R. K. Bachheti (ed.), *Nanotechnology in the Pulp and Paper Industry*, Smart Nanomaterials Technology, https://doi.org/10.1007/978-981-95-3473-9_1

industry, highlighting its potential to improve paper properties like strength, printability, glossiness, and gas barrier capabilities. It emphasizes how nanotechnology can promote sustainability and efficiency by optimizing resource utilization and minimizing environmental impact.

Keywords Nanofibers · Nanotechnology · Paper · Pulp · Wood industry

1 Introduction

The new scientific and technological field of nanotechnology will completely transform the utilization of materials in the twenty-first century. Highly effective and ecologically friendly nanotechnologies will replace the comparatively simple and undeveloped technologies that we presently rely on [144]. Novel nanosystems and nanostructured materials can be developed thanks to the discovery of new materials, processes, and phenomena at the nanoscale (usually 1 to 100 nanometers) and the creation of new experimental and theoretical study techniques [22]. Policymakers and business executives have immediately placed nanotechnology on their agendas, and by 2024, the global nanotechnology market is expected to reach approximately $125 billion [74]. Several factors highlight the significance of nanotechnology: (1) it is smaller, faster, uses less material, and uses less energy, (2) it allows for new properties and phenomena, (3) it is the most effective length scale for manufacturing, and (4) it makes it easier for living and nonliving systems to interact [69]. Numerous ultra-small-scale materials, such as metal oxides, ceramics, and polymeric materials, as well as various processing methods, including those that involve molecular-scale "self-assembly," are either in use or being prepared for commercial use [101]. The global market for nanotechnology in environmental applications is anticipated to reach $20 billion by 2025, with other examples, which are in various stages of development, focusing on pollution prevention and treatment [16, 74]. Since the beginning of the new millennium in 2000, Malaysia has been developing its microtechnology and nanotechnology, and this endeavor has continued to gain momentum [51]. Government financing for nanotechnology started in 2001, and the Eighth Malaysia Plan (8MP) classified it as a Strategic Research (SR) program under the IRPA [126]. Since nanotechnology was acknowledged as a new and growing field in the Third Industrial Master Plan (2005–2020), the Malaysian government has taken significant steps to advance the technology in the nation [52]. Since the establishment of the National Nanotechnology Initiatives in 2006, Malaysia has taken an active role in several nanotechnology-related forums, contributing to the global conversation in this revolutionary sector [40]. Since 2006, several countries in Latin America, as well as China, India, South Korea, and the United States, have been actively involved in conferences related to nanotechnology. Argentina, Brazil, Chile, and Uruguay are among these countries [73]. Hong Kong, Singapore, Thailand, Indonesia, and Malaysia are among the participating nations in the Asia Nano Forum [73]. This chapter explores the application of nanotechnology in the pulp and paper industry.

It focuses on how nanotechnology can enhance paper properties, such as strength, printability, gloss, and gas barrier capabilities. This sentence emphasizes the role of nanotechnology in promoting sustainability and efficiency. The chapter also highlights the development and application of nanoadditives that enhance product quality and alter production processes.

2 Nanomaterials in Pulp and Paper

The pulp and paper industry is undergoing significant changes due to the increasing use of nanomaterials, which enhance product sustainability and performance. Nanotechnology advancements, such as the application of cellulose nanofibers, are anticipated to enhance mechanical properties and reduce environmental impact, supporting the industry's transition to eco-friendly and biodegradable materials [9].

Because they are plentiful, renewable, have a nanofibrillar structure, can be made multifunctional, and self-assemble into distinct architectures, cellulose and lignocellulose have enormous potential as nanomaterials [140]. Since cellulosic nanofibers have a large surface area, the most crucial factor to regulate for nanocomposite applications is their adherence characteristics [140]. The contact angle measurements demonstrated that when acetylated, the surface properties of nanofibers changed from hydrophilic to more hydrophobic [66]. According to the microscopy investigation, the acetylation caused the cell wall of the kenaf fiber to enlarge, and the isolated nanofibers had diameters ranging from 5 to 50 nm. Nanotechnology has a great deal to offer the paper industry. This adaptable technology can also be applied to fillers for wet-end applications in papermaking [70].

In addition to enhancing the mechanical and functional qualities of pulp and paper products, the addition of these nanoparticles helps the sector transition to more creative and sustainable solutions. The potential uses and advantages of nanoparticles in the pulp and paper industry are anticipated to grow considerably as research advances [50].

2.1 Types of Nanomaterials Used for Pulp and Paper

2.1.1 Nanocellulose (NCC, NFC)

NCC, which is made by hydrolyzing cellulose fibers, is powerful, low in density, and has excellent mechanical properties. Products are papered with it to increase their strength and barrier properties [136]. Nanofibrillated cellulose (NFC), which is produced mechanically or chemically, is a network of cellulose fibrils that can improve the tensile strength, flexibility, and printability of paper. NFC can also make coatings and inks thicker [104].

Papermaking additives are sometimes known as papermaking chemicals. Papermaking auxiliaries can impart several exceptional features to the paper and are essential to the pulping, papermaking, and paper processing processes [83]. Paper additives can significantly improve paper performance even when they are only added in very small amounts. The most widely used papermaking auxiliaries include sizing agents, retention aids, filter aids, enhancers, defoamers, and dispersants [37].

NC is a significant byproduct of cellulose, a naturally occurring polymer with strong degradability, biocompatibility, and swelling rate. The properties of its molecular structure suggest that it has intrinsic benefits as an additive for papermaking [14]. In addition to giving cellulose fibers their strong structure through hydrogen bonding, the structural hydroxyl group (-OH) found in the glucose unit also makes cellulose more hygroscopic [102]. With Young's modulus of roughly 138 GPa, NC possesses high specific surface area, aspect ratio, and tensile strength. It also contains many hydroxyl groups on its surface that can create hydrogen bonds with the fibers [89]. At the very least, NC has retention, drainage, and enhancement effects. Additives frequently have several actions at once. Research on papermaking additives, particularly their status as enhancers, retention aids, and filter aids—all of which are covered in more detail in the sections that follow—has been the foundation for the papermaking and packaging industries' explosive growth and development [83].

2.1.2 Nanozeolite

Together with pulp, nanozeolite filler materials are utilized to improve the characteristics of paper and asphalt [45]. The addition of microzeolite also aids in reducing gas emissions within the bounds of high-quality paper. This is due to their crystal structure, which consists of millions of channels and cavities with a relatively huge surface area [59, 60]. Furthermore, natural nanozeolite has a strong tendency to attract moisture and can return absorbed water without changing its crystal structure or ability to absorb moisture [103]. The paper industry uses activated nanozeolites extensively as desiccants due to these characteristics. Natural zeolites, however, have several benefits over synthetic ones, such as being more readily available to the paper industry, more economical, and possessing more resources [103].

2.1.3 Nanoclays

Layered silicate minerals, such as kaolinite and montmorillonite, known as nanoclays, can enhance the mechanical and barrier properties of paper. When added to paper compositions, nanoclays improve the paper's resistance to gases and moisture, increase its strength, and reduce its porosity. They also have an impact on the paper's brightness and opacity [120].

Nanoclay's low gas permeability is one of its most intriguing properties when added to the paper industry, particularly for nylon paper [148]. It promised to provide a CO_2 barrier and a 100-fold reduction in O_2 permeability, extending the paper's overall shelf life [88]. Traditionally, multilayer polyethylene terephthalate bottles and containers for cheese and meat have utilized nanoclay as a barrier layer. In high-density polyethylene containers and food packaging, nanoclay is also utilized in place of ethylene glycol copolymer [148]. Consequently, nanoclay can serve as a waterproof barrier once it is used in the paper sector [111]. It provided a better substitute for widely used acrylic acid and traditional polyethylene, and it has a lot of potential for use in paper beverage technology [106].

2.1.4 Carbon Nanotubes (CNTs)

Cylindrical nanostructures, known as carbon nanotubes, have remarkable thermal, electrical, and mechanical qualities. CNTs can significantly increase the electrical conductivity and tensile strength of paper. This qualifies them for use in electronic paper and smart packaging [58]. Research on carbon nanotubes (CNTs) in the pulp and paper industry suggests their potential to improve the mechanical properties, electrical conductivity, and sustainability of paper products. Studies have shown that incorporating carbon nanotubes (CNTs) can significantly enhance paper strength and functionality while also enabling innovative applications such as smart papers and efficient recycling methods [7]. Studies have also shown that CNTs can enhance a paper's tensile strength, flexibility, and resistance to tearing. Furthermore, CNTs can make paper electrically conductive, enabling the creation of "smart" or electronic papers and more efficient recycling processes [71].

2.1.5 Metallic and Metal Oxide Nanoparticles

Metallic Nanoparticles: By incorporating silver, gold, and copper nanoparticles into paper, items can acquire antibacterial properties that extend their shelf life and inhibit the growth of microorganisms [55]. Metal Oxide Nanoparticles: Due to their photocatalytic qualities, zinc oxide (ZnO) and titanium dioxide (TiO_2) nanoparticles are utilized to enhance the whiteness and brightness of paper. They can also increase the longevity of paper goods and offer UV protection [125].

For supporting items on paper sheets, nano-TiO_2 may be utilized as a retention agent. Two methods for using nano-TiO_2 as a retention agent are wet end addition, which involves depositing nano-TiO_2 onto individual fibers prior to sheet formation, resulting in a bulk distribution of the TiO_2 loading, and coating the sheet surface with a nano-TiO_2 suspension [108] (Fig. 1). For instance, the surface characteristics of nano-TiO_2 with hexadecanoic acid showed a significant improvement in the mixture's wetting and dispersion. Compared to the empty ones, the paper containing nano-TiO_2 showed a greater dynamic elastic modulus [108]. This promotes the development of a cutting-edge technique for creating paper with a superhydrophobic surface, achieved by adding modified nano-TiO_2 to cellulosic fibers [75]. Nano-TiO_2 was dispersed using a superb speed homogenizer, and 3-(trimethoxysilyl) propyl methacrylate was utilized as a coupling agent to modify the surface [108]. According to the data, the

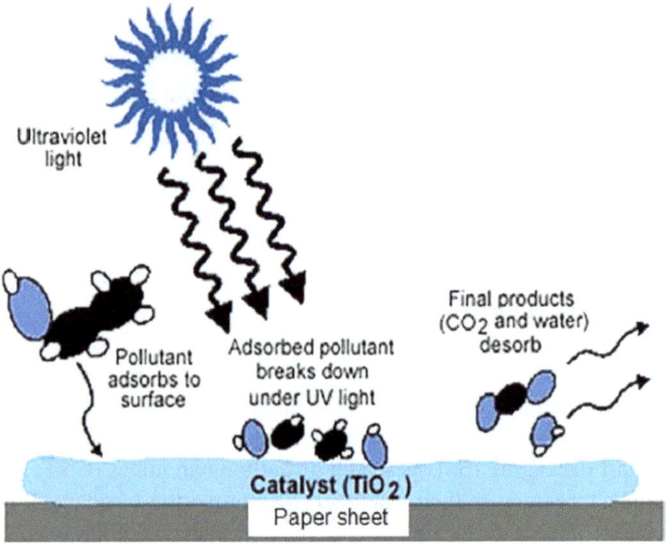

Fig. 1 TiO_2 paper sheet in photocatalysis system [113]

water contact angles of the changed paper increased from 126.5 to 154.2, indicating an improvement in hydrophobicity.

In the latter approach, a binder can be used to increase the TiO_2 loading and dispersion in the paper while also protecting the fiber from radiation. As a result, the produced paper shows exceptional stability when exposed to solar-like light [128]. For instance, study on sludge recovered from paper ash that contained nano-TiO_2 formed by acid treatment indicated that the anatase's crystal structure and size were 10–20 nm [113]. The stability and degradability properties of nano-TiO_2 were evaluated by removing acetaldehyde from the paper sludge. When subjected to UV and solar light, the resultant paper was found to be capable of completely breaking down acetaldehyde into CO_2 [84] (Table 1).

2.2 *Sources and Extraction Methods*

The primary natural sources of nanocellulose include wood pulp, agricultural waste (such as wheat straw and sugarcane bagasse), and other plant-based biomass that contains cellulose [124, 145]. Conversely, clay deposits worldwide yield naturally occurring layered silicate minerals known as nanoclays. These minerals occur in common forms, including bentonite, kaolinite, and montmorillonite [141]. Carbon nanotubes (CNTs) and other synthetic nanomaterials are often created from carbon-rich precursors in regulated laboratory or industrial settings. Lastly, metallic and

Table 1 Types of nanomaterials used in the pulp and paper industry

Nanomaterial	Source	Properties	Applications	Key references
Cellulose nanofibers	Wood pulp	High strength, lightweight, biodegradable	Reinforcement in paper, barrier coatings	[97]
Nanoclay	Natural clay minerals	High aspect ratio, mechanical strength	Paper reinforcement, moisture barrier	[98]
Silver nanoparticles	Silver compounds	Antimicrobial, conductive	Antimicrobial coatings for paper products	[12]
Titanium dioxide	Titanium ores	UV-blocking, photocatalytic	Coatings for paper to enhance durability and whiteness	[33]
Graphene oxide	Graphite	High strength, electrical conductivity	Conductive paper, sensors	[87]
Lignin nanoparticles	Lignin from wood	Biodegradable, UV-absorbing	Additive for paper strength and barrier properties	[132]

metal oxide nanoparticles are usually produced rather than retrieved using metal salts or oxides as precursors [86].

Four methods are available for producing cellulose fibers at the nanoscale: (1) bacterial cellulose nanofibers, (2) electrospun cellulose nanofibers, (3) microfibrillated cellulose plant cell fibers, and (4) cellulose whiskers or nanorods [39]. However, there are a number of methods for creating and describing nanofibers. The structure and properties of nanocellulose are determined mainly by its shape, crystalline structure, surface features, chemical and physical properties, and qualities in liquid suspension [112]. The research investigates the impact of various pretreatments on the structure and chemical composition of the isolated nanofibers, as well as their size and properties [107].

One common method for producing nanocrystalline cellulose that is distinct from native cellulose is acid hydrolysis. There have also been studies on the production of nanocellulose using other methods, such as TEMPO-mediated oxidation [133] and ionic liquids [95], which have been investigated for nanocellulose preparation. An application of cellulase enzyme has also been reported [53, 137].

The degree of cellulose polymerization rapidly drops during acid hydrolysis, but it stabilizes at a point known as the level-off degree of polymerization [116]. The quick hydrolysis of amorphous regions by the acid may be the cause of this behavior. Because of its large volume, the acid preferentially targets the amorphous region as hydrolysis begins, hydrolyzing the readily accessible glycosidic link [10]. Following the hydrolysis of the readily accessible bond, more hydrolysis takes place at the reducing end of the glucose chain and on the surface of the crystalline area, but considerably more slowly [36]. Several variables, including acid type, acid concentration, temperature, and time, influence acid hydrolysis [10, 11, 146]. Variations in temperature, hydrolysis duration, and acid content impact the shape of the nanocrystalline cellulose. For instance, shorter nanocrystalline cellulose is produced by increasing the hydrolysis duration [23] and the acid concentration [20] causes shorter nanocrystalline cellulose. In contrast, high temperatures cause cellulose to completely transform into glucose [15].

While nanofibrillated cellulose (NFC) is made mechanically by high-pressure homogenization, grinding, or ultrasonication, and occasionally in conjunction with enzymatic or chemical pretreatments to promote fibrillation, nanocrystalline cellulose (NCC) is usually extracted by acid hydrolysis, frequently employing sulfuric acid, which breaks down the amorphous regions of cellulose fibers to isolate the crystalline nanodomains [152]. Using purification methods such as sedimentation, centrifugation, and chemical treatments, contaminants are removed to obtain nanoclays. Then, by employing exfoliation methods like mechanical stirring or ultrasonic treatment, the layers are delaminated into nanoscale platelets that can be added to paper matrices [17]. Carbon nanotubes (CNTs) are produced via high-energy techniques such as arc discharge and laser ablation, as well as Chemical Vapor Deposition (CVD), which decomposes hydrocarbon gasses at high temperatures over metal catalysts. Purification techniques used after synthesis are crucial for removing amorphous carbon and catalyst particles [139]. The final techniques for producing metallic and metal oxide nanoparticles are the sol-gel process, which yields metal oxides such as TiO_2 and ZnO, chemical reduction, which forms nanoparticles by reducing metal ions in solution,and green synthesis, which uses microorganisms or plant extracts to produce nanoparticles in an environmentally friendly manner [109].

3 Applications of Nanotechnology in Pulp and Paper

3.1 Enhanced Paper Properties

3.1.1 Mechanical Strength

Studies have indicated that adding nanocellulose, including NCC and NFC, to paper products greatly increases their mechanical strength. According to studies, adding just 1–5% nanocellulose can increase tensile strength by up to 50%, allowing for lighter and thinner paper without compromising durability. The fiber network is reinforced by the high aspect ratio and potent hydrogen bonding properties of nanocellulose, which are responsible for this improvement.

When nanocellulose is added to papermaking, the result is a better-performing paper with increased strength [82] and density [96]. Also, it reduces porosity [18]. When beaten fibers and nanostructured cellulose are combined, the paper's characteristics are comparable [25]. Theoretically, adding wet and dry strength agents to paper can increase its strength [5], as well as the addition of functionalized fibers [100] and beating [57]. The enhancement results in fiber bonding ability [25]. According to Boufi et al. [25], nanocellulose acts as a connection between fibers, increasing the bound area and promoting fiber-to-fiber bonding. Additionally, the increased bonding capability may also be a consequence of various networks formed within the fiber. Due to their micrometer length, nanocellulose can serve as a bridge to connect nearby fibers, thereby strengthening the network [25]. The paper's strength is increased by the network of mutual bonds that fiber and nanocellulose provide. As a nanoadditive, nanocellulose strengthens internal bonds, increasing dry tensile strength, decreasing air permeability and opacity, and increasing density [119]. Because nanocellulose is a substance at the nanoscale and has a large surface area, hydrogen bonds can form more successfully. It is utilized as a wet-end addition to improve retention and dry strength. Nanofibrillated cellulose is a type of nanocellulose that is frequently used as a dry-strength addition [64].

3.1.2 Barrier Properties (Moisture, Gas, Oil)

It has been demonstrated that nanomaterials, especially nanoclays and nanocellulose, enhance paper's ability to withstand moisture, gasses, and oils [134]. According to research, papers treated with nanoclays exhibit reduced permeability to oxygen and water vapor, making them suitable for use in packaging materials [42]. For example, adding 2–10% nano clay to the paper can increase the shelf life of food materials packed in paper by reducing water vapor transmission rates by up to 70% [114].

3.1.3 Printability and Optical Properties

Additionally, the use of nanoparticles can enhance the optical qualities and printability of paper. According to studies, the addition of metallic nanoparticles, including titanium dioxide (TiO_2), can enhance print quality by increasing brightness and whiteness [1]. Additionally, nanocellulose can enhance the paper's surface smoothness, which is crucial for high-quality printing applications, as it produces sharper images and improved ink absorption. The high surface area and special qualities of nanocellulose enhance fiber bonding. Paper becomes stronger and more durable due to conversation of paper in nanocellulose [82].

Nanoparticles can influence a paper's opacity and glossiness. Enhanced opacity reduces show-through, making it suitable for double-sided printing. The glossiness of the paper can be adjusted. Manufacturers can tailor the paper's finish to meet specific printing requirements by varying the concentration and type of nanoparticles used [72].

Recent advancements in the field have explored the use of nanoparticles. These nanoparticles, including silica and zinc oxide, can enhance the optical properties of paper. These nanoparticles have the potential to improve print quality and visual effects by improving light absorption and dispersion. Functional nanoparticles can impart antimicrobial properties to paper. This makes paper suitable for applications in packaging and hygiene products [46].

3.2 NC in Waterproof Paper

Along with the standard writing function, some unique additions have been tried to make the paper moisture-proof and hydrophobic. Many natural polymers, such as chitosan, whey protein isolate, starch, and its derivatives, have been shown to exhibit antibacterial and water-proof properties recently. In order to create waterproof and antibacterial bifunctional paper, silver nanoparticles are frequently added to beeswax emulsions. CNC has been tested as an Ag dispersion, where it stabilized Pickering emulsions, to increase the silver-loaded CNC's compatibility with beeswax [90]. Then, the paper could be treated with a mixture of Ag, CNC, and beeswax to produce a multipurpose paper that has antibacterial and waterproof properties. A simple dip coating method was employed to construct silver nanowires (AgNWs) and hydrophobic inorganic ceramics on pure NC paper, resulting in an electromagnetic interference shielding paper that is both mechanically flexible and water resistant [127]. According to research, the produced hydrophobic AgNW/cellulose (H-AgNW/CNF) composite paper maintained its good waterproof performance after 1,000 peeling tests and innumerable bending tests [127].

3.3 Nanocoatings for Improved Performance

The possibility of applying nanocellulose as a coating is increased by its low cost, availability, and renewability. It is a bio-based substance that can be used as a coating ingredient in place of other environmentally friendly substitutes that are not as financially viable as their petroleum-based equivalents [56]. Nanocellulose's high specific surface area, strong chemical reactivity, and favorable rheological characteristics allow it to exhibit exceptional effectiveness in paper coatings [81]. By selecting the type of effect that the nanocellulose will have on the coated paper, the coating qualities can be adjusted. By strengthening the network of hydrogen bonds that have developed on the paper's surface, the fundamental attribute of increasing mechanical paper stability will be accomplished. When cellulose nanofibers or nanocrystals entangle and interact with paper, mechanical improvement results. Additionally, coating effects, such as the inclusion of gloss modifiers or pigments in the coating or its barrier properties, may also affect the appearance of the paper [138].

Advanced coatings that improve the functionality of paper goods can be developed thanks to nanotechnology. Studies have demonstrated that nanocoatings can offer antibacterial, oil-repellent, and water-resistant qualities [149]. Papers coated with nanostructured materials, for instance, are perfect for food packaging because of their high resistance to water and grease. By applying these coatings in thin layers, performance can be maximized while reducing material consumption [130].

By creating a network of hydrogen bonds with paper fibers, CNFs, and CNCs may both stick to paper surfaces. In fact, applying a layer of CNFs or CNCs on paper provides a number of advantages: improving the surface's smoothness will increase the barrier against ink penetration during printing operations [76], while the intertwined hydrogen-bonded network spontaneously produced by CNFs or CNCs self-assembly on the paper surface will enhance its mechanical properties [76]. The coating composition will significantly improve print sharpness and water fastness while also enhancing regular paper's gas barrier and water-repellency qualities [129].

In general, cellulose nanofibers are a more effective coating layer than CNCs. According to Li et al., CNFs form a more closely entangled network than CNCs due to their larger aspect ratio [85]. On the other hand, cellulose nanocrystals will arrange themselves in a stack. The incapacity of CNCs to form entangled networks and the absence of an energy-dissipating amorphous phase make their self-assembled nanostructure (such as film) extremely fragile [91]. This conclusion is widely accepted for freestanding nanocellulose sheets, sometimes known as "nanopaper." However, it is genuinely debatable for paper coating, as research indicates that CNC coating can also produce favorable results. Since mechanical qualities are often derived from studies conducted by independent laboratories based on available data, these comparisons are quite challenging to conduct systematically [48, 147].

3.4 *Nanofibers in Composite Reinforcement*

Researchers in composite science are increasingly interested in cellulose micro/nanofibril (CMNF) as a reinforcing material for composites due to its potential for high strength and low weight [143]. According to Nakagaito and Yano [110], microfibrillated cellulose derived from kraft pulp was used to create a novel high-strength composite. According to a study by Iwamoto et al. [62], nanofibers show considerable promise as a reinforcing material for optically transparent composites. The theoretical and practical advantages of employing nanofibers in composite filter media were investigated by Graham et al. [49], who concluded that nanofibers can enhance filter efficiency without significantly increasing the filter pressure drop. They have been shown to improve the performance of air filters for mining vehicle personnel cabins and extend the life of filters in pulse-clean cartridge systems for dust collection. Bhatnagar and Sain [21] created a polymer/nanofibers matrix with 10% nanofiber reinforcement in 90% PVA. They discovered that even 10% nanofibres provide significant reinforcing potential. Using electrospinning technology, [123] produced uniformly smooth, nonwoven mats composed of PVA reinforced with cellulose nanocrystals. The thermomechanical properties of these fiber nanocomposites were enhanced. Zimmermann et al. [150] state that the strength of the composites reinforced with nanofibers is comparable to that of magnesium alloy or mild steel. When microfibrillated cellulose is used in place of such alloys, a significant weight reduction in automobiles can be accomplished because of its low weight. Nevertheless, the application of biopolymers has been limited due to their often subpar mechanical and barrier properties, which can be enhanced by the addition of reinforcing substances (fillers), thereby creating composites [131]. The filler-matrix interactions and the final material's performance are favored by the nanoparticles' comparatively higher surface area compared to their microscale counterparts. When added to a polymer, nanoparticles can perform additional tasks besides nano-reinforcements, including biosensing, enzyme immobilization, and antibacterial activity. Cellulose nanofibers (CNFs) were added to chitosan and mango puree-based edible films in varying quantities to create nanocomposite edible films [13].

3.5 Environmental and Sustainability Aspects

3.5.1 Reduction in Raw Material Usage

The integration of nanotechnology in pulp and paper production can lead to a significant reduction in raw material usage [105]. Research indicates that the enhanced properties of nanomaterials allow for the production of stronger and lighter paper, which can reduce overall fiber consumption by up to 30%. This not only lowers production costs but also minimizes the environmental impact associated with raw material extraction and processing [134].

3.5.2 Improved Recyclability and Biodegradability

Nanotechnology can enhance the recyclability and biodegradability of paper products. Studies have shown that the use of biodegradable nanomaterials, such as nanocellulose, can enhance the overall biodegradability of paper, facilitating its breakdown in natural environments [151]. Additionally, the incorporation of nanomaterials can enhance the recycling process by improving the quality of recycled fibers, leading to higher-quality end products [2].

3.5.3 Nanotechnology for Wastewater Treatment in Paper Mills

The application of nanotechnology in wastewater treatment processes within paper mills has shown promising results [68]. Research indicates that nanomaterials, such as titanium dioxide and activated carbon nanoparticles, can effectively remove contaminants from wastewater, including dyes and organic pollutants [94]. These advanced treatment methods not only improve water quality but also reduce the environmental footprint of paper production, contributing to more sustainable manufacturing practices [93].

3.6 Strength-Reinforcing Agent

Numerous published articles demonstrated the subsequent evolution of paper in the form of cellulose nanofiber sheets. According to Ioelovich [61], current research has indicated that adding nanocellulose particles or nanofibrillated cellulose to the composition of paper may improve its strength [6]. It has also been shown that cellulose nanofibrils have the potential to act as a strength-increasing agent. The nanofibrils are used with a cationic polyelectrolyte, poly(amideamine) epichlorohydrin (PAE), to enhance paper strength. It is concluded that the paper strength experiments and fundamental interaction studies have shown that the use of nanofibrils as a strength-reinforcing agent, together with PAE, leads to superior improvements in strength properties. According to Nogi et al. [117], cellulose nanofibers, rather than the micrometer-sized cellulose pulp fibers seen in regular paper, are the source of a new type of paper. Nanofibrils with varying molar masses and nanopapers with varying porosities are made. High mechanical nanofibril performance and nanofibrillar network structure are linked to the high toughness of extremely porous nanopaper [54]. Higher fibril strength, better interfibril adhesion properties, and considerably smaller and more uniformly distributed flaws (voids) are what make nanopapers excellent [35]. More reliable mass manufacturing of the nanofibers can be accomplished, since fibrillation processes can be used on any natural fiber resource that contains lignin, such as flax, sugarcane bagasse, and wheat straw [142], next to woody lignocellulosic components. For another millennium, nanofiber paper may take the place of or surpass conventional paper in terms of information storage and transmission [19].

4 Manufacturing Processes and Techniques

The use of nanoparticle systems in papermaking has evolved into a more "properties" management system, both within the mill at converting operations and printers, as well as outside of it. Advances in nanotechnology in the fields of fiber science, minerals, and other additives will enable papermakers to impart structure and order to sheet designs [101]. Magnetic papers have been prepared using both lumen loading and in situ synthesis techniques. Papers made using the lumen loading procedure have better magnetic qualities than those made using the in situ synthesis approach. This is because the particles created by the chemical coprecipitation process have a limited (nm) size and exhibit super paramagnetic activity. To reduce their surface energy, finite-size particles tend to aggregate and firmly adhere to one another, resulting in a low rate of detachment throughout the washing and papermaking processes [29]. The strength of the paper was negatively impacted by the magnetic particles that were deposited on the fiber surface [30]. In the process of creating new paper materials, Agarwal et al. [4] have successfully integrated scales from nano to micro to macroscale (nanocoating–microfibers–macropaper). In the process of creating new paper materials, Agarwal et al. [4] have successfully integrated scales from nano to micro to macroscale (nanocoating–microfibers–macropaper) [92]. According to recent research, sucrose, a nanoadditive, can prevent the paper from losing its strength over time by incorporating inorganic fillers such as kaolin [41]. The strength of the resulting paper nanocomposites is increased by combining sucrose with the cellulose fibers' nanoporous structure, which causes the fibers to be beaten.

4.1 Integration of Nanotechnology in Paper Mills

The integration of nanotechnology into the pulp and paper industry has the potential to revolutionize traditional papermaking processes by enhancing the properties of paper products [67]. Nanomaterials, such as nanocellulose, nanoclay, and nanoparticles, can be incorporated into various stages of paper production to enhance strength, durability, and functionality [34].

Nanocellulose can be effectively incorporated into the pulp and paper manufacturing process during the refining stage, where its high surface area and unique properties significantly enhance the mechanical strength and barrier characteristics of the resulting paper [121]. Additionally, nanoparticles can be utilized in coating applications to enhance surface properties, such as water resistance, printability, and gloss, by employing techniques like spray coating or dip coating for optimal results [3]. Furthermore, nanomaterials can be mixed with fiber suspensions prior to the sheet formation process, ensuring uniform distribution and leading to improved paper properties. They can also function as functional additives, imparting specific benefits such as antimicrobial effects or UV protection, thereby enhancing the overall value of paper products [63]. Research findings support these advancements, demonstrating that papers reinforced with nanocellulose exhibit higher tensile strength and lower porosity, which contribute to better print quality and durability. Collectively, these techniques illustrate the transformative potential of nanotechnology in enhancing the quality and functionality of paper products [47]

4.2 Challenges and Solutions

Notwithstanding the pulp and paper industry's apparent advantages, several obstacles need to be overcome before nanotechnology may be widely used. Scaling up manufacturing is a significant technological challenge, as the efficacy of nanomaterials depends on their consistent quality and performance when transitioning from laboratory-scale to industrial-scale applications [9]. Additionally, achieving uniform dispersion of nanomaterials within pulp and paper matrices can be challenging due to the tendency of nanoparticles to agglomerate, which diminishes their effectiveness. Ongoing research focuses on developing improved dispersion techniques and stabilizers to overcome this issue [77]. Furthermore, the high production costs of nanomaterials can limit their economic feasibility for some paper mills, prompting research into cost-effective synthesis methods and the exploration of by-products from other industries as alternative sources of nanomaterials [27]. Operational integration also poses challenges, as incorporating nanotechnology into existing manufacturing processes necessitates modifications to equipment and workflows, often requiring significant capital investment [31]. Pilot projects and collaborations between academia and industry have shown promise in demonstrating the feasibility and benefits of these integrations. Studies indicate that the use of nanocellulose can enhance paper strength by up to 50% while reducing weight, thereby improving overall product quality and functionality [122]. The integration of nanotechnology into the pulp and paper industry presents exciting opportunities. Addressing these technical and economic challenges is essential for the successful implementation of nanotechnology in this sector, and ongoing research and collaboration among stakeholders will be crucial in overcoming these barriers and realizing the full potential of nanotechnology in this sector [99] (Table 2).

5 Health, Safety, and Regulatory Aspects

Health and safety concerns regarding nanotechnology in the pulp and paper industry primarily revolve around the potential risks associated with exposure to nanomaterials, which can pose health hazards to workers and the environment [135]. Nanomaterials are unique due to their large surface area and compact size. Increased reactivity and bioavailability resulting from these characteristics may have adverse effects on human health and the environment [80].

During several phases of production, employees in the pulp and paper industry may come into contact with nanoparticles. Managing raw materials, processing, and utilizing nanomaterials in coatings and additives are among these steps. Skin contact may cause irritation or allergic reactions, while inhaling airborne nanoparticles can lead to respiratory problems [101]. Long-term exposure to nanomaterials can cause severe health effects. These effects include potential carcinogenicity and neurotoxicity, which necessitate thorough risk assessments and monitoring [26]. The environmental impact of nanomaterials when used in pulp and paper is a significant concern. Their release into ecosystems can disrupt biological processes and harm aquatic and terrestrial organisms [79]. Thorough safety procedures are essential to minimize risks. Personal protection equipment (PPE), appropriate handling techniques, and regular training on potential nanomaterial risks should all be part of these processes [8].

Table 2 Advantages and challenges of nanotechnology integration in pulp and paper manufacturing

Aspect	Advantages	Challenges	Key references
Enhanced strength	Nanotechnology can enhance the mechanical properties of paper, leading to stronger and more durable products	The cost of nanomaterials can be high, impacting overall production costs	[28]
Improved barrier properties	Nanoparticles can enhance the barrier properties of materials against moisture, gases, and oils, resulting in improved packaging materials	Potential environmental and health risks associated with the use of nanomaterials	[38]
Reduced chemical usage	Nanotechnology can reduce the need for chemical additives, leading to more sustainable manufacturing processes	Integration into existing processes may require significant changes in equipment and technology	[78]
Better printability	Nanomaterials can enhance the printability and surface properties of paper, improving the quality of printed products	Limited understanding of long-term stability and performance of nanomaterials in paper products	[28]
Biodegradability	Nanotechnology can facilitate the development of biodegradable paper products, reducing environmental impact	Regulatory challenges and the need for standardized testing methods for nanomaterials	[32]
Cost-effectiveness	Potential for reduced production costs through improved efficiency and reduced waste	Initial investment in research and development may be substantial	[65]

5.1 Health and Safety Concerns

The incorporation of nanomaterials in the pulp and paper industry raises several health and safety concerns due to the unique properties of these materials at the nanoscale. Potential risks include inhalation or dermal exposure to nanoparticles, which may lead to respiratory issues, skin irritation, or other health complications. Research has shown that specific nanomaterials can exhibit toxicological effects, including cytotoxicity and genotoxicity, depending on their composition, size, and surface characteristics [24]. To mitigate these risks, it is crucial to establish and implement effective safety protocols and best practices. This includes conducting thorough risk assessments, implementing engineering controls such as ventilation systems, and providing personal protective equipment (PPE) for workers. Training programs should also be developed to educate employees on the safe handling of nanomaterials and the importance of adhering to established safety guidelines. Ongoing research into the long-term effects of nanomaterial exposure is crucial for refining these protocols and ensuring a safe working environment [115].

5.2 Regulatory Framework

Some countries have enacted laws and regulations governing the use of nanotechnology in the pulp and paper industry to address the unique challenges posed by nanomaterials. The Environmental Protection Agency (EPA) and the Food and Drug Administration (FDA) have developed guidelines for assessing and managing nanomaterials in the United States, with a focus on their impacts on human health and the environment [44]. In a similar vein, manufacturers are required to submit safety data for nanomaterials under the REACH (Registration, Evaluation, Authorization, and Restriction of Chemicals) legislation created by the European Union. Maintaining awareness of these rules, conducting regular audits, and ensuring that all products containing nanomaterials meet the necessary safety requirements are all essential components of compliance strategies for the pulp and paper sector. To exchange best practices and stay ahead of changing regulations, businesses should also participate in industry collaborations and communicate proactively with regulatory agencies. Research findings indicate that companies that prioritize compliance not only enhance their reputation but also reduce the risk of legal liabilities and improve overall operational efficiency [118]. By adopting a comprehensive approach to regulatory compliance, the pulp and paper industry can effectively navigate the complexities of nanotechnology while ensuring the health and safety of workers and the environment [43].

6 Conclusion and Future Directions

By enhancing the usability, sustainability, and quality of paper products, nanotechnology has the potential to revolutionize the pulp and paper industry. It has been demonstrated that incorporating nanomaterials into paper, such as cellulose nanofibers, nanoclays, and silver nanoparticles, can improve its mechanical properties, barrier effectiveness, and antibacterial capabilities. By reducing the need for chemical additives and enhancing paper's recyclability, these developments not only help produce stronger and more resilient paper products but also promote environmental sustainability due to their distinctive nanofibrillar structure, availability, and renewability. Cellulose and lignocellulosic materials, in particular, offer considerable potential as nanomaterials. Because they can be made versatile and self-assemble into well-defined forms, they are ideal candidates for a range of uses in the papermaking process. Numerous aspects of this process can be enhanced by nanotechnology, such as the development of novel paper types or the application of nanofibers and other nanomaterials (such as coating minerals, fillers, and sizing) to raise the caliber of the final product. Through mechanical and chemical processes, cellulose nanofibers can be effectively produced from a variety of fibrous raw materials. They have demonstrated considerable promise as reinforcing materials in composites, particularly in enhancing biopolymers to create environmentally friendly, biodegradable films. Industry participants must collaborate and invest in the development and marketing of nanotechnology applications in the pulp and paper industry as research advances. Malaysian research and development efforts in nanotechnology have already garnered sufficient attention and action plans, demonstrating international interest in utilizing these cutting-edge materials for a more sustainable future industry.

Competing Interests The authors declare that they have no competing interest.

Declaration of generative AI and AI-assisted technologies in the writing process During the preparation of this work, the book chapter author(s) have not taken any assistance from AI, and the book chapter author(s) reviewed and edited the content as needed and take(s) full responsibility for the content of the publication.

References

1. Abd El-Thalouth JI, Abdelslam S, Omar AK, Raslan WM (2024) Effect of treatment with laser/titanium dioxide nanoparticles on printability of cellulose acetate fabric with disperse and natural dyes. J Text Color Polym Sci 21(1):221–230
2. Abd Elkodous M et al (2022) Cutting-edge development in waste-recycled nanomaterials for energy storage and conversion applications. Nanotechnol Rev 11(1):2215–2294
3. Abu Jarad N, Imran H, Imani SM, Didar TF, Soleymani L (2022) Fabrication of superamphiphobic surfaces via spray coating; a review. Adv Mater Technol 7(10):2101702
4. Agarwal M, Lvov Y, Varahramyan K (2006) Conductive wood microfibres for smart paper through layer-by-layer nanocoating. Nanotechnology 17(21):5319
5. Ahola S, Österberg M, Laine J (2008) Cellulose nanofibrils—adsorption with poly (amideamine) epichlorohydrin studied by QCM-D and application as a paper strength additive. Cellulose 15:303–314
6. Ahola S, Salmi J, Johansson L-S, Laine J, Österberg M (2008) Model films from native cellulose nanofibrils. Preparation, swelling, and surface interactions. Biomacromolecules 9(4):1273–1282
7. Ali A, Koloor SSR, Alshehri AH, Arockiarajan A (2023) Carbon nanotube characteristics and enhancement effects on the mechanical features of polymer-based materials and structures–a review. J Mater Res Technol 24:6495–6521
8. Amoabediny G et al (2009) Guidelines for safe handling, use and disposal of nanoparticles. In: Journal of physics: conference series. IOP Publishing, pp 012037
9. Antony Jose S et al (2025) A comprehensive review on cellulose nanofibers, nanomaterials, and composites: manufacturing, properties, and applications. Nanomaterials 15(5):356
10. Araki J (2013) Electrostatic or steric?–preparations and characterizations of well-dispersed systems containing rod-like nanowhiskers of crystalline polysaccharides. Soft Matter 9(16):4125–4141
11. Araki J, Wada M, Kuga S, Okano T (1998) Flow properties of microcrystalline cellulose suspension prepared by acid treatment of native cellulose. Coll Surf A: Physicochem Eng Aspects 142(1):75–82
12. Arman Kandirmaz E, Ozcan A (2019) Antibacterial effect of Ag nanoparticles into the paper coatings. Nordic Pulp Paper Res J 34(4):507–515

13. Azeredo HM et al (2009) Nanocomposite edible films from mango puree reinforced with cellulose nanofibers. J Food Sci 74(5):N31–N35
14. Aziz T, Li W, Zhu J, Chen B (2024) Developing multifunctional cellulose derivatives for environmental and biomedical applications: Insights into modification processes and advanced material properties. Int J Biol Macromol 134695
15. Azizi Samir MAS, Alloin F, Sanchez J-Y, El Kissi N, Dufresne A (2004) Preparation of cellulose whiskers reinforced nanocomposites from an organic medium suspension. Macromolecules 37(4):1386–1393
16. Bachheti RK, Bachheti A, Husen A (eds) (2023) Nanomaterials for environmental and agricultural sectors. Springer, Singapore. https://doi.org/10.1007/978-981-99-2874-3
17. Bahranowski K et al (2017) Influence of purification method of Na-montmorillonite on textural properties of clay mineral composites with TiO_2 nanoparticles. Appl Clay Sci 140:75–80
18. Balea A et al (2020) Industrial application of nanocelluloses in papermaking: a review of challenges, technical solutions, and market perspectives. Molecules 25(3):526
19. Barhoum A, Samyn P, Öhlund T, Dufresne A (2017) Review of recent research on flexible multifunctional nanopapers. Nanoscale 9(40):15181–15205
20. Beck-Candanedo S, Roman M, Gray DG (2005) Effect of reaction conditions on the properties and behavior of wood cellulose nanocrystal suspensions. Biomacromolecules 6(2):1048–1054
21. Bhatnagar A, Sain M (2005) Processing of cellulose nanofiber-reinforced composites. J Reinf Plast Compos 24(12):1259–1268
22. Bhushan B (2017) Introduction to nanotechnology. Springer handbook of nanotechnology. Springer, pp 1–19
23. Bondeson D, Mathew A, Oksman K (2006) Optimization of the isolation of nanocrystals from microcrystalline cellulose by acid hydrolysis. Cellulose 13:171–180
24. Borm PJ et al (2006) The potential risks of nanomaterials: a review carried out for ECETOC. Particle Fibre Toxicol 3:1–35
25. Boufi S et al (2016) Nanofibrillated cellulose as an additive in papermaking process: a review. Carbohydr Polym 154:151–166
26. Boyes WK, van Thriel C (2020) Neurotoxicology of nanomaterials. Chem Res Toxicol 33(5):1121–1144
27. Charitidis CA, Georgiou P, Koklioti MA, Trompeta A-F, Markakis V (2014) Manufacturing nanomaterials: from research to industry. Manuf Rev 1:11
28. Chauhan VS, Chakrabarti SK (2012) Use of nanotechnology for high performance cellulosic and papermaking products. Cellul Chem Technol 46(5):389
29. Chia C, Zakaria S, Ahamd S, Abdullah M, Jani SM (2006) Preparation of magnetic paper from kenaf: lumen loading and in situ synthesis method. Am J Appl Sci 3(3):1750–1754
30. Chia C, Zakaria S, Nguyen K, Abdullah M (2008) Utilisation of unbleached kenaf fibers for the preparation of magnetic paper. Indus Crops Prod 28(3):333–339
31. Cooper KP, Wachter RF (2011) Challenges and opportunities in nanomanufacturing. Instrum Metrol Stan Nanomanuf Opt Semicond V 8105:7–12
32. Corsi I et al (2018) Ecofriendly nanotechnologies and nanomaterials for environmental applications: key issue and consensus recommendations for sustainable and ecosafe nanoremediation. Ecotoxicol Environ Saf 154:237–244
33. Das R, Ambardekar V, Bandyopadhyay PP (2021) Titanium dioxide and its applications in mechanical, electrical, optical, and biomedical fields. Titanium Dioxide-Adv Appl
34. Dhiman J, Anupam K, Kumar V, Saruchi (2024) Varieties, characteristics, and applications of nanofillers in pulp and paper industry. Handbook of nanofillers. Springer, pp 1–32
35. Dong J et al (2023) Mechanically strong nanopapers based on lignin containing cellulose micro-and nano-hybrid fibrils: lignin content-fibrils morphology-strengthening mechanism. Carbohyd Polym 311:120753
36. Dong XM, Revol J-F, Gray DG (1998) Effect of microcrystallite preparation conditions on the formation of colloid crystals of cellulose. Cellulose 5:19–32
37. Dulany MA, Batten Jr GL, Peck MC, Farley CE (2000) Papermaking additives. Kirk-Othmer Encycl Chem Technol 1–28

38. Edo GI et al (2025) Advancing sustainable food packaging: the role of green nanomaterials in enhancing barrier properties. Food Eng Rev 1–35
39. Eslahi N, Mahmoodi A, Mahmoudi N, Zandi N, Simchi A (2020) Processing and properties of nanofibrous bacterial cellulose-containing polymer composites: a review of recent advances for biomedical applications. Polym Rev 60(1):144–170
40. Ezema IC, Ogbobe PO, Omah AD (2014) Initiatives and strategies for development of nanotechnology in nations: a lesson for Africa and other least developed countries. Nanoscale Res Lett 9:1–8
41. Fahmy TY, Mobarak F (2009) Advanced nano-based manipulations of molasses in the cellulose and paper discipline: Introducing a master cheap environmentally safe retention aid and strength promoter in papermaking. Carbohydr Polym 77(2):316–319
42. Fatyeyeva K, Chappey C, Marais S (2017) Biopolymer/clay nanocomposites as the high barrier packaging material: recent advances. Food Packag 425–463
43. Festus-Ikhuoria IC, Obiuto NC, Adebayo RA, Olajiga OK (2024) Nanotechnology in consumer products: a review of applications and safety considerations. World J Adv Res Rev 21(3):2050–2059
44. Foster B, Fazarro D, Trybula W, Tate J, Hanks C (2017) The world of nanotechnology. Nano-safety: what we need to know to protect workers. De Gruyter, Boston, pp 1–40
45. Fukahori S, Ichiura H, Kitaoka T, Tanaka H (2003) Photocatalytic decomposition of bisphenol A in water using composite TiO_2-zeolite sheets prepared by a papermaking technique. Environ Sci Technol 37(5):1048–1051
46. Gaur J et al (2024) Current trends: zinc oxide nanoparticles preparation via chemical and green method for the photocatalytic degradation of various organic dyes. Hybrid Adv 5:100128
47. Ghamari M et al (2024) Valorization of diverse waste-derived nanocellulose for multifaceted applications: a review. Int J Biol Macromol 136130
48. Gicquel E, Martin C, Garrido Yanez J, Bras J (2017) Cellulose nanocrystals as new bio-based coating layer for improving fiber-based mechanical and barrier properties. J Mater Sci 52:3048–3061
49. Graham K et al (2002) Polymeric nanofibers in air filtration applications. In: Fifteenth annual technical conference & expo of the American filtration & separations society, Galveston, Texas, pp 9–12
50. Gupta GK, Sailwal M, Shukla P (2024) Sustainable nanotechnology based techniques for mitigating the pollutants from pulp and paper industry. ACS Omega 9(49):47904–47919
51. Hashim U, Nadia E, Salleh S (2009) Nanotechnology development status in Malaysia: industrialization strategy and practices. Int J
52. Hashim U, Nadia E, Shahrir S (2009) Nanotechnology development in Malaysia: current status and implementation strategy. Int J Nanoelectron Mater 2(1):119–134
53. Henriksson M, Berglund LA (2007) Structure and properties of cellulose nanocomposite films containing melamine formaldehyde. J Appl Polym Sci 106(4):2817–2824
54. Henriksson M, Berglund LA, Isaksson P, Lindström T, Nishino T (2008) Cellulose nanopaper structures of high toughness. Biomacromolecules 9(6):1579–1585
55. Hoseinnejad M, Jafari SM, Katouzian I (2018) Inorganic and metal nanoparticles and their antimicrobial activity in food packaging applications. Crit Rev Microbiol 44(2):161–181
56. Hossain L et al (2020) Structure and swelling of cross-linked nanocellulose foams. J Coll Interface Sci 568:234–244
57. Hubbe MA (2006) Bonding between cellulosic fibers in the absence and presence of dry-strength agents–a review. BioResources 1(2):281–318
58. Ibrahim KS (2013) Carbon nanotubes-properties and applications: a review. Carbon Lett 14(3):131–144
59. Ichiura H, Kitaoka T, Tanaka H (2003) Removal of indoor pollutants under UV irradiation by a composite TiO_2–zeolite sheet prepared using a papermaking technique. Chemosphere 50(1):79–83
60. Ichiura H, Kubota Y, Wu Z, Tanaka H (2001) Preparation of zeolite sheets using a papermaking technique Part I dual polymer system for high retention of stock components. J Mater Sci 36:913–917

61. Ioelovich M (2008) Cellulose as a nanostructured polymer: a short review. BioResources 3(4):1403–1418
62. Iwamoto S, Nakagaito AN, Yano H, Nogi M (2005) Optically transparent composites reinforced with plant fiber-based nanofibers. Appl Phys A 81:1109–1112
63. Jalali E, Erasmus E, Schutte-Smith M, Visser HG (2024) Fixation of nanoparticles on fabric: applications in general health management. Mater Today Commun 41:110577
64. Jasmani L, Rusli R, Khadiran T, Jalil R, Adnan S (2020) Application of nanotechnology in wood-based products industry: a review. Nanoscale Res Lett 15:1–31
65. Jena AB, Philipson TJ (2008) Cost-effectiveness analysis and innovation. J Health Econ 27(5):1224–1236
66. Jonoobi M, Harun J, Mathew AP, Hussein MZB, Oksman K (2010) Preparation of cellulose nanofibers with hydrophobic surface characteristics. Cellulose 17:299–307
67. Julkapli NM, Bagheri S (2016) Developments in nano-additives for paper industry. J Wood Sci 62:117–130
68. Kamali M, Khodaparast Z (2015) Review on recent developments on pulp and paper mill wastewater treatment. Ecotoxicol Environ Saf 114:326–342
69. Kamel S (2007) Nanotechnology and its applications in lignocellulosic composites, a mini review. eXPRESS Polym Lett 1
70. Karimi S, Tahir PM, Karimi A, Dufresne A, Abdulkhani A (2014) Kenaf bast cellulosic fibers hierarchy: a comprehensive approach from micro to nano. Carbohyd Polym 101:878–885
71. Karthika R, Jayanthi B, Aruna A, Selvankumar T (2022) Nanocellulosic materials for paper-making and paper coating industry, handbook of nanocelluloses: classification, properties, fabrication, and emerging applications. Springer, pp 1001–1033
72. Kasmani JE, Mahdavi S, Alizadeh A, Nemati M, Samariha A (2013) Physical properties and printability characteristics of mechanical printing paper with LWC. BioResources 8(3)
73. Kay L, Shapira P (2009) Developing nanotechnology in Latin America. J Nanoparticle Res 11:259–278
74. Khatoon UT, Velidandi A (2025) An overview on the role of government initiatives in nanotechnology innovation for sustainable economic development and research progress. Sustainability 17(3):1250
75. Kim H-I, Moon G-H, Monllor-Satoca D, Park Y, Choi W (2012) Solar photoconversion using graphene/TiO_2 composites: nanographene shell on TiO_2 core versus TiO_2 nanoparticles on graphene sheet. J Phys Chem C 116(1):1535–1543
76. Kontturi E et al (2018) Advanced materials through assembly of nanocelluloses. Adv Mater 30(24):1703779
77. Korayem AH, Tourani N, Zakertabrizi M, Sabziparvar A, Duan W (2017) A review of dispersion of nanoparticles in cementitious matrices: nanoparticle geometry perspective. Constr Build Mater 153:346–357
78. Krajnik P et al (2016) Transitioning to sustainable production–part III: developments and possibilities for integration of nanotechnology into material processing technologies. J Cleaner Prod 112:1156–1164
79. Kumar CV et al (2022) The impact of engineered nanomaterials on the environment: release mechanism, toxicity, transformation, and remediation. Environ Res 212:113202
80. Lead JR et al (2018) Nanomaterials in the environment: behavior, fate, bioavailability, and effects—an updated review. Environ Toxicol Chem 37(8):2029–2063
81. Lee K et al (2021) Double-crosslinked cellulose nanofiber based bioplastic films for practical applications. Carbohyd Polym 260:117817
82. Lengowski EC, Bonfatti Júnior EA, Kumode MMN, Carneiro ME, Satyanarayana KG (2019) Nanocellulose in the paper making. Sustain Polym Compos Nanocompos 1027–1066
83. Li A et al (2021) Overview of nanocellulose as additives in paper processing and paper products. Nanotechnol Rev 10(1):264–281
84. Li LaiSheng LL, Zhu WanPeng ZW, Zhang PengYi ZP, Chen ZhongYin CZ, Han WenYa HW (2003) Photocatalytic oxidation and ozonation of catechol over carbon-black-modified nano-TiO_2 thin films supported on Al sheet

85. Li Q, Wang S, Jin X, Huang C, Xiang Z (2020) The application of polysaccharides and their derivatives in pigment, barrier, and functional paper coatings. Polymers 12(8):1837
86. Lian JZ, Balapa V, Goetheer E, Cucurachi S (2024) Production of carbon nanotubes from captured carbon: an ex-ante life cycle assessment case study. Chem Eng J 502:158007
87. Lin X et al (2012) Fabrication of highly-aligned, conductive, and strong graphene papers using ultralarge graphene oxide sheets. ACS Nano 6(12):10708–10719
88. Ling Y, Luo Y, Luo J, Wang X, Sun R (2013) Novel antibacterial paper based on quaternized carboxymethyl chitosan/organic montmorillonite/Ag NP nanocomposites. Indus Crops Prod 51:470–479
89. Liu DY, Sui G, Bhattacharyya D (2014) Synthesis and characterisation of nanocellulose-based polyaniline conducting films. Compos Sci Technol 99:31–36
90. Liu S-D, Li D-S, Yang Y, Jiang L (2019) Fabrication, mechanical properties and failure mechanism of random and aligned nanofiber membrane with different parameters. Nanotechnol Rev 8(1):218–226
91. Liu W et al (2022) Cellulose nanopaper: fabrication, functionalization, and applications. Nano-Micro Lett 14(1):104
92. Lu Z et al (2007) Layer-by-layer nanoparticle coatings on lignocellulose wood microfibers. Coll Surf A: Physicochem Eng Aspects 292(1):56–62
93. Ma X et al (2018) Energy and carbon coupled water footprint analysis for Kraft wood pulp paper production. Renew Sustain Energy Rev 96:253–261
94. Mahlambi MM, Ngila CJ, Mamba BB (2015) Recent developments in environmental photocatalytic degradation of organic pollutants: the case of titanium dioxide nanoparticles—a review. J Nanomater 2015(1):790173
95. Man Z et al (2011) Preparation of cellulose nanocrystals using an ionic liquid. J Polym Environ 19:726–731
96. Manninen M, Kajanto I, Happonen J, Paltakari J (2011) The effect of microfibrillated cellulose addition on drying shrinkage and dimensional stability of wood-free paper. Nordic Pulp Paper Res J 26(3):297–305
97. Mazhari Mousavi SM, Afra E, Tajvidi M, Bousfield D, Dehghani-Firouzabadi M (2017) Cellulose nanofiber/carboxymethyl cellulose blends as an efficient coating to improve the structure and barrier properties of paperboard. Cellulose 24:3001–3014
98. Merah N, Ashraf F, Shaukat MM (2022) Mechanical and moisture barrier properties of epoxy–nanoclay and hybrid epoxy–nanoclay glass fibre composites: a review. Polymers 14(8):1620
99. Mersha AK, Finina BF, Gebreslassie G (2023) Applications, opportunities and challenges of nanotechnology in the food industry. Nanomater Environ Agric Sectors 255–273
100. Minor J, Atalia R, Harten T (1993) Improving interfibre bonding of recycled fibres. J Pulp Paper Sci 19(4)
101. Mohieldin S, Zainudin E, Paridah M, Ainun Z (2011) Nanotechnology in pulp and paper industries: a review. Key Eng Mater 471:251–256
102. Mokhothu TH, John MJ (2015) Review on hygroscopic aging of cellulose fibres and their biocomposites. Carbohyd Polym 131:337–354
103. Montalvo SJ, Guerrero LE, Borja R (2014) Improvement in nitrification through the use of natural zeolite: influence of the biomass concentration and inoculum source. Int J Environ Sci Technol 11:43–52
104. Morais FP, Carta AM, Amaral ME, Curto JM (2021) Micro/nano-fibrillated cellulose (MFC/NFC) fibers as an additive to maximize eucalyptus fibers on tissue paper production. Cellulose 28(10):6587–6605
105. Mousa SA, Abdallah H, Khairy S (2025) The use of green synthesized TiO_2/MnO_2 nanoparticles in solar power membranes for pulp and paper industry wastewater treatment. Sci Rep 15(1):2102
106. Muñoz P, Juárez M, Morales M, Mendívil M (2013) Improving the thermal transmittance of single-brick walls built of clay bricks lightened with paper pulp. Energy Build 59:171–180
107. Nagarajan K et al (2021) A comprehensive review on cellulose nanocrystals and cellulose nanofibers: pretreatment, preparation, and characterization. Polym Compos 42(4):1588–1630

108. Naguib M et al (2014) One-step synthesis of nanocrystalline transition metal oxides on thin sheets of disordered graphitic carbon by oxidation of MXenes. Chem Commun 50(56):7420–7423
109. Nair GM, Sajini T, Mathew B (2022) Advanced green approaches for metal and metal oxide nanoparticles synthesis and their environmental applications. Talanta Open 5:100080
110. Nakagaito AN, Yano H (2005) Novel high-strength biocomposites based on microfibrillated cellulose having nano-order-unit web-like network structure. Appl Phys A 80:155–159
111. Narvestad H, Gregersen ØW, Kure K-A (2013) Filler clay induced discolouration of bleached mechanical pulp. Nordic Pulp Paper Res J 28(1):68–81
112. Nasir M, Hashim R, Sulaiman O, Asim M (2017) Nanocellulose: preparation methods and applications, cellulose-reinforced nanofibre composites. Elsevier, pp 261–276
113. Naskar S, Pillay SA, Chanda M (1998) Photocatalytic degradation of organic dyes in aqueous solution with TiO_2 nanoparticles immobilized on foamed polyethylene sheet. J Photochem Photobiol A: Chem 113(3):257–264
114. Nath D, Santhosh R, Pal K, Sarkar P (2022) Nanoclay-based active food packaging systems: a review. Food Packag Shelf Life 31:100803
115. Nazbeen N, Wang Q (2016) Guidelines and best practices for safe handling of nanomaterials in research laboratories and industries. A project of ARCI, Department of Science & Technology, Government of India, Hyderabad, India, p 25
116. Nickerson RF, Habrle J (1947) Cellulose intercrystalline structure. Indus Eng Chem 39(11):1507–1512
117. Nogi M et al (2013) High thermal stability of optical transparency in cellulose nanofiber paper. Appl Phys Lett 102(18)
118. Nwachukwu G, Oladepo O, Avickson EK (2024) Quality control in financial operations: best practices for risk mitigation and compliance. World J Adv Res Rev 24(1):735–749
119. Othman SH, Nordin N, Azman NAA, Tawakkal ISMA, Basha RK (2021) Effects of nanocellulose fiber and thymol on mechanical, thermal, and barrier properties of corn starch films. Int J Biol Macromol 183:1352–1361
120. Padil VV et al (2022) Sustainable and safer nanoclay composites for multifaceted applications. Green Chem 24(8):3081–3114
121. Padzil FNM et al (2022) Nanocellulose composites in the pulp and paper industry. Indus Appl Nanocellulose Nanocompos 375–395
122. Pereira B, Arantes V (2020) Production of cellulose nanocrystals integrated into a biochemical sugar platform process via enzymatic hydrolysis at high solid loading. Indus Crops Prod 152:112377
123. Peresin MS, Habibi Y, Zoppe JO, Pawlak JJ, Rojas OJ (2010) Nanofiber composites of polyvinyl alcohol and cellulose nanocrystals: manufacture and characterization. Biomacromolecules 11(3):674–681
124. Pires JR, Souza VG, Fernando AL (2019) Valorization of energy crops as a source for nanocellulose production–current knowledge and future prospects. Indus Crops Prod 140:111642
125. Puri N, Gupta A (2023) Water remediation using titanium and zinc oxide nanomaterials through disinfection and photo catalysis process: a review. Environ Res 227:115786
126. Rajalingam K (2013) A case study and explorative analysis of the development of nanotechnology in Malaysia. J Sustain Dev Stud 5(1)
127. Ren F et al (2019) Highly bendable and durable waterproof paper for ultra-high electromagnetic interference shielding. Polymers 11(9):1486
128. Ren L, Zeng YP, Jiang D (2007) Fabrication of gradient pore TiO_2 sheets by a novel freeze–tape-casting process. J Am Ceramic Soc 90(9):3001–3004
129. Rentzhog M, Fogden A (2005) Rheology and surface tension of water-based flexographic inks and implications for wetting of PE-coated board. Nordic Pulp Paper Res J 20(4):399–409
130. Ruzi M, Celik N, Onses MS (2022) Superhydrophobic coatings for food packaging applications: a review. Food Packag Shelf Life 32:100823

131. Sadasivuni KK et al (2020) Recent advances in mechanical properties of biopolymer composites: a review. Polym Compos 41(1):32–59
132. Sadeghifar H, Ragauskas A (2020) Lignin as a UV light blocker—a review. Polymers 12(5):1134
133. Saito T, Kimura S, Nishiyama Y, Isogai A (2007) Cellulose nanofibers prepared by TEMPO-mediated oxidation of native cellulose. Biomacromolecules 8(8):2485–2491
134. Samyn P, Barhoum A, Öhlund T, Dufresne A (2018) Nanoparticles and nanostructured materials in papermaking. J Mater Sci 53:146–184
135. Shatkin JA, Kim B (2015) Cellulose nanomaterials: life cycle risk assessment, and environmental health and safety roadmap. Environ Sci Nano 2(5):477–499
136. Shen P, Tang Q, Chen X, Li Z (2022) Nanocrystalline cellulose extracted from bast fibers: preparation, characterization, and application. Carbohyd Polym 290:119462
137. Siqueira G, Tapin-Lingua S, Bras J, da Silva Perez D, Dufresne A (2010) Morphological investigation of nanoparticles obtained from combined mechanical shearing, and enzymatic and acid hydrolysis of sisal fibers. Cellulose 17:1147–1158
138. Spagnuolo L, D'Orsi R, Operamolla A (2022) Nanocellulose for paper and textile coating: the importance of surface chemistry. ChemPlusChem 87(8):e202200204
139. Syduzzaman M et al (2025) Carbon nanotubes: structure, properties and applications in the aerospace industry. Results Mater 25:100654
140. Tayeb AH, Amini E, Ghasemi S, Tajvidi M (2018) Cellulose nanomaterials—binding properties and applications: a review. Molecules 23(10):2684
141. Uddin F (2008) Clays, nanoclays, and montmorillonite minerals. Metall Mater Trans A 39(12):2804–2814
142. Wang L et al (2021) Review on nonconventional fibrillation methods of producing cellulose nanofibrils and their applications. Biomacromolecules 22(10):4037–4059
143. Wang S, Cheng Q, Rials TG, Lee S-H (2006) Cellulose microfibril/nanofibril and its nanocompsites. In: Proceedings of the 8th Pacific rim bio-based composites symposium, pp 301–308
144. Wegner TH, Jones EP (2009) A fundamental review of the relationships between nanotechnology and lignocellulosic biomass. Nanosci Technol Renew Biomater 1:1–41
145. Worku LA, Bachheti A, Bachheti RK, Rodrigues Reis CE, Chandel AK (2023) Agricultural residues as raw materials for pulp and paper production: overview and applications on membrane fabrication. Membranes 13(2):228
146. Worku LA, Bachheti RK, Bachheti A, Milessi TS, Chandel AK (2024) Understanding the biochemical changes at molecular level during biomass pretreatment: a comprehensive analysis. Cellulose 31:7281–7312 https://doi.org/10.1007/s10570-024-06081-7
147. Zhang L et al (2024) Novel diacid–superbase ionic liquids for efficient dissolution of cellulose and simultaneous preparation of multifunctional cellulose materials. Green Chem 26(15):8794–8807
148. Zhen ZH, Wang ZS (2013) Study of coating binder migration and affected parameters. Adv Mater Res 790:45–48
149. Zhu Q et al (2022) Recent advances in nanotechnology-based functional coatings for the built environment. Mater Today Adv 15:100270
150. Zimmermann MV et al (2016) Drying techniques applied to cellulose nanofibers. J Reinf Plastics Compos 35(8):682–697
151. Zinge C, Kandasubramanian B (2020) Nanocellulose based biodegradable polymers. Eur Polym J 133:109758
152. Zulaikha W, Hassan MZ, Ismail Z (2022) Recent development of natural fibre for nanocellulose extraction and application. Mater Today: Proc 66:2265–2273

Nanocellulose: Properties, Synthesis, Characterization, and Applications

Altaf H. Basta and Vivian F. Lotfy

Abstract The biodegradability of nanocellulose (NC) promotes its use as an eco-precursor for several purposes. The extraction of this nanomaterial from undesirable wastes, besides achieving green disposal of these wastes, will boost the eco-friendly nature of the products. This chapter highlights on the methods of preparing the nature-based NCs, and bacterial-based NCs, techniques of their characterization, as well as how the synthesis route will affect the properties of the nanoproduct, and summarized the strategies to be followed for enhancing their performance in different applications, such as sustaining release of drugs (e.g., medical application), thermal stability for application in HDPE-composites, dispersing/adsorbent of dyes, as well as in electronics and energy purposes. This chapter aims to provide a comprehensive overview of the synthesis methods, structural and physicochemical characterization, and performance enhancement strategies of nanocellulose derived from both natural and bacterial sources. It also seeks to demonstrate the relationship between synthesis methods and final material properties, and to highlight its emerging applications across medical, environmental, and industrial sectors.

Keywords Nanocellulose · Synthesis · Characterization · Sustainable applications · Waste valorization

1 Introduction

Nanocellulose has diverse applications across various fields [1]. In packaging, it enhances barrier properties and biodegradability, making it a sustainable alternative to plastics. In optics, it is used in transparent films and coatings. In medicine, nanocellulose serves in drug delivery, wound dressings, and tissue engineering. In food, it improves packaging by preventing spoilage and enhancing shelf life. In electronics, it is utilized in flexible displays and sensors. In aerospace, its lightweight and strong properties contribute to advanced materials. Additionally, nanocellulose

A. H. Basta (✉) · V. F. Lotfy
Cellulose and Paper Department, National Research Centre, Dokki, Giza, Egypt
e-mail: altaf_halim@yahoo.com; altaf_basta2004@yahoo.com

R. K. Bachheti (ed.), *Nanotechnology in the Pulp and Paper Industry*, Smart Nanomaterials Technology, https://doi.org/10.1007/978-981-95-3473-9_2

is biodegradable, making it an eco-friendly option in many applications [7, 8, 19–21, 25, 32, 57, 79, 80, 103, 106, 118, 135–138]. The synthesis of these biodegradable nanoparticles from undesirable wastes will add eco-effective performance to the products, and at the same time achieve green disposal of these wastes.

Nanocellulose refers to cellulose at the nanoscale level. The precursors used for isolating nanocellulose (NC) can be derived from wood, nonwood sources, and bacterial strains. Depending on the synthesis method, nanocellulose exhibits different morphologies, including cellulose nanocrystals (CNCs), cellulose nanofibers (CNFs), and bacterial cellulose [2, 17, 73]. In light of environmental legislation, the valorization of agricultural waste as nonwood sources is crucial, as these materials serve not only as precursors for nanocellulose but also in the production of porous carbons (activated carbons and carbon nanotubes), cellulose derivatives that act as bioactive compounds to inhibit cancer cell proliferation, and functional lignocellulosic products such as carbon adsorbents. Additionally, nanocellulose is utilized in the development of hydrogels for agricultural and pharmaceutical applications, optical materials, and has been the subject of extensive research by many scientists [22, 23, 41, 42, 61, 113]. This chapter emphasizes the significant features of nanocelluloses, as summarized in Table 1. It covers the properties and methods of synthesis of nanocellulose, the types of precursors used for its production, as well as the characterization techniques and their various applications.

2 Properties of Nanocelluloses

Nanocelluloses have been considered a more attractive subject by many researchers than cellulose: this is due to its interesting features such as renewability, relatively high physicochemical property, resistance to thermal degradation, crystallinity, in addition to strength properties and its behavior to use in high-performance application [39, 110, 114, 121, 133]. Cellulose- or microorganism-based sources are used as precursors for production of NCs. With regard to cellulose-based NCs, many factors will affect the properties of the resulting NC (thermal stability, crystallinity), for example, precursor, mechanical route, or chemical for extraction, where it is extracted as sulfate included nanocrystalline cellulose (NCC) or oxidized cellulose (NFs). For example, using mechanical treatment or HCl or H_3PO_4 during the extraction process provided high thermal stability and crystalline NC than that produced from H_2SO_4 [34, 46, 48, 59, 77, 83, 96, 102, 126, 127, 131].

Regarding the bacterial cellulose (BC), it is pure cellulose, with above crystallinity index (>60%) than cellulose-NC. When compared to plant cellulose, the average DP of BC is 16,000–6000 [39, 123, 124]. For plant-based cellulose, it is 13,000 to 14,000. Moreover, BC is predominantly as found as cellulose Iα, while the plant cellulose is Cellulose Iβ [40].

Various factors significantly impact the properties of bacterial cellulose (BC), as detailed in Table 1. Firstly, culture conditions play a crucial role; for instance, the crystallinity and degree of polymerization of BC are greater in static processes compared

Table 1 Nanocellulose: properties, synthesis, characterization, and applications

Focus area	From plant precursors	From microorganism
Properties of nanocelluloses	• Particle size less than 100 nm (nanoscale) • Renewable and based on biomass • Low density, cost, and energy consumption • High mechanical and thermal stability • In aqueous dispersion media, it has self-assembly property • Reinforcing capabilities • Long-lasting nature [114, 121]	• Fibrils in nanoscale (diameter ~ 10–100 nm) • No chemical treatments for removing lignin, hemicellulose, extractives • High tensile strength • Has crystalline cellulose I • Light weight • Remarkable capacity to hold water • Directly synthesized the cellulose in the form of membrane during biosynthesis • Direct modification of cellulose during assembly [39, 110, 133]
Type and synthesis process	Type: nanocrystalline cellulose (CNCs) and nanofibers (CNFs) Synthesis methods: • Mechanical: high-pressure homogenization, microfluidization, grinding, cryocrushing, high intensity ultrasonication, ball milling, and ultrafine grinding processes [110, 131] • Chemical: Acid hydrolysis (H_2SO_4, HCl, H_3PO_4, phosphotungstic acid., as well as sodium periodate [59, 77] • TEMPO, APS, $KMnO_4$ and hybrid treatment [45, 46, 48, 50, 96, 108, 132, 134, 144]	Influencing factors [15, 39, 40, 58, 98, 123, 124] • Culture conditions: Static and agitation • Strain used, Carbon and nitrogen sources • Additives for improving the production yield • Temperature and pH value • Oxygen and carbon dioxide pressures
Precursors of nanocelluloses	• Natural cellulose (wood and nonwood) e.g., cotton, hemp, flax, wheat straw, mulberry bark, ramie, avicel, tunicin, cellulose from algae and bacteria [23, 24, 44, 60, 69, 75, 85, 90, 117]	Bacterial strains *Acetobacter, Achromobacter, Aerobacter, Agrobacterium, Alcaligenes,* • *Pseudomonas, Rhizobium sarcina* [15, 18, 27, 40, 47, 55, 62, 71, 81, 107, 120]

(continued)

Table 1 (continued)

Focus area	From plant precursors	From microorganism
Feature	CNCs: Rod-like structure, high dispersion in organic solvents (DMSO, ethylene glycol), variable length and diameter depending on the source material: Cotton (length: 200–300 nm; diameter: 10–30 nm); Ascidian (diameter: >1 μm) Plants (length 100–250 nm; diameter: 5–70 nm), Algae and bacteria: 100 nm to several micrometers CNFs: Long, flexible, and interconnected fibrils, contains amorphous and crystalline chains, high aspect ratio, specific surface area, high mechanical properties, and percolation threshold 5–60 nm in diameter and several micrometers in length [117]	• Ultrapurity • High crystallinity (60–90%), and average degree of polymerization (2000–8000) • Water-holding capacity reaches > 100 times/dry weight) • Reticulate fibrous structure, optical transparency, and bio-compatibility • Possess high surface area, mechanical strength, and thermal stability • Polyfunctionality, moldability, and nontoxicity [15, 58, 98]
	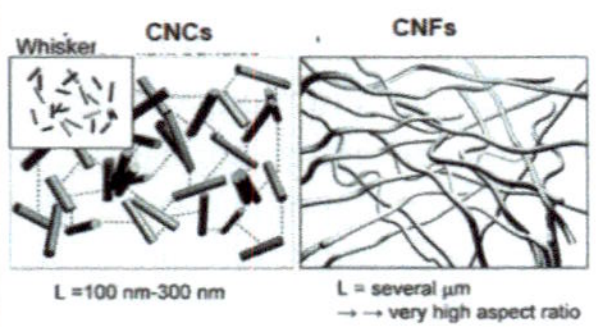	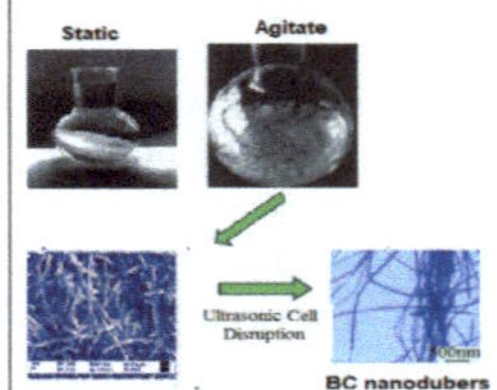

(continued)

Table 1 (continued)

Focus area	From plant precursors	From microorganism
Characterization techniques	• TEM and AFM to estimate the dimensions, and particle dimension distribution (PDD, aspect ratio • Crystallinity: by XRD and FTIR technique • TGA-techniques for determining the thermal stability, moreover evidenced the crystalline structure [28, 29, 31, 37, 49, 63, 68, 79, 86, 92, 100, 105]	• TEM and AFM to estimate the dimensionsand particle dimension distribution (PDD, aspect ratio • *Crystallinity: by XRD and FTIR technique,* • TGA-techniques [28, 31, 43, 78, 92, 100, 105]
Applications	In general, the application of cellulose-/BC-based NCs are * medical application, * improving the synthetic polymer, composites, *dispersing agent of dyes, * paper production, and food application [5, 24, 33, 53, 54, 64, 65, 76, 80, 84, 87, 99, 109, 125, 143] 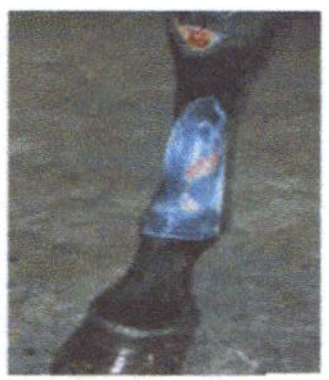 Application as dressing material on wounds of horses. 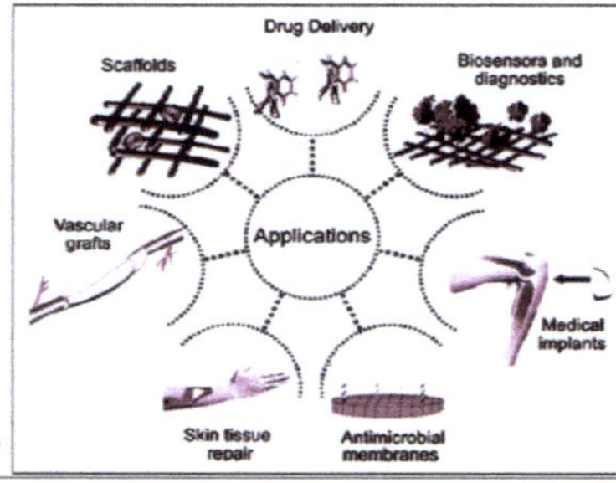 	

to agitated ones [122]. Additionally, agitated BC tends to have a smaller particle size, resulting in a higher water-holding capacity and a lower Young's modulus than static bacterial cellulose. Furthermore, the type of bacterial strain used and the additives present in the culture medium also influence the properties of BC, as illustrated in Tables 2 and 3.

Table 2 Effect of microorganisms on cellulose producers [39]

Organisms (genus)	Cellulose produced	Biological role
Acetobacter	Extracellular pellicle	To keep in aerobic
Achromobacter	Cellulose ribbons	Environment
Aerobacter	Cellulose fibrils	Flocculation in wastewater
Agrobacterium	Cellulose fibrils	Flocculation in wastewater
Alcaligenes	Short fibrils	Attachment of plant tissues
Pseudomonas	Cellulose fibrils	Flocculation in wastewater
Rhizobium	No distinct fibrils	Flocculation in wastewater
Sarcina	Short fibrils	Attached to most plants
Zoogloea	Amorphous cellulose	Unknown
	Not well defined	Flocculation in wastewater

Table 3 Effect of carbon sources on the production of cellulose by *Acetobacter xylinum* [39]

Source	Yield, %	Source	Yield, %
Monosaccharides		*Alcohols*	
D-Glucose	100	Ethanol	4
D-Fructose	92	Ethylene glycol	1
D-Galactose	15	Diethylene glycol	1
D-Mannose	3	Propylene glycol	8
D-Xylose	11	Glycerol	93
1-Arabinose	14	Myo-inositol	17
L-Sorbose	11	D-Arabitol	620
		D-Mannitol	380
Disaccharides		*Organic acids*	
Lactose	16	Citrate	20
Maltose	7	L-Malate	15
Sucrose	33	Succinate	12
Cellobiose	7–11		
Polysaccharides		*Others*	62
Starch	18	Glucono-lactone	0.5
		O-methyl-glucose	2

3 Type and Synthesis Process of Nanocelluloses

3.1 Mechanical Methods

The techniques involve high-pressure homogenization, micro-fluidization, grinding, cryo-crushing, high-intensity focused ultra-sonication, ball milling, and ultrafine grinding processes to produce cellulose nanofibers [110], or using the multistages: steam-explosion, followed by bleaching with hydrogen peroxide, blending under high-speed, finally ultrasonic treatments [131]. The last process provided nanocellulose fibers available to use as reinforcing agents in polymer materials at high temperature, due to their possess high thermal stability.

3.2 Chemical Methods

3.2.1 Cellulose Nanocrystals

The chemical approaches proceed via acid hydrolysis or acid free reagents. The acid hydrolysis was performed by using sulfuric acid (H_2SO_4), hydrochloric acid (HCl), hydrobromic acid (HBr), and phosphoric acids (H_3PO_4). The thermal stability of nanocellulose is significantly influenced by the hydrolyzed acid. Henrique et al. [59] found that the highest thermal stability of NCs from different sources (purified mango seeds and Kraft pulping) was higher in the case of HCl when compared to H_2SO_4. Lowering the concentration of acid is preferred to prepare thermal stable nanocellulose, where lowering the concentration of sulfuric acid from 64% to 20–47% led to produce high thermally stable NC and 35% sulfuric acid provided higher thermal stability than that prepared by 47% [34]. According to Roman and Winter [102], the drawback of high concentration of acid, especially sulfuric acid, is to the removal of the protective waxes and lignin layers from the fiber that make the fiber stronger. This view is also supported by Jintian and Yabin [67]. Tang et al. [111] and Lu and Hsieh [83] confirmed the beneficial role of phosphoric acid in the production of stable nanofibers from old corrugated container fibers.

One of the successful trials to overcome the drawback from sulfuric acid hydrolysis was the hydrolysis using oxalic acid [146]. Chen et al. [30] synthesized the high carboxylated-content nanocellulose from a pulp of eucalyptus kraft pulp and oxalic acid (50–70 wt%). Blends of oxalic, maleic, p-toluenesulfonic, and benzene-sulfonic acids, or phosphotungstic acid were successful in providing thermal stable NCs when compared to mineral acids. This can be ascribed to the relatively lower damage to the crystalline regions and surface structure during the mild conditions of hydrolysis process [77].

3.2.2 Cellulose Nanofibers (CNFs)

CNFs are synthesized by acid-free reagents; many reagents are used, such as 2,2,6,6-tetramethylpiperidine-1-oxyl radical (TEMPO), ammonium persulfate (APS), potassium permanganate. These reagents can be used to synthesize oxidized nanocelluloses. The TEMPO was investigated in the 1990s. This route is more advantageous than acid hydrolysis, because it is easy, cheap, and consumes very low energy [46, 48, 50, 108]. Recently, ammonium persulfate (APS) was used as strong oxidizing agents for obtaining carboxylayted nanocelluloses in the context of reducing the wastes remaining from acid. Many researchers studied the role of cellulose isolation methods (soda, hydrogen peroxide, and multistep) on the nanocellulose characteristics by the APS [96, 132]. The less common oxidizing agents and low-cost approach for preparing NCs-COOH are achieved by $KMnO_4$ and $(COOH)_2$ [45, 134, 144].

Combining the mechanical with chemical treatments is also reported in the production of nanocelluloses, for example, combining the ultrasound during APS oxidation and ultrasonic homogenizing method together with ammonium persulfate [45, 134].

3.3 *Bacterial Cellulose (BC)*

This type of nanocellulose is produced by various bacterial genera, including *Acetobacter xylinum, Acetobacter hansenii, Acetobacter pasteurianus*, and *Komagataeibacter europaeus*. It shares a similar structure with plant-derived cellulose but features a network of fine fibers with a diameter of approximately 0.1 μm. Additionally, bacterial cellulose exhibits higher crystallinity and an average degree of polymerization compared to plant cellulose [39, 123, 124]. Changes in the cellulose crystal structure and degree of polymerization occur based on the culture conditions, whether static or agitated. In agitated cultures, there is a reduction in crystallinity, resulting in a more disordered structure and a lower degree of polymerization. This structural disorder, combined with a smaller particle size, contributes to a higher water-holding capacity [15, 40]. Research indicates that additives such as carboxymethyl cellulose (CMC) and sodium lignosulfonate can significantly enhance the productivity of bacterial nanocellulose (BNC). For instance, the addition of lignosulfonate has been shown to improve BNC yield by approximately 57%, while CMC also contributes positively to fermentation processes [58, 98].

4 Precursors of Nanocelluloses

4.1 *Nanocellulose-Based Natural Origins*

These precursors can be obtained from wood, nonwood (agriculture wastes), and their pulps, or pre-prepared microcrystalline cellulose (MCC). Wood-based NC is usually synthesized by homogenization or via pretreatment by crushing in liquid nitrogen and refining then subjected to homogenizing [60, 69, 117]. Sulfite pulp of wood was used as a precursor for isolating the CNC. The increase of H_2SO_4/pulp ratio and the longer time of hydrolysis gives rise to shorter CNCs with low dispersity index [90, 128].

The utilization of nonwood wastes (flax, hemp, sisal, kenaf, and agricultural crops) as precursors of NCs as potential route to green disposal and increasing their value was also studied by many researchers. Cotton has a high cellulose content, does not need the purification step and gives a high yield. Hence it is a popular precursor for the synthesis of NCs [44, 75, 89]. The reagent used to purify the agriculture from noncellulosic components by pulping process (alkali, acidic or neutral) also effects on behaviour of NC, alters in the particle size distribution and thermal behavior of the resulted NCs (CNC and CNFs) [23, 24].

On using MCC precursor and mild acidic solution of 1-butyl-3-methylimidazolium hydrogen sulphate, as ionic liquid, the yield of the produced CNC is $48 \pm 2\%$.

4.2 *Nanocellulose-Based Bacterial Strains*

Nanocellulose (BC) is produced using various bacterial genera under static or agitation conditions, resulting in either a gelatinous pellicle or gel. To reduce production costs, these bacteria can be isolated from rotten fruits or cultivated with industrial wastes, including corn steep liquor, konjac powder, and fruit juices. Other sources such as maple syrup, thin stillage, wheat straw, and crude glycerol from biodiesel production can also be utilized. These alternatives effectively replace glucose or yeast, supplying essential carbon and nitrogen sources in the culture medium [18, 27, 55, 62, 71, 81, 120]. It is interesting to note that The isolated *Komagataeibacter intermedius MO* from rotten Guava and replacing half of yeast (nitrogen source) by Cheese whey improved the properties of BC pellicles (crystallinity, thermal stability and strength properties). While, on replacing the N and C sources as a whole by cheese by-product resulted high elastic BC pellicles and yield equivalent to that produced on using Hestrin-Schramm (high cost medium) [81]. The literature also reported the genetic engineering route to improve the production yield of BC and reduce the nutrient broth price. An example of enzyme engineering the sucrose phosphorylases has effective potential for bio-catalytic conversion of common sugar into products with attractive properties [47, 107].

5 Characterization Techniques

5.1 *Microscopic Techniques*

Various microscopic techniques (SEM, TEM, and AFM) are used to characterize the NCs. The TEM images showed the rod-like-shape particles of NC. Their particle dimension distribution (PDD) and aspect ratio was estimated, using software ImageJ. Program (Java-based public domain image processing), from length and diameter measures at different horizontal and vertical projections [72, 86]. The study recommended the relation between the method of extraction reagents (pulping and converting cellulose to nanoform) on PDD of the NCs, which will reflect on its applications as stabilizing or adsorbent agent of Indigo dye. The NC of type CNCs with length (204.4 ± 107.8 nm) and diameter (6.7 ± 2.3 nm), from alkaline pulping of rice straw (NaOH-AQ reagent) is beneficial as stabilizing agent of dye; while the CNFs from using the RS-neutral sulfite pulp precursor (PDD: 4.98 ± 1.6 and 90.5 ± 3.14) was most efficiency as dye adsorbent. Figure 1 illustrates the transmission electron microscopy (TEM) images along with particle size distribution data (PDD) of nanocrystals obtained through different methods, including acid hydrolysis (CNC) and oxidation using permanganate (K-ONCs) and ammonium persulfate (APS-ONCs). Additionally, the use of alkali-reduced sulfite (RS) pulp with NaOH-AQ-BH reagent as a precursor for nanocellulose resulted in CNCs with a low aspect length-to-width ratio [23, 24]. Balsa wood-CNCs with suitable crystallinity and aspect ratios can be used in producing polymer nanocomposites [90].

The atomic force microscopy (AFM) technique is essential for analyzing the surface morphology of cellulose nanocrystals (CNCs) extracted using sulfuric acid (H_2SO_4) and bacterial cellulose (BC) precursors. It provides detailed insights into the size, shape, and structural characteristics of these nanomaterials [75, 78]. This technique is also employed to estimate the cross-sectional dimensions of nanocellulose. Figure 2 presents an AFM image comparing mulberry branch bark-derived CNCs with bacterial cellulose (BC). The CNCs exhibit a diameter of approximately 20 nm, a length of 300–400 nm, and an aspect ratio ranging from 15 to 20. In contrast, the reticulated structure of BC features microfibrils, bundles, and ribbons with diameters of 17 nm, 37 nm, and 62 nm, respectively. These nanoscale dimensions further support the nanostructured nature of BC.

5.2 *Crystallinity*

The X-ray diffractometer was utilized to record the X-ray diffraction (XRD) patterns of nanocellulose samples, allowing for the determination of their degree of crystallinity using equations reported by Scherrer [105] and Doumeng et al. [37]. Table 4 presents the crystallinity parameters of naturally sourced cellulose nanocrystals (NCs) in comparison with bacterial cellulose nanocrystals (BC-NCs).

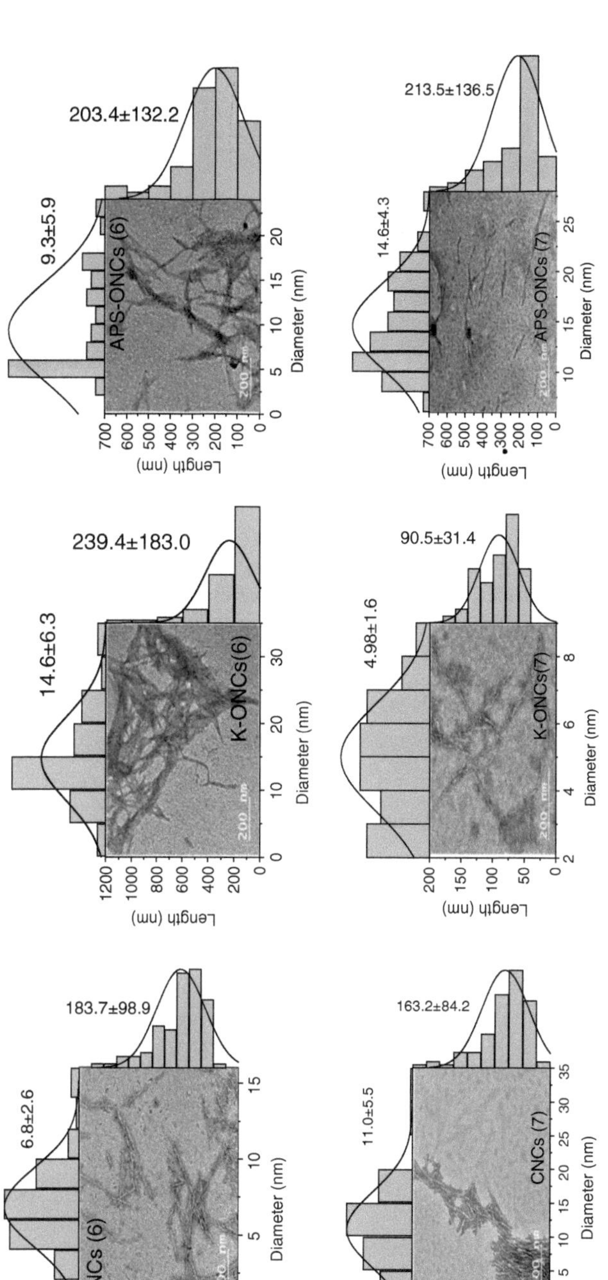

Fig. 1 TEM images of nanocelluloses from acid hydrolysis (CNCs), permanganate oxidation (K-ONCs) and APS oxidation (APS-ONCs), and their PDD [66]

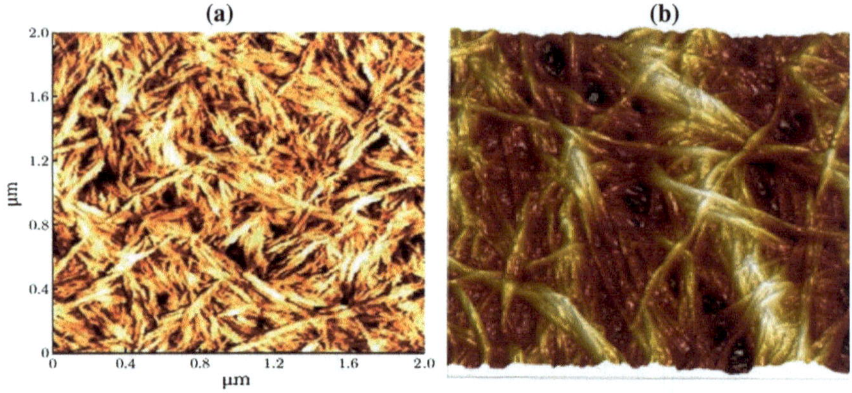

Fig. 2 AFM images of **a** mulberry branch bark-based CNC [65] and **b** BC [80]

Table 4 Crystallinity index of different wood-based, agriculture-based NCs in comparison with various drying BCs

	Precursor	Type	CI	References
Wood precursor	Rubber wood	Unbleached fibers	46	[69]
		Bleached pulp	64	
		CNFs	70	
	Wood (needle fir)	Original fibers	56	[29]
		Chemical purified cellulose	73.2	
		CNFs	71.0	
Agriculture precursor	Rice husks	Untreated	46.8	[68]
		Alkali treated	50.2	
		Bleached	56.5	
		CNCs	59.0	
	Rice straw	Raw	57.42	[23]
		Alkali-pulp-CNCs	62.8	
		Neutral pulp-CNCs	57.6	
		Raw	57.42	
		Alkali-pulp-CNFs	51.64	
		Neutral pulp-CNFs	63.35	
Bacterial cellulose	BC	Dried by lyophilized	68.20%	[43]
	BC	Dried by oven	96.54%	

5.3 Fourier-Transform Infrared Spectroscopy (FTIR)

The Fourier-transform infrared spectroscopy (FTIR) test indicates the presence of functional groups in the nanocrystals (NCs) and reveals their structural characteristics. Notably, there are observed increases in the intensities of bands associated with carbonyl and carboxyl groups in the range of 1700–1750 cm^{-1}. In the case of H_2SO_4-CNC, additional bands appear at 1030, 1170, and 1350 cm^{-1}, which are attributed to S=O vibrations of the formed SO_3H groups. Nelson and O'Connor [92] estimated the crystallinity index based on the ratio of absorbance bands at 1420 cm^{-1}.

5.4 Thermal Stability

Brown [28] estimated the thermal stability of compounds using thermogravimetric analysis (TGA) by examining changes in their physical properties as a function of temperature. Several researchers, including [49, 63], Coats and Redfern [31], and [100], developed equations to estimate the activation energy of degradation from TGA curves.

Similar to the impact role of precursor type and synthesized method of plant-NCs on their PDD, crystallinity they also effective on their thermal stability. For BC-NCs, Biyik and Coban and other researchers found that the thermal stability is affected by strain, medium and culture conditions [27, 72]. Table 5 summarizes the thermal stability of RS-NCs (as example of natural cellulose) versus type of pulp, in comparison with BC- NC in different media.

6 Applications of Nanocellulose

NCs represent a class of organic nanomaterials that have unique characteristics (biodegradable, biocompatible, low toxicity, environmental friendliness, cost-effectiveness, and ease of synthesis from renewable biomass). Due to this feature they offer a promising platform in the following applications (as examples), and most of applications are summarized in Table 6.

6.1 Biomedical Applications

This application is one topic field for using NCs. Drug delivery systems (DDS), tissue Engineering (TE) and cosmetic are examples of this field. The DDS is investigated at 1952, and it based on formation of NC-hydrogels, loaded by drugs, Basta and Lotfy [21], used chitosan and metal cations, as organic and inorganic chelating the NCs.

Table 5 TGA measurements of cellulose-based and BC-based NCs [22, 81]

	Code	Step	Temp. range (°C)	Peak maximum of DTG curve (°C)	Order "n"	R^2	Ea kJ/mol
Rice straw-based NC	Ex. RS	2nd	200.46–381.89	331.78	0.50	0.96	146.00
		3rd	381.89–450.94	410.75	2.00	0.92	298.21
	Alkali RS-pulp	2nd	209.5–339.09	233.57	1.5	0.843	9.5398
		3rd	339.09–446.75	389.12	1.5	0.914	155.606
	Neutral sulfite RS-pulp	2nd	212.03–430.07	267.4	2.0	0.944	113.921
	Acidic sulfite RS-pulp	2nd	218.25–306.01	266.99	0.5	0.953	172.488
		3rd	306.01–399.08	332.28	1.5	0.907	161.08
BC-based NCs	*Komagataeibacter intermedius MO*	2nd	117–217	146	1.0	0.975	67.23
		3rd	230–325	278	1.0	0.975	102.4
		4th	336–434	386	1.0	0.989	131.2
	Replacing carbon source of seed by cheese whey	2nd	240–316	281	1.0	0.898	171.7
		3rd	342–450	417	1.0	0.941	143.6
	Replacing the yeast of seed medium by cheese whey	2nd	235–329	294	1.0	0.857	159.1
		3rd	357–503	414	1.0	0.993	106.8

They also Conjugated the cellulose with amino acids for resulting the nanobioactive compounds to inhibite the growth of cancer cells) [79]. Another approach was investigated by [76], based on coating of NC surface by protein and plasma, or creating the surface charge. The DDS systems are capable of prolonging the release of drug, due to its ability to interact with hydrogel formulation [54, 99, 109, 143]. Based on the high affinity of nanostructure cellulose films to hold the liquid (e.g., water), together with excellent mechanical durability and high flexibility, it is used in wound dressing. The gelatinous membrane of the agitated-BC as an artificial skin was superior to conventional quaze toward temporary covering of wounds [64]. In addition, it has significant role in tissue engineering (regeneration and repair of osteoblast cells of bone). The main factors including the cultivation process during tissue engineering are shown in Fig. 3.

It was found that the impact factor for this application was based on specific sizes and shapes of the nanocellulose, which facilitates the cells to adopt more densely oriented nanocrystals surface. Also, blending nanocellulose with electrospun nanofibers can be used as supporting scaffold to culture human cells (Fig. 4).

For the application in cosmetic purpose, it was back to affinity of NCs to disperse in water, together with high aspect ratio property, which facilitate the gelation at low temperature, and further promote in cosmetic as biocounterpart, and green approach

Table 6 Application of NCs from cellulose-based and bacterial-based precursors

Field	Application	References
Medical and tissue	– Sustaining release of drugs – Water absorbents – Antimicrobial films and tampons, sanitary – Wound dressing – Cosmetic for laser therapy, mesotherapy	[24, 38, 53, 54, 64, 65, 76, 79, 80, 84, 94, 99, 109, 125, 130, 143]
Water treatment	– Industrial activities – Agronomic loaded irrigated water	[5, 6, 11, 26, 33, 56, 66, 122]
Cosmetic	Composite coating agent for nails, hair, eyebrows, or eyelashes	[35, 84]
Electronic devices	Sensor, electronic displays, and windows	[3, 74, 129, 142, 145]
Construction	– Sensors to monitor stress levels in bridge	[51]
Dyes	Dispersing agent and adsorbents	[23]
Paper products	Fire retardant paper Grease-proof paper	[18, 119]
Food products	Additives Packaging for oxygen barrier Edible films	[9, 12, 13, 97, 119, 139, 140]
Oil recovery	Fracturing fluid in oil recovery	[14]

(continued)

Table 6 (continued)

Field	Application	References
Composites	High-performance polymer composites – Improve the mechanical properties of polymer matrix	[26, 82, 91]
Automobile	Lightweight and high strength components such as bumpers, side panels, and dashboards	[93]

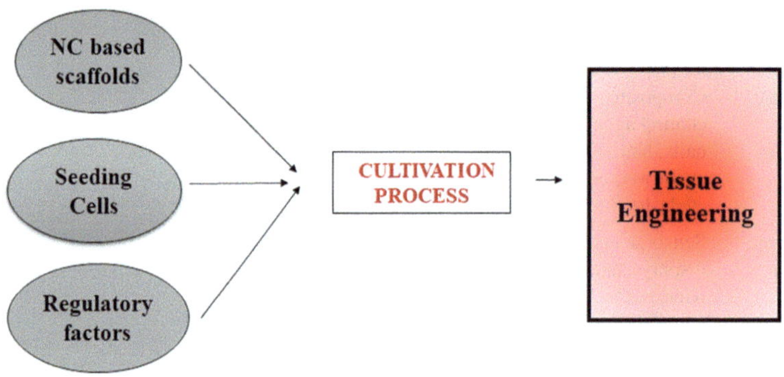

Fig. 3 Diagrammatic representation of basic factors involved in the cultivation process during tissue engineering [65]

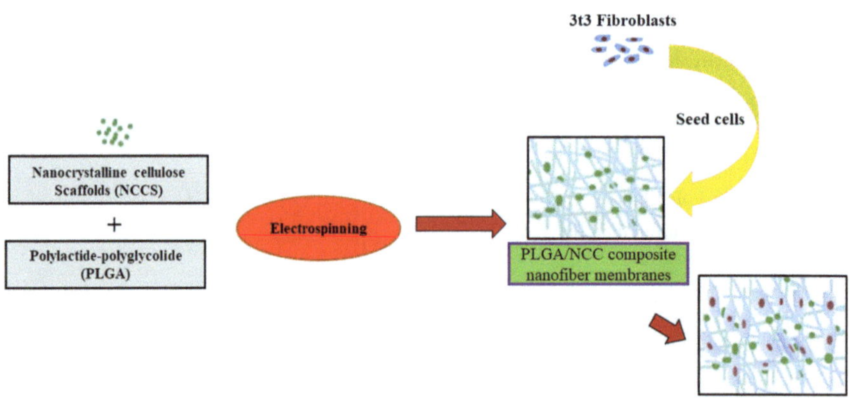

Fig. 4 Representation of nanocellulose-based skin scaffold [53, 125]

to avoid the nonhealth effect from using nondegradable plastic microbeads (decoratives and parabens as preservatives), when its used as mask packs and other specific cosmetics. On other side, the NCs can be applied as carrier to proteolytic enzymes, for cosmetics of peeling, laser therapy, mesotherapy, and atopic facial changes [84].

6.2 Environmental Applications

Environmental applications of nanocrystals (NCs) in water purification and reducing the limitations of synthetic polymers in packaging. Water pollution from industrial wastewater and fertilizer leaching poses risks to health and the environment. Traditional treatment methods are often inefficient and costly, despite the effectiveness of carbon nanostructure-based adsorbents [87]. Developing the application of Cellulose-in nano size (CNFs, CNCs, as well as cellulose-metal nanoparticles and nanocomposite hydrogels) in wastewater purification is carried out by techniques of effective membrane filtration, flocculation, sorption and catalytic degradation, systems [16, 33]. Recently, the electrospun nanofiber membranes are a novel way to wastewater treatment. The advantage of these membranes is ascribed to high surface/volume ratio and porosity [5, 11]. Promoting the nanocellulose membranes/filters in the system of water filtration is achieved by in situ functionalization of membranes and 3D printing [66, 122].

Regarding the environmental impact of soluble fertilizers, due to the CNCs have excellent behavior in mitigating the fast release of any active materials it succeeded in solving the eco-problem from loss of plant soil system, via limit the leaching of fertilizers and any agronomic compounds included the disposed irrigated water, and consequently achieves sustainable enhanced efficiency fertilizers [6, 56].

Substituting a part of synthetic polymer by biodegradable NCs will not only have a positive impact on the environment, but will also confer beneficial properties on the resulting polymer composites, compared to original polymers (less density, high mechanical, and thermal stability). CNFs is proved as the most benefit in applying which is return to its morphology hairy nanofibrils, and more flexibility. The study proved that the method of pulping process for preparation of CNs precursor have the positive role on their performance effect on properties of HDPE [26, 82, 91].

6.3 Food Additives and Edible Films

The application of NCs in food is assessed as they are used as additives to enhance the performance of biodegradable food packaging, films (chitosan, soy protein, and tragacanth, or as alternative to synthetic polymers in active food packaging). The route for their application is directly applied on surface of food (by spray/dipping). In both routes, the impact of NCs is positive on controlling the release of active loading compounds [4, 9, 12, 13]. The safety packaging is also investigated from nanoclay and polylactic acid (PLA) matrix, which further improves the water and oxygen barrier properties [116].

The incorporation of nanoparticles, both cellulose and metallic substrate, has been reported to improve film performance. These included metal nanoparticles, such as silver nanoparticle, and inorganic nanoparticles, such as silica [104]. The formulation of films from PLA/PVA-NC-AgNPs results in bioactive films suitable for food packaging, as AgNPs serve as agents that inhibit the growth of both positive and negative bacteria [13]. The presence of carboxyl (COO) groups in cellulose nanofibers (CNFs) enhances the production of antibacterial films for food packaging applications, attributed to the binding of COO groups with silver ions [36]. The plant extracts, which include the polyphenols, are also studied as additives to accept the film's functional properties (including antioxidant and antibacterial capacities). The essential oils from plants are also a great class of compound additives, to accept the film strong antibacterial capability [10, 88, 139, 140].

The BC has a great potential to apply in food purpose; examples of popular BC foods are Philippine desert, Nata de Coco, Chinese Kombuchar, or Manchurian Tea that are obtained by growing yeast and *Acetobacter* in a medium containing tea extract and sugar. On the contrary, it is applied as a functional food additive (e.g., thickener, texturizer, and/or calorie reducer) in ice cream, salad dressing, and weight reduction base. Due to disorder structure of agitated culture-BC, it is preferred to apply as emulsifiers than that from static culture, because the former has a disordered structure [52, 97, 141].

6.4 Electronics and Energy

The nanomaterials have advantageous role in the electronic devices in decreasing the power consumption, decrease their weight with the small size of transistors, and increase density of memory. For CNF composites with less roughness (smoothness) and possess good dimensional stability is able to apply in printed electronic, replacing replace oil-based plastic substrates [115]. Due to the behavior of nanocelluloses as liquid crystals materials, it also used in electro-optical devices [95, 112]. Figure 5 summarizes the beneficial applications of NC [101].

For application of NCs in area of energy storage and conversion is studied, as their adhesive, gel electrolyte, and in 3D Printing machine. For CNFs-adhesive is replaced Polyvinylidene fluoride in binding the production of batteries and supercapacitor electrodes in the production of batteries and supercapacitor electrodes to effectively binding its components, with forming uniform layer safety contact to the current collector [74, 142] (Fig. 6).

The gel electrolyte's network structure effectively traps water molecules, preventing evaporation while maintaining high ionic conductivity. This characteristic enhances its application in zinc batteries, supercapacitors, and other batteries. Polyvinyl alcohol is preferred over sodium carboxymethylcellulose due to its ability

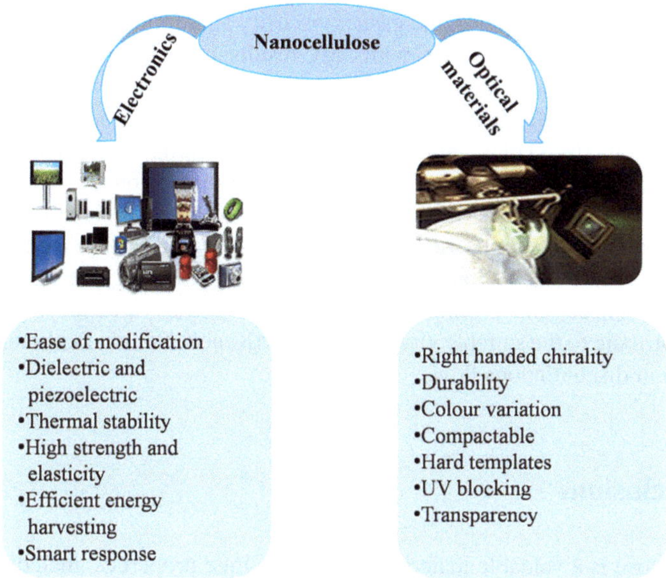

Fig. 5 Nanocellulose-based products for sustainable application in electronics and optical materials [101]

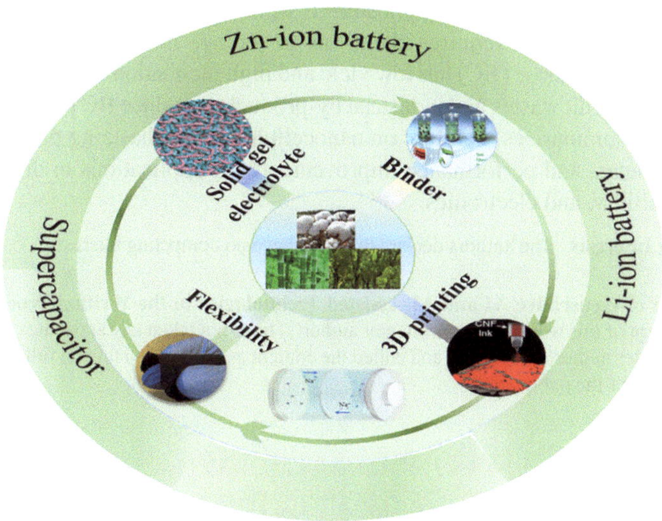

Fig. 6 Diagram for NCs in energy storage devices [70, 74, 129]

to form more hydrogen bonds, increasing viscosity and improving electrolyte–electrode contact [3]. The blend of NC with CMC results in an electrolyte with high ionic conductivity, mechanical strength, and low free water content, enabling a high-rate and long cycling life for battery [129].

Nanocrystals (NCs) play a significant role in 3D printing, where machines utilize a streamlined process to create high-resolution three-dimensional objects efficiently. To enhance the capabilities of MXene-based inks with low MXene loading, [145] developed controllable cellulose nanofibers (CNFs) via a TEMPO-mediated oxidation process. CNFs act as an intercalation agent, preventing the aggregation of MXene nanosheets and promoting a hierarchical porous structure. Additionally, CNFs contribute to the suitable viscosity of the MXene/CNF ink, facilitating droplet formation in discontinuous flow.

7 Conclusions

Nanocellulose is a valuable material due to its unique properties, including surface chemistry, renewability, biodegradability, and nontoxicity, making it suitable for various applications such as medical, food products, dye dispersants, synthetic polymer composites, and paper packaging. Although its effective characterization is established, industrial-scale application is limited by production costs. Isolation from natural precursors requires chemicals to remove noncellulosic components, while bacterial cellulose (BC) has low yield and high incubation costs. Recent efforts focus on using fruit wastes and industrial by-products to reduce BC production costs. This chapter summarizes literature on nanocelluloses, emphasizing precursor roles, synthesis routes, and performance improvements for applications in drug delivery, thermal stability, and electronics.

Competing Interests The authors declare that they have no competing interest.

Declaration of Generative AI and AI-Assisted Technologies in the Writing Process During the preparation of this work, the book chapter author(s) have not taken any assistance from AI, and the book chapter author(s) reviewed and edited the content as needed and take(s) full responsibility for the content of the publication.

References

1. Abate L, Bachheti A, Bachheti RK, Husen A, Pandey DP, Megra MB (2022) Coir fiber-based cellulose, nanocellulose, and their cutting-edge applications. In: Jawaid M (ed) Coir fiber and its composites. Elsevier Inc., United States, pp 309–325
2. Abdel-Hadi AK, Hosny WM, Basta AH, El-Saied H (1994) Metal Chelates with some cellulose derivatives. Pt I Polym Plast Technol Eng 33(6):781–791

3. Abral H, Atmajaya A, Mahardika M, Hafizulhaq F, Kadriadi, Handayani D, Sapuan SM, Ilyas RA (2020) Effect of ultrasonication duration of polyvinyl alcohol (PVA) gel on characterizations of PVA film. J Mater Res Technol 9:2477–2486
4. Ahankari SS, Subhedar AR, Bhadauria SS, Dufresne A (2021) Nanocellulose in food packaging: a review. Carbohydr Polym 255:117479
5. Ahmed SF, Mofijur M, Ahmed B, Mehnaz T, Mehejabin F, Maliat D, Hoang AT, Shafiullah GM (2022) Nanomaterials as a sustainable choice for treating wastewater. Environ Res 214(1):113807
6. Aikpokpodion PE, Hsiao BS, Dimkpa CO (2024) Mitigation of nitrogen losses in a plant-soil system through incorporation of nanocellulose and zinc-modified nanocellulose. J Agric Food Chem 72:17295–17305
7. Alavijeh MS, Rad I, Hatamie S (2021) Magnetic nanocomposite's mechanism of action during the hyperthermia treatment of breast cancer. Appl Nanosci 11:2739–2746
8. Alruwashid FS, Dar MA, Alharthi NH, Abdo HS, Almotairy S (2021) The synthesis and characterization of graphene-based cobalt ferrite nanocomposite materials and its electrochemical properties. Appl Nanosci 11:2661–2668
9. Amara C, El Mahdi A, Medimagh R, Khwaldia K (2021) Nanocellulose-based composites for packaging applications. Curr Opin Green Sustain Chem 31:100512
10. Atarés L, Chiralt A (2016) Essential oils as additives in biodegradable films and coatings for active food packaging. Trends Food Sci Technol 48:51–62
11. Amin MT, Alazba AA, Manzoor U (2014) A review of removal of pollutants from water/wastewater using different types of nanomaterials. Adv Mater Sci Eng Article 825910. https://doi.org/10.1155/2014/825910
12. Ayhan Z (2019) Packaging and the shelf life of fruits and vegetables. In: Reference module in food science. Elsevier. https://doi.org/10.1016/b978-0-08-100596-5.22649-8
13. Azeredo HMC, Rosa MF, Mattoso LHC (2017) Nanocellulose in bio-based food packaging applications. Ind Crops Prod 97:664–671
14. Barati R, Liang JT (2014) A review of fracturing fluid systems used for hydraulic fracturing of oil and gas wells. J Appl Polym Sci 131:40735(1–11)
15. Barja F (2021) Bacterial nanocellulose production and biomedical applications. J Biomed Res 14, 35(4):310–317
16. Bachheti RK, Bachheti A, Husen A (eds) (2023a) Nanomaterials for environmental and agricultural sectors. Springer, Singapore. https://doi.org/10.1007/978-981-99-2874-3
17. Basta AH, Girgis AS, El-Saied H (2002) Fluorescence pyridinecarbonitrile containing compounds and their application in security paper. Dyes Pigments 54(1):1–10
18. Basta AH, El-Saied H (2009) Performance of improved bacterial cellulose application in the production of functional paper. J Appl Microbiol 107(6):2098–2107
19. Basta AH, Lotfy VF, Mahmoud K, Abdelwahed NAM (2020) Synthesis and evaluation of protein-based biopolymer in production of silver nanoparticles as bioactive compound versus carbohydrates- based biopolymers. R Soc Open Sci 7(10):200928
20. Basta AH, Lotfy VF, Salem AM (2022) Valorization of biomass pulping waste as effective additive for enhancing the performance of films based on liquid crystal hydroxypropylcellulose nanocomposites. Waste Biomass Valoriz 13(4):2217–2231
21. Basta AH, Lotfy VF (2023) Performance assessment of rice-by product-derived methylene blue nano-carrier. Chem Data Collect 42:100957
22. Basta AH, Lotfy VF (2023) Essential role of pulping route on thermal stability performance of rice straw-based nanocelluloses. Chem Eng Technol 46(5):837–847
23. Basta AH, Lotfy VF (2023) Impact of pulping routes of rice straw on cellulose nanoarchitectonics and their behavior toward Indigo dye. Appl Nanosci (Switzerland) 13(6):4455–4469
24. Basta AH, Lotfy VF (2023) The synergistic route for enhancing rice by-product derived nanoparticles in sustained release of bioactive compound. Cellulose 30(18):11473–11491
25. Basta AH, Lotfy VF (2024) New HCHO-scavenger for enhancing the HCHO-based adhesive in production environmentally friendly agro-based composites. Int J Adhes Adhes 1333:103756

26. Bhatnagar A, Sain M (2005) Processing of cellulose nanofiber-reinforced composites. J Reinf Plast Compos 24:1259–1268
27. Biyik H, Coban EP (2017) Evaluation of different carbon, nitrogen sources and industrial wastes for bacterial cellulose production. Eur J Biotechnol Biosci 5:74–80
28. Brown ME (1988) Introduction to thermal analysis: techniques and applications. Springer Netherlands
29. Chen W, Yu H, Liu Y, Chen P, Zhang M, Hai Y (2011) Individualization of cellulose nanofibers from wood using high-intensity ultrasonication combined with chemical pretreatments. Carbohydr Polym 83:1804–1811
30. Chen L, Zhu JY, Baez C, Kitinb P, Elderc T (2016) Highly thermal-stable and functional cellulose nanocrystals and nanofibrils produced using fully recyclable organic acids. Green Chem 18:3835–3843
31. Coats A, Redfem J (1964) Kinetic parameters from thermogravimetric data. Nature 201:68–69
32. Danish M, Ayub H, Sandhu ZA, Shoaib A, Akram Z, Najeeb J, Naeem S (2021) Synthesis of cerium oxide/cadmium sulfide nanocomposites using inverse microemulsion methodology for photocatalytic degradation of methylene blue. Appl Nanosci 11:503–2515
33. Darabitabar F, Yavari V, Hedayati A, Zakeri M, Yousefi H (2020) Novel cellulose nanofiber aerogel for aquaculture wastewater treatment. Environ Technol Innov 18:100786
34. Das K, Ray D, Bandyopadhyay NR, Ghosh T, Mohanty AK, Misra M (2009) A study of the mechanical, thermal and morphological properties of microcrystalline cellulose particles prepared from cotton slivers using different acid concentrations. Cellulose 16:783–793
35. Dhar N (2010) Novel cellulose nanoparticles for potential cosmetic and pharmaceutical applications. MS. thesis of applied science, chemical engineering, Waterloo, Ontario, Canada
36. Dong C, Zhang H, Pang Z, Liu Y, Zhang F (2013) Sulfonated modification of cotton linter and its application as adsorbent for high-efficiency removal of lead(II) in Effluent. Bioresour Technol 146:512–518. https://doi.org/10.1016/j.biortech.2013.07.108
37. Doumeng M, Makhlouf L, Marsan O, Delbé K, Denape J, Chabert F (2021) A comparative study of the crystallinity of polyetheretherketone by using density, DSC, XRD, and Raman spectroscopy techniques. Polym Test 93:106878
38. El-Saied H, Basta AH, Abdel-Hadi AK, Hosny WM (1994) Metal Chelates with some cellulose derivatives, Pt II Polym Int 35(1):27–33
39. El-Saied H, Basta AH, Gobran RH (2004) Research progress in friendly environmental technology for the production of cellulose products (Bacterial cellulose and its application). Polym Plast Technol Eng 43:797–820
40. El-Saied H, El-Diwany AI, Basta AH, Atwa NA, El-Ghwas DE (2008) Production and characterization of economical bacterial cellulose. BioResources 3(4):1196–1217
41. El-Saied H, El-Hady OA, Basta AH, El-Dewiny CY, Abo-Sedera SA (2016) Bio-chemical properties of sandy calcareous soil treated with rice straw-based hydrogels. J Saudi Soc Agric Sci 15(2):188–194
42. Fathy NA, Basta AH, Lotfy (2019) Novel trends for synthesis carbon nanostructures from agricultural wastes. Carbon nanomaterials for agri-food and environmental applications, pp 59–74
43. Feng X, Ullah N, Wang X, Sun X, Li C, Bai Y, Chen L, Li Z (2015) Characterization of bacterial cellulose by *Gluconacetobacter hansenii* CGMCC 3917. J Food Sci. https://doi.org/10.1111/1750-3841.13010
44. Ferrer A, Filpponen I, Rodríguez A, Laine J, Rojas OJ (2012) Valorization of residual empty palm fruit bunch fibers (EPFBF) by microfluidization: production of nanoflbrillated cellulose and EPFBF nanopaper. Bioresour Technol 125:249–255
45. Filipova I, Fridrihsone V, Cabulis U, Berzins A (2018) Synthesis of nanofibrillated cellulose by combined ammonium persulphate treatment with ultrasound and mechanical processing. Nanomater 8:640
46. Filipova I, Serra F, Tarrés Mutjé P, Delgado-Aguilar M (2020) Oxidative treatments for cellulose nanofibers production: a comparative study between TEMPO-mediated and ammonium persulfate oxidation. Cellulose 27:10671–10688

47. Franceus F, Desmet T (2020) Sucrose phosphorylase and related enzymes in glycoside hydrolase family 13: discovery, application and engineering. Int J Mol Sci 21(7):2526
48. Fraschini C, Chauve G, Bouchard J (2017) TEMPO-mediated surface oxidation of cellulose nanocrystals (CNCs). Cellulose 24:2775–2790
49. Freeman ES, Carroll B (1958) The application of thermos-analytical techniques to reaction kinetics: the thermogravimetric evaluation of the kinetics of the decomposition of calcium oxalate monohydrate. J Phys Chem 62:394–397
50. Fukuzumi H, Saito T, Okita Y, Isogai A (2010) Thermal stabilization of TEMPO-oxidized cellulose. Polym Degrad Stab 95:1502–1508
51. Gao D, Luan Y, Wang Q, Liang Q, Qi Q (2015) Construction of cellulose-utilizing *Escherichia coli* based on a secretable cellulase. Microb Cell Fact. https://doi.org/10.1186/s12934-015-0349-7
52. Gayathry G (2015) Production of Nata de Coco—a natural dietary fibre product from mature coconut water using *Gluconacetobacter xylinum.* Natl J Food Ferment Technol 5(2):231–235
53. Ghafari R, Jonoobi M, Amirabad LM, Oksman K, Ahmad Reza Taheri AR (2019) Fabrication and characterization of novel bilayer scaffold from nano-cellulose based aerogel for skin tissue engineering applications. Int J Biol Macromol 136(1):796–803
54. Gopi S, Amalraj A, Thomas S (2016) Effective drug delivery system of biopolymers based on nanomaterials and hydrogels a review. Drug Design 5:1e7
55. Greser AB, Avcioglu NH (2022) Optimization and physicochemical characterization of bacterial cellulose by *Komagataeibacter nataicola* and *Komagataeibacter maltaceti* strains isolated from grape, thorn apple and apple vinegars. Arch Microbiol 204:465
56. Gomez-Maldonado D, Phillips SG, Vaidya SR, Bartley PC, White JC, Fairbrother DH, Peresin MS (2024) Modifying soluble NPK release with hydrophobized nanocellulose-based hydrogels for sustainable enhanced efficiency fertilizers. Environ Sci Nano 11:529–545
57. Han L, Cui S, Yu H, Song M, Zhang H, Huang C, Kim D, Tam KMC (2019) Self-healable conductive nanocellulose nanocomposites for biocompatible electronic skin sensor system self-healable conductive nanocellulose nanocomposites for biocompatible electronic skin sensor system. ACS Appl Mater Interfaces 11:44642–44651
58. Heßler N, Klemm D (2009) Alteration of bacterial nanocellulose structure by insitu modification using polyethylene glycol and carbohydrate additives. Cellulose 16:899–910
59. Henrique MA, Neto WPF, Silvério HA, Martins DF, Gurgel LVA, Barud HDS, Pasquin D (2015) Kinetic study of the thermal decomposition of cellulose nanocrystals with different polymorphs, cellulose I and II, extracted from different sources and using different types of acids. Ind Crops Prod 76:128–140
60. Herrick FW, Casebier RL, Hamilton JK, Sandberg KR (1983) Microfibrillated cellulose: morphology and accessibility. J Appl Polym Sci Appl Polym Symp 37:797–813
61. Ho PH, Lotfy V, Basta A, Trens P (2021) Designing microporous activated carbons from biomass for carbon dioxide adsorption at ambient temperature. A comparison between bagasse and rice by-products. J Clean Prod 294:126260
62. Hong F, Qiu K (2008) An alternative carbon source from konjac powder for enhancing production of bacterial cellulose in static cultures by a model strain *Acetobacteraceti* subsp. *xylinus* ATCC 23770. Carbohydr Polym 72:545–549
63. Horowitz HW, Metzger G (1963) A new analysis of thermogravimetric traces. Anal Chem 35:1464–1468
64. Horue M, Silva JM, Berti IR, Brandão LR, da Barud HS, Castro GR (2023) Bacterial cellulose-based materials as dressings for wound healing. Pharmaceutics 15(2):424
65. Jacoba S, Reshmy Rb R, Antonyc S, Madhavand A, Sindhue B, Awasthif MK, Kuddusg M, Pilla S, Varjanii S, Pandey A, Binod P (2022) Nanocellulose in tissue engineering and bioremediation: mechanism of action. BioEng 13(5):12823–12833
66. Jin H, Kettunen M, Laiho A, Pynnonen H, Paltakari J, Marmur A, Ikkala O, Ras RHA (2011) Superhydrophobic and superoleophobic nanocellulose aerogel membranes as bioinspired cargo carriers on water and oil. Langmuir 27:1930–1934

67. Jintian H, Yabin L (2014) Performance of salix cellulose in the process of micro/nano. J Appl Sci 14:592–595
68. Johar N, Ahmad I, Dufresne A (2012) Extraction, preparation and characterization of cellulose fibres and nanocrystals from rice husk. Ind Crops Prod 37:93–99
69. Jonoobi M, Khazaeian A, Tahir PM, Azry SS, Oksman K (2011) Characteristics of cellulose nanofibers isolated from rubberwood and empty fruit bunches of oil palm using chemo-mechanical process. Cellulose 18:1085–1095
70. Kobayashi Y, Saito T, Isogai A (2014) Aerogels with 3D ordered nanofiber skeletons of liquid-crystalline nanocellulose derivatives as tough and transparent insulators. Angew Chem Int Ed Engl 53:10394–10397
71. Kurosumi A, Sasaki C, Yamashita Y, Nakamura Y (2009) Utilization of various fruit juices as carbon source for production of bacterial cellulose by *Acetobacter xylinum* NBRC 13693. Carbohydr Polym 76:333–335
72. Kusdianto K, Diastatic W, Shimada M, Qomariyah L, Winardi S (2019) Fabrication of ZnO-SiO_2 nanocomposite materials prepared by a spray pyrolysis for the photocatalytic activity under UV and sunlight irradiations. IOP Conference Series: Materials Science Engineering, vol 778, 26th regional symposium on chemical engineering (RSCE 2019) 30–31, Kuala Lumpur, Malaysia
73. Lavanya D, Kulkarni P, Dixit M, Raavi PK, Krishna LNV (2011) Source of cellulose derivatives and their applications. Int J Drug Dev Res 2:19–38
74. Li R, Tian D, Chen L, Zhuang B, Feng H, Li Q, Yu L, Ling Y (2023) The application of cellulose nanofibrils in energy systems. Rev Batteries 9:399
75. Liu L, Yao J (2012) Properties of biocomposite fibers from cellulose nanowhiskers and cellulose matrix. J Fiber Bio-eng Inform 5:207–215
76. Lin N, Dufresne A (2014) Nanocellulose in biomedicine: current status and future prospect. Eur Polym J 59:302–325
77. Liu Y, Wang H, Yu G, Yu Q, Li B, Mu X (2014) A novel approach for the preparation of nanocrystalline celluloseby using phosphotungstic acid. Carbohydr Polym 110:415–422
78. Liu X, Cao L, Wang S, Huang L, Zhang Y, Tian M, Li X, Zhang J (2023) Isolation and characterization of bacterial cellulose produced from soybean whey and soybean hydrolyzate. Sci Rep 13:16024
79. Lotfy VF, Basta AH (2020) Optimizing the chitosan-cellulose based drug delivery system for controlling the ciprofloxacin release versus organic/inorganic crosslinker, characterization and kinetic study. Int J Biol Macromol 165:1496–1506
80. Lotfy VF, Hassan SS, Khalf-Alla PA, Basta AH (2020) The role of side chain of amino acid on performance of their conjugates with carboxymethyl cellulose and their Pd (II) complexes as bioactive agents. Int J Polym Mater Polym Biomater 69(1):21–31
81. Lotfy VF, Basta AH, Abdel-Monem MO, Abdel-Hamed GZ (2021) Utilization of bacteria in rotten Guava for production of bacterial cellulose from isolated and protein waste. Carbohydr Polym Technol Appl 2:100076
82. Lotfy VF, Basta AH, Shafik ES (2023) Synergic effect of cellulose-and carbon-based nanostructure derived from agricultural biowaste on the performance of polyester–HDPE nanocomposite. Appl Nanosci 13(9):6503–6518
83. Lu F-F, Yu H-Y, Zhou Y, Yao J-M (2016) Spherical and rod-like dialdehyde cellulose nanocrystals by sodium periodate oxidation: optimization with double response surface model and templates for silver nanoparticles. Express Polym Lett 10:965–976
84. Ludwicka K, Jedrzejczak-krzepkowska M, Kubiak K, Kolodziejczyk M, Pankiewicz T, Bielecki S (2016) Medical and cosmetic applications of bacterial nanocellulose. In: Gama M, Dourado F, Bielecki S (eds). Elsevier, Amsterdam, pp 145–165. https://doi.org/10.1016/B978-0-444-63458-0.00009-3
85. Mao J, Osorio-Madrazo A, Laborie MP (2013) Preparation of cellulose I nanowhiskers with a mildly acidic aqueous ionic liquid: reaction efficiency and whiskers attributes. Cellulose 20:1829–1840

86. Mazzoli A, Favoni O (2012) Particle size, size distribution and morphological evaluation of airborne dust particles of diverse woods by Scanning Electron Microscopy and image processing program. Powder Technol 225:65–71
87. Miandad AR, Waqas M, Gehany F, Barakat MA (2019) Remediation of wastewater using various nano-materials. Arab J Chem 12(8):4897–4919
88. Mir SA, Dar BN, Wani AA, Shah MA (2018) Effect of plant extracts on the techno-functional properties of biodegradable packaging films. Trends Food Sci Technol 80:141–154
89. Morais JPS, Rosa MDF, De Souza Filho MDSM, Nascimento LD, Do Nascimento DM, Cassales AR (2013) Extraction and characterization of nanocellulose structures from raw cotton linter. Carbohydr Polym 91:229–235
90. Morelli CL, Marconcini JM, Pereira FV, Bretas RES, Branci-forti MC (2012) Extraction and characterization of cellulose nanowhiskers from balsa wood. Macromol Symp 319:191–195
91. Nair SS, Yan N (2015) Effect of high residual lignin on the thermal stability of nanofibrils and its enhanced mechanical performance in aqueous environments. Cellulose 22:3137–3150
92. Nelson ML, O'Connor RT (1964) Relation of certain infrared bonds to cellulose crystallinity and crystal lattice type. J Appl Polym Sci 8(3):1235–1341
93. Nurazzi NM, Jenol MA, Kamarudin SH, Aisyah HA, Hao LC, Yusuff SM, Nur Amira MR, Harussani MM, Norrahim MNF, Ilyas RA, Norli AA (2022) Nanocellulose composites in the automotive industry. Industrial applications of nanocellulose and its nanocomposites. Woodhead Publishing Series in Composites Science and Engineering, pp 439–467
94. Otoni CG, Espitia PJP, Avena-Bustillos RJ, McHugh TH (2016) Trends in antimicrobial food packaging systems: emitting sachets and absorbent pads. Food Res Int 83:60–73
95. Oulachgar H, Bolduc M, Chauve G, Desroches Y, Beaupre P, Bouchard J, Galarneau P (2015) Fabrication and electro-optical characterization of a nanocellulose- based spatial light modulator. MRS Adv. https://doi.org/10.1557/adv.2015.34
96. Oun AA, Rhim J-W (2018) Isolation of oxidized nanocellulose from rice straw using the ammonium persulfate method. Cellulose 25:2143–2149
97. Phisalaphong M, Chiaoprakobkij N (2016) Applications and products. Nata De Coco, In Book, Bacterial NanoCellulose, Ch. 7, pp 143–155. https://doi.org/10.1201/b12936-8
98. Premjet S, Ohtani Y, Sameshima K (1994) The contribution of high molecular lignosulfonate to the powerful bacterial cellulose production system with *Acetobacter xylinum* ATCC 10245. Sen'i Gakkaishi 50(10):458–463
99. Rajwade JM, Paknikar KM, Kumbhar JV (2015) Applications of bacterial cellulose and its composites in biomedicine. Appl Microbiol Biotechnol 99:2491–2511. https://doi.org/10.1007/s00253-015-6426-3
100. Reich L, Stivala SS (1978) Kinetic parameters from thermogravimetric curves. Thermochim Acta 24(1):9–16
101. Reshmy R, Philip E, Sherely Annie Paul SA (2020) Nanocellulose-based products for sustainable applications—recent trends and possibilities. Rev Environ Sci Bio/Technol 19:779–806
102. Roman M, Winter WT (2004) Effect of sulfate groups from sulfuric acid hydrolysis on the thermal degradation behavior of bacterial cellulose. Biomacromol 5:1671–1677
103. Rusmirovic JD, Ivanovi JZ, Pavlovi VB, Raki VM, Rancic MP, Djokic V, Marinković AD (2017) Novel modified nanocellulose applicable as reinforcement in high-performance nanocomposites. Carbohydr Polym 164:4–74
104. Sarwar MS, Niazi MBK, Jahan Z, Ahmad T, Hussain A (2018) Preparation and characterization of PVA/nanocellulose/Ag nanocomposite films for antimicrobial food packaging. Carbohydr Polym 184:453–464
105. Scherrer P (1918) Nachrichten von der Gesellschaft der Wissenschaften zu Göttingen. Mathematisch- Physikalische Klasse 2:98–100
106. Shu J, Tang D (2020) Recent Advances in photoelectrochemical sensing: from engineered photoactive materials to sensing devices and detection modes. Anal Chem 92:363–377
107. Singhania RR, Patel AK, Tsai M-L, Chen C-W, Dong CD (2021) Genetic modification for enhancing bacterial cellulose production and its applications. Bioengineering 12(1):6793–6807

108. Sun X, Wu Q, Ren S, Lei T (2015) Comparison of highly transparent all-cellulose nanopaper prepared using sulfuric acid and TEMPO-mediated oxidation methods. Cellulose 22:1123–1133
109. Sun F, Nordli HR, Pukstad B, Kristofer Gamstedt R, Chinga-Carrasco G (2017) Mechanical characteristics of nanocellulose-PEG bionanocomposite wound dressings in wet conditions. J Mech Behav Biomed Mater 69:377–384. https://doi.org/10.1016/j.jmbbm.2017.01.049
110. Suryanegara L, Nakagaito AN, Yano H (2010) Thermo-mechanical properties of microfibrillated cellulose-reinforced partially crystallized PLA composites. Cellulose 17:771–778
111. Tang Y, Shen X, Zhang J, Guo D, Kong F, Zhang N (2015) Extraction of cellulose nanocrystals from old corrugated container fiber using phosphoric acid and enzymatic hydrolysis followed by sonication. Carbohydr Polym 125:360–366
112. Teters C, Kong W, Taylor M, Simonsen J, Lerner M, Plant T, Evans G (2007) The cellulose nanocrystal electro-optic effect. In: International conference on nanotechnology for the forest products industry, Knoxville, TN
113. Thakur VK, Thakur MK, Gupta RK (2013) Development of functionalized cellulosic biopolymers by graft copolymerization. Int J Biol Macromol 62:44–51
114. Thomas B, Raj MC, Athira KB, Joy J, Moores A, Drisko GL, Sanchez C (2018) Nanocellulose, a versatile green platform: from biosources to materials and their applications. Chem Rev 118(24):11575–11625
115. Torvinen K, Sievänen J, Hjelt T, Hellén E (2012) Smooth and flexible filler-nanocellulose composite structure for printed electronics applications. Cellulose 19:821–829
116. Trifol J, Plackett D, Sillard C, Szabo P, Bras J, Daugaard AE (2016) Hybrid poly(lactic acid)/nanocellulose/nanoclay composites with synergistically enhanced barrier properties and improved thermomechanical resistance. Polym Int 65:988–995
117. Ullah MW, Manan S, Ul-Islam M, Vasilyevich RV, Thomas S, Yang G (2021) Ch. 1 Introduction to nanocellulose. World Scientific Publishing Europe Ltd. https://doi.org/10.1142/978 1786349477_0001
118. Vaikundam M, Shanmugam S, Aldawood S, Arunkumar P, Santhanam A (2023) Preparation of biopolymer pectin fascinate hydroxyapatite nanocomposite for biomedical applications. Appl Nanosci 13:6039–6051
119. Valera N, Chaussy D, Passas R (2004) Barrier properties of grease proof papers. Cellul Chem Technol 38(3):245–254
120. Vazquez A, Foresti ML, Cerrutti P, Galvagno M (2013) Bacterial cellulose from simple and low cost production media by *Gluconacetobacter xylinus*. J Polym Environ 21:545–554
121. Wang B, Sain M, Oksman K (2007) Study of structural morphology of hemp fiber from the micro to the nanoscale. Appl Compos Mater 14:89–103
122. Wang K, Abdalla AA, Khaleel MA, Hilal N, Khraisheh MK (2017) Mechanical properties of water desalination and wastewater treatment membranes. Desalination 401:190–205
123. Watanabe K (1997) Research progress in production of bacterial cellulose by aeration and agitation culture and its application as a new industrial material. Biosci Biotechnol Biochem 61:219–224
124. Watanabe K, Tabuchi M, Morinaga Y, Yoshinaga F (1998) Structural feature and properties of bacterial cellulose produced in agitated culture. Cellulose 5:187–200
125. Wei Z, Wu C, Li R, Yu D, Ding Q (2021) Nanocellulose based hydrogel or aerogel scaffolds for tissue engineering. Cellulose 28(12):7497–7520
126. Worku LA, Bachheti RK, Tadesse MG (2023) Preparation and characterization of carboxylated cellulose nanocrystals from *Oxytenanthera abyssinica* (Ethiopian lowland bamboo) cellulose via citric acid anhydrous hydrolysis catalyzed by sulfuric acid. Biomass Conv Bioref. https://doi.org/10.1007/s13399-022-03718-0
127. Worku LA, Bachheti RK, Bachheti A, Milessi TS, Chandel AK (2024) Understanding the biochemical changes at molecular level during biomass pretreatment: a comprehensive analysis. Cellulose 31:7281–7312. https://doi.org/10.1007/s10570-024-06081-7
128. Worku LA, Tadesse MG, Bachheti RK, Bachheti A, Husen A (2024) Synthesis of carboxylated cellulose nanocrystal/ZnO nanohybrids using *Oxytenanthera abyssinica* cellulose and zinc

nitrate hexahydrate for radical scavenging, photocatalytic, and antibacterial activities. Int J Biol Macromol 267:131228. https://doi.org/10.1016/j.ijbiomac.2024.131228
129. Xu L, Meng TT, Zheng XY, Li TY, Brozena AH, Mao YM, Zhang Q, Clifford BC, Rao JC, Hu LB (2023) Nanocellulose-carboxymethylcellulose electrolyte for stable, high-rate zinc-ion batteries. Adv Funct Mater 33:2302098
130. Yadav S, Illa MP, Rastogi T, Sharma CS (2016) High absorbency cellulose acetate electrospun nanofibers for feminine hygiene application. Appl Mater Today 4:62–70
131. Yang W, Feng Y, He H, Yang Z (2018) Environmentally-friendly extraction of cellulose nanofibers from steam-explosion pretreated sugar beet pulp. Mater 11(7):1160
132. Yang H, Zhang Y, Kato Z, Rowan SJ (2019) Preparation of cellulose nanofibers from Miscanthus x. Giganteus by ammonium persulfate oxidation. Carbohydr Polym 212:30–39
133. Yano H, Sugiyama J, Nakagaito AN, Nogi M, Matsuura T, Hikita M, Handa K (2005) Optically transparent composites reinforced with networks of bacterial nanofibers. Adv Mater 17:153–155
134. Yoshinaga F, Tonouchi N, Chen W, Yu H, Liu Y, Chen P, Zhang M, Hai Y (2011) Individualization of cellulose nanofibers from wood using high-intensity ultrasonication combined with chemical pretreatments. Carbohydr Polym 83:1804–1811
135. Yu Z, Gong H, Xum J, Li Y, Xue F, Zeng Y, Lium X, Tang D (2022) Liposome-embedded Cu2–x AgxS nanoparticle-mediated photo- thermal immunoassay for daily monitoring of cTnI protein using a portable thermal imager. Anal Chem 94:7408–7416
136. Yu Z, Gong H, Xu J, Li Y, Zeng Y, Liu X, Tang D (2022) Exploiting photoelectric activities and piezoelectric properties of $NaNbO_3$ semiconductors for point-of-care immunoassay. Anal Chem 94:3418–3426
137. Zeng R, Wang W, Cai G, Huang Z, Tao J, Tang D, Zhu C (2020) Single-atom platinum nanocatalyst-improved catalytic efficiency with enzyme-DNA supermolecular architectures. Nano Energy 74:104931
138. Zeng R, Wang W, Chen M, Wan Q, Wang C, Knopp D, Tang D (2020) CRISPR-Cas12a-driven MXene-PEDOT: PSS piezoresistive wireless biosensor. Nano Energy 82:105711
139. Zhang W, Jiang W (2020) Antioxidant and antibacterial chitosan film with tea polyphenols-mediated green synthesis silver nanoparticle via a novel one-pot method. Int J Biol Macromol 155:1252–1261
140. Zhang W, Li X, Jiang W (2020) Development of antioxidant chitosan film with banana peels extract and its application as coating in maintaining the storage quality of apple. Int J Biol Macromol 154:1205–1214
141. Zhang Y, Yang S, Tang H, Wan S, Qin W, Zeng Q, Huang J, Yu G, Feng Y, Li J (2022) Depletion stabilization of emulsions based on bacterial cellulose/carboxymethyl chitosan complexes. Carbohydr Polym 2022(297):119904
142. Zhong X, Han J, Chen L, Liu W, Jiao F, Zhu H, Qin W (2021) Binding mechanisms of PVDF in lithium ion batteries. Appl Surf Sci 553:149564
143. Zhao D, Zhuo RX, Cheng SX (2012) Alginate modified nanostructured calcium carbonate with enhanced delivery efficiency for gene and drug delivery. Mol Biosyst 8:753e759
144. Zhou L, Li N, Shu J, Liu Y, Wang K, Cui X, Yuan Y, Beibei D, Geng Y, Wang Z, Duan Y, Zhang J (2018) One-pot preparation of carboxylated cellulose nanocrystals and their liquid crystalline behaviors. ACS Sustain Chem Eng 6(9):12403–12410
145. Zhou G, Li MC, Liu C, Wu Q, Mei C (2022) 3D printed $Ti_3C_2T_x$ MXene/cellulose nanofiber architectures for solid-state supercapacitors: ink rheology, 3D printability, and electrochemical performance. Adv Funct Mater 32:2109593
146. Zoppe JO, Johansson L-S, Seppälä J (2015) Manipulation of cellulose nanocrystal surface sulfate groups toward biomimetic nanostructures in aqueous media. Carbohydr Polym 126:23–31

Nanocellulose-Based Paper Reinforcement

Abhay Shankar, Monali Maiti, S. N. Raju Kutcherlapati, Jayaramudu Jarugala, and Arindam Chakrabarty

Abstract Nanocellulose, derived from renewable sources such as cellulose fibers, has garnered significant attention as a reinforcement material in the paper industry due to its exceptional mechanical properties, high surface area, and sustainability. This chapter explores the dual role of nanocellulose as an additive and coating agent for enhancing the physical, mechanical, and barrier properties of paper. As an additive, nanocellulose creates a dense, hydrogen-bonded network inside the fiber matrix, significantly increasing the internal bonding and tensile strength of paper. As a coating agent, nanocellulose can provide a dense, smooth layer that improves surface properties like gloss and printability while adding mechanical strength, air and water resistance properties. Moreover, nanocellulose can be used as a composite material with synthetic coating ingredients to improve the properties as paper coatings. It enhances mechanical properties such as durability and tensile strength while offering better air and moisture resistance. These developments pave the way for paper industries to be more environmentally friendly, which is required to meet the rising demand for eco-friendly and sustainable alternatives to nondegradable plastic films.

Keyword Cellulose nanofiber · Paper coating · Mechanical strength · Water resistance · Bio-degradable

---

A. Shankar · S. N. R. Kutcherlapati · J. Jarugala · A. Chakrabarty (✉)
Department of Polymers and Functional Materials, CSIR-Indian Institute of Chemical Technology, Hyderabad, India
e-mail: arindam.iict@csir.res.in

Academy of Scientific and Innovative Research (AcSIR), Ghaziabad, India

M. Maiti
R.B.V.R.R. Women's College, Hyderabad, India

R. K. Bachheti (ed.), *Nanotechnology in the Pulp and Paper Industry*, Smart Nanomaterials Technology, https://doi.org/10.1007/978-981-95-3473-9_3

1 Introduction

Paper is a thin sheet substance generated by chemical and mechanical processing of cellulose pulp fibers obtained from rags, woods, and grasses. For centuries, paper has been created from different substances like bamboo, cotton, wood, wheat straw, hemp, linen rags, and sugar cane waste. However, the two prime sources of paper production are wood and recycled paper products.

Paper goods find application throughout a wide range of human endeavors, such as for writing purposes, printing books, newspapers, packaging, and wrapping objects, and in manufacturing tickets, currency, cheques, and documents [6, 51]. Transferring and archiving information have long been vital functions of paper materials, serving as the backbone of record-keeping and knowledge preservation throughout history. From ancient manuscripts to modern documents, the ability to store and retrieve information has shaped societies, facilitated communication, and fostered learning. In an era where digitalization is rapidly transforming how we manage data, the role of paper still holds significance, particularly for its tangibility and permanence. Electronic material often only lasts for several years to several decades, while paper documents, books, and manuscripts can be preserved for centuries without specific conservation measures. However, carelessness, deliberate harm, or the unavoidable deterioration brought on by past manufacturing techniques (such as the use of aluminum salts), improper storage, and regular human use all have a negative impact on paper products [30, 57]. The necessity to prolong the life of paper products meant to convey information is becoming increasingly recognized, as is the need to preserve cultural treasures. This holds true for both recent ones and ones that were made in the past.

Over the past 20 years, Express delivery and takeout have seen a significant uptick in business due to the quick growth of online purchasing. Additionally, online buying has increased dramatically because of the lockdown and restrictions on retail outlets during the Coronavirus (COVID-19) epidemic. Fuel-based plastics continue to be the most often utilized materials in express packaging because of their good mechanical and waterproofing qualities at a low cost. They gradually contaminate the land and oceans due to inappropriate disposal. Thus, biodegradable plastic was used instead of fuel-based plastics, particularly in disposable packaging. The last several years have seen a rise in interest in a number of biodegradable plastics, like poly(butyleneadipate-co-terephthalate), polylactic acid, and polycaprolactone [29, 36]. However, most biodegradable plastics have limited barrier qualities, poor mechanical strength, and high costs [1].

In its raw form, paper often exhibits poor mechanical properties and high hydrophilicity, which can erode interfibrillar hydrogen bonding, causing the material to lose its strength in watery or highly humid conditions, making it inappropriate for applications like printing books, documentation, and packaging. Hence, reinforcement is essential to enhance its performance [1]. Several studies have been conducted to strengthen the paper. For instance, a conventional and simple method to improve the wet strength of paper is to mechanically beat or refine the pulp [9, 13, 16]. The specific surface area of fibrillating fibers can be increased by beating or refining,

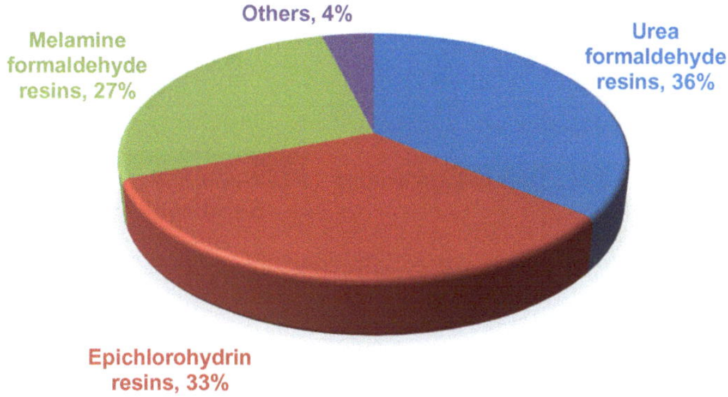

Fig. 1 Global consumption of synthetic resins as additives for paper coating

and fines can be produced on the pulp fiber surface, which improves the hydrogen bonding between fibers [12]. Adding wet strength additives/ sizing agents to pulp or coating paper surfaces is another way to increase the wet strength and reinforce the paper. The majority of wet-strength additives are cationic polymeric compounds that are soluble in water and have high reactivity with paper fibers. Additives for wet strength are synthetic resins and other substances appropriate for various paper goods. Figure 1 shows the global consumption of these additives [17].

The coating methods involve the use of laminate of low-density polyethylene (LDPE) and biopolymers such as polysaccharides, alginate, proteins, lipids, or waxes to improve the strength and hydrophobicity [22, 28, 39, 46]. The use of bio-based materials such as chitosan, cellulose nanofiber (CNF), cellulose nanocrystal (CNC), soybean oil, and castor oil in the coating composition has been the focus of recent paper coating research [38]. In this regard, chitosan and cellulose were found to be quite promising because of their biodegradability and functionality for further chemical modification [21, 49]. Recently, Shankar et al. have reported a drastic increase in tensile strength by using acrylic/nanocellulose composite latex as the coating material for kraft paper [44]. They also demonstrated the application of crosslinked alkyl chitosan as a coating material to achieve improved mechanical properties and hydrophobicity in coated kraft papers [45]. A library of natural and synthetic polymers is being used for several applications (Fig. 2). In terms of paper coating application, the degradable polymers from both petro-based and bio-based sources play a major role.

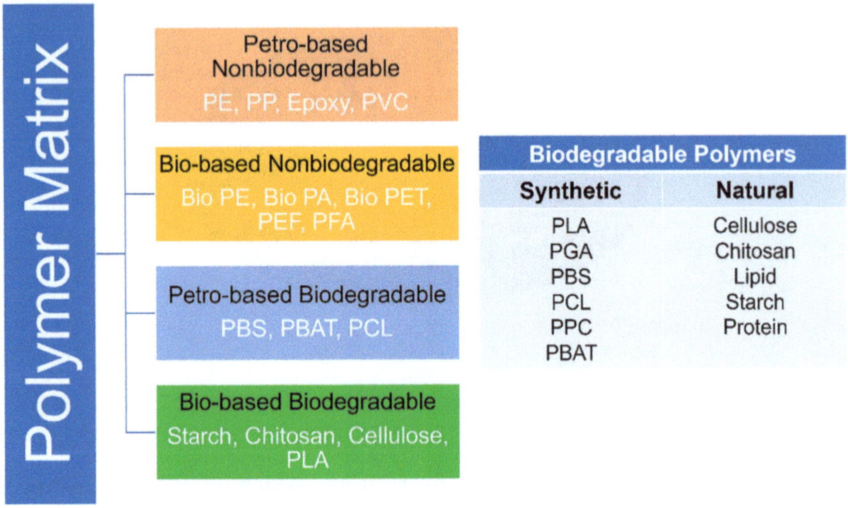

Fig. 2 Classification of degradable and nondegradable polymers obtained from natural and synthetic sources

2 Paper Reinforcement

Paper reinforcement refers to techniques and materials used to enhance the strength, durability, and longevity of paper products. Traditionally, paper is vulnerable to tearing, wear, and environmental factors like moisture. Reinforcement strategies aim to improve its resilience and functionality, making it suitable for various applications.

3 Process of Paper Reinforcement

The process of paper reinforcement falls into two primary areas: adding wet strength additives/internal sizing agents to pulp slurry and surface sizing agents, which are coated to the paper surface once the paper sheet has formed [11]. Among the most widely used and conventional internal sizing agent is rosin. The primary internal sizing chemicals utilized in the wet end of papermaking include polymeric compounds, rosin, alkyl ketene dimer (AKD), and alkenyl succinic anhydride (ASA). These additives influence the dewatering and retention of fillers and fibers in the paper sheet by increasing the hydrophobicity of the paper [58]. Styrene copolymers are the source of most synthetic copolymers used for surface sizing. These substances prolong the life of paper, since they are incorporated into the cellulose network and do not require the addition of fixing agents, such as alum, which is used to fix rosin [14]. Synthetic copolymer sizing agents have also reportedly been shown to offer the added benefit of increasing paper strength [50]. Similarly, wet strength

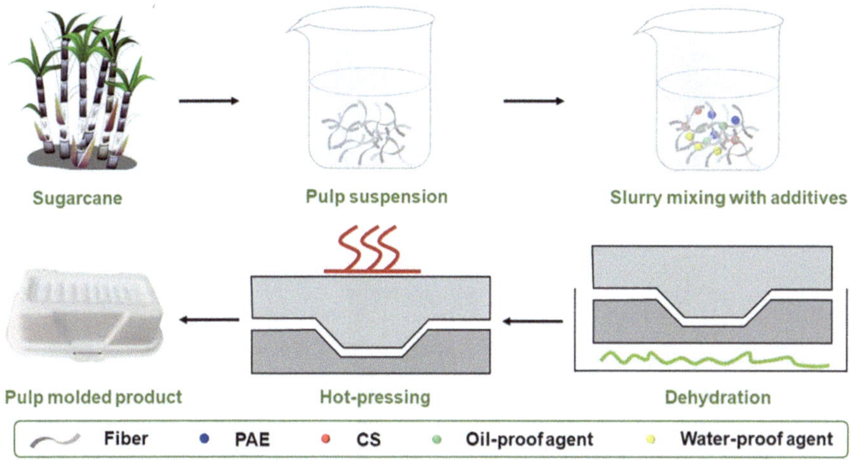

Fig. 3 Schematic representation of the process used in making the pulp-molded product polyamide epichlorohydrin (PAE) and cationic starch (CS) [35]

additives such as polyamide amine-epichlorohydrin, urea–formaldehyde resins, and melamine formaldehyde resins are crucial for enhancing the durability and performance of paper products, especially in applications where moisture resistance is essential [43]. Wet additives improve the paper's ability to maintain its strength when wet. Studies have shown that the wet strength of filter paper can be increased by using a phosphoric acid-urea solution [54]. The procedure for combining additives in pulp slurry involved mixing cationic starch (CS) and polyamide epichlorohydrin (PAE) with the pulp suspension after adding water- and oil-proof agents. The pulp molded product that was produced after additional dehydration and hot pressing. The complete process from pulp source to pulp molded product is shown in Fig. 3.

Paper can also be reinforced by applying coating over the paper surface. There are several techniques used for coating such as extrusion, solvent coating, and dispersion coating. The most well-known method for applying coatings over the surface of the paper is extrusion, as shown in Fig. 4, which is used in industry for a number of reasons, including continuous processing, consistent coating, low probability of cracks and pinholes, and solvent-free application. However, there are certain drawbacks to extrusion coatings as well. For example, high coat weight is necessary to provide the necessary qualities, polymer melt instability, and other material-related issues that sometimes restrict coating speed and effectiveness. Because of this, natural thermoplastic biopolymers like polylactic acid (PLA) and poly-3-hydroxybutyrate (PHB) have not been used much in industry and can only be used after being modified in bulk to facilitate processing [7, 37]. PHB and PLA have a tendency to experience poor melt stability and thermal degradation above their melting temperatures, which is one of the main disadvantages of processing them. To increase PHB and PLA's melt stability, a number of approaches have been tried, including blending with plasticizers, adding additives, combining with other thermally stable polymers,

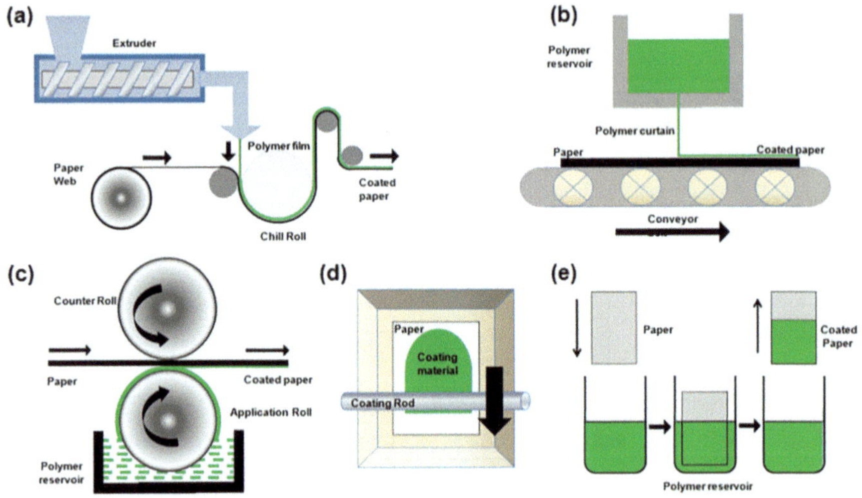

Fig. 4 Various coating techniques used for paper coating, **a** extrusion coating; **b** curtain coating; **c** size press coating; **d** bar coating; and **e** dip coating [38]

or chemically altering the polymer backbone by adding copolymers [19]. However, further work needs to be done to enhance other required properties such as flexibility, permeability to gas and vapor, and compatibility with natural fibers as reinforcement. A Paper substrate can be coated with dispersions and solvent-based coatings using various methods, including surface sizing, curtain coating, dip coating, and bar or rod coating.

4 Synthetic Polymers for Paper Reinforcement

4.1 Synthetic Polymers as Additive for Paper

Some commonly used synthetic polymer additives are polyacrylamide (PAM), melamine formaldehyde (MF), glyoxylated polyacrylamide (GPAM), polyvinyl amine (PVAm), polyamide amine-epichlorohydrin (PAE), polyethyleneimine (PEI), and polycarboxylic acids that are predominantly utilized. Paper can be made to have the necessary wet-strength qualities by using these synthetic polymer additives alone or in combination. The polyamide-epichlorohydrin (PAE) resin is the most often utilized as a wet strength agent in the pulp and paper industry. This synthetic resin has a linear chain with the active functional group 3-hydroxy-azacyclobutane ring and a molecular structure made up of polyamide and epichlorohydrin. It can cross-link cellulose fibers, by reacting with carboxyl groups to create a robust network that prevents paper from breaking down when it comes into contact with water [47].

However, the composite paper with PAE still has drawbacks such as poor dimensional instability due to the limited anchoring sites of the linear structure and the pulp recyclability [55, 56]. The generation of absorbable chlorinated organics may arise from the breakdown of PAE, which has the potential to seriously affect both human health and the environment [18].

4.2 Synthetic Polymers as Coating

Synthetic polymers such as paraffin wax, polyethylene, polyvinylidene chloride, and acrylic polymers are increasingly being used as paper substrate coatings to enhance their mechanical strength and durability. Low-density polyethylene (LDPE) is the most popular polymer for paper coating due to its cost-effectiveness, good strength, and mild chemical resistance [40]. However, LDPE is a synthetic nondegradable polymer. After end-use, these papers become difficult to recycle and biodegrade, because it is difficult to separate the coated polymer layer from the paper [27]. In order to avoid the use of such nondegradable polymers, per- and poly-fluoroalkyl substances (PFASs) are utilized in the manufacturing of numerous paper-based goods, like boxes, plates, and cups, in the paper industry [3, 23]. However, some fluorinated substances are poisonous and cause long-term harm to the ecosystem [48]. The paper coated with fluorochemicals, has raised questions about possible exposure during manufacture, use, and disposal [5]. Epidemiological research has indicated a potential link between prolonged human exposure to the long-chain perfluorooctanoic acid (PFOA) and perfluorooctanesulfonic acid (PFOS) compounds and a number of illnesses, including kidney disease, thyroid disease, malformations, and problems with male reproduction, pregnancy-related problems, and immune system dysfunction in children [2]. Thus, in recent years, research on paper coatings has concentrated on fluorine-free polymers. Paraffin waxes use paper coating, lubricants, printing ink, agricultural products, rheology modifiers, and other applications [32, 41]. Paraffin waxes are frequently used in packaging to create grease and moisture barriers for various goods, including cereal, baked goods, snacks, and fast food. However, paraffin waxes hinder the paper repulping process by creating stickies and are nonbiodegradable and nonrenewable [10, 33]. Our group has recently revealed that employing acrylic/nanocellulose composite latex as the coating material for kraft paper has produced a notable rise in tensile strength [44]. Recently, Rabnawaz and coworkers have synthesis water-dispersible poly(dimethylsiloxane) (PDMS) for paper coating [15]. However, since PDMS and acrylic are both nondegradable synthetic polymers, they will affect the paper degradation after end-use. To overcome these limitations, Rabnawaz et al. have synthesized degradable polymeric waxes composed of stearate ethylene glycol lactide acrylate as a coating material for paper substrate [53]. These polymeric waxes were synthesized using a four-step procedure. Kraft paper was coated with these synthesized waxes using both solvent-based and solvent-free (hot-press) techniques. However, the synthesized waxes could improve only oil- and water-resistant properties.

5 Bio-based Materials in Paper Reinforcement

The need for eco-friendly paper reinforcement additives and coatings has increased as environmental concerns have grown. In this regard, bio-based materials derived from renewable sources are gaining attention as sustainable alternatives to synthetic polymers. These materials, such as chitosan, starches, alginate, proteins, lipids, and nanocellulose, enhance paper strength and durability while reducing environmental impact.

5.1 Bio-based Materials as Additives for Paper

Recently Zhou et al. have reported a water-soluble hyperbranched wet strength agent based on the cross-linking of polyamines and oxidized amylopectin from waxy corn, an industrial crop [60], illustrated in Fig. 5. This polymer has multidirectional binding sites due to its hyperbranched structure, which is produced from oxidized amylopectin. Rich aldehyde and amine functional groups at the end of the polymer increase binding strength through co-crosslinking and self-crosslinking sites, hence enhancing the wet strength of paper. This synthesized polymer was added to the pulp slurry during the paper-making process to improve the wet strength of paper. Salam et al. developed soy-based polymer as a dry-strength agent for paper. [42] The chemical modification of soy flour has derived this dry strength additive for pulp. Bhardwaj et al. showed the effect of chitosan and cationic starch as wet-end additives on increasing the mechanical properties of paperboard [4].

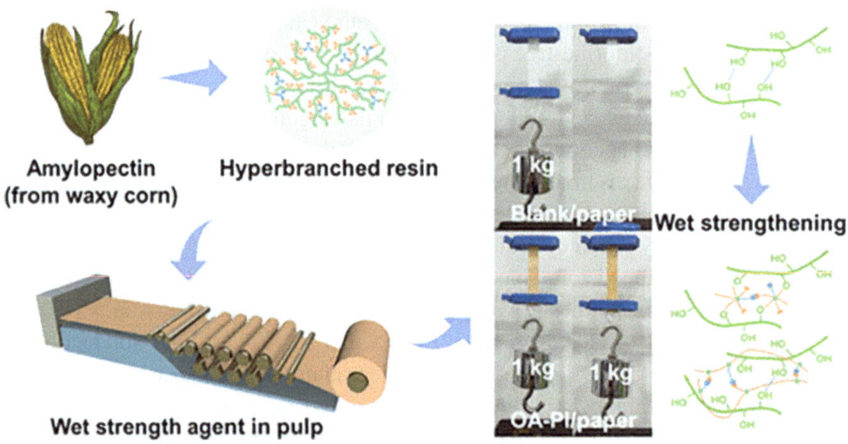

Fig. 5 Hyperbranched polyamines from amylopectin to improve the wet strength [60]. (Reproduced with the permission from the American Chemical Society)

5.2 *Bio-based Materials as a Coating Material*

In order to enhance the durability and mechanical strength of paper substrates, bio-based materials such as chitosan, castor oil, soybean oil, and starch are being used more frequently as coatings. Recently, our group has reported crosslinked alkyl chitosan-based coating material for improving the mechanical properties and water-repelling properties of coated paper [45], in which the Michael addition reaction did the crosslinking of chitosan with triacrylate functionality like pentaerythritol triacrylate followed by the reaction with octadecylamine addition. The mechanism of paper reinforcement by crosslinked alkyl chitosan is shown in Fig. 6. Rabnawaz et al. developed waterborne acrylated epoxidized soybean oil (AESO) emulsion to coat kraft paper [20]. The waterborne AESO emulsion was cast onto kraft paper and was then photo-cured. The mechanical properties of coated paper were found to be ~90% of the uncoated paper. Tian et al. introduced the dispersion of AESO, which is stabilized by modified nanocrystalline cellulose for the paper coating to improve the barrier properties of coated paper [49]. Rabnawaz et al. demonstrated the synthesis of bio-based coating material chitosan-graft-castor oil (CHI-*g*-CO) copolymer for paper coating application [24]. Castor-oil-capped isocyanate (CO-capped-NCO) was chemically grafted onto a chitosan polymer backbone to produce CHI-*g*-CO. The obtained CHI-*g*-CO copolymer was then employed as a waterborne coating for kraft paper to improve both water- and oil-resistant properties. Zhu and coworkers have demonstrated that the wet and dry tensile strength of paper could be significantly enhanced after coating with regenerated cellulose [61].

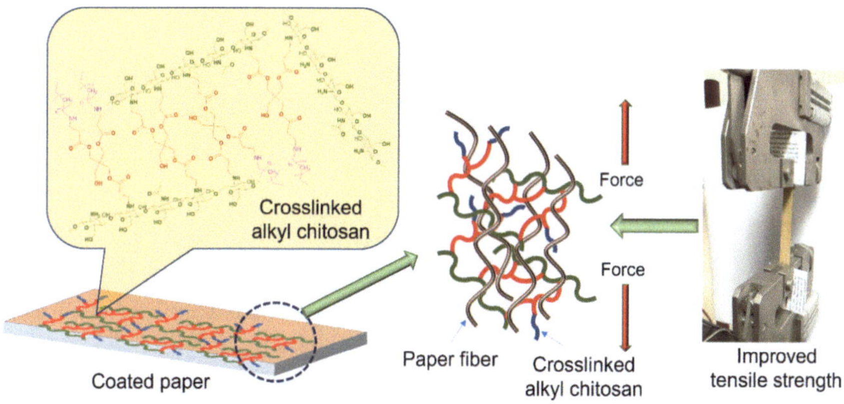

Fig. 6 Schematic presentation of the interaction between paper fibers and crosslinked alkyl chitosan responsible for the improvement in the mechanical properties of the coated paper [45]. (Reproduced with the permission from the American Chemical Society)

5.3 Nanocellulose in Paper Reinforcement

Cellulose is a renewable bio-based substance that occurs abundantly in nature, performs well, degrades quickly, and is ideal for sustainable development trends. This linear natural polysaccharide polymer's hydrophilicity, degradability, and chemical diversity are determined by the many intramolecular hydrogen bonds between *D*-glucose and β-1,4-glucoside linkages [31]. Besides its many excellent properties, cellulose also possesses a special hierarchical structure that helps prepare nanoscale derivatives in the context of rapidly advancing nanotechnology. Chemical, biological, or physical/mechanical processes are used to prepare cellulose nanoscale materials called nanocellulose (NC).

The researchers found tremendous potential in NC for making functional composites with synthetic polymers. Figure 7 summarizes the application areas of such composites [25]. The largest possible user of cellulose and NC is probably the paper sector. Throughout the paper-making process, NC has been used in several stages (such as forming, pressing, drying, and refining). Since NC is the same chemical substance that forms the bonds in the structure of paper, it is used as a strengthening agent in paper [34]. The NC surface contains many free hydroxyl groups. When NC is added to pulp slurry, it will be dispersed throughout the fiber gaps or on the fiber surface and closely blended with the pulp fibers: strengthening the adhesion between the fibers, filling in the spaces in the paper, and producing the effect of increasing the paper strength [8, 59]. Recently, Liu et al. demonstrated the improved mechanical properties under wet and dry conditions of paper sheets after adding cellulose nanofibers (CNFs) to pulp addition [26].

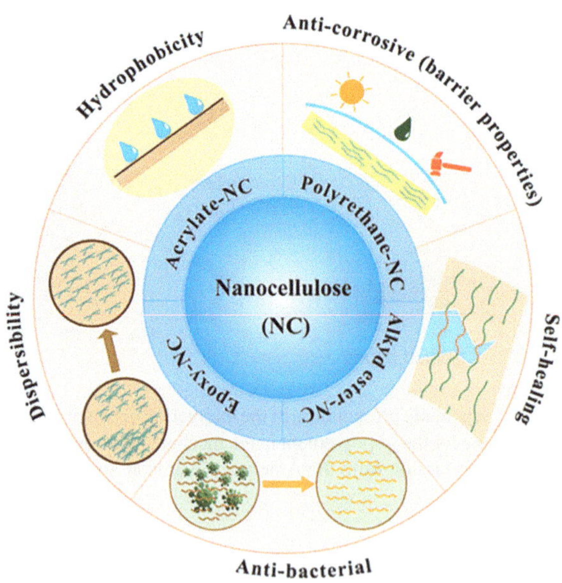

Fig. 7 Nanocellulose-based composites for functional coating application [25]. (Reproduced with the permission from the American Chemical Society)

5.4 *Nanocellulose as a Coating Material*

Nanocellulose as a coating material can improve the durability and mechanical strength of paper, making it more resistant to tearing and folding. Recently, Xiang et al. has reported superhydrophobic cellulose nanocrystals (CNCs) are produced by reacting octadecylamine (ODA) on the surface of CNCs through the self-polymerization of tannic acid and Michael addition/Schiff base reaction between ODA and polytannic acid-coated CNCs (PTA@CNCs) [52] as shown in Fig. 8. The obtained coating material was dispersed in ethanol and coated onto the surface of filter paper via the spray coating method. The coated filter paper showed improved wet strength.

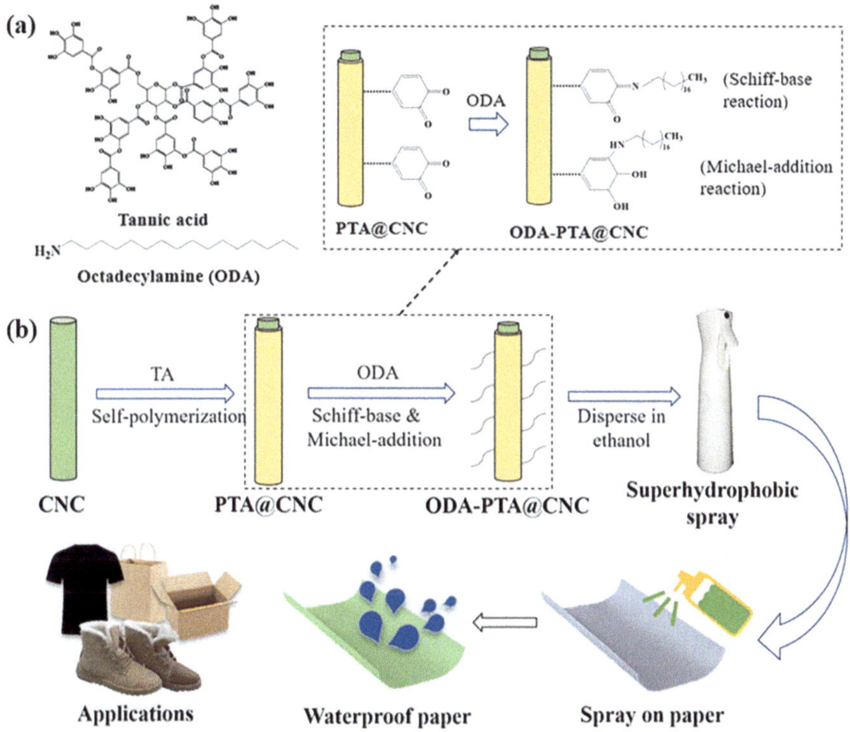

Fig. 8 **a** Mechanism of the reaction for preparing superhydrophobic coating **b** diagram illustrating the simple synthesis of sustainable superhydrophobic ODA-PTA@CNCs and their application [52]

6 Conclusion

The search for sustainable alternatives to synthetic additives and coatings for paper substrates has gained momentum in recent years. Among the most promising candidates is nanocellulose, a material derived from cellulose, which is the most prevalent natural polymer on the planet. Its unique properties and environmental benefits make it an exceptional choice for enhancing the performance of paper products while addressing the growing concerns about sustainability and environmental impact.

Cellulose is a fundamental component of plant cell walls, making it widely available from various sources such as wood, agricultural by-products, and even certain types of algae. This abundance means that nanocellulose can be sourced sustainably, providing a renewable alternative to synthetic materials that rely on finite fossil fuels. By leveraging this natural resource, industries can contribute to a circular economy, promoting the use of renewable materials and reducing reliance on nonrenewable resources.

Nanocellulose exhibits excellent barrier properties and mechanical properties. With high tensile strength and flexibility, nanocellulose can significantly enhance the mechanical properties of paper substrates. When incorporated into paper products, it improves tear resistance, tensile strength, and overall durability, allowing for thinner and lighter products without sacrificing performance. This is especially beneficial in packaging applications where barrier properties, weight, and strength are critical for maintaining product integrity during transportation and storage.

7 Future Challenges and Applications

Sustainability and renewability are taken into account in the early phases of scientific study in a period of growing environmental concern. Therefore, as a mechanical reinforcing agent, nanocellulosic materials can displace synthetic-based materials. However, biopolymer development as additives/coatings for paper in desired applications is still in the early stages. Poor mechanical properties, barrier qualities, and high processing costs impede industrial-scale commercialization compared to traditional petroleum-based polymers. Industrializing biopolymer as an additive and coatings for pulp and paper requires additional research, which can improve the overall cost-effectiveness.

Competing Interests The authors declare that they have no competing interest.

Declaration of Generative AI and AI-Assisted Technologies in the Writing Process During the preparation of this work, the book chapter author(s) have not taken any assistance from AI, and the book chapter author(s) reviewed and edited the content as needed and take(s) full responsibility for the content of the publication.

References

1. Alava MJ, Niskanen KJ (1997) Performance of reinforcement fibres in paper. Fundam Papermak Mater 2:1177–1213. https://doi.org/10.15376/frc.1997.2.1177
2. Barry V, Winquist A, Steenland K (2013) Perfluorooctanoic acid (PFOA) exposures and incident cancers among adults living near a chemical plant. Environ Health Perspect 121:1313–1318. https://doi.org/10.1289/ehp.1306615
3. Begley TH, Hsu W, Noonan G, Diachenko G (2008) Migration of fluorochemical paper additives from food-contact paper into foods and food simulants. Food Addit Contam Part A 25:384–390. https://doi.org/10.1080/02652030701513784
4. Bhardwaj S, Bhardwaj NK, Negi Y (2016) Effect of chitosan and cationic starch as wet-end additives to enhance the strength properties of paperboard using OCC recycled pulp. IPPTA Q. J Indian Pulp Pap Tech Assoc 28:164–172
5. Biege LB, Hurtt ME, Frame SR, O'Connor JC, Cook JC (2001) Mechanisms of extrahepatic tumor induction by peroxisome proliferators in male CD rats. Toxicol Sci 60:44–55. https://doi.org/10.1093/toxsci/60.1.44
6. Campano C, Miranda R, Merayo N, Negro C, Blanco A (2017) Direct production of cellulose nanocrystals from old newspapers and recycled newsprint. Carbohydr Polym 173:489–496. https://doi.org/10.1016/j.carbpol.2017.05.073
7. Cheng HY, Yang YJ, Li SC, Hong JY, Jang GW (2015) Modification and extrusion coating of polylactic acid films. J Appl Polym Sci 132:1–8. https://doi.org/10.1002/app.42472
8. Dai D, Fan M (2013) Green modification of natural fibres with nanocellulose. RSC Adv 3:4659–4665. https://doi.org/10.1039/c3ra22196b
9. Dunlop-Jones N (1991) Wet-strength chemistry. Pap Chem. https://doi.org/10.1007/978-94-011-6474-0_6
10. Dwivedi P, Mishra PK, Mondal MK, Srivastava N (2019) Non-biodegradable polymeric waste pyrolysis for energy recovery. Heliyon 5:e02198. https://doi.org/10.1016/j.heliyon.2019.e02198
11. Ferreira PJ, Gamelas JA, Moutinho IM, Ferreira AG, Gómez N, Molleda C, Figueiredo MM (2009) Application of FT-IR-ATR spectroscopy to evaluate the penetration of surface sizing agents into the paper structure. Ind Eng Chem Res 48:3867–3872. https://doi.org/10.1021/ie801765c
12. Gharehkhani S, Sadeghinezhad E, Kazi SN, Yarmand H, Badarudin A, Safaei MR, Zubir MNM (2015) Basic effects of pulp refining on fiber properties—a review. Carbohydr Polym 115:785–803. https://doi.org/10.1016/j.carbpol.2014.08.047
13. Gigac J, Fišerová M (2008) Influence of pulp refining on tissue paper properties. Tappi J 7:27–32. https://doi.org/10.32964/tj7.8.27
14. Ginebreda A, Guillén D, Barceló D, Darbra RM (2012) Additives in the paper industry. Handb Environ Chem 18:11–34. https://doi.org/10.1007/698_2011_109
15. Hamdani SS, Elkholy HM, Alford A, Jackson K, Naveed M, Wyman I, Wang Y, Li K, Haider SW, Rabnawaz M (2024) Synthesis of water-dispersible poly(dimethylsiloxane) and its potential application in the paper coating industry as an alternative for PFAS-coated paper and single-use plastics. Polymers (Basel) 16:1006. https://doi.org/10.3390/polym16071006
16. Hassan ML, Mathew AP, Hassan EA, El-Wakil NA, Oksman K (2012) Nanofibers from bagasse and rice straw: process optimization and properties. Wood Sci Technol 46:193–205. https://doi.org/10.1007/s00226-010-0373-z
17. Hubbe MA, Nanko H, McNeal MR (2009) Retention aid polymer interactions with cellulosic surfaces and suspensions: a review. BioResources 4:850–906. https://doi.org/10.15376/biores.4.2.850-906
18. Kamali M, Gameiro T, Costa MEV, Capela I (2016) Anaerobic digestion of pulp and paper mill wastes—an overview of the developments and improvement opportunities. Chem Eng J 298:162–182. https://doi.org/10.1016/j.cej.2016.03.119

19. Kervran M, Vagner C, Cochez M, Ponçot M, Saeb MR, Vahabi H (2022) Thermal degradation of polylactic acid (PLA)/polyhydroxybutyrate (PHB) blends: a systematic review. Polym Degrad Stab. https://doi.org/10.1016/j.polymdegradstab.2022.109995
20. Khan A, Kumar V, Anulare J, Dam M, Wyman I, Mayekar P, Rabnawaz M (2024) Sustainable packaging with waterborne acrylated epoxidized soybean oil. ACS Sustain Resour Manag 1:879–889. https://doi.org/10.1021/acssusresmgt.3c00093
21. Khattak S, Wahid F, Liu L, Jia S, Chu L, Xie Y, Li Z (2019) Applications of cellulose and chitin/chitosan derivatives and composites as antibacterial materials: current state and perspectives. Appl Microbiol Biotechnol 103:1989–2006. https://doi.org/10.1007/s00253-018-09602-0
22. Kopacic S, Walzl A, Zankel A, Leitner E, Bauer W (2018) Alginate and chitosan as a functional barrier for paper-based packaging materials. Coatings 8:235. https://doi.org/10.3390/coatings8070235
23. Langberg HA, Arp HPH, Breedveld GD, Slinde GA, Høiseter Å, Grønning HM, Jartun M, Rundberget T, Jenssen BM, Hale SE (2021) Paper product production identified as the main source of per- and polyfluoroalkyl substances (PFAS) in a Norwegian lake: source and historic emission tracking. Environ Pollut 273:116259. https://doi.org/10.1016/j.envpol.2020.116259
24. Li Z, Rabnawaz M, Khan B (2020) Response surface methodology design for biobased and sustainable coatings for water- and oil-resistant paper. ACS Appl Polym Mater 2:1378–1387. https://doi.org/10.1021/acsapm.9b01238
25. Liu E, Meng H, Zhang H, Liang H, Wu Z (2024) Nanocellulose, a renewable biobased functional/reinforcement additive for composite coating materials: a review. ACS Appl Polym Mater 6:28–48. https://doi.org/10.1021/acsapm.3c02088
26. Liu H, Yano H, Abe K (2023) Reinforcement of dry and wet paper sheets by cellulose nanofibers. Cellulose 30:211–222. https://doi.org/10.1007/s10570-022-04888-w
27. Marsh K, Bugusu B (2007) Food packaging—roles, materials, and environmental issues: scientific status summary. J Food Sci. https://doi.org/10.1111/j.1750-3841.2007.00301.x
28. Matsui KN, Larotonda FDS, Paes SS, Luiz DB, Pires ATN, Laurindo JB (2004) Cassava bagasse-Kraft paper composites: analysis of influence of impregnation with starch acetate on tensile strength and water absorption properties. Carbohydr Polym 55:237–243. https://doi.org/10.1016/j.carbpol.2003.07.007
29. Moshood TD, Nawanir G, Mahmud F, Mohamad F, Ahmad MH, AbdulGhani A (2022) Sustainability of biodegradable plastics: new problem or solution to solve the global plastic pollution? Curr Res Green Sustain Chem 5:100273. https://doi.org/10.1016/j.crgsc.2022.100273
30. Motelica L, Popescu A, Răzvan AG, Oprea O, Trușcă RD, Vasile BS, Dumitru F, Holban AM (2020) Facile use of zno nanopowders to protect old manual paper documents. Materials (Basel) 13:1–20. https://doi.org/10.3390/ma13235452
31. Oksman K, Aitomäki Y, Mathew AP, Siqueira G, Zhou Q, Butylina S, Tanpichai S, Zhou X, Hooshmand S (2016) Review of the recent developments in cellulose nanocomposite processing. Compos Part A Appl Sci Manuf 83:2–18. https://doi.org/10.1016/j.compositesa.2015.10.041
32. Pashova S (2023) Application of plant waxes in edible coatings. Coatings 13:911. https://doi.org/10.3390/coatings13050911
33. Pawar PA, Purwar AH (2013) Bioderadable polymers in food packaging. Am J Eng Res 151–164
34. Perdoch W, Cao Z, Florczak P, Markiewicz R, Jarek M, Olejnik K, Mazela B (2022) Influence of nanocellulose structure on paper reinforcement. Molecules. https://doi.org/10.3390/molecules27154696
35. Qin C, Li J, Wang W, Li W (2022) Improving mechanical strength and water barrier properties of pulp molded product by wet-end added polyamide epichlorohydrin/cationic starch. ACS Omega 7:22173–22180. https://doi.org/10.1021/acsomega.1c07369
36. Rai P, Mehrotra S, Priya S, Gnansounou E, Sharma SK (2021) Recent advances in the sustainable design and applications of biodegradable polymers. Bioresour Technol 325:124739. https://doi.org/10.1016/j.biortech.2021.124739

37. Rasal RM, Janorkar AV, Hirt DE (2010) Progress in polymer science poly (lactic acid) modifications. Prog Polym Sci 35:338–356. https://doi.org/10.1016/j.progpolymsci.2009.12.003
38. Rastogi VK, Samyn P (2015) Bio-based coatings for paper applications. Coatings 5:887–930. https://doi.org/10.3390/coatings5040887
39. Rhim JW, Lee JH, Hong SI (2006) Water resistance and mechanical properties of biopolymer (alginate and soy protein) coated paperboards. Lwt 39:806–813. https://doi.org/10.1016/j.lwt.2005.05.008
40. Rojas K, Canales D, Amigo N, Montoille L, Cament A, Rivas LM, Gil-Castell O, Reyes P, Ulloa MT, Ribes-Greus A, Zapata PA (2019) Effective antimicrobial materials based on low-density polyethylene (LDPE) with zinc oxide (ZnO) nanoparticles. Compos Part B Eng 172:173–178. https://doi.org/10.1016/j.compositesb.2019.05.054
41. Saji VS (2020) Wax-based artificial superhydrophobic surfaces and coatings. Coll Surf A Physicochem Eng Asp 602:125132. https://doi.org/10.1016/j.colsurfa.2020.125132
42. Salam A, Lucia LA, Jameel H (2015) A new class of biobased paper dry strength agents: synthesis and characterization of soy-based polymers. ACS Sustain Chem Eng 3:524–532. https://doi.org/10.1021/sc500764m
43. Schäfer JL, Schölch S, Prucker O, Brandstetter T, Rühe J, Stockert AR, Meckel T, Biesalski M (2021) Accessibility of fiber surface sites for polymeric additives determines dry and wet tensile strength of paper sheets. Cellulose 28:5775–5791. https://doi.org/10.1007/s10570-021-03817-7
44. Shankar A, AM K, Narayan R, Chakrabarty A (2023) Progress in Organic Coatings Emulsion polymerized styrene acrylic/nanocellulose composite coating to improve the strength and hydrophobicity of kraft paper. Prog Org Coat 182:107634. https://doi.org/10.1016/j.porgcoat.2023.107634
45. Shankar A, Areekkadan AM, Chakrabarty A (2024) Water-repelling and reinforced Kraft paper by biobased coating with robustness, compostability, and mulching potential. ACS Sustain Chem Eng 12:14657–14669. https://doi.org/10.1021/acssuschemeng.4c03792
46. Shankar S, Bang YJ, Rhim JW (2019) Antibacterial LDPE/GSE/Mel/ZnONP composite film-coated wrapping paper for convenience food packaging application. Food Packag Shelf Life 22:100421. https://doi.org/10.1016/j.fpsl.2019.100421
47. Siqueira EJ, Salon MB, Belgacem MN, Mauret E (2015) Carboxymethylcellulose (CMC) as a model compound of cellulose fibers and polyamideamine epichlorohydrin (PAE)—CMC interactions as a model of PAE—fibers interactions of PAE-based wet strength papers 42144, pp 1–10. https://doi.org/10.1002/app.42144
48. Surma M, Wiczkowski W, Zielinski H, Cieslik E (2015) Determination of selected perfluorinated acids (PFCAs) and perfluorinated sulfonates (PFASs) in food contact materials using LC-MS/MS. Packag Technol Sci 28:789–799. https://doi.org/10.1002/pts
49. Tian X, Wu M, Wang Z, Zhang J, Lu P (2022) Journal of Colloid and Interface Science A high-stable soybean-oil-based epoxy acrylate emulsion stabilized by silanized nanocrystalline cellulose as a sustainable paper coating for enhanced water vapor barrier. J Colloid Interface Sci 610:1043–1056. https://doi.org/10.1016/j.jcis.2021.11.149
50. Wang Y, Guo R (2014) Polymers used in papermaking industry process, pp 1–72
51. Worku LA, Bachheti A, Bachheti RK, Rodrigues Reis CE, Chandel AK (2023) Agricultural residues as raw materials for pulp and paper production: overview and applications on membrane fabrication. Membranes 13(2):228
52. Xiang H, Wang B, Zhong M, Liu W, Yu D, Wang Y, Tam KC, Zhou G, Zhang Z (2022) Sustainable and versatile superhydrophobic cellulose nanocrystals. ACS Sustain Chem Eng 10:5939–5948. https://doi.org/10.1021/acssuschemeng.2c00311
53. Yadav S, Khan A, Hamdani SS, Rabnawaz M (2024) Degradable polymeric waxes for paper coating applications. ACS Appl Polym Mater 6:3263–3272. https://doi.org/10.1021/acsapm.3c03072
54. Yamamoto Y, Ichiura H, Ohtani Y (2019) Improvement of wet paper strength using a phosphoric acid–urea solution. Cellulose 26:5105–5116. https://doi.org/10.1007/s10570-019-02423-y

55. Yang D, Chen Y, Pelton RH (2019) Challenges to achieving strong but fully degradable adhesive joints between wet cellulose surfaces. Langmuir 35:13286–13291. https://doi.org/10.1021/acs.langmuir.9b02481
56. Yang D, Pelton RH (2017) Degradable microgel wet-strength adhesives: a route to enhanced paper recycling. ACS Sustain Chem Eng 5:10544–10550. https://doi.org/10.1021/acssuschemeng.7b02541
57. Zervos S, Alexopoulou I (2015) Paper conservation methods: a literature review. Cellulose 22:2859–2897. https://doi.org/10.1007/s10570-015-0699-7
58. Zhang H, Kannangara D, Hilder M, Ettl R, Shen W (2007) The role of vapour deposition in the hydrophobization treatment of cellulose fibres using alkyl ketene dimers and alkenyl succinic acid anhydrides. Colloids Surf A Physicochem Eng Asp 297:203–210. https://doi.org/10.1016/j.colsurfa.2006.10.059
59. Zhou Y, Fan M, Luo X, Huang L, Chen L (2014) Acidic ionic liquid catalyzed crosslinking of oxycellulose with chitosan for advanced biocomposites. Carbohydr Polym 113:108–114. https://doi.org/10.1016/j.carbpol.2014.06.081
60. Zhou Y, Huang Q, Wang L, Wang X (2024) Enhancing cross-linking network for superior wet strength of paper by sustainable hyperbranched polyimines. ACS Appl Mater Interfaces 16:15383–15393. https://doi.org/10.1021/acsami.4c01403
61. Zhu R, Liu X, Song P, Wang M, Xu F, Jiang Y, Zhang X (2018) An approach for reinforcement of paper with high strength and barrier properties via coating regenerated cellulose. Carbohydr Polym 200:100–105. https://doi.org/10.1016/j.carbpol.2018.07.069

Nanocoatings for Enhanced Paper Surface Properties

Sakshi Sahu, Jaydip D. Bhaliya, Neha Kesarwani, S. N. Raju Kutcherlapati, Teja Sai Telagumsetti, Jayesh Shankar Waghmare, Shivam Shailesh Kumar Joshi, Prashanthi Yarasani, Jayaramudu Jarugala, Arindam Chakrabarty, and Emmanuel R. Sadiku

Abstract The integration of nanotechnology in the paper industry represents a transformative advancement in enhancing the surface properties of paper products. This chapter examines the innovative potential of nanocoating, whose nanoscale dimensions and unique physicochemical properties offer significant enhancements in paper functionality and performance. Starting with an overview of the importance of surface properties in various paper applications, the chapter outlines the challenges and opportunities associated with the use of nanocoatings. A detailed exploration of the fundamentals of nanocoatings follows, including their classification into polymer-based, inorganic, and hybrid types, as well as an examination of the mechanisms by which these coatings enhance surface attributes. The chapter also covers the synthesis and application techniques of nanocoating, such as layer-by-layer (LBL) assembly, sol–gel processes, and plasma-enhanced chemical vapor deposition (PECVD), along with other advanced methodologies, for example, electrospinning and dip coating. The discussion then transitions to specific application methods that are tailored for the paper industry, emphasizing techniques such as high-speed blade coating, rod coating, spray coating, and inkjet printing. Various types of nanocoatings are examined, including hydrophobic, antimicrobial, barrier, conductive, and mechanical reinforcement coatings, each designed to enhance specific paper qualities. Environmental and economic considerations, including its sustainability, cost–benefit analysis, and regulatory aspects, are also addressed. The chapter concludes with a forward-looking

S. Sahu · J. D. Bhaliya · N. Kesarwani · S. N. R. Kutcherlapati (✉) · T. S. Telagumsetti · J. S. Waghmare · S. S. K. Joshi · J. Jarugala · A. Chakrabarty
Polymers and Functional Materials Division, CSIR-Indian Institute of Chemical Technology, Hyderabad, Telangana, India
e-mail: ksnraju.iict@csir.res.in

S. N. R. Kutcherlapati · S. S. K. Joshi · J. Jarugala · A. Chakrabarty
Academy of Scientific and Innovation Research (AcSIR), Uttar Pradesh, Ghaziabad, India

P. Yarasani
Department of Chemistry, Mahatma Gandhi University, Nalgonda, Telangana, India

E. R. Sadiku
Department of Chemical, Metallurgical and Materials Engineering (Polymer Division), Tshwane University of Technology, Pretoria, South Africa

R. K. Bachheti (ed.), *Nanotechnology in the Pulp and Paper Industry*, Smart Nanomaterials Technology, https://doi.org/10.1007/978-981-95-3473-9_4

perspective on emerging nanomaterials, multifunctional and smart coatings, and the integration of nanotechnology with digital and smart packaging, thereby offering valuable insights into future research and development in the field.

Keywords Nanocoating · Paper industry · Surface properties · Synthesis techniques

1 Introduction

Nanotechnology is a rapidly advancing field in science and technology that is set to transform material usage in the twenty-first century. The discovery of new materials, processes, and phenomena at the nanoscale level, along with the advancement of experimental and theoretical research techniques, opens up new possibilities for the creation of innovative nanosystems and nanostructured materials. Nanotechnology has quickly become a priority for policymakers and industry leaders. Nanotechnology is important because of the following points, viz: (1) lower space requirement, minimal material needed, less energy requirement, and faster than the conventional materials, (2) novel properties and phenomena, and (3) most efficient manufacturing and the ability of nanomaterials to interact with living or nonliving entities [29]. A variety of ultra-small-scale materials, such as metal oxides, ceramics, and polymers, along with a broad range of processing methods, for example, the molecular-scale self-assembly techniques, are either currently in use or being developed for commercial applications. Paper has been utilized for over 2000 years due to its widespread availability, ease of handling and processing, biocompatibility, and biodegradability. Its compatibility with the fast roll-to-roll printing and low cost further contribute to its pervasive use in various applications. Paper is particularly well suited as a substrate for affordable printed microelectronics, thereby offering applications that complement those of high-end silicon-based electronics [64].

The conventional paper is composed of cellulose fibers with diameters ranging from 20 to 50 μm. The surface roughness, porous texture, and lack of transparency of standard paper present challenges for integrating electronic devices onto its surface. Techniques such as super-calendaring to smooth the surface, applying coatings, or creating devices on flat substrates before transferring or laminating them onto paper can be employed to address these issues. Paper is a versatile and flexible material that finds diverse applications in daily life, including printing, household products, and food packaging. The versatility of paper is attributed to its properties, such as low density, good mechanical characteristics, excellent recyclability, and biodegradability [31]. However, the fibrous and porous structure of paper results in poor barrier properties, which is a drawback [40]. Consequently, plastic films with significantly higher barrier properties have replaced paper in many packaging applications and have become an integral part of daily life. Nevertheless, the poor degradability of plastic films has raised substantial environmental concerns, leading to renewed interest in the use of paper products for packaging applications [23].

Paper exhibits certain drawbacks, necessitating modifications to enhance its use in packaging applications and improve the key properties required for specific needs. For instance, in the food packaging domain, a crucial requirement is to enhance the barrier properties of paper. This is attributed to paper's inherent high porosity, leading to poor resistance to gases such as oxygen and water vapor. The most common approach to improve the barrier properties of paper is to apply a thin wax coating on the surface. However, this technique presents a challenge, as it can hinder the recyclability of the final products [76]. Previous research has demonstrated that nanocoating technology can significantly enhance the barrier properties, flame retardancy, and mechanical performance of coated substrates. Furthermore, nanocoatings are considerably thinner than traditional coatings, resulting in minimal cost concerns. Therefore, the primary goal of researchers is to enhance the barrier properties and flame retardancy of paper substrates for use in food packing [47]. Consequently, nanocoating is a key factor within the nanotechnology domain. By engineering the surfaces and layers at the nanoscale level, a wide array of functionalities and novel physical effects can be realized. Table 4.1 provides an overview of the various applications for nanolayers and coatings.

The surface chemistry of paper often varies depending on its grade and origin. Different surface treatments can modify the surface polarities of materials. Understanding and managing the physicochemical properties of paper, sizing agents, and inks is crucial for achieving the desired printing qualities. Most sizing materials used for paper surface treatment are polymeric, consisting of large linear macromolecules that provide the necessary properties, such as strength and flexibility, for effective

Table 4.1 Applications of nanocoating [1]

Surface properties	Application area
Mechanical properties (e.g., tribology, hardness, scratch resistance	Wear protection of machinery and equipment and mechanical protection of soft materials. (polymers, wood, textile, etc.), super-plasticity of ceramics
Wetting properties (e.g., antiadhesive, hydrophobic, hydrophobic)	Antigraffiti, antifouling, lotus effect, self-cleaning surfaces
Thermal and chemical properties (e.g., heat resistance and insulation, corrosion resistance)	Corrosion protection for machinery and equipment, heat resistance for turbines and engines, thermal insulation equipment, and building materials
Biological properties (biocompatibility, anti-infective)	Biocompatible implants, medical tools, wound dressings
Electronical and magnetic properties (e.g., magneto resistance, dielectric)	Ultrathin dielectrics for field-effect transistors, magneto-resistive sensors, and data memory
Catalytic efficiency	Good catalytic efficiency through high surface-to-volume ratio
Optical properties (e.g., antireflection, photochromic and electrochromic)	Photochromic and electrochromatic windows, antireflective screens, and solar cells

coating. The adhesion efficiency is influenced by the balance between adhesion and cohesion interactions. Issues may arise from mass transfer processes, including the penetration of adhesives into the paper substrate and their subsequent migration to the surface. These interactions are ultimately governed by factors such as the shape and size of paper pores, their wettability and swelling characteristics, as well as the viscosities and surface tension of the sizing materials. While the physicochemical properties of paper and lignocellulosic surfaces have been extensively studied, many factors influencing surface characteristics remain unclear [48]. Gardner and colleagues [10] employed dynamic contact angle analysis to explore the wettability of wood samples from various origins. They found that factors such as elemental and chemical composition, surface roughness, capillary-flow properties, sample storage conditions, and the speed of DCA measurements significantly impact wettability results. The authors highlighted methods suitable for assessing the surface energetics and chemistry of wood.

2 Fundamentals of Nanocoating

Nanocoating is an incredibly slender layer that is accumulated at the nanoscale to improve the performance of materials by utilizing special qualities, for example, enhanced surface area and quantum effects. These coatings are built up by using techniques such as chemical and physical vapor deposition, sol–gel, electrochemical deposition, and spray coating. In the following section, the fundamentals of nanocoating will be discussed in detail.

2.1 Definition and Characteristics of Nanocoating

An ultrathin layer of material, usually <100 nm thick, called nanocoating, is applied to a surface to improve the material's existing characteristics or add new features. By taking advantage of the special properties of materials at the nanoscale level, such as an enhanced surface area and quantum effects, these coatings enhance the material's performance in aspects including antibacterial activity, durability, corrosion resistance, optical clarity, and thermal insulation. A high surface area-to-volume ratio, which enhances reactivity and contact with the environment, is just a few of the distinctive qualities that nanocoatings exhibit. They provide exceptional resistance to wear and abrasion, while also greatly enhancing strength and durability. Because they offer anticorrosion protection, nanocoatings are the best choice for preventing oxidation and chemical degradation. Their antibacterial activity prevents the growth of bacteria and pathogens, which is beneficial in the food and medical industries. Meanwhile, their hydrophobic and hydrophilic qualities allow surfaces to self-clean. Furthermore, electrical conductivity, thermal stability, and optical qualities can all be improved by nanocoating [21].

2.2 *Classification of Nanocoatings*

The nanocoating is generally classified into several forms, out of which, the mainly used, are polymer coating, inorganic coating, hybrid coating, etc. The following section briefly describes each of these forms.

2.2.1 Polymer-Based Coatings

Polymer-based nanocoating is a coating whose characteristics are improved by integrating nanoparticles with polymers. These coatings are ideal for a wide range of applications due to their combination of robustness, durability, and flexibility. The polymer matrix's tensile strength, abrasion resistance, and chemical stability are all markedly enhanced by the addition of nanoparticles. Furthermore, polymer-based nanocoating offers superior chemical and moisture resistance, shielding surfaces from the environment. They are widely used in various industries, including packaging, textiles, automotive, and electronics. These coatings shield the electronics from corrosion, dust, and moisture. They are used on automobile components to improve their resilience to deterioration and longevity. They add qualities, such as water resistance and stain resistance, to textile products [44]. Furthermore, food packaging often utilizes polymer-based nanocoatings to create barriers against moisture and gases, thereby extending the shelf life of the goods. They are a well-liked option in different industries due to their increased performance and versatility. In Fig. 4.1, polymer coating was carried out on paper with polysaccharide solutions, while fillers were used for coating on paper because of their good mechanical strength and antibacterial properties.

2.2.2 Inorganic Coatings

Resources, such as metals, metal oxides, or ceramics, are used for the deposition of inorganic nanocoating, which provides remarkable chemical and thermal resiliencies. These coatings are perfect for specific demanding applications because of their exceptional hardness, wear resistance, and capacity to tolerate harsh environments. Inorganic nanocoatings has a high degree of stability and robustness, which make them helpful in shielding electrical components from heat-related corrosion. Figure 4.2 shows the inorganic coating by the layer-by-layer assembly on Mg alloy [19]. It improves the antireflective and UV protection qualities in optics, thereby enhancing the longevity and functionality of lenses and other optical equipment [25]. Inorganic nanocoatings are advantageous for medical equipment, because they are resistant to microbial development and are biocompatible.

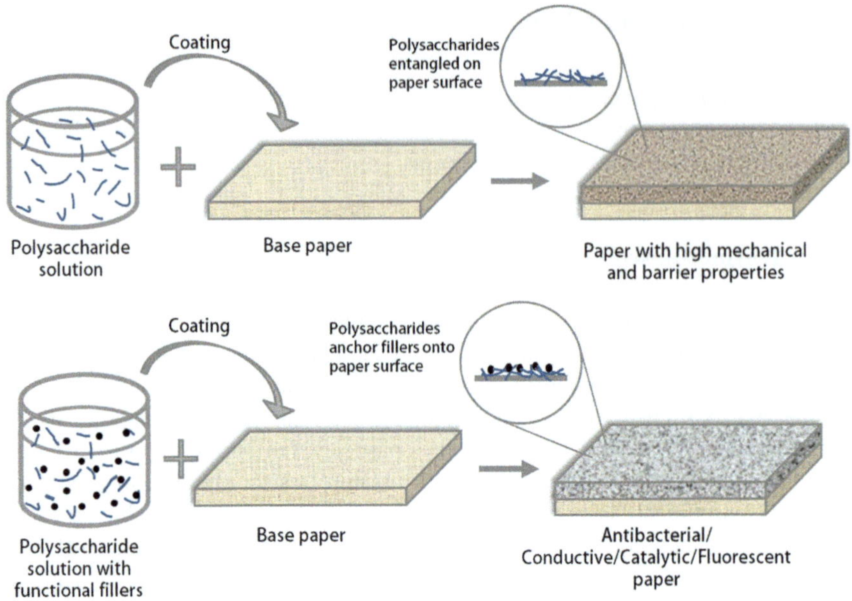

Fig. 4.1 Polymer coating on paper [36]. Copyright (2020). Reproduced with permission from MDPI. (Open access)

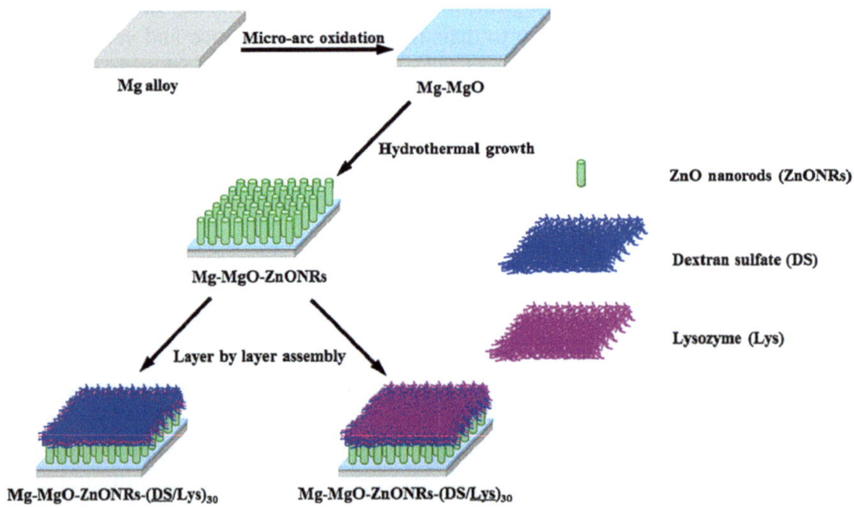

Fig. 4.2 Inorganic polymer coating [19]. Copyright 2020. Reproduced with permission from Elsevier

2.2.3 Hybrid Coatings

The benefits of inorganic and organic (polymer) materials are combined in hybrid nanocoating to provide coatings with improved and adaptable qualities. These coatings combine the toughness and stability of the inorganic components with the flexibility and endurance of polymers. It is possible to fine-tune mechanical strength, heat and chemical resistance, and overall performance through the integration of various materials [46]. The enhanced mechanical characteristics of the hybrid nanocoating, including greater flexibility and durability, define it. Furthermore, they exhibit excellent thermal stability, which qualifies them for use in applications subjected to high temperatures. These coatings' resistance to chemicals guarantees defense against deterioration and corrosion from a variety of substances [77]. Hybrid nanocoatings can also offer cutting-edge features, including antibacterial, UV protection, and self-healing capabilities. Applications for hybrid coatings are numerous. Due to their favorable structures, they have developed a range of characteristics that are utilized in various industries, including flexible displays, food packaging coatings, adhesive coatings, and membranes. These features include antireflective, antistatic, and barrier properties [19]. Nanocomposite coatings have been utilized as a means to control their optical properties, for example, refractive index. The increased mechanical strength of these coatings has proven beneficial due to their abrasion resistance and protective properties, which include applications for anticorrosion and protective coatings for metallic substrates, such as those used in aircraft.

2.3 Mechanisms of Surface Property Enhancement

Nanoscale processes enhance surface functionality and are involved in the mechanisms that govern surface property enhancement in nanocoatings. Using functional groups, nanoparticles can chemically alter surface characteristics, mechanically strengthen substrate durability, and establish physical barriers against moisture and impurities [59]. Properties, such as mechanical strength or optical clarity, can be further tailored with nanostructured designs. Furthermore, the interlinking of nanocoatings with biological systems can enhance their antibacterial effectiveness and biocompatibility. The mechanisms of surface property enhancement refer to the various processes and interactions at the nanoscale level that improve the functional characteristics of coated surfaces. These mechanisms include the following:

- **Physical Barrier**: The protective layer created by nanocoating serves as a barrier against impurities, gases, and moisture.
- **Chemical Modification**: By changing their surface chemistry, nanoparticles can improve certain characteristics, including hydrophobicity, adhesion, and resistance to corrosion.
- **Mechanical Reinforcement**: Coatings with incorporated nanoparticles provide better wear resistance, hardness, and toughness.

- **Surface Energy Modification**: Wetting characteristics and adhesion are impacted by the surface energy adjustments that are made by nanostructured coatings.

3 Synthesis and Application Techniques

Paper coatings are necessary to improve the surface attributes, such as printability, brightness, and durability of paper. Paper coatings are made by selecting binders (e.g., latex or starch), colors (such as clay, calcium carbonate, or titanium dioxide), and additives. These components recombine to create a homogeneous slurry. This slurry can be applied to paper through a variety of processes, including blade coating, roll coating, and curtain coating. Advanced techniques, such as dip-coating, which immerses paper in a coating solution; layer-by-layer assembly, which creates coatings by alternating layers of different materials; and electrospinning, which uses an electric field to create fine fibers that form a coating, are also used to achieve specific properties.

3.1 Layer-By-Layer (LBL) Assembly

The manufacturing of paper plays a crucial role in the economic infrastructure and has a significant impact on both personal and professional realms. These coating technologies apply a layer to the paper substrate during or after the sheet formation process, utilizing microfibers. Agarwal et al. [3] demonstrated a layer-by-layer nanoassembly of conductive polyelectrolyte thin films on lignocellulose microfibers, subsequently integrating them with paper. Nanotechnology, including the layer-by-layer nanoassembly, has the potential to enhance pulping, process chemistry, paper coating, and recycling procedures. The nanoassembly technique involves the sequential deposition of oppositely charged polyelectrolytes or nanoparticles on surfaces of diverse shapes and sizes. This distinctive characteristic of the layer-by-layer assembly has generated extensive interest in its application within the field of nanocoating. Over the past decade, layer-by-layer nanoassembly has emerged as a straightforward, practical, and adaptable approach, capable of producing ultrathin films over vast surfaces, microfibers, and cores with the desired compositions. LBL self-assembly was first proposed by Iler in 1966 for microparticles, and it received significant attention after its rediscovery and considerable development for polyelectrolytes. It is a multistep deposition procedure that may produce multilayered thin films of oppositely charged polyelectrolytes and/or nanoparticles on almost any solid surface [56]. LBL self-assembly involves immersing a charged substrate in an oppositely charged colloid solution to adsorb the first monolayer, followed by a washing cycle to remove the unbound material and prevent contamination of the next oppositely charged colloid. Finally, the coated substrate is submerged to deposit a second layer, forming the multilayered structure (Fig. 4.3). Some LBL techniques eliminate the

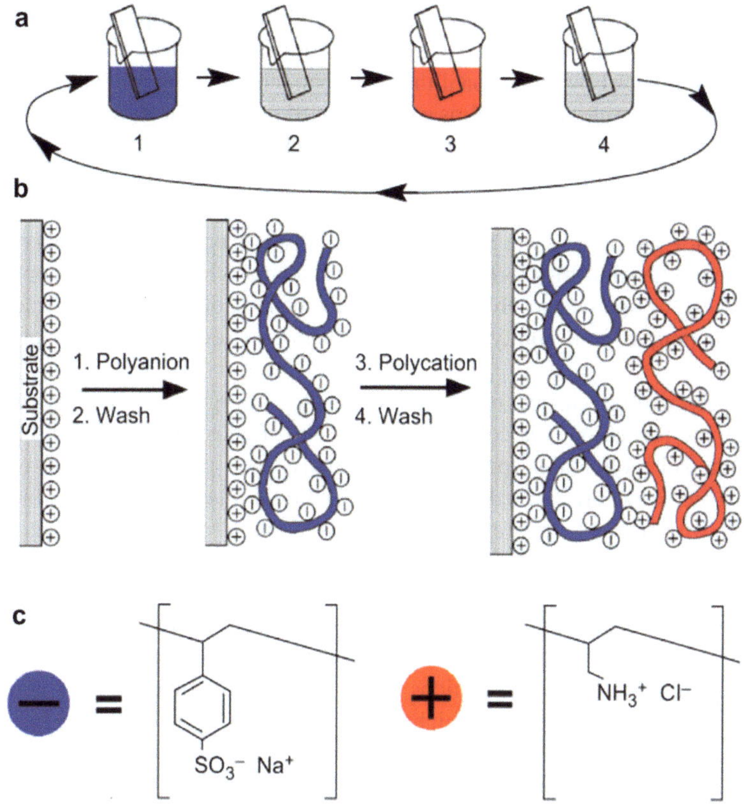

Fig. 4.3 The following are shown in a schematic diagram: **a** immersion LBL construction employing oppositely charged polymers on a planar substrate, **b** the charge properties of the films following each step of deposition, and **c** the polymer's chemical structure [56]. Copyright (2016). Reproduced with permission from ACS

need for washing cycles, resulting in faster assembly times [3]. When compared to more traditional coating techniques, LBL self-assembly methods offer several advantages. These include (1) ease of use and equipment simplicity, (2) suitability for coating a wide range of surfaces, (3) abundance of natural and synthetic colloids, (4) flexible application to objects with irregular shapes and sizes, (5) formation of stabilizing coats, and (6) control over the required multilayer thickness [33].

3.2 Sol–Gel Processes

The all-round mode for producing coatings that change from a solution (sol) to a solid (gel) state is known as the sol–gel process. In paper coatings, this approach yields thin layers that enhance the material's mechanical, thermal, and chemical properties.

The procedure forms a sol out of metal salts or alkoxides by dipping or spraying it on the paper, and then waiting for it to gel and dry. The features of the coating can be further enhanced by an optional thermal treatment, thereby producing uniform, long-lasting, and multipurpose films with improved functionalization, aesthetics, and barrier properties for a wide range of applications. In its most basic form, sol–gel dip coating involves removing a substrate from a fluid sol. A solid layer is then deposited as a result of further condensation processes, solvent evaporation, and gravitational draining. Sol–gel dip-coating requires significantly less equipment and may be less costly than other thin-film forming techniques, such as chemical vapor deposition, evaporation, or sputtering. However, the main benefit of sol–gel over traditional coating techniques is its ability to customize the microstructure of the film produced [54].

Sol–gel coatings can be made via two routes: inorganically and organically. Through the formation of a colloidal suspension (often oxides) and with the gelation of the sol (colloidal suspension of extremely small particles, with sizes of 1–100 nm), networks are formed in a continuous liquid phase by using the inorganic approach. In general, there are four steps involved in the formation of sol–gel. They are as follows: (a) hydrolysis, (b) condensation and polymerization of monomers to form chains and particles, (c) particle growth, and (d) agglomeration of the polymer structures, followed by the formation of networks that extend throughout the liquid medium, resulting in the gel thickening. Once the hydrolysis reaction has started, the condensation and hydrolysis reactions happen at the same time. Alcohol and water are examples of low-molecular-weight byproducts, produced during the hydrolysis and condensation processes. These tiny molecules are forced away by drying, and as more condensation takes place, the network gets smaller.

As shown in Fig. 4.4, the sol–gel process typically comprises four main phases: hydrolysis, where monomers undergo chemical reactions to form chains and particles; condensation and polymerization, where these chains and particles continue to grow; particle growth, where the particles increase in size; and finally, agglomeration of the polymer structures, leading to the formation of interconnected networks that permeate the liquid medium, resulting in the thickening and gelation of the solution [6]. In many applications, sol–gel materials have been utilized in conjunction with paper. Work has been reported, for instance, on coating paper with sol–gel compounds or adding them to paper structures. The paper was treated with standard sol–gel materials to strengthen its resilience to heat and moisture [57]. Recent observations have revealed that treating paper with a methyltrimethoxysilane oligomer increased its tensile strength and water repellency. A study revealed that the leaching stability of paper, coated with sol–gel silica layers, containing bound colors, was enhanced.

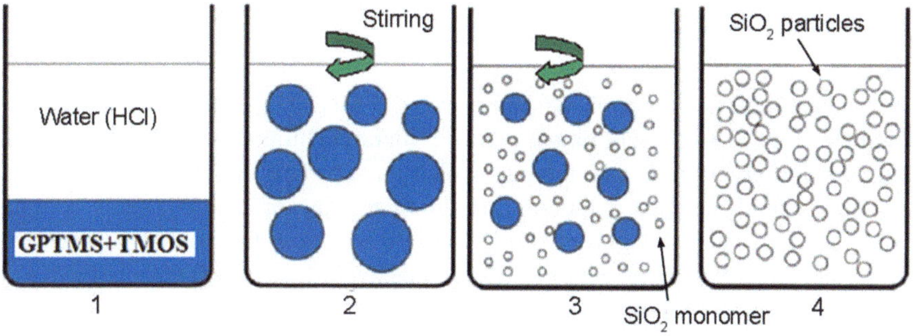

Fig. 4.4 Schematic diagram for sol–gel process [6]. Copyright (2016). Reproduced with permission from Springer Nature

3.3 *Plasma-Enhanced Chemical Vapor Deposition (PECVD)*

The deposition of thin-film dielectrics at low temperatures (usually between 100 and 500 °C) is of great interest. These films find application in the following areas: (a) surface passivates for device architectures, (b) interlevel dielectrics in integrated circuits, and (c) gate dielectrics, in field effect transistors. The deposition method needs to be tailored for each application, since each one has a unique set of material and process limitations. A remote PECVD method was applied to the deposition of dielectric thin films suitable for device applications [38]. Attention was focused on the connections among the deposition reactions, chemical bonding, and physical characteristics of the deposited layers. Many PECVD functional coating materials have been studied and explored for a wide range of applications, some of these materials are already being used in industrial settings. Figure 4.5 shows the growth of 2D materials by the PECVD growth mechanism.

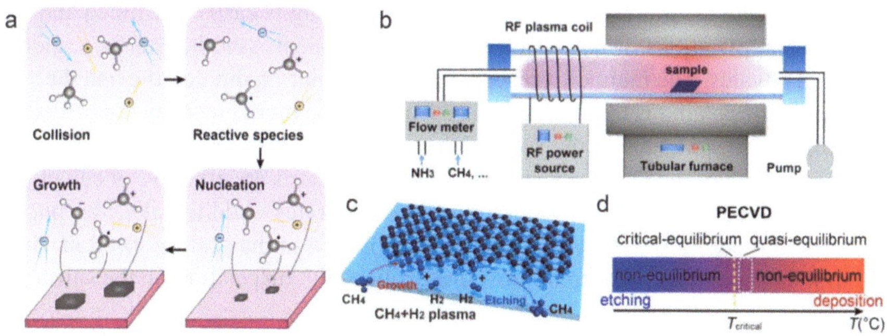

Fig. 4.5 Growth of 2D materials by PECVD: **a** growth mechanism; **b** PECVD setup; **c** competition processes in CH_4/H_2 PECVD; **d** different equilibrium states of PECVD [73]. Copyright (2021). Reproduced with permission from ACS

3.4 Other Techniques

To produce superhydrophobic cellulose paper, a modified surface is required that combines low surface energy materials with hierarchical microscale and nanoscale roughness, thereby imparting the paper with its superhydrophobic characteristics. To date, several methods, including plasma treatment, dip coating, spray coating, brush coating, chemical grafting, solution immersion, expansion of supercritical fluids, and electrospinning, have been developed to produce superhydrophobic paper. These techniques can employ polymers and/or nanoparticles, and they can include monolayer or multilayer deposition. Unfortunately, several disadvantages, including their intricacy, lack of adaptability, and restricted industrial use, are associated with some of the aforementioned methodologies [72]. By using an electric field, charged polymer solutions are drawn into ultrafine fibers for electrospinning coating, which is subsequently applied to paper. By using this method, a consistent, nanoscale fiber network is created, thereby improving the strength, hydrophobicity, and barrier effectiveness of the paper. The electrospinning method is ideal for sophisticated paper coating applications, because it provides the exact control required over fiber diameter and alignment.

A popular and flexible method for attaching thin films to a substrate is the dip coating technique (Fig. 4.6). A substrate must be submerged in a coating solution and then removed at a certain pace. A small layer of the coating solution sticks to the substrate's surface as it is removed. The surface tension, withdrawal speed, solution viscosity, and other variables all affect the coating's thickness and homogeneity. To harden the coating, the substrate is usually dried or cured after coating. Due to its ease of use, low cost, and ability to produce homogeneous coatings, the dip coating method is widely utilized across various sectors. Its applications range from coatings that are both beautiful and protective to those that are functional in electronics and medical equipment. Large surface areas and intricate forms can be coated with this technology, which makes it useful for both large-scale industrial production and small-scale laboratory studies. Because dip-coating is easily machined and has numerous applications, it has become a more prominent issue in interface science and fluid physics. The main difference between the solution dip-coating and the sol–gel-based dip-coating is that the deposited films can be more precisely controlled by controlling the synthesis of the sol and the ensuing gelation process. Additionally, improved vacuum- and spin-assisted dip-coating methods are effective at enhancing the infiltration of coating solutions into fiber substrates. Additionally, the photo-assisted dip-coating is used to regulate the evaporation of the coating solution, since the irradiation impact helps to improve the coating film disposition on the substrate.

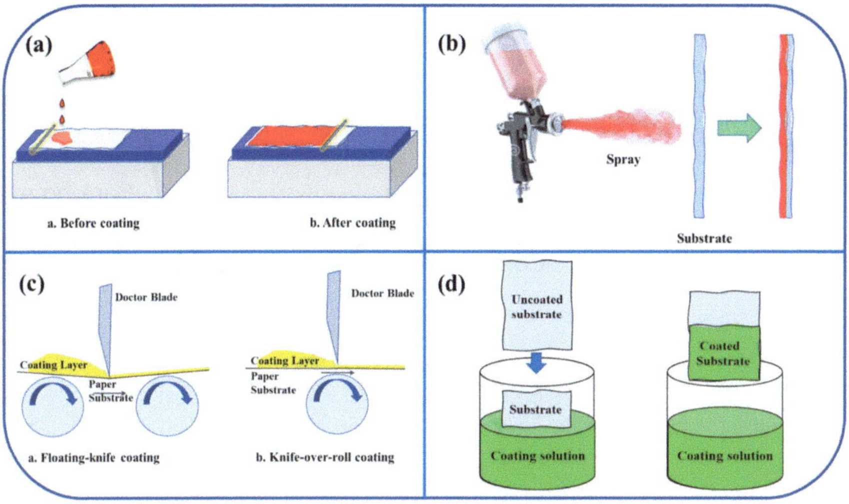

Fig. 4.6 Different methods for coating on the paper substrate: **a** bar coating, **b** spray coating, **c** knife coating, and **d** dip coating [34]. Copyright (2024) Reproduced with permission from Springer Nature

4 Application Methods for Nanocoating in the Paper Industry

4.1 High-Speed Blade Coating

The blade coating method is one of the most widely used techniques for applying uniform thin films of liquid onto moving webs, particularly, in industrial settings. In this process, a blade acts as a smoothing device, metering the coating by removing excess material that is applied upstream. The coating is pulled into the narrow channel between the blade and the moving substrate, thereby creating a shear stress that helps achieve a consistent film thickness. The geometry of the blade nip, where the coating is metered, plays a crucial role in determining the uniformity, thickness, and overall quality of the final coated film.

In the setup, a liquid layer flows through a shallow channel formed between a fixed blade and a mobile substrate, which is held against a roller (Fig. 4.7). The blade can be either rigid or flexible, and the substrate can vary in compressibility, depending on the specific application requirements. In the flexible blade coating, the blade may feature a mechanical component or an inflatable pressure hose near the blade tip, hence enhancing the control over the coating process. This technique is widely used in various industrial applications, including the application of oxide coatings on magnetic recording tapes, paperboard, and films. The main objective of blade coating in paper and paperboard manufacturing is to create a smooth surface for printing, which enhances optical properties and the printed ink density. However, achieving uniformity can be challenging due to the rough, porous, and compressible nature

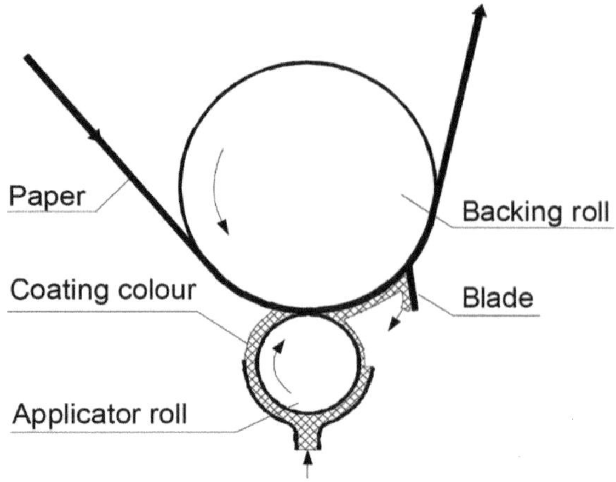

Fig. 4.7 Principle set up of blade coating with an applicator roll [4]. Copyright (2011). Reproduced with permission from WIT Press. (Open access)

of the substrates. The blade coating is a fundamental yet sophisticated technique, often compared to the spreading of butter on bread, where the blade's angle and its interaction with the substrate are critical to controlling the coat weight and achieving the desired finish [4].

4.2 Rod Coating

Carbon can form a variety of allotropes in its nonorange, one of which is carbon nanotube (CNT). It is further classified into two types: the first is single-walled carbon nanotube (SWCNT), and the other is multiwalled carbon nanotube (MWCNT). Based on the solution-based deposition method and direct growth, the CNT-based TCFs (CNT-TCFs) were produced. We offer various coatings based on solution-based deposition methods, including spray coating, dip coating, inkjet coating, spin coating, gel coating, and major rod coating, among others. Mayer road coating is one of the most widely used coatings, which can coat larger area films on substances with zero wastage of CNT ink and at low temperatures, making the coating effective (Fig. 4.8).

In the work of Meng et al. [28], a glass drawdown pad and a stainless-steel rod wrapped with stainless steel wire were used. The substrate was clamped onto the glass pad, and coating liquid was applied to the substrate in front of the coating rod by using a dropper. Excess liquid was removed during the rod coating process. The amount of liquid left on the substrate depended on the spacing between the high points of the wire. The geometry of this system resulted in a wet film thickness that was directly proportional to the wire diameter. Initially, the coating appeared as a series of stripes spaced according to the wire winding intervals. The thin film can be approximated as an ideal sinusoidal surface. The surface tension created a pressure differential across the liquid–air interface, thereby causing the liquid to flow from the peaks to the valleys of the coating [45].

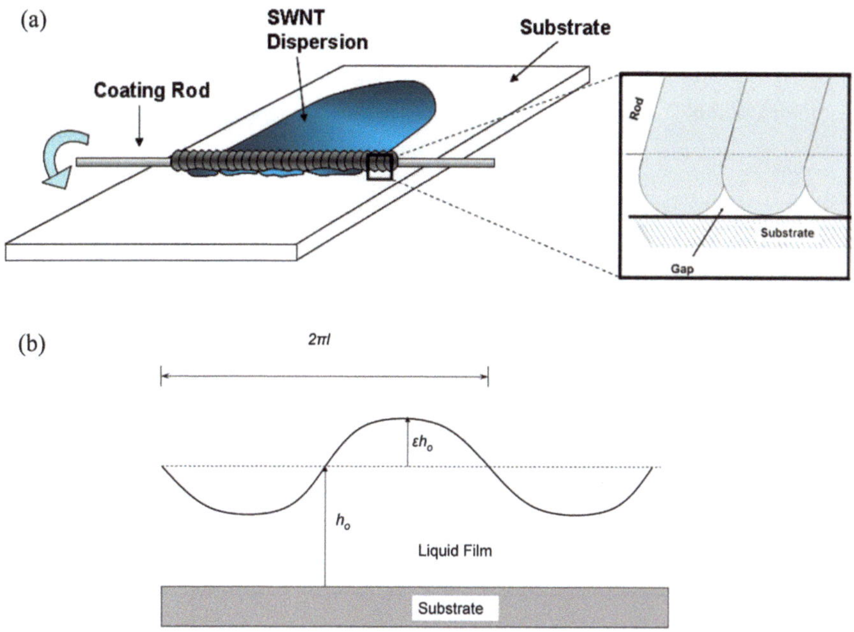

Fig. 4.8 **a** Schematic illustration of a draw-down rod coater with a wire-wound Mayer coating rod and **b** the free surface profile of a thin film, immediately after the Mayer rod coating [13]. Copyright (2009). Reproduced with permission from ACS

4.3 Spay Coating

Spray coating, also known as spin coating, is one of the coating techniques. This is the method with most of the applications with simple, uniform films and rapid techniques for depositing CNT films. It produces highly uniform CNT films [2]. In spray coating, the substrate is acquired, and then, the colloidal suspension is evenly by the spray method, which is followed by heating to evaporate the dispersing fluid. Hence, the coating is formed by the colloidal deposit.

In Fig. 4.9, the spray coating of the CNT film on the substrate can be seen, typically achieved by spraying the CNT solution onto a heated substrate using an air gun. Based on the solvent and the dispersion of the surfactant, the substrate temperature depends on the use of an air-atomizing nozzle spray coating, which will deposit the layers. To get an even and homogenous film, nozzles as well as ultrasonic spray nozzles were used, which give a fine degree of atomization [30]. Based on the size, geometry, and physical property of the liquid being atomized, spray nozzle performance depended on the drying of the droplet, which mainly depended on the boiling point and vapor pressure of the solvent during the formation of the layer. In Fig. 4.9, polydimethylsiloxane (PDMS) was related to CNT and sprayed on the glass substrate, which is set with temperature (temperature is set based on the solvent), and

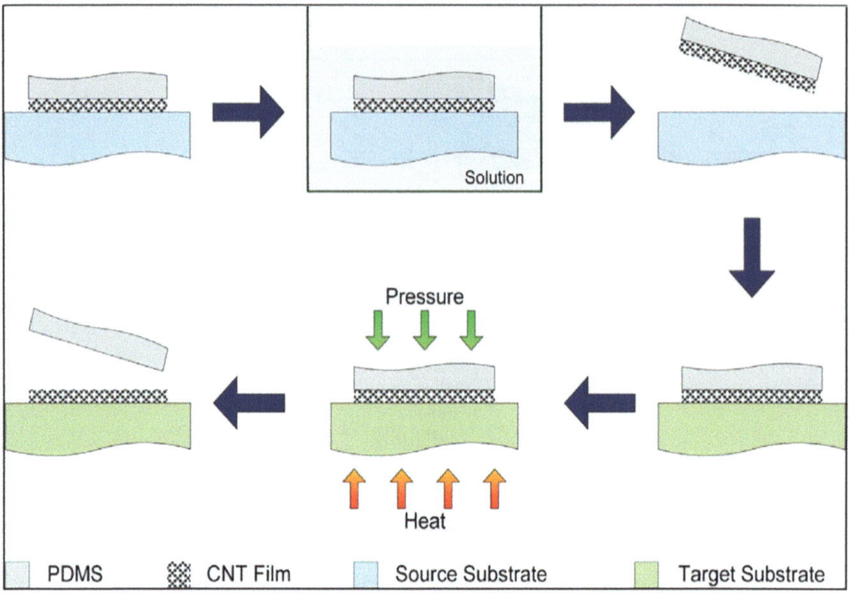

Fig. 4.9 Schematic of the major steps involved in the transfer printing process [2]. Copyright (2013). Reproduced with permission from Elsevier

then applied the pressure with atomizing gas. Furthermore, temperature and pressure are reduced to achieve low deposition [2].

4.4 Inkjet Printing-Pattern and Labels

The placement of active electronics on a plastic substrate was achieved by using inkjet printing. In carbon nanotubes, for feasible large-area fabrication, inkjet printing is the best method for thin film transistors (TFTs). Inkjet printing has a great influence on the industrial scale, because it targets specific film deposition [9]. It consists of ink cartridges, print head, droplet formation, and paper feed. It is used for various applications, such as conjugated polymers and gold nanoparticles (Fig. 4.10), among others. Initially, the cartridges were filled with liquid ink; then through the print head, the ink was sprayed onto the paper with the help of tiny nozzles. The nozzle size can be adjusted based on the printing model. The ink spews out in the form of a drop (droplet formation onto the paper).

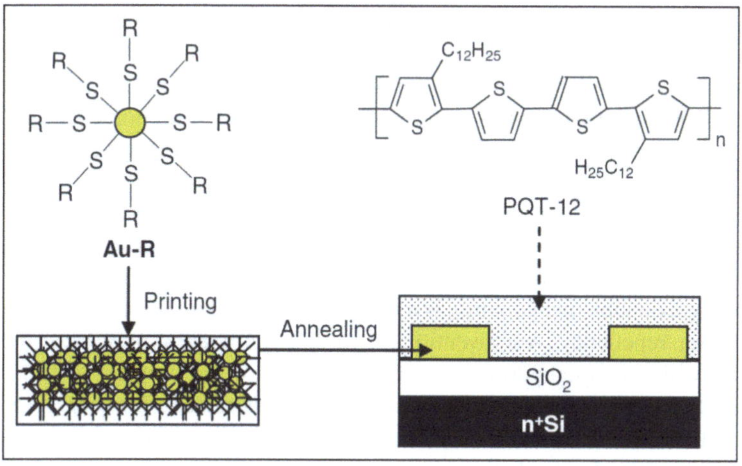

Fig. 4.10 Schematic depiction of printing gold source and drain electrode using gold nanoparticle ink for a bottom-contact OTFT device with PQT-12 semiconductor layer [71]. Copyright 2005. Reproduced with permission from Wiley

5 Types of Nanocoatings for Paper Enhancement

In the rapidly evolving field of materials science and nanotechnology, the application of nanocoatings to enhance traditional materials represents a groundbreaking breakthrough with far-reaching implications [41]. Paper's intrinsic limits have pushed the hunt for advancements in a variety of industries, including packaging and printing, as well as art and education. In this chapter, the numerous forms of nanocoating that can be applied to paper are critically examined.

5.1 Hydrophobic and Superhydrophobic Paper Coatings

Hydrophobic and superhydrophobic coatings stand out in the rapidly changing arena of paper improvement due to their potential to impart remarkable water-repellent qualities on paper surfaces. These coatings solve a basic weakness of regular paper, namely, its vulnerability to water damage. By generating paper surfaces that are either hydrophobic or superhydrophobic, these coatings significantly improve the functional lifespan of paper goods and widen their range of applications, particularly in areas where moisture resistance is key [60]. A contact angle greater than 90° suggests a hydrophobic surface, which allows water to bead up rather than spread out. Superhydrophobic surfaces have even higher water repellency, with contact angles surpassing 150° [14]. On such surfaces, water droplets create nearly spherical beads that readily roll away, thereby reducing contact area and preventing water absorption. These coatings can be used on a range of materials, including glass, metal, fabric, and wood. In Fig. 4.11, droplets of water are seen in different substrates. Hydrophobic nanocoatings function by altering the surface tension of the substance. Water has a high surface tension, which implies the fact that its molecules are attracted to one another. This attraction enables water to condense on a hydrophobic surface rather than to spread out. There are numerous advantages when employing hydrophobic nanocoatings. They can protect surfaces from water damage, corrosion, and discoloration. They may also make the surfaces, simpler to clean. Hydrophobic nanocoatings can also be utilized to make surfaces that clean themselves. These surfaces repel dirt and debris, hence making them easier to maintain clean over time.

Superhydrophobic nanocoating is a form of hydrophobic nanocoating that maximizes water repellency. A conventional hydrophobic coating may cause water to bead up, while a superhydrophobic coating causes water droplets to roll off the surface, almost perfectly spherical, with minimal constraints. This roughness forms air pockets between the water droplet and the surface, hence reducing contact and causing the water to bead up very tightly. A superhydrophobic surface has an extremely high contact angle, which is a measure of how far a droplet spreads out on a surface, usually exceeding 150°.

Superhydrophobic surfaces can be created by using a variety of methods, including polymer and nanoparticle deposition, layer-by-layer (LBL) procedures, and electrochemical processes. The LBL approach employs the alternating deposition of charged

Fig. 4.11 Images of droplets of water above the superhydrophobic coating deposited on various surfaces such as **a** gold film, **b** Si wafer, **c** tissue, and **d** cotton [32]. Copyright (2008). Reproduced with permission from Taylor & Francis

polyelectrolytes or nanoparticles to achieve regulated layer thickness and roughness via electrostatic interactions. After fluorinating the outer layers, these films exhibit outstanding superhydrophobicity [17]. Nanoparticles in block copolymer micelles can produce porous films that, when fluorinated, become superhydrophobic. Organic polymer solutions can be phase-separated to create sol–gel processed porous silica substrates that are, then, superhydrophobic. Sol–gel phase separation, for example, can produce superhydrophobic foams on their own. Furthermore, inorganic materials, such as ZnO and TiO_2, can generate superhydrophobic surfaces via the vertically oriented nanorods that expose the low-energy crystal planes [39].

5.2 Mechanical Reinforcement Coatings

Mechanical reinforcement coatings are designed to enhance the durability, strength, and longevity of materials by providing a protective layer. These coatings are essential in many industries, including construction, automotive, aerospace, and electronics, where materials are subjected to harsh climatic conditions, mechanical wear, and stress. These coatings are primarily, designed to improve substrate mechanical qualities, such as hardness, abrasion resistance, impact resistance, and tensile strength. These coatings defend the substrates against physical impacts, friction, and other mechanical forces, by providing a robust, protective barrier [24]. Additionally, they can enhance resistance to corrosion, chemical attacks, and UV degradation, thereby extending the lifespan of the coated materials. Mechanical reinforcement coatings can be made from a variety of materials, including metals, ceramics, polymers, and composites. These coatings can be applied using classic processes, such as painting and electroplating, or more modern techniques, including thermal spraying, chemical vapor deposition (CVD), and physical vapor deposition (PVD) [30]. The coating material and application method employed are determined by the specific requirements of the intended application, such as the type of substrate, environmental conditions, and the required mechanical properties [30]. Advanced coatings may also incorporate nanomaterials, such as nanocellulose or graphene, to enhance mechanical properties without adding weight or altering the texture of the paper. These coatings can be developed to be environmentally friendly and recyclable, in line with sustainable standards. Cellulose nanocrystals (CNCs) can significantly improve the mechanical properties of paper [15]. Their high aspect ratios and strong intermolecular hydrogen bonding contribute to increased tensile strength, stiffness, and resistance to tearing and folding [24]. This enhancement is particularly beneficial in applications that require robust and durable paper, such as in packaging, where reinforced paper can better withstand handling, transportation, and mechanical stresses [15].

5.3 *Antimicrobial and Antifouling Nanocoatings*

Antimicrobial and antifouling coatings are an advanced combination of materials science and microbiology, developed to address the persistent and widespread challenges in various sectors [67]. These coatings are engineered surface treatments designed to inhibit the growth and accumulation of microbes and organic materials on various substrates. Their importance extends across numerous industries, including healthcare, marine, and food processing, thereby providing many benefits, ranging from infection reduction to increased operational efficiency and equipment longevity [18]

5.3.1 Antimicrobial Coating

Antimicrobial coatings are designed to prevent or eliminate microorganisms, such as bacteria, viruses, fungi, and algae (Fig. 4.12). These coatings can activate a variety of mechanisms, including chemical, physical, and photocatalytic reactions. Chemical agents, such as silver ions, copper, and other metallic nanoparticles, damage microbial cell walls and interfere with critical cellular activities, thereby resulting in cell death. Physical agents use nanostructures or micropatterned surfaces to produce physical damage to microbial cells upon contact. Photocatalytic compounds, such as titanium dioxide (TiO_2), produce reactive oxygen species (ROS) under UV light, effectively neutralizing microorganisms. Nanocoatings with silver nanoparticles have emerged as a powerful solution in the field of paper coatings, providing remarkable antibacterial properties that enhance the hygiene, safety, and longevity of paper products. This innovative technology takes advantage of the unique properties of silver at the nanoscale level to make coatings that efficiently limit the growth of a wide range of bacteria, hence making them especially useful in areas such as healthcare, food packaging, and consumer products. Antimicrobial coatings have a wide range of usages and are crucial in sterile workplaces. In health-care settings, these coatings are used to protect medical devices, implants, and hospital surfaces from hospital-acquired infections and in packaging materials, which helps to extend shelf life and prevent contamination, thereby maintaining food safety and quality.

5.3.2 Antifouling Coating

Antifouling coatings, on the other hand, are intended to head off the growth of biofouling organisms, such as barnacles, algae, and mussels, often found on submerged surfaces. These coatings work by a variety of methods, including the release of poisonous chemicals and the formation of nonstick surfaces. Nonbiocidal antifouling coatings, including foul-release coatings, provide a slippery surface that prevents organisms from adhering strongly, as well as hydrophobic or hydrophilic

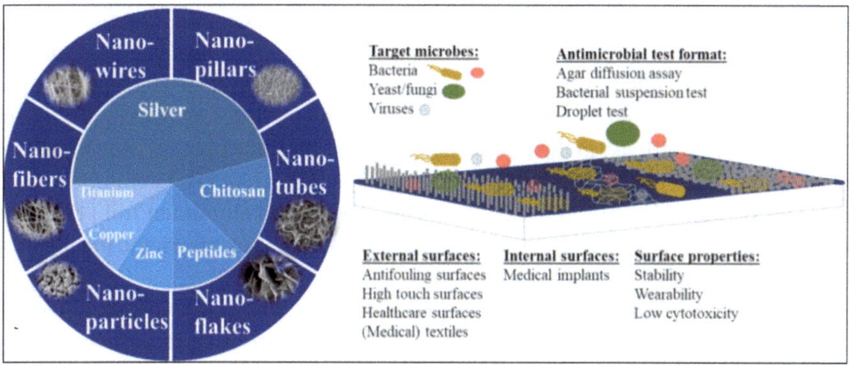

Fig. 4.12 Types of antimicrobial nanoparticles and the essential characteristics of antimicrobial surfaces based on nanomaterials. Antimicrobial agents on the left antimicrobial nanomaterials can take the shape of elongated nanostructures, which can be nanotubes, nanoparticles, nanowires, pillars, or other structures. Right side: structures are made of nanoparticles, nanopillars, or nanofibers that are embedded in or deposited on antimicrobial surfaces. Typical surface types, important physical and chemical characteristics, microbiological goals, and antimicrobial test formats are displayed [27]. Copyright (2022). Reproduced with permission from MDPI. (Open access)

coatings, which use surface chemistry to discourage adhesion. Biomimetic coatings also reduce biofouling by mimicking natural antifouling techniques, such as the shark skin's rough surface [62]. Antifouling coatings are used extensively in the marine industry. By avoiding biofouling on the ship hulls, these coatings reduce drag, thereby resulting in lower fuel consumption and operational expenses [37]. They are also used on underwater structures, such as oil rigs and pipelines, to extend their lifespans and reduce maintenance requirements. Antifouling coatings in aquaculture help to keep water flowing and decrease disease spread by preventing fouling organisms from growing [18].

5.4 *Barrier Coatings for Moisture and Gas*

Barrier coatings for moisture and gas are significant advances in the paper technology domain that aim to enhance the protective characteristics of paper products. These coatings form a barrier that prevents moisture, oxygen, carbon dioxide, and other gases from permeating the contents, thereby extending the shelf-life and preserving the quality of the products. Such coatings are especially useful in packaging applications, where they prevent food, pharmaceuticals, and other sensitive products from deteriorating due to environmental exposure. These coatings form a barrier that prevents moisture, oxygen, carbon dioxide, and other gases from permeating the contents, extending shelf life and preserving their quality. Such coatings are beneficial in packaging applications, where they prevent food, pharmaceuticals, and other sensitive products from deteriorating due to environmental exposure. Various techniques are employed to apply these barrier coatings. Extrusion coating involves melting a polymer and extruding it over a moving paper web, resulting in a continuous and uniform coating with superior barrier properties. Polyethylene (PE), polypropylene (PP), ethylene vinyl alcohol (EVOH), and polyethylene terephthalate (PET), are typical polymers utilized in this procedure. Lamination is the process of applying an adhesive to a film or foil and adhering it to a paper substrate. The commonly used materials include PE, PET, polyvinyl chloride (PVC), and aluminum foil [12]. This method provides outstanding barrier characteristics, and it increases the mechanical strength of the paper.

Solvent-based coatings require the application of a barrier material dissolved in a solvent to the paper, which evaporates to form a thin barrier layer. The commonly utilized materials include latex, acrylics, and starch-based compounds. The co-extrusion coating technique combines multiple layers of various polymers, extruded simultaneously, and attaches them to the paper substrate, resulting in exceptional barrier qualities suitable for packaging that requires high moisture and gas resistance. Nanocoating involves dispersing nanoparticles in a solvent or polymer matrix before applying them to the paper surface. Nanoclays, nitrocellulose, graphene oxide, and silver nanoparticles are the common materials often employed [65].

5.5 *Conductive and Antistatic Coatings*

Conductive and antistatic coatings have become vital materials in various high-tech industries due to their ability to impart electrical conductivity or prevent static charge accumulation. These coatings are crucial for safeguarding sensitive electronic components, enhancing aerospace safety, ensuring medical device reliability, and optimizing automotive and industrial manufacturing processes. Conductive coatings utilize elements, such as metals, carbon, or conductive polymers, to transfer electrons, whereas antistatic coatings employ moisture-attracting additives or ionic chemicals to discharge static charge [74]. Despite their numerous benefits, these coatings encounter hurdles in terms of durability, compatibility, and cost-efficiency. Nanotechnology, hybrid materials, and environmentally friendly formulations are tackling these concerns, thereby resulting in improved performance and versatility. Conductive and antistatic paper coatings have become increasingly important in the paper and packaging industries, especially as electronic packaging and printed electronics continue to expand [70]. Conductive coatings on paper frequently include ingredients, such as conductive polymers, carbon-based additives, or metal nanoparticles, to enhance electron transport. Antistatic coatings, on the other hand, disperse static charges with moisture-attracting or ionic chemical additions [74].

Since polyaniline (PANI) has a high electrical conductivity and tolerance to environmental changes, it is frequently employed in nanocoatings. PANI creates a continuous conductive network on paper to enhance its antistatic properties [74]. The mechanical properties and conductivity of other conductive polymers, such as polypyrrole and polythiophene, differ [53]. Metal nanoparticles, such as copper, gold, and silver, improve conductivity by forming a variety of conductive pathways. In addition to helping to dissipate static charges, carbon nanotubes (CNTs) have exceptional mechanical strength, thermal conductivity, and electrical conductivity. Because graphene is very mechanically strong, it creates strong conductive networks that remove static charges from nanocoatings.

Antistatic nanocoatings are designed to prevent the accumulation of static electricity on paper surfaces, a widespread concern that can lead to dust attraction, handling difficulties, and even damage to sensitive electronic components. The basic mechanism by which these nanocoatings work is to either dissipate static charges or prevent their accumulation. This is accomplished by including conductive nanoparticles or antistatic chemicals that promote the transfer of charges away from the surface, thereby neutralizing any static electricity that may accumulate. When static charges accumulate on the surface of the untreated paper, they might remain localized, resulting in high surface resistance. An antistatic nanocoating makes the paper's surface more conductive, allowing charges to disperse rather quickly. This is especially critical in circumstances where static discharge might cause harm, such as in packaging for electronic components or in cleanroom settings, where dust accumulation must be reduced.

6 Environmental and Economic Considerations

Nanocoating improves the paper properties. Before the widespread adoption, their environmental and economic impacts must be assessed. Nanoparticles in paper coatings can boost performance and functionality. However, environmental and economic considerations must be carefully considered. Nanomaterials can improve printability, mechanical strength, and barrier properties. This may reduce paper use and waste [8]. However, nanoparticles' environmental risks must be carefully assessed throughout the manufacture, use, and disposal cycles. Additionally, the economic viability of mass-producing nanomaterials must be considered, while considering material, processing, and regulatory costs. This section highlights the key variables for sustainable nanomaterial-based paper coating development and elucidates the factors necessary to preserve nanoparticle-based paper coatings [75].

6.1 Sustainability of Nanocoating Materials and Processes

The demand for durable and high-performance materials in industries and biomedicine, combined with global environmental concerns, has driven significant research into biopolymers. Renewable materials may solve the sustainability predicament. Surface alterations have increased product quality and reduced costs throughout the late twentieth and early twenty-first centuries. Material characteristics have been improved through the use of microscale and nanoscale coatings.

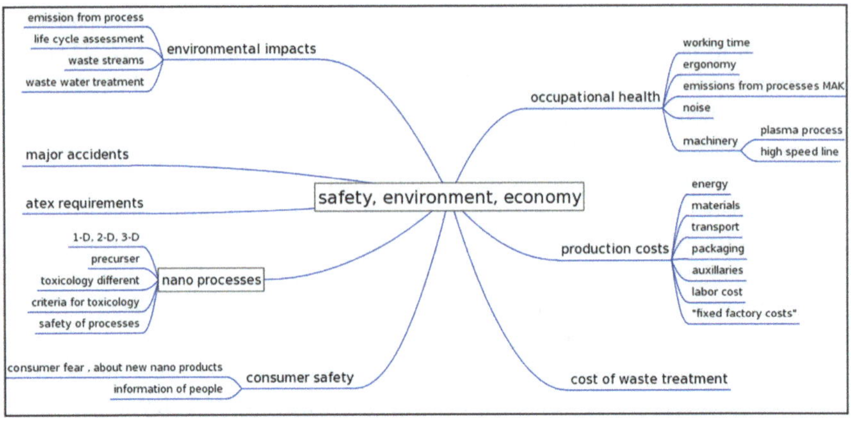

Fig. 4.13 A mind map illustrating the safety, environmental, and economic factors that need to be taken into consideration to ensure the sustainable manufacture of the new paper coating [43]. Copyright (2010). Reproduced with permission from DTU Library. (Open access)

Sustainable approaches have become increasingly important, as resources continue to be depleted. To reconcile innovation with environmental responsibility and meet the EU's preservation and sustainable development goals, academics and industry must collaborate (Fig. 4.13) [43].

Nanocoated paper possesses antibacterial and durability qualities; however, its environmental impact necessitates sustainable alternatives (Fig. 4.14). Traditional coatings are excellent barriers; however, they use nonbiodegradable polymers, thereby causing environmental problems. Due to these negative issues, the industry has been seeking green solutions. Biopolymer-based paper coatings, made from renewable sources such as cellulose and chitosan, show promise for sustainable and biodegradable paper coatings that retain performance. Biopolymers are divided into three categories, as per their origin and manufacturing (Fig. 4.15). Biopolymers from biomass start with polysaccharides (starches and cellulose), proteins, and lipids. The second category includes conventionally synthesized bio-based polymers. Polylactic acid, the third group, includes the microorganism-produced or microorganism-extracted polymers. This group includes PHA [28].

In the following section, the environmental and sustainability arising from each type of coating will be discussed.

6.1.1 Starch-Based Coatings

Renewable and biodegradable starch can replace synthetic polymers in the paper coatings domain. It improves the paper quality and reduces pollution. Starch is dispersed, modified, and applied to paper during coating. Although starch-based coatings increase paper characteristics, water sensitivity remains. Research is underway

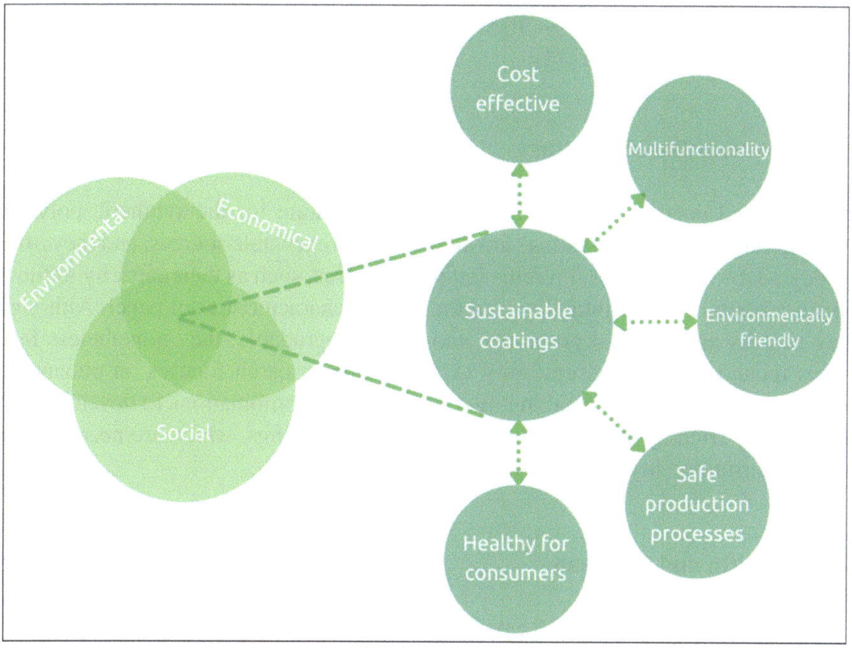

Fig. 4.14 This figure shows how multiple components aid sustainable product design [49]. Copyright (2021). Reproduced with permission from MDPI. (Open access)

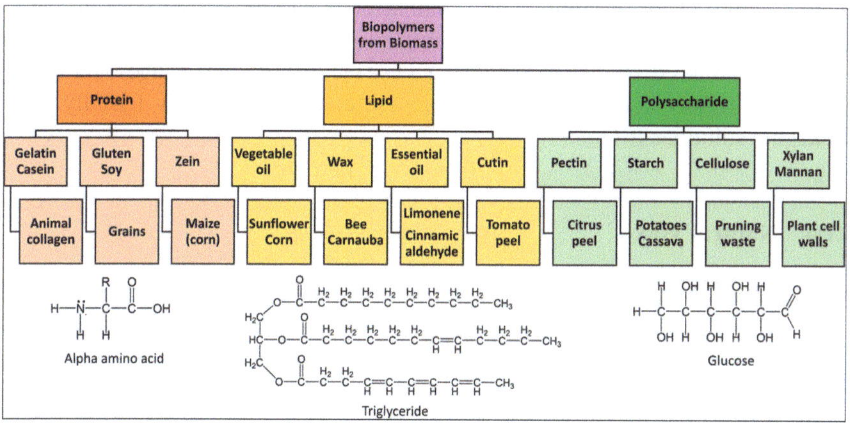

Fig. 4.15 This figure classifies biomass biopolymers into proteins, lipids, and polysaccharides. Representative examples and sources are in each category. Schematics of biopolymer chemical structures are also shown. Proteins are amino acids, lipids are triglycerides, and polysaccharides are long chains of monosaccharides like glucose [28]. Copyright (2024). Reproduced with permission from RSC. (Open access)

to enhance starch performance, making it a viable and eco-friendly option for paper manufacturing.

6.1.2 Protein-Based Coatings

Protein-based coatings provide a sustainable substitute for the synthetic polymer coatings on paper. These coatings are derived from renewable sources, and they offer biodegradable alternatives. Proteins improve features, such as their capacity to function as a barrier, their printability, and the surface characteristics of paper. Although they possess antibacterial properties and are biodegradable, they nonetheless face issues related to their susceptibility to water. The research focuses on enhancing protein-based coatings by making modifications and combinations to develop environmentally friendly and high-performance paper coatings, which are necessary to promote sustainability in the future.

6.1.3 Chitin and Chitosan

Chitin and chitosan generated from crustacean shells possess biodegradable, antimicrobial, and barrier qualities, which make them well suited for environmentally friendly paper coatings. The procedure involves the extraction of chitin, followed by deacetylation to obtain chitosan, and the preparation of a solution for applying a coating. Chitosan coatings improve the antibacterial, barrier, and mechanical characteristics of paper. Despite the presence of water solubility and mechanical constraints, current research is dedicated to enhancing these factors through changes and combinations, therefore facilitating the development of environmentally friendly paper packaging solutions [61].

6.1.4 Natural Wax

A. **Beeswax**: Beeswax, a naturally water-repellent substance, provides a sustainable option to synthetic coatings for paper. Throughout the millennia, humans have utilized beeswax for its water resistance and protective properties. The coating process involves the purification of beeswax, its emulsification, and subsequent application to paper. The beeswax-coated paper demonstrates good water resistance, resistance to grease, and enhanced aesthetics. Although beeswax is biodegradable, its high melting point and inability to mix with water present difficulties. Aranda-Ledesma et al.'s research centered on the development of novel beeswax compositions and techniques for applying them to produce environmentally friendly paper coatings that offer superior performance [7].
B. **Candelilla wax**: Candelilla wax, derived from desert bushes, provides a sustainable substitute for synthetic paper coverings. The water resistance, barrier qualities, and natural origin of this material make it a highly promising choice. The

coating process involves removing wax, refining it, and subsequently applying the refined wax to paper. Candelilla-wax-coated paper exhibits enhanced water resistance, improved printability, and enhanced aesthetics. Although the material is biodegradable, it still faces challenges, such as its high melting point and the risk of breaking. The research focuses on addressing these constraints to fully harness the potential of candelilla wax for environmentally friendly paper coverings [7].

6.1.5 Cellulosic Materials

Nanocellulose, a substance obtained from cellulose, possesses remarkable strength, biodegradability, and barrier characteristics, thereby rendering it a highly promising and sustainable option for paper coating. The production process entails the disintegration of cellulose fibers into nanocellulose particles, which are subsequently spread and placed onto paper. Nanocellulose coatings improve the strength, barrier qualities, and smoothness of paper. Although biodegradable and renewable, it is necessary to overcome problems related to production costs and coating processes. Ongoing research endeavors to enhance the efficiency of nanocellulose manufacturing and utilization, hence facilitating the development of cutting-edge, high-performance, and environmentally sustainable paper coatings.

6.1.6 Vegetable-Oil-Based Paper Coatings

Paper coatings made from vegetable oils are better sustainable than petroleum-based ones. They resist water, break down naturally, and may be renewed. The coating process involves oil refinement, modification, emulsification, and application to paper. Vegetable oil coatings increase paper's water, grease, and barrier properties. Despite being biodegradable, the material needs more research on its oxidation and water sensitivity. To circumvent these limitations, the paper industry can utilize vegetable oils to produce innovative, eco-friendly, and effective packaging coatings [51].

6.1.7 Marine Algae

Biopolymers from marine algae are suitable and sustainable for paper coatings. These aquatic creatures produce chemicals used in the production of paper coatings. Biopolymer extraction, purification, and coating formulation follow the algae culture or harvesting. The marine-algae-based coatings improve the paper's antibacterial, wettability, printability, and barrier qualities. The widespread adoption of algae culture, biopolymer extraction, and coating performance enhancement is possible; however, hurdles remain. Therefore, the potential of marine algae for sustainable paper coatings requires further study [16].

6.1.8 Sustainable Coating Process

Biopolymers, prepared via extrusion, dispersion, and solvent coating, protect paper and paperboard. These techniques employ polysaccharides, proteins, and lipids to wrap materials without the need for heating. Additionally, they can dissolve or disperse polymers in solvents. Figure 4.16a illustrates an extrusion procedure, a widely used method for coating paper that offers advantages such as uninterrupted processing, uniform coating, and a reduced occurrence of pinholes. Nevertheless, it encounters obstacles, such as instability in the polymer melt and the need for high coat weight. Commercial utilization of biopolymers such as PHB and PLA is seldom owing to heat degradation. However, the addition of certain additives can enhance their resistance to melting. Although there have been some advancements, problems, such as water vapor and gas permeability, elasticity, and fiber compatibility, continue to exist. The barrier dispersion and solvent-based coatings, although lightweight, may necessitate the application of two layers to address the pinholes and water vapor barriers. However, the usage of energy-intensive post-coating procedures restricts their industrial application. Coatings can be applied using several methods, including surface sizing, curtain coating, bar or rod coating, and dip coating. Each method has its benefits and limitations when it comes to controlling the thickness of the coating and the scale of application (Fig. 4.18b–e). The application procedures have a substantial influence on the form and features of the coated surface [55, 68].

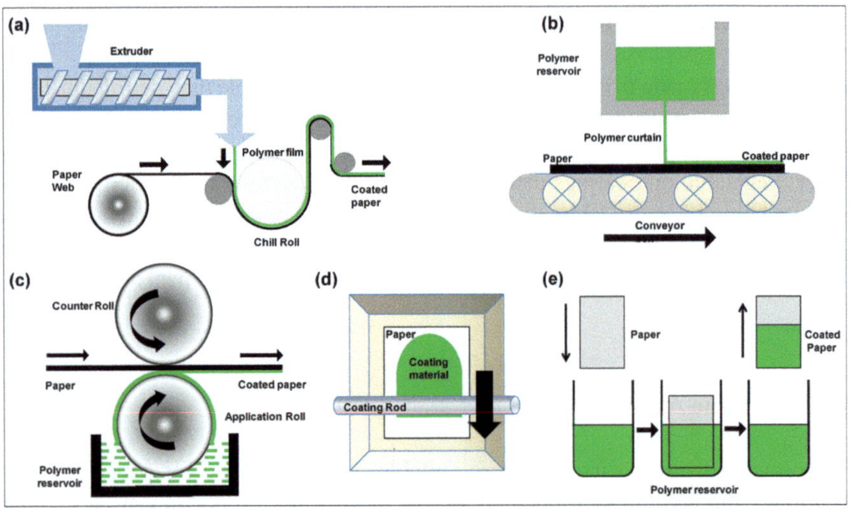

Fig. 4.16 Different coating processes for paper applications: **a** extrusion coating, **b** curtain coating, **c** size press coating, **d** bar coating, and **e** dip coating [55]. Copyright (2015). Reproduced with permission from MDPI. (Open access)

6.2 Cost–Benefit Analysis of Sustainable Nanocoatings in Paper Coatings

To determine whether it is economically feasible and environmentally sustainable to add sustainable nanocoatings to paper, a thorough cost–benefit analysis is necessary. This study should consider both the concrete aspects, such as manufacturing costs, material procurement, energy usage, and environmental consequences, as well as the intangible advantages, including enhanced paper characteristics, resource preservation, and waste reduction. An exhaustive evaluation of these parameters is crucial for determining the overall economic viability and environmental sustainability of the nanocomposite paper coatings. Table 4.2 presents a comprehensive analysis of the potential advantages of using sustainable nanocoatings in paper coatings, highlighting their impact on both the paper industry and the final product's performance [22, 66].

The integration of eco-friendly nanocoating into the paper production process incurs significant expenses. The initial capital allocated to research and development, coupled with the relatively higher price of sustainable nanoparticles in comparison to conventional materials, are factors that contribute to the overall expenditure. The expenses are further increased by the need to modify the current infrastructure for the application of nanocoating and the implementation of rigorous quality control systems.

Sustainable nanocoating has great initial investment and production costs, but long-term economic gains can balance the investment and production costs. Profits may rise with better paper, less waste, and environmental sustainability. Continuous research and production will likely lower the nanocoating costs, making them more

Table 4.2 Advantages of employing sustainable nanocoatings in paper coatings [35, 66]

Benefit	Description	Impact
Enhanced barrier properties	Improved resistance to moisture, grease, and gases	Reduces product spoilage, extends shelf life, and minimizes packaging waste
Improved Mechanical strength	Increased paper strength, stiffness, and tear resistance	Enables the use of lighter-weight paper, reducing material costs and transportation emissions
Enhanced printability	Improved ink adhesion and surface smoothness	This leads to higher-quality printing results and potentially reduces ink consumption
Biodegradability and sustainability	Derived from renewable resources and readily decompose	Aligns with eco-friendly packaging principles and reduces reliance on synthetic materials
Potential antibacterial properties	Certain nanomaterials exhibit inherent antibacterial activity	Contributes to improved food safety and shelf life of packaged products

economically viable. Industrial applications require a rigorous cost–benefit analysis of production costs, savings, and projected market value.

6.3 Regulatory and Safety Aspects of Nanocoating for Paper and Pulp Industry

To effectively work with sustainable nanocoatings for paper applications, it is crucial to have a thorough understanding of the regulatory and safety concerns associated with them. Although these materials have potential benefits in terms of sustainability and performance, their distinct properties necessitate a thorough assessment to ensure the safety of consumers and the preservation of the environment [42].

6.3.1 Regulatory Framework

The regulatory framework for nanomaterials is complex and constantly changing. The regulatory authorities worldwide have established precise criteria for goods that incorporate nanomaterials, aiming to strike a balance between innovation and safeguarding human health and the environment. Here (in Fig. 4.17) is a detailed examination of the participation of some key regulatory organizations:

By using the Toxic Compounds Control Act, the EPA oversees the import, manufacturing, processing, distribution, usage, and disposal of chemical compounds, including nanomaterials. Manufacturers and importers must provide risk information for newly produced nanomaterials, and the EPA can ban their use if they pose risks. The EU REACH legislation encompasses chemical registration, assessment, licensing, and limiting. Nanoparticle manufacturers and exporters must register with the European Chemicals Agency (ECHA) under the Registration, Evaluation, Authorization, and Restriction of Chemicals (REACH) regulation. They must also include detailed information on the properties, risks, and proper use of nanomaterials. The FDA regulates nanoparticles in food packaging, contact materials, and medical devices. The FDA requires manufacturers to demonstrate the safety and efficacy of nanoparticles in specific uses. MHRA regulates medicines, medical equipment, and blood products in the UK. They also guide the assessment of the safety of nanomaterials in these products. The Australian Therapeutic Products Administration (TGA) regulates pharmaceuticals, medical equipment, and cosmetics. There are strict nanomaterial use guidelines for these products. Japanese drugs and medical devices are regulated, evaluated, and approved by the Ministry of Health, Labour and Welfare (MHLW) and the Pharmaceuticals and Medical Devices Agency (PMDA). MHLW sets rules, whereas PMDA evaluates and approves items. They have recommendations for reviewing the safety of pharmaceutical and medical device nanomaterials [5, 50]. These regulations cover a wide range of elements:

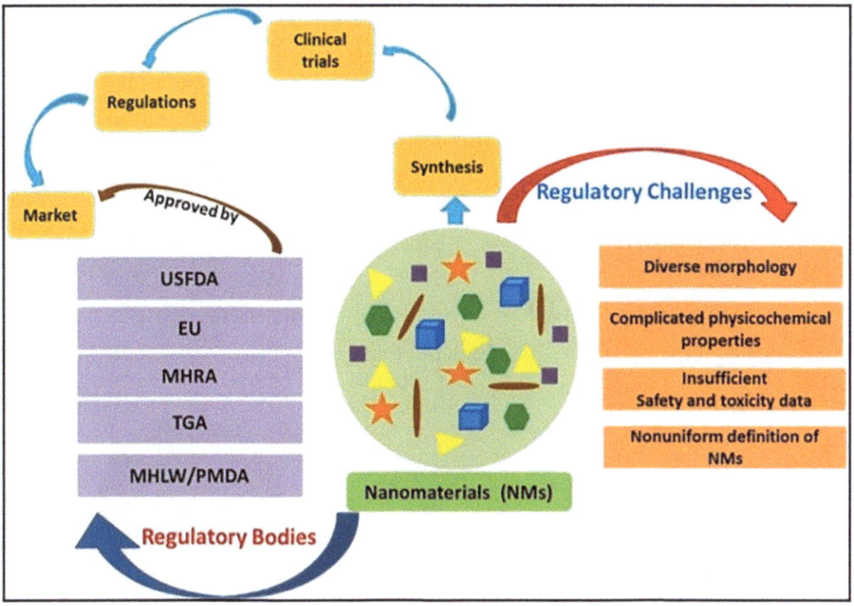

Fig. 4.17 Regulatory bodies and their challenges [5]. Copyright (2023). Reproduced with permission from Elsevier

- Registration is mandatory for manufacturers and importers of nanomaterials, who are required to register these materials with the appropriate authorities.
- A thorough analysis of the possible risks to human health and the environment must be conducted as part of the risk assessment process.
- Nanomaterial-containing products often require specialized labeling to provide customers and handlers with relevant information.

Safeguarding workers against nanoparticle toxicity requires occupational exposure limits. Safety evaluations are needed to detect nanocoated paper product production, usage, and disposal dangers due to inhalation, ingestion, and skin contact concerns (Fig. 4.18) [5, 11].

6.3.2 Strategies for the Mitigation of Risk

Several risk reduction measures may be adopted to handle the possible associated risks:

- By encapsulating nanoparticles within a matrix, it is possible to reduce the bioavailability of the nanoparticles and diminish their potential toxicity.
- Surface functionalization: Changes in the surface characteristics of a nanoparticle can change its characteristics and lessen the amount of contact it has with biological systems.

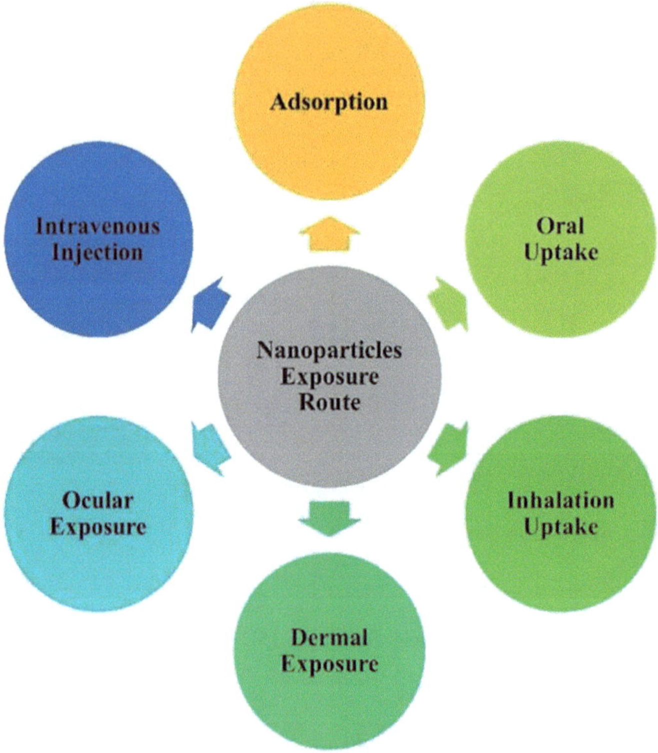

Fig. 4.18 Pathways through which exposure to nanocoatings might occur [52]. Copyright (2005). Reproduced with permission from Intech Open. (Open access)

- The use of appropriate handling and disposal techniques can reduce occupational and environmental exposures. This is referred to as the exposure control method [11].

Life cycle assessments (LCA) are essential for the assessment of the environmental consequences of nanocoated paper products at every stage of their production cycle. By complying with the regulatory rules and undertaking comprehensive risk evaluations, it is feasible to create nanocoatings that are both safe and environmentally friendly. It is crucial to have open and honest communication, which includes providing clear information about the presence of nanoparticles and the possible advantages and disadvantages they may bring to customers. Providing customers with information on the appropriate usage and disposal methods can reduce the customer's negative exposure [58].

7 Future Trends and Prospects of Paper Nanocoatings

Nanocoatings applied to paper have the potential to significantly improve its functional qualities, barrier performance, and sustainability. Food packaging enhances barrier characteristics, prolongs shelf life, and offers antibacterial protection. Enhanced protection against degradation and the controlled release of active chemicals are advantageous for pharmaceutical packaging. Additionally, nanocoatings enhance the electrical and thermal properties of paper-based substrates, thereby facilitating the development of flexible and eco-friendly electronic devices [50, 69]. Current developments are:

- The sector is being revolutionized by the use of multifunctional nanocoatings, which combine several nanomaterials to provide improved performance. An example of this is the combination of antibacterial nanoparticles with barrier qualities, which results in the creation of packaging that can preserve itself. The future of nanocoatings rests in their customization for specific applications, such as food, medicine, and electronics. This requires the exact optimization of the nanomaterial selection and the coating composition [51].
- Sustainable nanomaterials made from biological sources, such as nanocellulose and chitosan, are gaining popularity. These bio-based alternatives to synthetic nanomaterials are propelling the industry toward sustainability. Advanced characterization methods, such as atomic force microscopy (AFM), transmission electron microscopy (TEM), and X-ray diffraction (XRD), will enable researchers to better understand the structure–property correlations in nanocoatings [35].

Despite the exciting promise of paper nanocoatings, various obstacles must be overcome. The attainment of a homogeneous nanoparticle dispersion within the coating matrix is crucial for achieving optimal performance. Efficient and cost-effective manufacturing processes are essential for achieving commercial success. The prioritization of safety and regulatory compliance are critical for the protection of both customers and the environment.

Although there are difficulties to overcome, the potential advantages of paper nanocoatings are significant. Through persistent research and development efforts, the paper sector can overcome these challenges and discover new opportunities to produce innovative products that align with the requirements of a sustainable future (Table 4.3).

7.1 Emerging Nanoparticles for Paper Nanocoatings

The study of nanoparticles is undergoing continuous development, which opens up new avenues for enhancing paper coatings. The innovative nanoparticles feature several distinguishing qualities that have the potential to significantly improve the efficiency and capacities of paper-based products [63].

Table 4.3 Analysis of various nanomaterials for paper coatings [20]

Nanomaterial	Properties	Potential applications
Silica nanoparticles	High surface area, mechanical strength	Barrier, antifogging, flame retardancy
Titanium dioxide nanoparticles	UV protection, antimicrobial	Food packaging, healthcare
Carbon nanotubes	High conductivity, mechanical strength	Electronic packaging, sensors
Cellulose nanocrystals	Renewable, biodegradable, high strength	Barrier, mechanical reinforcement

7.1.1 Nanoparticles with Great Potential

The emerging materials show great potential for improving paper coatings. Metal–Organic Frameworks (MOFs) have a high surface area and configurable pore size, thereby resulting in increased gas barrier, fire resistance, and controlled release of functionalities. Two-dimensional materials, such as graphene and its variants, excel in the provision of strength, conductivity, and barrier qualities. Quantum dots provide illumination, counterfeit detection, and sensing capabilities. Clay minerals improve barrier characteristics, flame retardancy, and mechanical strength, thereby broadening the varied options for advanced paper coatings (Table 4.4).

7.2 Potentials for Multifunctional and Smart Coatings in Paper Nanocoatings

The incorporation of versatile and intelligent features into paper nanocoatings signifies cutting-edge development in the fields of packaging and information technology. By integrating the benefits of nanotechnology with the adaptability of paper,

Table 4.4 Analysis of the various nanomaterials for paper coatings [49]

Nanoparticle	Properties	Potential applications
Metal–organic frameworks (MOFs)	High surface area, tunable porosity, versatility	Gas barriers, flame retardants, drug delivery
Two-dimensional (2D) materials	High strength, conductivity, barrier properties	Conductive paper, sensors, flexible electronics
Quantum dots (QDs)	Tunable optical properties, luminescence	Security features, displays, sensors
Clay minerals	Barrier properties, flame retardancy, mechanical strength	Packaging, fire-resistant paper

it becomes feasible to generate novel materials with improved capabilities and reactivity [28].

7.2.1 Nanocoatings with Several Functions

Multifunctional nanocoatings enhance resource utilization by combining various capabilities within a single layer. The integration of antimicrobial nanoparticles with barrier materials results in coatings that effectively shield against microbial contamination while maintaining the pristine condition of the product. By combining nanocellulose or graphene with hydrophobic nanoparticles, the mechanical strength and barrier qualities of paper are enhanced, resulting in increased package performance and durability.

7.2.2 Intelligent Nanocoatings

Intelligent nanocoatings exhibit a reactive nature toward environmental stimuli, thereby allowing them to adjust to varying circumstances.

- Sensor-based Coatings: By integrating nanoparticles with sensing capabilities, such as conductive polymers or metal oxides, into paper coatings, it is possible to develop sensors that can sense temperature, humidity, or gas.
- Autonomous Repair Coatings: The incorporation of self-healing nanoparticles, such as microcapsules containing healing agents, can enable the paper to autonomously repair small defects autonomously, thereby prolonging its durability.

7.3 Integration of Paper Nanocoatings with Digital and Smart Packaging Technologies

The combination of nanotechnology, paper science, and digital technologies is propelling the advancement of creative and interactive packaging solutions. By combining paper nanocoatings with digital and smart packaging technologies, it is possible to develop intelligent and interactive packaging systems. These systems may enhance the customer experience, increase product traceability, and give real-time information [16, 68].

7.3.1 Methods for Integrating Different Components or Elements

- Radio-frequency identification (RFID) Integration: Conductive nanocoatings can be utilized to embed RFID tags in paper packaging, thereby enabling radio-frequency identification. These tags are used to store product information, check inventory, and allow customer involvement.
- Quick Response (QR) Code Integration: High-resolution printing on nanocoated paper allows for the incorporation of QR codes that can be scanned to obtain product information, nutritional data, and interactive content.
- Sensor Integration: Nanocoatings may be utilized to develop sensors that can effectively monitor and assess product parameters, including temperature, humidity, and light exposure. The transmission of this information can be achieved through the utilization of integrated wireless communication technologies [26].
- Nanocoatings can be used to facilitate Augmented Reality (AR) integration by providing a suitable surface for AR experiences. Consumers can engage with virtual items and information superimposed over physical packaging through the use of smartphones [8].

7.3.2 Obstacles and Possibilities

- Compatibility: It is essential to ensure that diverse technologies and materials can work together effectively for successful integrations.
- Cost-effectiveness: Ensuring that the benefits of digital and smart features outweigh the expenses of manufacturing, which is crucial for the commercial feasibility of a product.
- Consumer Adoption: It is essential to educate customers about the benefits of interactive packaging and promote its adoption to achieve the desired market success.
- Although there are obstacles, the combination of paper nanocoatings with digital and innovative packaging technologies presents substantial prospects:
- Improved Consumer Experience: Interactive elements have the potential to enhance consumer engagement and satisfaction.
- Enhanced supply chain management can be achieved through the use of real-time tracking and monitoring of items, resulting in optimized efficiency.
- Product authentication may be enhanced by the use of digital elements, which serve to deter counterfeiting and safeguard a brand's reputation [16].

7.3.3 Anticipated Developments

- Integration of the Internet of Things (IoT): By connecting paper packaging to the IoT, it will be possible to gather and analyze data in real time.
- Customized package: Utilizing consumer data to create personalized package experiences.

- Developing environmentally friendly nanocoatings and packaging materials to support sustainability.

The coupling of paper nanocoatings with digital and smart packaging technologies is a rapidly evolving field that has the potential to transform the packaging industry significantly. It is possible to develop innovative and long-lasting packaging solutions that meet the needs of both consumers and businesses by overcoming challenges and capitalizing on available opportunities.

8 Conclusion and Future Perspectives

Improvements in surface qualities are required to meet the growing demand for high-performance paper, and the introduction of nanotechnology in the paper sector is a big step forward. The purpose of these developments is to improve the performance and functionality of paper goods. This chapter has critically explored how the inclusion of nanocoating in paper substrates can improve their surface qualities. With their distinct physicochemical characteristics and nanoscale thickness, nanocoatings have demonstrated outstanding potential in enhancing the paper's qualities. This chapter begins with an introduction to nanotechnology in the paper industry, highlighting the vital role that surface qualities play in various paper applications and outlining the potential and challenges associated with this topic. Furthermore, the advancements in the fundamentals of nanocoatings and the various synthesis and application techniques of nanocoatings, such as layer-by-layer assembly, sol–gel, PECVD, and many more, were succinctly discussed. However, the paper has a fibrous structure. Hence, it has poor barrier properties, which is a drawback. Nanocoating technology can considerably enhance the barrier properties, flame retardancy, and mechanical performance of coated substrates. Additionally, nanocoating is significantly thinner than traditional coatings (such as wax coatings on paper), resulting in minimal cost. The application of a nanocoating, a very thin layer of material typically less than 100 nm thick, to a surface can enhance its current properties or introduce new characteristics. These coatings enhance performance in areas including antibacterial activity, durability, corrosion resistance, optical clarity, and thermal insulation by leveraging the unique properties, such as increased surface area and quantum effects, of materials at the nanoscale. A few unique properties of nanocoating include a high surface area-to-volume ratio, which promotes reactivity and contact with the environment.

Future research should focus on enhancing the durability of superhydrophobic characteristics, improving optical clarity, and developing thermal nanocoatings to enhance paper surface properties, particularly in terms of laundering resistance, to expand their applications in flexible paper coating materials. While various studies have demonstrated the successful synthesis of highly resistant superhydrophobic nanocoatings against mechanical stresses, the impact of laundering has been scarcely reported. To enable large-scale deployment, researchers need to devise eco-friendly, cost-effective, and scalable methods for preparing this interesting class of nanocoating materials to enhance the surface properties of paper. Investigations should also concentrate on the design and development of superhydrophobic surfaces, improved optical clarity, and thermally resistant nanocoatings for use in advanced applications, such as sustainable energy production and storage. The integration of additional properties with surface coatings to create multifunctional materials is expected to undergo significant advancements in the coming decades.

Competing Interests The authors declare that they have no competing interest.

Declaration of Generative AI and AI-Assisted Technologies in the Writing Process During the preparation of this work, the book chapter author(s) have not taken any assistance from AI, and the book chapter author(s) reviewed and edited the content as needed and take(s) full responsibility for the content of the publication.

References

1. Abdel-Karim RFA (2013) Nanocoatings. In: Modern surface engineering treatments. InTech. https://doi.org/10.5772/55776
2. Abdelhalim A, Abdellah A, Scarpa G, Lugli P (2013) Fabrication of carbon nanotube thin films on flexible substrates by spray deposition and transfer printing. Carbon N Y 61:72–79. https://doi.org/10.1016/j.carbon.2013.04.069
3. Agarwal M, Lvov Y, Varahramyan K (2006) Conductive wood microfibres for smart paper through layer-by-layer nanocoating. Nanotechnology 17:5319–5325. https://doi.org/10.1088/0957-4484/17/21/006
4. Alam P, Stoor S, Toivakka M (2011) Predicting the causes of operational anomalies in blade coating. In: WIT transactions on engineering sciences, pp 205–216. https://doi.org/10.2495/SECM110181
5. Ali F, Neha K, Parveen S (2023) Current regulatory landscape of nanomaterials and nanomedicines: a global perspective. J. Drug Deliv Sci Technol 80:104118. https://doi.org/10.1016/j.jddst.2022.104118
6. Amiri S, Rahimi A (2016) Hybrid nanocomposite coating by sol–gel method: a review. Iran Polym J (English Ed.) 25:559–577. https://doi.org/10.1007/s13726-016-0440-x
7. Aranda-Ledesma NE, Bautista-Hernández I, Rojas R, Aguilar-Zárate P, del Medina-Herrera NP, Castro-López C, Guadalupe Martínez-Ávila GC (2022) Candelilla wax: Prospective suitable applications within the food field. Lwt 159:113170. https://doi.org/10.1016/j.lwt.2022.113170
8. Arora R, Roy T, Adak P (2024) A review of the impact of nanoparticles on environmental processes. BIO Web Conf 86:01001. https://doi.org/10.1051/bioconf/20248601001
9. Beecher P, Servati P, Rozhin A, Colli A, Scardaci V, Pisana S, Hasan T, Flewitt AJ, Robertson J, Hsieh GW, Li FM, Nathan A, Ferrari AC, Milne WI (2007) Ink-jet printing of carbon nanotube thin film transistors. J Appl Phys. https://doi.org/10.1063/1.2770835
10. Bryne LE, Lausmaa J, Ernstsson M, Englund F, Wälinder MEP (2010) Ageing of modified wood. Part 2: determination of surface composition of acetylated, furfurylated, and thermally modified wood by XPS and ToF-SIMS. Holzforschung 64:305–313. https://doi.org/10.1515/HF.2010.062
11. Buzea C, Pacheco II, Robbie K (2007) Nanomaterials and nanoparticles: sources and toxicity. Biointerphases 2:MR17–MR71. https://doi.org/10.1116/1.2815690
12. Chen X, Xiao B, Yang Y, Jiang Y, Song X, Chen F, Wang W, Wu J, Meng Y (2024) A novel paper-based composite film with enhanced oxygen and water vapor barrier properties. Prog Org Coatings 186:108042. https://doi.org/10.1016/j.porgcoat.2023.108042

13. Dan B, Irvin GC, Pasquali M (2009) Continuous and scalable fabrication of transparent conducting carbon nanotube films. ACS Nano 3:835–843. https://doi.org/10.1021/nn8008307
14. Das S, Kumar S, Samal SK, Mohanty S, Nayak SK (2018) A review on superhydrophobic polymer nanocoatings: recent development and applications. Ind Eng Chem Res 57:2727–2745. https://doi.org/10.1021/acs.iecr.7b04887
15. Dastjerdi S, Naeijian F, Akgöz B, Civalek Ö (2021) On the mechanical analysis of microcrystalline cellulose sheets. Int J Eng Sci 166:103500. https://doi.org/10.1016/j.ijengsci.2021.103500
16. De Jesus Raposo MF, De Morais AMB, De Morais RMSC (2015) Marine polysaccharides from algae with potential biomedical applications. Mar Drugs 13:2967–3028. https://doi.org/10.3390/md13052967
17. De Villiers MM, Otto DP, Strydom SJ, Lvov YM (2011) Introduction to nanocoatings produced by layer-by-layer (LbL) self-assembly. Adv Drug Deliv Rev 63:701–715. https://doi.org/10.1016/j.addr.2011.05.011
18. Dong W, Li B, Wei J, Tian N, Liang W, Zhang J (2021) Environmentally friendly, durable and transparent anti-fouling coatings applicable onto various substrates. J Colloid Interface Sci 591:429–439. https://doi.org/10.1016/j.jcis.2021.02.014
19. Du M, Peng M, Mai B, Hu F, Zhang X, Chen Y, Wang C (2020) A multifunctional hybrid inorganic-organic coating fabricated on magnesium alloy surface with antiplatelet adhesion and antibacterial activities. Surf Coatings Technol 384:125336. https://doi.org/10.1016/j.surfcoat.2020.125336
20. Erkoc P, Ulucan-Karnak F (2021) Nanotechnology-based antimicrobial and antiviral surface coating strategies. Prosthesis 3:25–52. https://doi.org/10.3390/prosthesis3010005
21. Ganeshkumar S, Kumar A, Maniraj J, Suresh Babu Y, Ansu AK, Goyal A, Kareem Kadhim I, Saxena KK, Prakash C, Altuijri R, Ijaz Khan M, Hassan AM (2023) Exploring the potential of nano technology: a assessment of nano-scale multi-layered-composite coatings for cutting tool performance. Arab J Chem 16:105173. https://doi.org/10.1016/j.arabjc.2023.105173
22. Garg R, Garg R, Eddy NO (2024) Applications of nanocoatings, pp 28–56. https://doi.org/10.4018/979-8-3693-3136-1.ch002
23. Ghule K, Ghule AV, Chen BJ, Ling YC (2006) Preparation and characterization of ZnO nanoparticles coated paper and its antibacterial activity study. Green Chem 8:1034–1041. https://doi.org/10.1039/b605623g
24. Gicquel E, Martin C, Garrido Yanez J, Bras J (2017) Cellulose nanocrystals as new biobased coating layer for improving fiber-based mechanical and barrier properties. J Mater Sci 52:3048–3061. https://doi.org/10.1007/s10853-016-0589-x
25. Hernadi K, Ljubović E, Seo JW, Forró L (2003) Synthesis of MWNT-based composite materials with inorganic coating. Acta Mater 51:1447–1452. https://doi.org/10.1016/S1359-6454(02)00539-6
26. Huston M, Debella M, Dibella M, Gupta A (2021) Green synthesis of nanomaterials. Nanomaterials 11:2130. https://doi.org/10.3390/nano11082130
27. Ivask A, Ahonen M, Kogermann K (2022) Antimicrobial nano coatings. Nanomaterials 12:10–13. https://doi.org/10.3390/nano12234338
28. Jahangiri F, Mohanty AK, Misra M (2024) Sustainable biodegradable coatings for food packaging: challenges and opportunities. Green Chem 26:4934–4974. https://doi.org/10.1039/D3GC02647G
29. Kamel S (2007) Nanotechnology and its applications in lignocellulosic composites, a mini review. Express Polym Lett 1:546–575. https://doi.org/10.3144/expresspolymlett.2007.78
30. Kang K, Bae G, Won J, Lee C (2012) Mechanical property enhancement of kinetic sprayed Al coatings reinforced by multi-walled carbon nanotubes. Acta Mater 60:5031–5039. https://doi.org/10.1016/j.actamat.2012.05.034
31. Khwaldia K, Arab-Tehrany E, Desobry S (2010) Biopolymer coatings on paper packaging materials. Compr Rev Food Sci Food Saf 9:82–91. https://doi.org/10.1111/j.1541-4337.2009.00095.x

32. Kim SH (2008) Fabrication of superhydrophobic surfaces. J Adhes Sci Technol 22:235–250. https://doi.org/10.1163/156856108X305156
33. Köklükaya O, Carosio F, Grunlan JC, Wågberg L (2015) Flame-retardant paper from wood fibers functionalized via layer-by-layer assembly. ACS Appl Mater Interfaces 7:23750–23759. https://doi.org/10.1021/acsami.5b08105
34. Kunam PK, Ramakanth D, Akhila K, Gaikwad KK (2024) Bio-based materials for barrier coatings on paper packaging. Biomass Convers Biorefinery 14:12637–12652. https://doi.org/10.1007/s13399-022-03241-2
35. Lengowski EC, Bonfatti Júnior EA, Coelho Simon L, Bolzon de Muniz GI, Sulato de Andrade A, Neves Leite A, de Miranda S, Leite EL (2023) Nanocellulose coating on kraft paper. Coatings 13:1705. https://doi.org/10.3390/coatings13101705
36. Li Q, Wang S, Jin X, Huang C, Xiang Z (2020) The application of polysaccharides and their derivatives in pigment, barrier, and functional paper coatings. Polymers (Basel). https://doi.org/10.3390/POLYM12081837
37. Li X, Yan J, Yu T, Zhang B (2022) Versatile nonfluorinated superhydrophobic coating with self-cleaning, anti-fouling, anti-corrosion and mechanical stability. Colloids Surfaces A Physicochem Eng Asp 642:128701. https://doi.org/10.1016/j.colsurfa.2022.128701
38. Lucovsky G, Tsu DV, Kim SS, Markunas RJ, Fountain GG (1989) N film diele~Cs by remote plasma-enhanced Che~Cal-vapor deposithon (remote Pecvd). Appl Surf Sci 39:33–56
39. Luo G, Teh KS, Liu Y, Zang X, Wen Z, Lin L (2015) Direct-write, self-aligned electrospinning on paper for controllable fabrication of three-dimensional structures. ACS Appl Mater Interfaces 7:27765–27770. https://doi.org/10.1021/acsami.5b08909
40. Mahadeva SK, Walus K, Stoeber B (2015) Paper as a platform for sensing applications and other devices: a review. ACS Appl Mater Interfaces 7:8345–8362. https://doi.org/10.1021/acsami.5b00373
41. Makhlouf ASH, Tiginyanu I (2011) Introduction. Nanocoatings Ultra-Thin Film Technol Appl 15–20. https://doi.org/10.1016/B978-1-84569-812-6.50017-8
42. Makino H (2018) Environmental and safety issues with nanoparticles. In: Nanoparticle technology handbook. Elsevier, pp 365–395. https://doi.org/10.1016/B978-0-444-64110-6.00007-X
43. Markert F, Breedveld L, Lahti J, Vangeneuenden D (2010) Development of sustainable paper coatings using nanoscale industrial surface processing
44. Mathai S, Shaji PS (2022) Polymer-based nanocomposite coating methods: a review. J Sci Res 14:973–1002. https://doi.org/10.3329/jsr.v14i3.58338
45. Meng Y, Xu XB, Li H, Wang Y, Ding EX, Zhang ZC, Geng HZ (2014) Optimisation of carbon nanotube ink for large-area transparent conducting films fabricated by controllable rod-coating method. Carbon N Y 70:103–110. https://doi.org/10.1016/j.carbon.2013.12.078
46. Michael FM, Khalid M, Walvekar R, Siddiqui H, Balaji AB (2017) Surface modification techniques of biodegradable and biocompatible polymers, biodegradable and biocompatible polymer composites: processing, properties and applications. Elsevier Ltd. https://doi.org/10.1016/B978-0-08-100970-3.00002-X
47. Moggridge GD, Lape NK, Yang C, Cussler EL (2003) Barrier films using flakes and reactive additives. Prog Org Coatings 46:231–240. https://doi.org/10.1016/S0300-9440(02)00180-7
48. Mohammed M, Rahman R, Mohammed AM, Adam T, Betar BO, Osman AF, Dahham OS (2022) Surface treatment to improve water repellence and compatibility of natural fiber with polymer matrix: Recent advancement. Polym Test 115:107707. https://doi.org/10.1016/j.polymertesting.2022.107707
49. Nartita R, Ionita D, Demetrescu I (2021) Sustainable coatings on metallic alloys as a nowadays challenge. Sustain 13:10217. https://doi.org/10.3390/su131810217
50. Nizam NUM, Hanafiah MM, Woon KS (2021) A content review of life cycle assessment of nanomaterials: current practices, challenges, and future prospects. Nanomaterials 11:3324. https://doi.org/10.3390/nano11123324
51. Nurchi C, Buonvino S, Arciero I, Melino S (2023) Sustainable vegetable oil-based biomaterials: synthesis and biomedical applications. Int J Mol Sci 24:2153. https://doi.org/10.3390/ijms24032153

52. Oberdörster G, Maynard A, Donaldson K, Castranova V, Fitzpatrick J, Ausman K, Carter J, Karn B, Kreyling W, Lai D, Olin S, Monteiro-Riviere N, Warheit D, Yang H (2005) Principles for characterizing the potential human health effects from exposure to nanomaterials: elements of a screening strategy. Part Fibre Toxicol 2:8. https://doi.org/10.1186/1743-8977-2-8
53. Ohtsuka T (2012) Corrosion protection of steels by conducting polymer coating. Int J Corros. https://doi.org/10.1155/2012/915090
54. Owens GJ, Singh RK, Foroutan F, Alqaysi M, Han CM, Mahapatra C, Kim HW, Knowles JC (2016) Sol-gel based materials for biomedical applications. Prog Mater Sci 77:1–79. https://doi.org/10.1016/j.pmatsci.2015.12.001
55. Rastogi VK, Samyn P (2015) Bio-based coatings for paper applications. Coatings 5:887–930. https://doi.org/10.3390/coatings5040887
56. Richardson JJ, Cui J, Björnmalm M, Braunger JA, Ejima H, Caruso F (2016) Innovation in layer-by-layer assembly. Chem Rev 116:14828–14867. https://doi.org/10.1021/acs.chemrev.6b00627
57. Samyn P, Barhoum A, Öhlund T, Dufresne A (2018) Review: nanoparticles and nanostructured materials in papermaking. J Mater Sci 53:146–184. https://doi.org/10.1007/s10853-017-1525-4
58. Simamora J, Wiloso EI, Yani M (2023) Life cycle assessment of paper products based on recycled and virgin fiber. Glob J Environ Sci Manag 9:89–106. https://doi.org/10.22034/gjesm.2023.09.SI.07
59. Solberg A, Zehner J, Somorowsky F, Rose K, Korpela A, Syverud K (2023) Material properties and water resistance of inorganic–organic polymer coated cellulose paper and nanopaper. Cellulose 30:1205–1223. https://doi.org/10.1007/s10570-022-04925-8
60. Song J, Rojas OJ (2013) Approaching super-hydrophobicity from cellulosic materials: a review. Nord Pulp Pap Res J 28:216–238. https://doi.org/10.3183/npprj-2013-28-02-p216-238
61. Song Z, Li G, Guan F, Liu W (2018) Application of chitin/chitosan and their derivatives in the papermaking industry. Polymers (Basel) 10:389. https://doi.org/10.3390/polym10040389
62. Sun L, Wang N, Wang Y, Chen P, Liu H, Cao S, Liu Y, Gao C (2023) Fabrication of an anti-fouling coating based on epoxy resin with a double antibacterial effect via an in situ polymerization strategy. Prog Org Coatings 184:107837. https://doi.org/10.1016/j.porgcoat.2023.107837
63. Thipe VC, Karikachery AR, Çakılkaya P, Farooq U, Genedy HH, Kaeokhamloed N, Phan DH, Rezwan R, Tezcan G, Roger E, Katti KV (2022) Green nanotechnology—an innovative pathway towards biocompatible and medically relevant gold nanoparticles. J Drug Deliv Sci Technol 70:103256. https://doi.org/10.1016/j.jddst.2022.103256
64. Tobjörk D, Österbacka R (2011) Paper electronics. Adv Mater 23:1935–1961. https://doi.org/10.1002/adma.201004692
65. Tyagi P, Salem KS, Hubbe MA, Pal L (2021) Advances in barrier coatings and film technologies for achieving sustainable packaging of food products—a review. Trends Food Sci Technol 115:461–485. https://doi.org/10.1016/j.tifs.2021.06.036
66. Uttarwar RG (2013) Multiscale design and life-cycle based sustainability assessment of polymer nanocomposite coatings
67. Vartiainen J, Motion R, Kulonen H, Rättö M, Skyttä E, Ahvenainen R (2004) Chitosan-coated paper: effects of nisin and different acids on the antimicrobial activity. J Appl Polym Sci 94:986–993. https://doi.org/10.1002/app.20701
68. Vasile C (2018) Polymeric nanocomposites and nanocoatings for food packaging: a review. Materials (Basel). https://doi.org/10.3390/ma11101834
69. Vasile C (2018) Polymeric nanomaterials: recent developments, properties and medical applications. In: Polymeric nanomaterials in nanotherapeutics. Elsevier, pp 1–66. https://doi.org/10.1016/B978-0-12-813932-5.00001-7
70. Wouters MEL, Wolfs DP, Van Der Linde MC, Hovens JHP, Tinnemans AHA (2004) Transparent UV curable antistatic hybrid coatings on polycarbonate prepared by the sol-gel method. Prog Org Coatings 51:312–319. https://doi.org/10.1016/j.porgcoat.2004.07.020
71. Wu Y, Li Y, Ong BS, Liu P, Gardner S, Chiang B (2005) High-performance organic thin-film transistors with solution-printed gold contacts. Adv Mater 17:184–187. https://doi.org/10.1002/adma.200400690

72. Xiang Q, Ma YM, Yu DG, Jin M, Williams GR (2013) Electrospinning using a Teflon-coated spinneret. Appl Surf Sci 284:889–893. https://doi.org/10.1016/j.apsusc.2013.08.030
73. Yi K, Liu D, Chen X, Yang J, Wei D, Liu Y, Wei D (2021) Plasma-enhanced chemical vapor deposition of two-dimensional materials for applications. Acc Chem Res 54:1011–1022. https://doi.org/10.1021/acs.accounts.0c00757
74. Youssef AM, El-Samahy MA, Abdel Rehim MH (2012) Preparation of conductive paper composites based on natural cellulosic fibers for packaging applications. Carbohydr Polym 89:1027–1032. https://doi.org/10.1016/j.carbpol.2012.03.044
75. Zahra Z, Habib Z, Hyun S, Sajid M (2022) Nanowaste: another future waste, its sources, release mechanism, and removal strategies in the environment. Sustainability 14:2041. https://doi.org/10.3390/su14042041
76. Zhang D, Xiao H (2013) Dual-functional beeswaxes on enhancing antimicrobial activity and water vapor barrier property of paper. ACS Appl Mater Interfaces 5:3464–3468. https://doi.org/10.1021/am400585m
77. Zvonkina I, Soucek M (2016) Inorganic-organic hybrid coatings: Common and new approaches. Curr Opin Chem Eng 11:123–127. https://doi.org/10.1016/j.coche.2016.01.008

Nanoparticles Used in Pulp and Paper Functionalization

Mehak and Trishita Saha

Abstract Nanotechnology has the potential to significantly impact economic development through the use of various materials, devices, and systems at the nanoscale. The focus of this chapter is on the application of nanoparticles in the pulp and paper industry, highlighting various functionalization techniques that can enhance the existing properties of paper products. Nanoparticles, such as silver, zinc oxide, and titanium dioxide, in addition to carbonaceous materials like graphene oxide and carbon nanotubes, can improve the mechanical, barrier, antimicrobial, and smart properties of paper. The chapter covers various approaches for applying nanoparticles to paper products, including techniques such as impregnation, in situ synthesis, and blending with biopolymer matrices, as well as the associated issues of uniform dispersion, recycling, and biodegradability. Despite progress, significant research gaps remain, particularly in understanding the long-term environmental impacts of nanoparticle-functionalized papers and the scale-up of their applications. This work highlights the importance of interdisciplinary research in assessing the safety and effectiveness of such materials in real-world applications, contributing to a broader understanding of how to improve the sustainability and performance of paper products within the context of a circular economy. We have demonstrated the transformative potential of nanotechnology to enable lightweight, multifunctional, and environmentally friendly paper products.

Keywords Nanoparticles · Functionalization · Pulp and paper · Antimicrobial properties · Challenges · Limitations

Mehak · T. Saha (✉)
Cellulose and Paper Discipline, Forest Products Division, Forest Research Institute, Dehradun, India
e-mail: trishitasaha1@gmail.com

R. K. Bachheti (ed.), *Nanotechnology in the Pulp and Paper Industry*, Smart Nanomaterials Technology, https://doi.org/10.1007/978-981-95-3473-9_5

1 Introduction

Nanotechnology can be one significant element of global trajectory and economic growth in the near future. Already, this broad multidisciplinary discipline has discovered interesting novel abilities. This broad multidisciplinary discipline is already revealing glimpses of fascinating new possibilities, allowing materials, devices, and systems to be investigated, developed, and produced at the nanoscale. Using nanotechnology to produce nanomaterials with unique properties controllably is expected to revolutionize technology and industry. The purpose of nanotechnology is to control and manipulate materials to obtain a special function. Nanotechnology is a new tool added in recent years with a wide range of applications.

Nanotechnological applications are the highest in the packaging segment, followed by electronics, pharmaceuticals, plastics, cosmetics, and others, including paper coating. Paper is a material made of cellulosic fibers, derived mainly from wood, rags, and certain grasses, processed into flexible sheets or rolls by deposit from an aqueous suspension. Cellulose is a high-molecular-weight, stereoregular, and linear polymer of repeating beta-D-glucopyranose units. It is the principal structural element and a major constituent of the cell walls of trees and plants. The empirical formula for cellulose is $(C_6H_{10}O_5)_n$, where "n" is degree of polymerization.

By carefully incorporating nanoparticles to provide synergistic effects with the cellulose fiber network, rather than just adding them as fillers, the functionalization process is a sophisticated approach to material design. Physical entrapment of nanoparticles inside the fiber matrix, chemical interaction with cellulosic hydroxyl groups, or surface coating applications are some of the ways through which its integration can occur [1].

Metal nanoparticles, including silver (Ag), zinc oxide (ZnO), and titanium dioxide (TiO_2), are acknowledged in the Ababu et al. (2024) study for their antibacterial, UV-protective, and self-cleaning qualities, respectively. Likewise, the materials derived from nanocellulose, particularly cellulose nanocrystals (CNCs) and cellulose nanofibrils (CNFs), offer improved mechanical performance, reduced weight, and flexibility, leading to sheets with stronger qualities (Feldhusen, 2025). Carbonaceous nanoparticles, such as graphene oxide (GO) and carbon nanotubes (CNTs), enable the production of conductive sheets for use in flexible electronics and smart packaging (Sundararaman et al., 2025). In particular, food packaging applications might benefit from the incorporation of silica and clay nanoparticles by improving barrier properties and resistance to gases, oils, and moisture (Ahmadi et al., 2024).

This chapter is designed to help readers gain knowledge on how nanotechnology is impacting the pulp and paper industry through systematic functionalization methods. We will discuss the types of particles included, the various types of nanoparticles, the specific contributions of nanoparticles to paper properties, and the methods used to incorporate nanoparticles into pulp and paper. We will also discuss some of the practical limitations and challenges to consider when using these technologies, which influence the commercial viability of papers developed with nanotechnology-based novel products (Fig. 1).

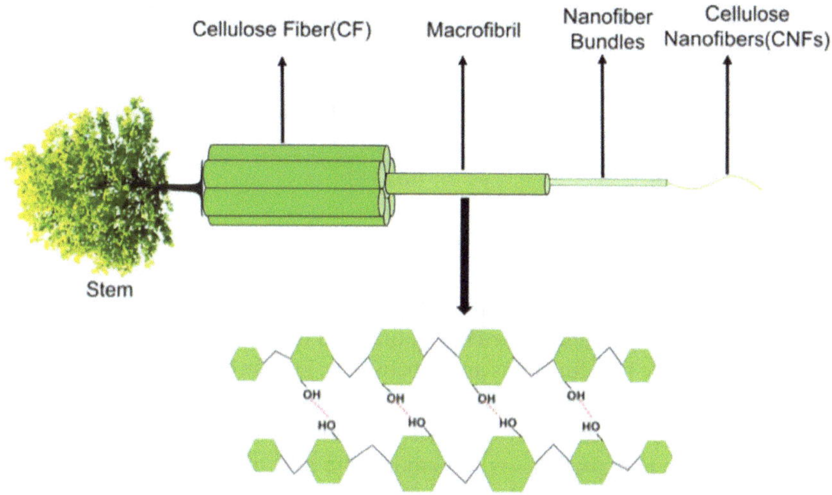

Fig. 1 Hierarchical structure of cellulose from plant stem to nanofiber [2]

2 Types of Nanoparticles Used in Pulp and Paper Functionalization

Nanoparticles that are used in paper engineering are classified as inorganics, organics, and hybrid nanoparticles.

2.1 *Inorganic Nanoparticles*

Inorganic nanoparticles have received significant interest in the pulp and paper sector due to their ability to improve paper quality, increase process efficiency, and provide additional functionality. The porous nature of the structural makeup, combined with the hydrophilicity of the fiber surface, allows for the effective absorption of nanoparticle suspensions through capillary forces and dry nanoparticle loading. The paper serves as an inexpensive substrate, and the small quantities of nanoparticles are deposited, providing efficient optical, antibacterial, anticounterfeiting, Surface Enhanced Raman Scattering (SERS), and Surface Plasmon Resonance (SPR) functionality when using TiO_2, gold, and silver nanoparticles. Paper has numerous benefits associated with being biodegradable, biocompatible, and renewable, with its structural morphology and surface chemistry readily engineered for fabrication [3].

Nanoparticles can be successful nanofillers added to wood pulp, in addition to the general use of nanomaterials as nanopigments for coatings. Major steps in the papermaking process involve either filling the fiber lumen, altering or modifying the fiber walls, or Layer-by-Layer depositing of nanomaterials in order to promote in situ impregnation and treatment of paper fibers. Producing advanced functional paper products with advanced functionalization requires an understanding of the interactions among nanomaterials, cellulose matrices, functional additives, and processes to achieve targeted functionalization of paper while minimizing waste and energy costs [4].

2.1.1 Metal and Metal Oxides

Pulp and paper are increasingly employing metal and metal oxide nanoparticles [such as silver, gold, copper, zinc oxide (ZnO), titanium dioxide (TiO_2), and ferric oxide (Fe_2O_3)] to provide new functions and capabilities. For the development of functionalized paper products, nanoparticles should be applied to the pulp and paper through impregnation, coating, or in situ synthesis. The primary objective of functionalizing paper with these nanoparticles is to demonstrate added value, including mechanical strength, UV protection, catalytic activity, barrier properties, and antibacterial activity. A study has reported that adding ZnO nanoparticles in situ to bamboo paper results in high antibacterial activity and UV protection, which may be advantageous in food packaging or medical documents [5]. As discussed by Tang et al. [6], nanoparticles of noble metals, for example, gold and silver, are directly synthesized on pulp fibers to enhance their antibacterial characteristics as well as color induced by plasmonic effects. A number of environmental applications involving gas purification and pollution degradation are possible as a result of catalytic papers, containing copper or platinum nanoparticles [7]. By utilizing copper, platinum, gold, or silver nanoparticles, the pulp and paper industry is producing sustainable, flexible, and value-added products that meet current demands (Table 1).

2.2 *Organic Nanoparticles*

Recently, organic nanoparticles, particularly eco-friendly organic nanoparticles prepared from natural polymers such as cellulose, chitosan, and polysaccharides, have gained significant interest in the field of sustainable materials science. Organic nanoparticles are more beneficial than their inorganic counterparts, because they are biodegradable/biocompatible and are generally associatively less toxic, are less harmful to the environment, and offer options for environmentally friendly processing methods. In the pulp and papers sector, organic nanoparticles provide an additional unique means for added performance of materials while also meeting the sustainability agenda and circular economy [4].

One of the most well-studied and utilized organic nanoparticles or natural nanoreinforcements is cellulose nanocrystals (CNCs) or cellulose nanofibers (CNFs), which can be sourced from either plant biomass and, more importantly, from paper waste in the form of cellulose feedstock. Cellulose nanocrystals offer significant specific improvements to the mechanical properties of paper (i.e., tensile strength

Table 1 Metal and metal oxide nanoparticles in pulp and paper functionalization

Nanomaterial	Type of functionalization	Methodology	Application	References
Zinc oxide (ZnO)	Surface/internal	In situ chemosynthesis	UV shielding, antimicrobial	[8]
Silver (Ag)	Surface	Green synthesis + coating	Antifungal, antibacterial	[9]
Ferric oxide (Fe_2O_3)	Surface	Green synthesis + coating	Antifungal	[9]
Copper (Cu)	Catalytic	In situ synthesis on ZnO paper	Catalytic reactions	[7]
Platinum (Pt)	Catalytic	In situ synthesis on ZnO paper	Catalytic efficiency	[7]
Gold (Au)	Surface/catalytic	In situ synthesis	UV protection, color, catalytic	[6]
Titanium dioxide (TiO_2)	Surface	Nanoparticle dispersion	Barrier properties	[10]
Aluminum oxide (Al_2O_3)	Surface	Nanoparticle dispersion	Dielectric enhancement	[10]
Silicon dioxide (SiO_2)	Surface	Coating/ layer-by-layer	Barrier, ink adhesion	[10]
Copper oxide (CuO)	Catalytic	Paper-structured composite	Catalytic degradation	[7]
Tantalum oxide	Reinforcement	Surface grafting	Mechanical reinforcement	[11]
Zirconium oxide	Reinforcement	Surface grafting	Mechanical reinforcement	[11]

and stiffness) when combined into the pulp as a function of both their crystallinity and crystalline aspect ratio. Also, surface chemistry can be engineered to add specific functional responses (e.g., antimicrobial, UV resistance, or catalytic properties).

Types of Organic Nanoparticles (Nanocellulose Based)

2.2.1 Nanofibrillated Cellulose (NFC)

The concept of nanocellulose or nanofibrillated cellulose (NFC) was initially introduced by Turbak, Snyder, and Sandberg in the 1970s. Cellular nanofiber (CNF), microfibrillated cellulose (MFC), nanofibrous cellulose, or nanofibrillar cellulose (NFC) are interchangeable terminologies frequently utilized in scientific literature [12]. The diversity in nanofibrillated cellulose characteristics stems from cellulosic sources and preparation methodologies, which significantly impact the fiber length, diameter, and yield. Nanofibrillated cellulose (NFC) is a type of nanocellulose, which comprises of long, flexible fibrils that are derived from cellulose plant material by mechanical processing (e.g., grinding, homogenization, and microfluidization). Through mechanical processing, cellulose fibers are defibrillated to smaller diameters from a micrometer to nanometer scale (i.e., typically 20–30 nm) [13]. With the development of new chemical and enzymatic pretreatments (e.g., TEMPO, periodate, carboxymethylation, and endoglucanase) and the ongoing work to improve commercial production of NFC due to tremendous reductions in energy demands after pretreatment (i.e., energy demands were reduced from 500 to 1500 kWh per ton) [14].

When NFC is incorporated into paper formulations, it greatly enhances the inter-fiber bonding of virgin and recycled pulp. In effect, the tensile, burst, and tear strength values improved as the bonds created a dense matrix of hydrogen-bonded fibrils within the paper sheet, even in recycled pulp. For example, 10% NFC was added to recycled pulp, resulting in a nearly 97% improvement in tensile strength, a nearly 133% improvement in burst strength, and a 104% improvement in tear resistance compared to untreated papers [13]. Additionally, it reduces paper porosity and improves apparent density,NFC can provide smoothness and improved printability. These reasons make NFC properties even more beneficial in the manufacture of specialty paper, quality packaging, and barrier coatings. NFC can help improve paper recyclability by recovering the bondability loss in recycled fibers, enabling sustainable and circular economy practices in papermaking.

2.2.2 Cellulose Nanocrystals (CNCs)

Initially, the application of acid hydrolysis for extracting cellulose nanocrystals (CNCs) was introduced by Nickerson and Habrle in 1947. Subsequent pioneering work by Rånby in 1949 was, however, initially disregarded; it was later rediscovered in the 1970s following the research of Derek Gray at McGill University, Canada. Cellulose nanocrystals, also referred to as nanocrystalline cellulose (NCC), exhibit a highly crystalline structure (approximately 90%). With the chemical treatment, the amorphous areas of cellulose are carefully dissolved, while crystalline regions stay intact. Crystalline regions exhibit a uniform shape in three dimensions, characterized by high stiffness and a high aspect ratio. The surface chemistry of the CNC may contain sulfate groups (resulting from the acid hydrolysis step) and confer colloidal stability in an aqueous suspension. Unlike NFC, CNCs do not create continuous networks but act as discrete reinforcing elements for composites and paper materials. Their dimensions range from 3 to 35 nm in diameter and 100 to 1000 nm in length [15].

In pulp and paper functionalization, CNCs are critical for improving the stiffness, dimensions stability, and barrier properties of paper. The high crystallinity and rigidity of CNCs improve the thermal and mechanical stability of paper, making it suitable for more demanding applications such as food packaging and technical papers [16]. CNC's decrease in gas and moisture permeability of paper improves some of the protection it can provide in packaging. Furthermore, their inclusion may result in lower coefficient of thermal expansion for the paper while also improving optical properties. The use of CNCs is being increasingly tested in engineered coatings and functional films to enhance performance and durability [17]. Just like NFC, CNCs contribute to the creation of higher-value and more sustainable paper products from recycled fibers, as well as reduced use of synthetic additives.

2.2.3 Clay and Silica Nanoparticles

Papermaking utilizes kaolin nanoparticles, which originate from the kaolinite clay minerals commonly found in sedimentary rocks rich in clay. For transparency, smoothness, brightness, and opacity in papermaking, kaolin nanoparticles can be utilized as pigments or as fillers. With the advancement of nanotechnology, scientists and manufacturers can produce kaolin at the nanolevel, thereby increasing its surface area and reactivity with other materials. Different forms of kaolin nanoparticles include pure kaolin nanoclays, organically modified kaolin, and kaolin–silica core–shell composites, which can be used to improve the mechanical properties, barrier performance, and print quality of paper products. Additionally, when kaolin nanoparticles were combined with specific polymers (e.g., polyacrylic acid) or cellulose fibers, the authors found increased paper tensile strength, burst index, folding endurance, and water absorption, which can ultimately be utilized in specialty or absorbent grades of paper [18].

Montmorillonite is closely related to kaolin, which is a layered smectite clay mineral. Montmorillonite nanoparticles are being added to pulp and paper formulations primarily due to their high aspect ratio, swelling, and large surface area. Montmorillonite is capable of improving the barrier properties, lowering air permeability, and enhancing the moisture resistance of paper and paperboard products, which is particularly necessary in food packaging and specialty papers [4]. Montmorillonite nanoparticles can be functionalized with polymers or biopolymers to enhance their adherence to cellulose fibers, while also acquiring flame retardant or antimicrobial properties.

Silica nanoparticles are a further significant inorganic additive in pulp and paper functionalization. They are commercially available in various forms, including colloidal silica, fumed silica, and mesoporous silica. They are used as structures to enhance paper strength, improve filler retention, and improve barrier properties. Silica nanoparticles reduce air permeability and porosity when applied as a coating, making them advantageous in papers requiring moisture and/or gas barrier performance [19]. Silica generated from kaolin sources has also been used as a source for creating colloidal silica, which is applied to enhance drainage, retention, and uniformity of the paper sheet during papermaking [20].

In combination, kaolin, montmorillonite, and silica nanoparticles are all multifunctional components that can maximize the sustainable, high-performance, value-added aspects of the paper products produced [21].

3 Functionalization of Pulp and Paper Using Nanoparticles

3.1 Mechanical Property Enhancement

Nanoparticles are increasingly being used as additives to enhance the functional properties of pulp and paper, primarily to improve mechanical strength, flame resistance, and other desirable properties. Graphene oxide (GO) has excellent potential to improve the paper and pulp sector due to its enhanced strength, reinforcement, and improved paper products. Depending on the various parameters, GO papers with the incorporation of nanoparticles such as silica have shown improved mechanical performance. For example, the GO paper reinforced with 10% functionalized silica nanoparticles showed an approximately 51% improvement in tensile strength, a 317% increase in elastic modulus, and a 69% increase in toughness. These improvements can be mainly attributed to strong hydrogen bonding and covalent interactions that build a structure throughout the paper matrix between GO sheets and Silica nanoparticles, allowing for better energy dissipation and load transfer during the loading process [22].

In addition to mechanical improvements, GO-based papers also provide enhanced flame resistance, attributing to the formation of protective char layers made from reduced graphene oxide and silica, which act as barriers against heat and flame propagation. Such flame-resistant nanocomposite papers have potential applications in flexible electronics, fire-safe packaging, and thermal insulation materials where conventional paper would fail to provide adequate protection [22].

3.2 Antimicrobial Activity Improvement

AgNPs, or silver nanoparticles, play a crucial role in enhancing the antimicrobial properties of pulp and paper products, particularly in the packaging sector. AgNPs have been utilized in paper coatings and mixed with paper matrices to impart excellent antibacterial properties, which will aid in developing food safety and prolonging product shelf life. Instead of heavily reliant methodologies, eco-friendly processes have been developed to use paper-based materials for food packaging, utilizing G-AgNPs produced through environmentally friendly methods [23]. These coatings, which are often added to natural polymer matrices such as starch and chitosan, produce a barrier against harmful microorganisms, including *Salmonella enteritidis, Staphylococcus aureus, and Escherichia coli* [24].

AgNPs coatings not only provide antimicrobial protection but also improve the mechanical properties, water resistance, and oil barrier properties of paper packaging. Chitosan-silver nanoparticles coatings have shown to have enhanced tensile strength, water resistance, and grease resistance as packaging for direct contact food packaging [25]. Coatings based on cellulose nanofibers with silver nanoparticles have similarly shown improved water vapor permeability and tensile properties, along with the ability to inhibit bacterial growth [26].

In conclusion, the end products, AgNP-functionalized paper packaging, offer numerous benefits compared to regular paper packaging. They prevent contamination during the storage and transport of food products, which extends the shelf life of products and reduces food waste. Furthermore, these antimicrobial papers can help to reduce the amount of chemical preservatives needed in food packaging as consumers increasingly demand cleaner and safer food packaging options [27].

3.3 *Barrier Properties*

3.3.1 Water and Oil Resistance Barrier Properties Through Nanoparticles

Functionalizing paper with nanoparticles is a viable approach to making it water- and oil-resistant, especially for packaging where contact with grease and water is routine. The integration of nanoparticles, such as silica (SiO_2), zinc oxide (ZnO), cellulose nanocrystals (CNCs), and nanoclays, greatly enhances these barrier properties by minimizing surface energy and blocking or filling the pore network of the paper. For example, hydrophobically modified silica nanoparticles produce a superhydrophobic effect through the formation of a hierarchical rough surface, which efficiently repels water droplets [3]. Similarly, the application of cellulose nanocrystals blended with biopolymers such as starch or chitosan yields biobased coatings that are not only water-repellent but also compostable and biodegradable [28].

For resistance to oils, nanoparticles assist by creating thick, continuous coats that impede the capillary absorption of non-polar liquids, such as fats and oils. Montmorillonite and kaolin nano clays, for instance, produce tortuous pathways that resist the flow of oils, increasing the kit value of the paper—a measure used in the industry to assess oil resistance. These films are especially useful in food packaging, including wraps, takeaway boxes, and snack packaging, where both oil and water resistance are required. In addition, ZnO nanoparticles have been researched with dual functionality—they not only offer oil resistance but also exhibit antimicrobial action, making them suitable for use in perishable foods [29].

3.3.2 Oxygen and Moisture Barrier Properties for Packaging Applications

Nanoparticle coatings are highly successful functional agents for improving barrier properties against oxygen and moisture in paper packaging. The use of barriers is critical to protect sensitive products, particularly in food, pharmaceutical, and electronic packaging, due to the potential for oxygen and water vapor contact that could spoil, degrade, or reduce the shelf life of a product. Nanoclay, silica, cellulose nanocrystals (CNCs), and metal oxide nanoparticles are the most frequently used nanomaterials for this purpose. By creating tortuous diffusion pathways—providing complicated, long paths for gas or vapor molecules to follow—nanoparticles can reduce oxygen and matter permeability [30].

In moisture barrier applications, nanoparticles are typically incorporated into a hydrophilic or hydrophobic polymer matrix (e.g., montmorillonite (MMT) clay and silica with polyvinyl alcohol (PVA), starch, or chitosan) as thin but dense films that substantially reduce the water vapor transmission rate (WVTR), improving moisture resistance while achieving flexibility and biodegradability for the paper [31]. For instance, PVA/MMT nanocomposites form layered structures with improved crystallinity and moisture resistance, making them suitable for packaging dry foodstuffs or pharmaceuticals.

In the oxygen barrier regime, cellulose nanocrystals (CNCs) and nanofibrillated cellulose (NFC) are being used more and more, because they are renewable and have higher barrier properties. These nanocelluloses, when employed as coatings or thin films, form a densified network that minimizes free volume and hence inhibits the diffusion of oxygen molecules. Moreover, CNCs may be mixed with graphene oxide or layered silicates to further enhance oxygen impermeability while ensuring paper recyclability [28]. These bio-coatings are considered sustainable alternatives to traditional plastic coatings for oxygen-sensitive products, such as coffee, snacks, and processed foods.

Zinc oxide (ZnO) and titanium dioxide (TiO_2) nanoparticles not only add barrier efficiency but also offer UV protection and antimicrobial activity. They are multivalent, serving various benefits and being particularly suited for packaging perishables, such as dairy products or meat. The synergy between biopolymers and nanoparticles enables the design of coatings that are both high-performance and biodegradable, a key requirement of the current green market.

3.4 Self-Cleaning and Superhydrophobic Surface Properties

Functionalization of paper surfaces with nanoparticles to render them superhydrophobic and self-cleaning has been of rising interest, particularly for specialty papers, high-performance packaging, and antifouling.

A superhydrophobic surface will also exhibit a water contact angle greater than 150° such that water droplets will roll off and bead up while removing dirt and contaminants [32]. Such self-cleaning, or Lotus Effect, is realized through the combination of low-surface-energy materials and microlevel and nanolevel roughness [33]. The most widely used nanoparticles to produce such a dual-scale structure on paper supports include silica (SiO_2), titaniaum dioxide (TiO_2), zinc oxide (ZnO), and fluorinated materials. For example, hydrophobized silica nanoparticles deposited through spray-coating or dip-coating methods create porous, water-repellent coatings that mimic the natural lotus leaves, allowing paper to repel water and even some oily contaminants [34]. Furthermore, TiO_2 nanoparticles have a photo-induced self-cleaning function.

TiO_2, when irradiated by UV light, produces reactive oxygen species with the capability to degrade organic contaminants on the paper surface and is hence found to be suitable for antibacterial food wraps, wallpaper, and hygienic products. Furthermore, by combining cellulose nanocrystals (CNCs) or chitosan with nanoparticles such as ZnO or SiO_2, scientists have created biodegradable and eco-friendly superhydrophobic coatings for paper-based products used in smart packaging and outdoor applications. They not only enhance the water resistance but also durability, stain resistance, and print quality of wet paper products. The addition of nanoparticles makes functionalized paper lightweight, breathable, and recyclable without affecting the purposes of sustainable material development for the packaging and paper industry [35].

3.5 Conductivity and Smart Paper Applications

Nanoparticles, particularly graphene and carbon nanotubes (CNTs), have revolutionized the dynamics of pulp and paper functionalization, enabling the transformation of paper to accommodate additional applications beyond those of conventional paper. With nanomaterials, paper can be utilized as an active component in electronic, sensing, and thermal applications while retaining all its desirable attributes, including flexibility, low weight, and cost-effectiveness. The unique characteristics of graphene electrical conduction and the mechanical integrity of CNTs create a highly effective conductive framework within a paper structure [36].

Graphene and CNT-based papers exhibit remarkable properties that enable a high level of electromagnetic interference (EMI) shield; therefore, they can be useful for electronic applications requiring protection from external electrical noise. For example, a CNT composite paper exhibited shielding effectiveness of -18 to -37 dB, and its applicability will vary across different frequency ranges, making it suitable for packaging small, high-value electronics or sensitive instruments [37]. These papers are also helpful as flexible electrodes (or electronic components) in paper-based electronics such as transistors, heating elements, and sensors.

Smart paper sensors, produced using graphene and CNT coatings, can sense temperature fluctuations, humidity, or even gases like hydrogen, providing warnings in hazardous conditions. For instance, using layer-by-layer assembly methodology, special coatings made from functionalized graphene oxide and CNT were applied to paper, allowing it to detect flames within seconds, making it useful for fire detection in packaging or construction materials [38]. Similarly, carbon nanotube-based humidity and gas sensors have been developed on paper substrates, offering low-cost solutions for environmental monitoring and disposable sensor applications [39].

The final graphene- and CNT-functionalized papers surpass the performance characteristics of conventional papers by offering enhanced electrical conductivity, mechanical strength, thermal stability, and additional functionality. These smart papers can enable flexible and portable electronics, safer packaging, and responsive systems, as well as support investors in developing scalable and sustainable production processes. This technology represents a significant step toward developing high-performance, operationally sustainable materials for the next generation of devices.

4 Methods for Incorporation of Nanoparticles in Pulp and Paper

4.1 In Situ Synthesis on Pulp Fibers

In situ synthesis involves the fabrication of nanoparticles directly on pulp fiber surfaces during the pulp processing itself. The pulp fibers are first chemically modified, which can be achieved through oxidation processes (e.g., TEMPO oxidation, 2,2,6,6-tetramethylpiperidine-1-oxyl) that produce pulp with active functional groups (e.g., carboxyl and aldehyde groups) on the cellulose chains, which can be easily transformed into active sites for nanoparticle growth. The treated pulp is then immersed in a solution of metal precursor (e.g., a gold or zinc salt solution), which helps to facilitate a reaction with the functional groups on the cellulose of the pulp. The functional groups will reduce metal ions in the solution to nanoparticles on the pulp surface, often without the need to add another reducing agent or invoke heat. The resulting paper will comprise nanoparticles covalently bonded to the network of cellulose fibers, ensuring that the nanoparticles will not leach out and can enable the paper to exhibit different properties, such as antimicrobial activity, UV protection, or catalytic capability. For example, gold nanoparticles synthesized on TEMPO-oxidized pulp were shown to enable catalysis for the reduction of 4-nitrophenol, while pulp with zinc oxide nanoparticles sized on it demonstrated UV-shielding properties and antibacterial properties [8, 40]. The process yields functional papers directly synthesized during paper production, eliminating the need to synthesize nanoparticles separately and making them more efficient and arguably more environmentally friendly (Figs. 2 and 3).

4.2 Mixing Nanoparticle Dispersions into Pulp Slurry

The method discussed involves the introduction of presynthesized nanoparticles into the pulp slurry at the wet end of the papermaking process. The nanoparticles will be evenly dispersed in water or another medium and can be one of the following: chitosan-complexed starch, titanium dioxide, or silver, among others. The dispersion will then be added to the pulp slurry at a very late stage in the process before sheet

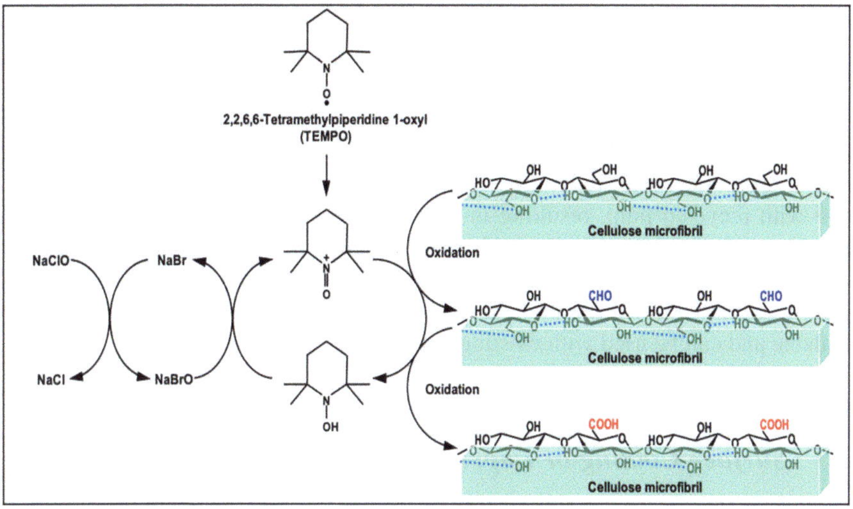

Fig. 2 Schematic illustration of the TEMPO oxidation process of cellulose pulp fibers Reference [40]

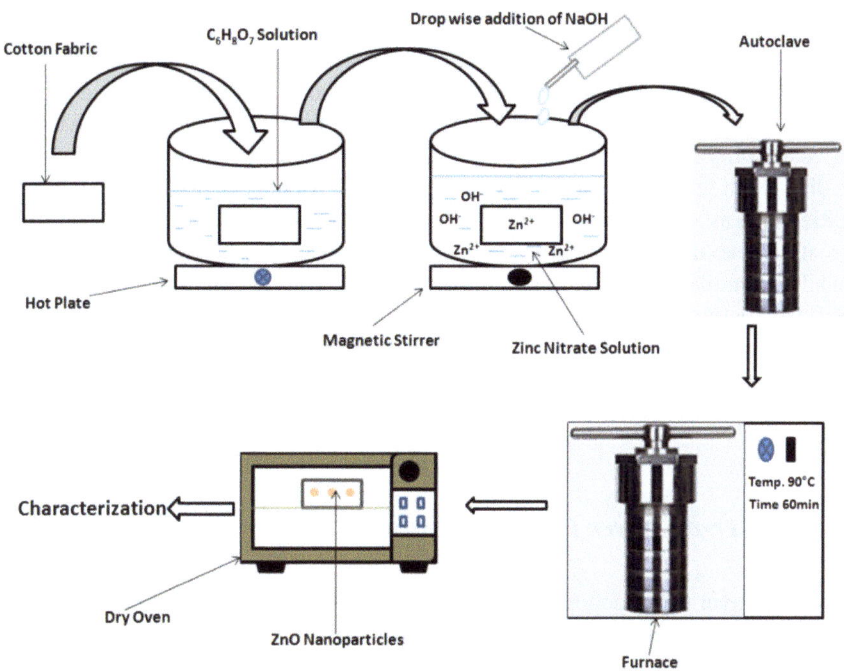

Fig. 3 Schematic diagram of in-situ synthesis of ZnO-nano particles on cellulosic fibres [41]

formation occurs. Mixing must be carefully performed to ensure uniform distribution of nanoparticles within the fibers. The mixture will be formed into sheets using standard papermaking equipment before being pressed and dried. The nanoparticles act as nanofillers, opening up more sites for interfiber adhesive bonding and improving specific mechanical properties, such as tensile strength, burst strength, and barrier properties. One study utilized new chitosan-starch nanoparticles in conjunction with recycled pulp, resulting in increased tensile and burst strength, as well as improved surface properties [42]. Another study mixed TiO_2 nanoparticles into sugarcane bagasse pulp, allowing for both barrier and antimicrobial functionality, with no significant deterioration in mechanical performance [43]. This method is scalable and can be used with existing papermaking facilities.

4.3 Surface Coating or Impregnation

In this method, nanoparticle suspensions are used to coat a paper sheet that has been produced prior to the synthesis of the nanoparticles. For example, silver, zinc oxide, or titanium dioxide nanoparticles are first synthesized (chemical, green, or physical to obtain the nanoparticles, such as laser ablation). A stable suspension of these nanoparticles in water or another carrier solvent is prepared, and the paper sheets are immersed, sprayed, or roll-coated with the nanoparticle suspension so that the nanoparticles are uniformly dispersed across the surface. The paper sheet can then be dried; in some cases, a thermal curing is utilized to set the nanoparticles. This method is directed at applying coating to the surface of the paper sheet to produce functionalities such as antimicrobial activity, UV protection, or barrier properties, while not greatly impacting bulk properties of the paper, for example, the potential efficiencies of silver nanoparticle-coated papers as an antibacterial with low water leaching [44]. Surface coatings are particularly valuable for applications in packaging, hygiene products, and antimicrobial papers, where surface interaction may be the key factor in product utilization. This process is flexible, with no disruption in sheet formation, so that it can be considered as a postprocessing step for functionalization Fig. 4.

4.4 Layer-by-Layer (LbL) Deposition

Layer-by-layer (LbL) deposition is a highly controlled technique that constructs thin films of nanoparticles on pulp fibers or paper surfaces, utilizing alternating layers of positively and negatively charged species. The first step is to dip or spray the paper or fiber with a solution containing cationic species (e.g., chitosan, cationic nanoparticles) and then rinse. Next, treat the paper or fiber with an anionic solution (e.g., carboxymethylcellulose, anionic nanoparticles). These steps can be repeated for as many layers as are desired, with rinses to eliminate loosely bound material between layers. The process allows for control of the placement of nanoparticles, the

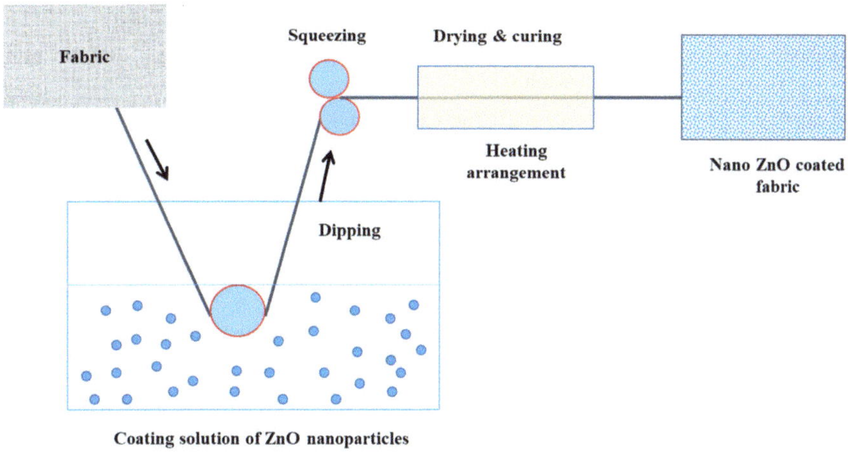

Fig. 4 The sequence of thermo-fixation coating of nanoparticles on the fabric surface [46]

thickness of the layer, and the composition of the coating or layer. The application of LbL has been used to improve mechanical strength, enhance barrier properties, or impart specific functionalities such as flame retardancy or water resistance [4]. LbL also allows the use of multiple nanoparticles (clay, silica, and TiO_2) to create functional multilayered structures for future advanced paper applications. The deposition and growth of Au NPs have been achieved through electrostatic assembly on wood cellulosic fibers that were alternately dipped in polyelectrolytes and a colloidal nanoparticle solution [45]. The deposition and growth of Au NPs have been achieved by electrostatic assembly on wood cellulosic fibers that were alternately dipped in polyelectrolytes and a colloidal nanoparticle solution [34]. While LbL processing can be more challenging, its advantageous ability for precision and customizability warrants its application here.

4.5 Green Synthesis and Incorporation

Green nanoparticle synthesis has been defined as the preparation of nanoparticles using environmentally friendly methods, such as plant extracts or physical methods like laser ablation. For example, the aqueous extracts of *Syzygium cumini* leaves can be used as reducing and stabilizing agents to produce silver and ferric oxide nanoparticles. Alternatively, laser ablation generates a mixture of nanoparticles in suspension without the use of chemicals to assist the process, as the ablation process can occur only in water combined with pure metals. These nanoparticles can be mixed with the pulp slurry or coated onto the paper surface. This has the advantage of eliminating toxic chemicals and replaced with nanomaterials that will measure functional properties including antimicrobial properties, UV protection, and barrier properties [9, 44].

Table 2 Methods of nanoparticle incorporation for pulp and paper functionalization

Method	Type of functionalization	Nanoparticles used	Functional role	References
In situ chemosynthesis	Internal functionalization	Zinc oxide (ZnO)	UV protection, antimicrobial activity	[8]
Nanofillers in slurry	Bulk reinforcement	Bark, sawdust nanoparticles	Strength enhancement	[47]
Surface impregnation	Surface functionalization	Silver (Ag)	Antimicrobial	[44]
Chitosan-starch nanocomposite slurry	Bulk reinforcement	Chitosan-starch nanoparticles	Strength, gloss improvement	[42]
Layer-by-layer (LbL) assembly	Coating functionalization	Various organic/inorganic NPs	Controlled barrier, strength	[4]
Green synthesis + coating	Surface functionalization	Silver (Ag), ferric oxide (Fe_2O_3)	Antifungal, antibacterial	[9]
TiO_2 in bagasse paper	Bulk functionalization	TiO_2	Barrier, antimicrobial	[43]
Gold nanoparticle in situ synthesis	Internal catalytic functionalization	Gold (Au)	Catalytic, functional paper	[40]
Ferroelectric nanoparticle loading	Bulk electrical functionalization	Barium titanate Strontium titanate	Dielectric properties	[48]
10.Wastepaper valorization coating	Surface functionalization	Biogenic silver (Ag)	Antimicrobial packaging	[27]

Generally, the process consists of nanoparticle synthesis → mixing/impregnation → sheet or coated surface → drying. Green synthesis employs eco-friendly methods that meet sustainability goals while remaining functional, making it viable for applications in food packaging and hygiene, as well as for an eco-friendly line of products (Table 2).

5 Challenges in Pulp and Paper Functionalization Using Nanoparticles

The fibrous, porous, and chemically heterogeneous nature of paper substrates presents challenges in achieving a uniform dispersion of nanoparticles on these substrates. As noted by [3], the authors have cited these differences as being a source of variability, which addresses issues related to the uniformity of nanoparticle adsorption and aggregation, which ultimately affects the printability, antibacterial properties, barrier properties, and other properties. The fibers' variable surface chemistry inevitably affects the efficiency of nanoparticle incorporation regardless of efforts to control the size and charge of the particles during functionalization.

The synthesis and application methods of nanoparticles are both expensive and energy-intensive, which limits the ability to integrate nanoparticles into more commercial applications. The different technologies that involve the chemical synthesis of metal nanoparticles or the mechanical fibrillation of nanocellulose typically involve very expensive reagents, specialized equipment, and considerable energy input [49]. Furthermore, commercializing the incorporation of nanoparticles and the functionalization of fibers into existing manufacturing systems may, in itself, add complexity to the production process, which can translate to higher operating costs—especially in applications where cost is competitive factor, such as packaging or disposable products.

Recyclability issues are a significant limitation of nanoparticle-enhanced papers. The incorporation of nanoparticles could affect standard recycling technology, including fiber recovery, de-inking, and the quality of secondary fiber products made after recycling [49]. As we consider the circular economy, there is concern about how recycling rates and material reuse factors impact the effective recyclability of products enhanced with nanoparticles, which can create significant barriers to environmental resilience and sustainability.

There are several uncertainties regarding the environmental fate and potential toxicity of nanoparticles, particularly those composed of metals such as silver and titanium dioxide. While some metallic nanoparticles may withstand environmental degradation and exhibit ecotoxicological properties, comprehensive life-cycle assessments and toxicological studies are yet to be identified [50]. Regarding this aspect of the material(s), the variable nature of nanoparticles raises regulatory questions concerning safety in use, recycling, and disposal. The issues surrounding the control of nanoparticle release upon use, recycling, or disposal can raise safety concerns.

Transitioning nanoparticle functionalization processes to large-scale industrial applications is a technically challenging task. Many techniques developed on an experimental scale, such as layer-by-layer assembly or in situ synthesis, cannot be easily transferred to the fast, operational, and continuous production methods used in contemporary papermaking processes [51]. Developing and transferring these approaches at the industrial scale requires significant alteration to the process or new technology investments, which can result in a commercialization lag.

The inconsistency in characterizing nanoparticle distribution, adhesion, and functional performance on paper substrates is hindered by a lack of standardized methodologies, resulting in unpredictable quality control and hesitant industry adoption. The absence of a standardization and characterization approach makes it impossible to compare nanoparticle-functionalized papers across studies or define standards for commercial applications [3]. This lack of standardization hinders the development of consistent and reproducible products based on nanoparticle functionalities on paper substrates that are suitable for following a regulated supply chain.

Regulatory uncertainty and the acceptance of new markets introduce additional challenges. For example, in the case of types of paper with nanomaterials for food

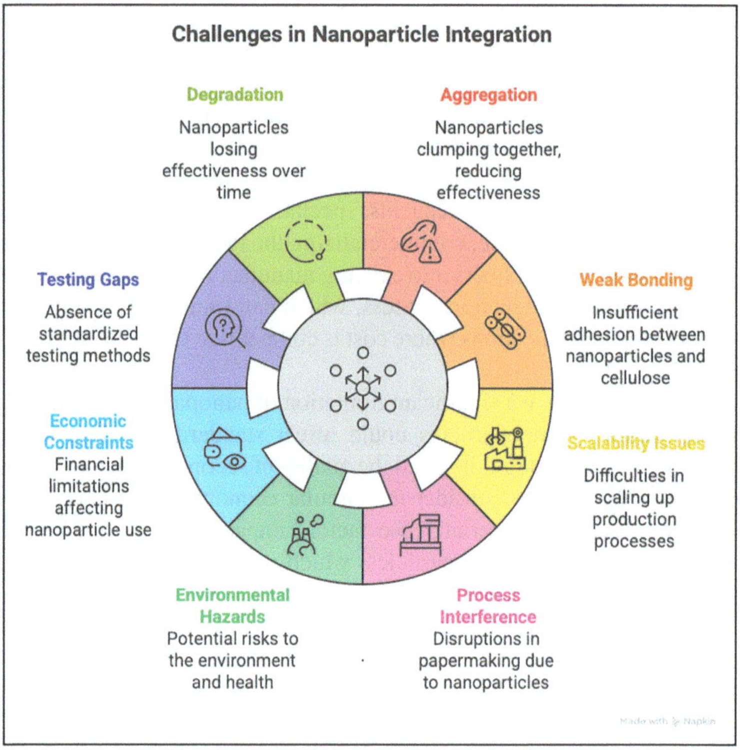

Fig. 5 Illustration of various challenges in nanoparticle incorporation

packaging, medical, or other sensitive uses, companies must undergo a level of scrutiny by regulators. Currently, guidelines and testing methods for nanomaterial-functionalized paper have not been established, and a level of uncertainty remains for manufacturers to confidently enter their paper products into the market [51] (Fig. 5).

6 Conclusion

The incorporation of nanoparticles has revolutionized the pulp and paper industry by imparting advanced functionalities, including enhanced mechanical strength, antimicrobial activity, flame resistance, and barrier properties. Nanomaterials like graphene oxide (GO), silver nanoparticles (AgNPs), silica (SiO_2), zinc oxide (ZnO), titanium dioxide (TiO_2), carbon nanotubes (CNTs), cellulose nanocrystals (CNCs), and nanoclays are increasingly integrated into paper matrices. These materials, due to their nanoscale size and large surface area, enhance interactions with cellulose fibers, resulting in improved performance. For example, GO combined with functionalized silica nanoparticles significantly improves tensile strength and elasticity, while AgNPs embedded in biopolymer coatings boost hydrophobicity and antimicrobial properties—key for food packaging. TiO_2 and silica offer self-cleaning capabilities, while CNTs and graphene enhance electrical conductivity for paper-based electronics. Various methods, such as in situ synthesis, green synthesis, impregnation, and blending with biopolymers, are employed to incorporate these nanoparticles. However, achieving proper dispersion and strong adhesion to the paper substrate while retaining recyclability and biodegradability remains a critical challenge. Despite progress, several research gaps persist. These include assessing long-term environmental and health impacts, especially in food-contact applications, developing sustainable end-of-life strategies, scaling up nanoparticle integration at industrial levels, and achieving uniform dispersion at higher loadings.

Furthermore, most studies focus on single-function improvements, whereas multifunctional, sustainable formulations are needed. In conclusion, nanoparticle functionalization has opened new horizons for creating lightweight, multifunctional, and eco-friendly paper products. To fully harness these benefits, interdisciplinary research addressing environmental safety, scalability, and lifecycle performance is essential. Such efforts will support the transition toward circular economy principles and greener manufacturing in the pulp and paper industry.

Competing Interests The authors declare that they have no competing interest.

Declaration of Generative AI and AI-Assisted Technologies in the Writing Process During the preparation of this work, the book chapter author(s) have not taken any assistance from AI, and the book chapter author(s) reviewed and edited the content as needed and take(s) full responsibility for the content of the publication.

References

1. Lavoine N, Desloges I, Dufresne A, Bras J (2012) Microfibrillated cellulose–Its barrier properties and applications in cellulosic materials: a review. Carbohyd Polym 90(2):735–764
2. Sun J, Yang X, Bai Y, Fang Z, Zhang S, Wang X, Yang Y, Guo Y (2024) Recent advances in cellulose nanofiber modification and characterization and cellulose nanofiber-based films for eco-friendly active food packaging. https://doi.org/10.3390/foods13243999
3. Ngo YH, Li D, Simon GP, Garnier G (2011) Paper surfaces functionalized by nanoparticles. Adv Coll Interface Sci 163(1):23–38. https://doi.org/10.1016/j.cis.2011.01.004
4. Samyn P, Barhoum A, Öhlund T, Dufresne A (2017) Review: nanoparticles and nanostructured materials in papermaking. J Mater Sci 53:146–184
5. Sobri Z, Ainun Z, Zainudin E (2018) Distribution of zinc oxide nanoparticles on unbleached and bleached bamboo paper via in-situ approaches. IOP Conf Ser Mater Sci Eng 368:012046
6. Tang B, Sun L, Li J, Kaur J, Zhu H, Qin S, Yao Y, Wu C, Wang X (2015) Functionalization of bamboo pulp fabrics with noble metal nanoparticles. Dyes Pigm 113:289–298
7. Koga H, Kitaoka T (2011) On-paper synthesis of metal nanoparticles for catalytic applications. Sen'i Gakkaishi 67(7):141–152
8. Li Y, Zhang Y, Wanru D, Yue J, Xu M, Shi S (2018) Preparation and properties of pulp fibers treated with zinc oxide nanoparticles by in situ chemosynthesis. Holzforschung 72:923–931

9. Elshaer MA, Abd-Elraheem MA, Taha AS et al (2024) Green synthesis of silver and ferric oxide nanoparticles using Syzygium cumini leaf extract and their antifungal activity when applied to oak wood and paper pulp from Imperata cylindrica grass biomass. Waste Biomass Valorization
10. Tabari K, Hosseinpour S, Parashos P, Khozestani PK, Rahimi H (2017) Cytotoxicity of selected nanoparticles on human dental pulp stem cells. Iran Endod J 12:137–142
11. Furman B, Rawls HR, Wellinghoff S, Dixon H, Lankford J, Nicolella D (2000) Metal-oxide nanoparticles for the reinforcement of dental restorative resins. Crit Rev Biomed Eng 28(3–4):439–443
12. Moon RJ, Martini A, Nairn J, Simonsen J, Youngblood J (2011) Cellulose nanomaterials review: structure, properties and nanocomposites. Chem Soc Rev 40(7):3941–3994. https://doi.org/10.1039/C0CS00108B
13. Viana LC, Potulski DC, Bolzón de Muñiz GI, Andrade AS, da Silva EL (2018) Nanofibrillated cellulose as an additive for recycled paper. Cerne 24(2):251–260. https://doi.org/10.1590/01047760201824022518
14. Gupta GK, Shukla P (2020) Lignocellulosic biomass for the synthesis of nanocellulose and its eco-friendly advanced applications
15. Kumar V, Pathak P, Bhardwaj NK (2020) Waste paper: an underutilized but promising source for nanocellulose mining. Waste Manage 102:281–330
16. Giannelli R, Giannelli R, Babudri F, Operamolla A (2021). Nanocellulose-based functional paper. https://doi.org/10.1016/B978-0-12-822350-5.00003-5
17. He Y, Li H, Fei X, Peng L (2021) Carboxymethyl cellulose/cellulose nanocrystals immobilized silver nanoparticles as an effective coating to improve barrier and antibacterial properties of paper for food packaging applications. Carbohyd Polym 252:117156
18. Ibrahim M, El-Zawawy WK, Nawwar G (2012) Modified kaolin and polyacrylic acid-g-cellulosic fiber and microfiber as additives for paper properties improvements. Carbohyd Polym 88:1009–1014
19. Hollertz R, Pitois C, Ariza D, Wågberg L (2015) Dielectric response of kraft paper from fibres modified by silica nanoparticles. IEEE CEIDP 2015:459–462
20. Afshar M, Alipour A, Norouzbeigi R (2023) Kaolin-based stable colloidal nano-silica: peptization factors and stability assessments via designed experiments. Res Eng
21. Murray H (2006) Kaolin applications. In: Developments in clay science, vol 2, pp 85–109
22. Yu Z-R, Li S-N, Zang J, Zhang M, Gong L-X, Song P, Zhao L, Zhang G-D, Tang L (2019) Enhanced mechanical property and flame resistance of graphene oxide nanocomposite paper modified with functionalized silica nanoparticles. Compos B Eng. https://doi.org/10.1016/J.COMPOSITESB.2019.107347
23. Trotta F, Da Silva S, Massironi A, Mirpoor SF, Lignou S, Ghawi SK, Charalampopoulos D (2024) Advancing food preservation: sustainable green-AgNPs bionanocomposites in paper-starch flexible packaging for prolonged shelf life. Polymers 16(7):941. https://doi.org/10.3390/polym16070941
24. Nechita P (2017) Active-antimicrobial coatings based on silver nanoparticles and natural polymers for paper packaging functionalization. Nord Pulp Pap Res J 32(3):452–458. https://doi.org/10.3183/npprj-2017-32-03-p452-458
25. Jung J, Kasi G, Seo J (2018) Development of functional antimicrobial papers using chitosan/starch-silver nanoparticles. Int J Biol Macromol 112:530–536. https://doi.org/10.1016/j.ijbiomac.2018.01.155
26. Amini E, Azadfallah M, Layeghi M, Talaei-Hassanloui R (2016) Silver-nanoparticle-impregnated cellulose nanofiber coating for packaging paper. Cellulose 23:557–570
27. Nwabor O, Singh S, Ontong JC, Vongkamjan K, Voravuthikunchai S (2020) Valorization of wastepaper through antimicrobial functionalization with biogenic silver nanoparticles, a sustainable packaging composite. Waste Biomass Valorization 12:3287–3301. https://doi.org/10.1007/s12649-020-01237-5
28. Hubbe MA, Ferrer A, Tyagi P, Yin Y, Salas C, Pal L, Rojas OJ (2017) Nanocellulose in thin films, coatings, and plies for packaging applications: a review. BioResources 12(1):2143–2233

29. Kunam PK, Ramakanth D, Akhila K, Gaikwad KK (2024) Bio-based materials for barrier coatings on paper packaging. Biomass Convers Biorefinery 14(12):12637–12652
30. Lagaron JM, Catalá R, Gavara R (2004) Structural characteristics defining high barrier properties in polymeric materials. Mater Sci Technol 20(1):1–7
31. Rhim JW, Park HM, Ha CS (2013) Bio-nanocomposites for food packaging applications. Prog Polym Sci 38(10–11):1629–1652
32. Gao X, Jiang L (2004) Water-repellent legs of water striders. Nature 432(7013):36–36
33. Li XM, Reinhoudt D, Crego-Calama M (2007) What do we need for a superhydrophobic surface? A review on the recent progress in the preparation of superhydrophobic surfaces. Chem Soc Rev 36(8):1350–1368
34. Li H, Fu S, Peng L (2013) Surface modification of cellulose fibers by layer-by-layer self-assembly of lignosulfonates and TiO_2 nanoparticles: Effect on photocatalytic abilities and paper properties. Fibers Polym 14:1794–1802
35. Ghosh M, Stebe KJ (2010) Spreading and retraction as a function of drop size. Adv Coll Interface Sci 161(1–2):61–76
36. Tang Y-J, Gou J (2010) Synergistic effect on electrical conductivity of few-layer graphene/multi-walled carbon nanotube paper. Mater Lett 64:2513–2516. https://doi.org/10.1016/J.MATLET.2010.08.035
37. Pang Z (2014) Preparation and EMI shielding performance of carbon nanotubes conductive paper. J Synth Cryst
38. Chen W, Liu P, Liu Y, Wang Q, Duan W (2018) A temperature-induced conductive coating via layer-by-layer assembly of functionalized graphene oxide and carbon nanotubes for a flexible, adjustable response time flame sensor. Chem Eng J. https://doi.org/10.1016/J.CEJ.2018.07.110
39. Han J-W, Kim B, Li J, Meyyappan M (2012) Carbon nanotube based humidity sensor on cellulose paper. J Phys Chem C 116:22094–22097. https://doi.org/10.1021/JP3080223
40. Azetsu A, Koga H, Yuan L, Kitaoka T (2013) Direct synthesis of gold nanocatalysts on TEMPO-oxidized pulp paper containing aldehyde groups. BioResources 8(3):3706–3717
41. Javed A, Wiener J, Tamulevičienė A, Tamulevičius T, Lazauskas A, Saskova J, Račkauskas S (2021) One step In-Situ synthesis of zinc oxide nanoparticles for multifunctional cotton fabrics. Materials 14(14):3956. https://doi.org/10.3390/ma14143956
42. Salam A, Lucia L, Jameel H (2013) Synthesis, characterization, and evaluation of chitosan-complexed starch nanoparticles. ACS Appl Mater Interfaces 5(21):11029–11037
43. Jain P, Tripathi S, Deshmukh RK, Gaikwad KK, Singh S (2023) Functionalization of sugarcane bagasse–based paper with amla pomace/titanium dioxide nanoparticles. Biomass Convers Biorefinery
44. Schlemmer W, Fischer W, Zankel A et al (2018) Green procedure to manufacture nanoparticle-decorated paper substrates. Materials 11(12)
45. Pinto RJ, Marques PA, Martins MA, Neto CP, Trindade T (2007) Electrostatic assembly and growth of gold nanoparticles in cellulosic fibres. J Colloid Interface Sci 312(2):506–512
46. Tania IS, Ali M (2021) Coating of ZnO nanoparticle on cotton fabric to create a functional textile with enhanced mechanical properties. Polymers 13(16):2701. https://doi.org/10.3390/polym13162701
47. Laka M, Vikele L, Rozenberga L, Janceva S (2016) Nanoparticle fillers obtained from wood processing wastes for reinforcing of paper. AIP Conf Proc 1736:020112
48. Omari H, Zyane A, Belfkira A et al (2016) Dielectric properties of paper made from pulps loaded with ferroelectric particles. J Nanomater 2016(3982572)
49. Balea A, Fuente E, Monte MC, Merayo N, Campano C, Negro C, Blanco Á (2020) Industrial application of nanocelluloses in papermaking: a review of challenges, technical solutions, and market perspectives. Molecules 25(3):526. https://doi.org/10.3390/molecules25030526
50. Manda BMK, Blok K, Patel MK (2012) Innovations in papermaking: an LCA of printing and writing paper from conventional and high yield pulp. Sci Total Environ 439:307–320. https://doi.org/10.1016/j.scitotenv.2012.09.022
51. Gupta GK, Sailwal M, Shukla P (2024) Sustainable nanotechnology-based techniques for mitigating the pollutants from pulp and paper industry. ACS Omega. https://doi.org/10.1021/acsomega.4c06022

Nanotechnology in Paper Coating and Printing

Unnati Chaudhary, Rohit Gusain, Gyanesh Joshi, and Vikas Rana

Abstract Nanotechnology is an emerging field of science and technology that is transforming material science in the twenty-first century by offering unique and valuable functionalities. In the pulp and paper industry, the use of nanotechnology has increased, because it has the potential to enhance product performance, since processes can become more efficient through nanotechnological interventions. The use of nanoparticles and nanostructured materials in papermaking initially aimed at improving process operations and reducing raw material usage. However, its scope has now expanded significantly, which has allowed the integration of a wide variety of nanomaterials into paper structures, bestowing it with new and multifunctional properties. Nanotechnology has enabled improvements, particularly in barrier properties, printability, mechanical and optical characteristics, leading to significant advancements in paper coating and printing. The ability to modify surface features at the nanoscale opens up new opportunities for creating high-performance, value-added paper products. In this chapter, the basic principles of nanotechnology along with its various roles and applications in paper coating and printing are discussed. Nanotechnology is used to enhance coating formulations, and recent literature also highlights developments in printing quality. To fully realize the potential of nanotechnology in the pulp and paper industry, collaboration between academic researchers and policymakers is essential. Such cooperation can address current challenges and drive sustainable innovation in paper-based materials.

Keywords Paper · Nanotechnology · Printing · Coating · Nanomaterials

U. Chaudhary (✉) · R. Gusain · G. Joshi · V. Rana
Cellulose and Paper Discipline, Forest Products Division, Forest Research Institute, Dehradun, India
e-mail: unnatichaudhary7@gmail.com

R. K. Bachheti (ed.), *Nanotechnology in the Pulp and Paper Industry*, Smart Nanomaterials Technology, https://doi.org/10.1007/978-981-95-3473-9_6

1 Introduction

Nanotechnology, which involves controlling matter on an incredibly small scale (usually less than 0.1 μm or 100 nm), has revolutionized various industries by allowing us to develop new matter with cutting-edge uses [13]. Nanotechnology is one of the fastest-growing sciences where different processes and technologies are used to evolve molecules/products at the nanoscale range with improved properties and functionalities. This transformative technology has emerged as one of the most rapidly evolving scientific disciplines, such as electronics, healthcare, material science, and polymer engineering, particularly in paper production, where it offers promising advancements in the coating and printing process (Fig. 1).

The pulp and paper industries hold high prominence due to their influence on the local economy and their evolving focus on sustainability and performance enhancement. According to statistical data, the production of paper and paperboard globally was estimated to reach approximately 400 million metric tons in 2023 [37]. Additionally, the global paper and pulp market size is expected to increase from US$351.53 billion in 2021 to US$373 billion by 2029 [38]. With the increasing demand for high-performance, sustainable, and multifunctional paper products, integrating nanotechnology into paper coating and printing processes has become a key innovation pathway.

Paper sheets, due to their highly porous and hygroscopic nature, are generally incompatible with many coating colors and printing inks. To overcome this limitation and enhance their printing performance, various external additives are incorporated to improve ink penetration control, surface smoothness, dimensional stability, and optical qualities. Nonfibrous additives used in papermaking can be broadly categorized into two types: (i) Functional Additives, which include coating pigments, fillers, both wet and dry strength, as well as sizing agents and retention aids, which

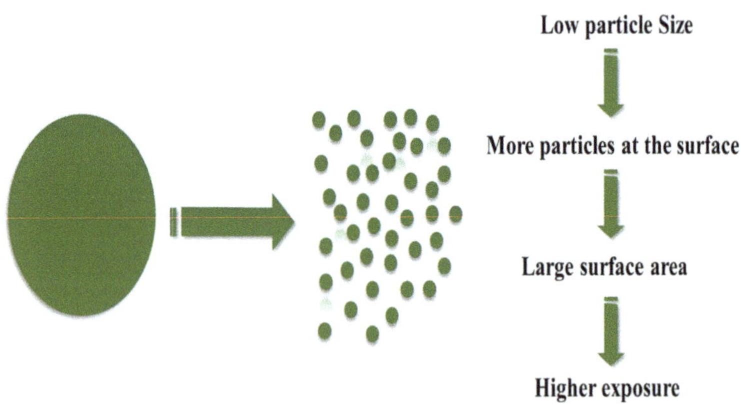

Fig. 1 Basic concepts of nanomaterials

enhance the end-use properties of paper such as brightness, opacity, and printability. (ii) Control Additives that comprise pitch control agents, defoamers, retention and drainage aids, biocides, and slimicides. Their role is to optimize the papermaking process, particularly at the wet end, by improving filler retention, accelerating drainage, removing contaminants, minimizing foam formation, and preventing microbial growth [82]. To optimize surface properties and address printing challenges, fillers are commonly added at the wet-end stage of papermaking, typically in the range of 10–20%, using retention aids to distribute them effectively within the fiber matrix. These fillers serve multiple functions, such as improving formation of the sheet, optical properties, reducing air permeability, increasing drainage efficiency and machine speed, improving printability, and lowering production costs by partially replacing more expensive fibers. In addition, coating pigments are applied to further enhance paper smoothness and quality. These coatings fill surface crevices, leading to the production of a compact, plane surface with a superior finish compared to the sized material or blending with refined fibers. The resulting surface characteristics of paper are majorly influenced by the particle shape, size, and distribution of the fillers and pigments used [32]. The two main strategies for integrating nanoparticles into paper are wet-end addition and surface treatment. The former involves the incorporation of nanoparticles during the papermaking process and direct bonding to fibers before sheet formation, while the latter is treating preformed paper sheets post production, either by immersion in nanoparticle-containing solutions or through surface coating techniques, to enhance functionality [75].

Modern paper products require more than essential mechanical strength and printability. Industries now seek advanced properties like enhanced barrier properties, smoothness with controlled ink absorption, improved opacity, and vibrant print quality through more sustainable and cost-effective processes [86, 96]. Traditional coating materials like clay and calcium carbonate, while effective to a degree, often fall short of achieving this balance. Nanomaterials offer promising solutions to achieve these goals due to their high surface area, tunable characteristics, and versatility.

A wide range of nanoadditives, including nanofibrils, nanoparticles, and nanogels, are increasingly integrated into paper coating and printing applications to impart distinctive properties that affect paper quality. For example, nano and hydrated calcium silicate as well as calcium carbonates have shown notable success in enhancing the opacity of newspaper print and reducing ink spread, thereby improving print clarity [27, 56]. To make superhydrophobic high-opacity papers, researchers use titanium dioxide (TiO_2) nanoparticles and, when these are paired with zinc oxide (ZnO), they also bring in antimicrobial benefits [20, 91, 93]. Nano silica (SiO_2) stands for its ability to combat bacteria, enhance print performance, and provide greater opacity [64, 98]. Nanotechnology is revolutionizing the printing sector, paving the way for innovative and smart printing applications. Materials like graphene, carbon nanotubes, and metal nanoparticles are key to the advancement of printed electronics, paving the way for innovations including flexible circuits and smart packaging [66]. Moreover, nanotechnology plays a crucial role in security printing by introducing anti-counterfeiting polymeric inks with print-marking and fingerprinting applications [1].

Moreover, nanopigments (e.g., nanokaoline) and nanopolymers as binding agents have been investigated as coating agents in paper manufacturing, enhancing water barrier properties and mechanical attributes [85]. Nanocoated papers are now evolving into multifunctional materials with diverse applications from water and gas barrier papers for food packaging to pollutant-absorbing sheets for indoor pollutants, and printing packaging materials enhanced with nanoclays and SiO_2 nanoparticles [14, 31, 73]. Nevertheless, low levels and high aspect ratios of nanoparticles or nanoadditives can significantly influence paper properties, particularly for packaging material [55]. However, despite their potential, the small particle size that gives nanoadditives their unique properties also presents challenges in ensuring they are uniformly distributed and effectively embedded within the fiber matrix and also raises concerns about their potential toxicity and environmental impact [54, 90]. Nanotechnology has opened new frontiers for paper coating and printing by enabling enhanced optical, mechanical, and barrier properties. From improved print quality to specialized functional papers, the integration of nanomaterials is paving the way for innovative, high-performance, and sustainable paper products.

2 The Applications of Nanotechnology in Paper Coating

Nanotechnology is offering new advancements in papermaking by introducing materials and processes at the nanoscale to overcome traditional limitations of paper products. The complex structure of paper comprises cellulose molecules, fibers, their networks, and fibrils that play a critical role in determining key properties like (i) suitability for printing, which includes the capillary phenomenon of ink, its absorption, and drying, (ii) visible appearance in terms of gloss, surface smoothness, and color, (iii) barrier characteristics viz. absorption of water and infiltration of grease and gas, and (iv) operability in reference to friction and abrasion [3, 21, 26, 51]. Researchers have unlocked new development possibilities through the application of nanotechnology in various fields, including medical/biomedical, pharmaceutical, automobile, agriculture, construction, food packaging, renewable energy, and environmental protection. However, its application in papermaking is still in its infancy. While nanotechnology has demonstrated transformative potential across diverse industries, its integration into papermaking—particularly in the field of paper coating—offers significant advancements in both functionality and performance. Using nanoscale materials like nanoclays, nanosilica, and nanocellulose in coating formulations can boost paper products by making them smoother, better for printing, and with superior barrier properties against moisture, oils, and gases [36, 63]. Nanocoatings enable paper to replace traditional plastic or foil laminations, providing eco-friendly options for packaging and specialty paper uses. The integration of nanotechnology into paper coating not only boosts product quality but also meets the industry's growing need for sustainable and efficient materials.

Typical coatings often include substances like kaolin, calcium carbonate, titanium dioxide, polyethylene, polyvinyl alcohol, and fluorocarbon. While these coatings work well, they do pose some environmental issues, mainly because of poor biodegradability or recyclability. On the other hand, cellulose nanofibers (CNFs) are being recognized as an eco-friendly alternative to enhance the strength, barrier properties, and printing quality of paper [2, 60, 76]. Given these points, innovations in paper coatings that leverage nanotechnology serve as a vital link between traditional paper-making processes and the rising need for high-quality, eco-friendly paper products.

2.1 Cellulose-Based Nanomaterials

Recently, there has been a lot of interest in cellulose-based nanomaterials (CNMs), especially in paper coatings due to their large surface area, strong mechanical properties, and biodegradability. Among these, cellulose nanocrystals (CNCs) and cellulose nanofibrils (CNFs) are commonly used to enhance paper properties. In coatings, CNMs serve as functional additives and provide structural support. They help create barriers against water, grease, and gases, while keeping the coating recyclable and sustainable. The hydroxyl groups on CNMs form strong hydrogen bonds with the paper base and other coating components. This bonding results in coatings that are dense, even, and mechanically strong. Additionally, chemical modifications of CNMs can give them special features like hydrophobicity or antimicrobial activity, making them more useful for high-performance coated papers. Recent innovations in the field of CNMs open new pathways for their application as promising, eco-friendly alternatives to petroleum-based synthetic coatings in the papermaking industry.

In the work by [10], CNF coating was used that significantly reduced the permeability of air of paper by creating a high-density surface stratum. Studies also indicate that CNF combined with shellac reduced the permeability of water vapor, air, and oxygen [41]. CNF-clay composites improve ink absorption and print density, while CNF-based coatings decrease water permeability in coated paper [7, 22]. Composites of nanofibrillated cellulose and silver nanoparticles were prepared via the electrostatic-interaction-based assembly of silver nanoparticles (aqueous colloids) onto the nanofibrillated cellulose, which were then examined for their antibacterial feature toward microorganisms like *S. aureus* and *K. pneumoniae*. Subsequently, the antibacterial nanocomposites were employed as fillers in starch-based coating systems applied to paper substrates from *Eucalyptus globulus* [59]. A hybrid nanopaper was fabricated by introducing cellulose nanofibrils into the cellulose nanowhiskers matrix that achieved high optical transparency while maintaining iridescent characteristics under polarized light [94]. Despite these, the application of CNF for making paper is often constrained due to its extremely fine particle size, which can reduce retention efficiency. This limitation hinders its effective integration into the paper's three-dimensional fiber network, thereby affecting overall

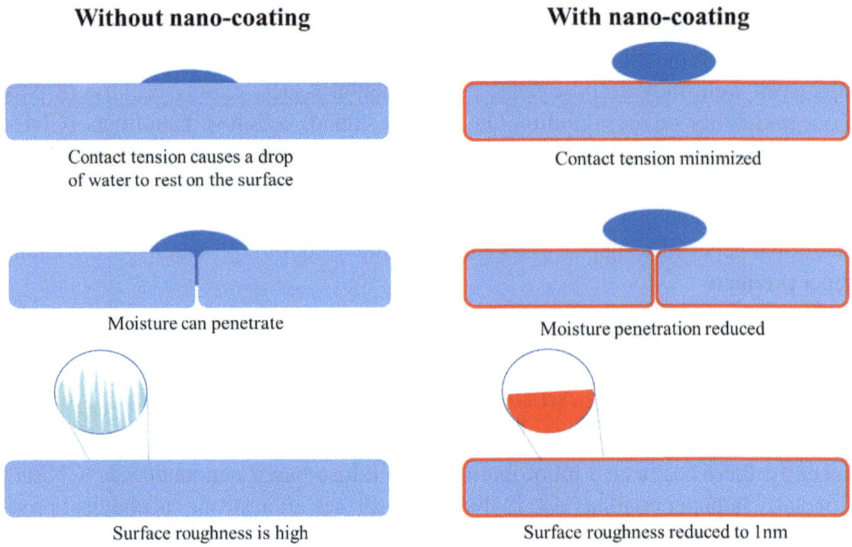

Fig. 2 Nanocoating versus conventional coating (Recreated from [61])

cost-efficiency [80]. Figure 2 illustrates the difference between conventional coatings and nanocoatings.

2.2 *Nanosilica and Other Inorganic Fillers*

Inorganic mineral fillers are one of the most vital components in the papermaking process, with fillers widely used to reduce production costs, lower energy demands, and improve optical as well as printing characteristics of paper. Common fillers include ground and precipitated $CaCO_3$ (GCC and PCC), hydrous and calcined kaolin, precipitated silica and silicate, talc, and TiO_2. While fillers improve brightness and opacity, especially in printing and packaging applications, however, it may also harm the interfiber bonding that occurs within the different strata of paper, decreasing the strength of the paper, decreasing sizing efficiency, poor filler retention, increased wire abrasion, and dusting, particularly at high addition levels [24]. To address these limitations, various preflocculation and filler modification strategies have been developed using both natural and synthetic polymers in addition to inorganic compounds and even surfactants or emulsions as chemicals. Furthermore, the incorporation of nanoparticles has enabled the introduction of novel functionalities in uncoated and coated papers, with key parameters being particle dimensions and distribution profile, length-to-width ratio, surface functionalization, and stabilization.

Adding nanosilica and other inorganic fillers like nano–calcium carbonate into coating formulations has led to considerable developments in improving paper's barrier properties, mechanical strength, and appearance. Nanosilica, with its large surface area and low particle size, makes the coating layer denser and smoother. Including nanosilica helps reduce air permeability, increase ink absorption uniformly, and enhance gloss, important for high-quality printing. Adding about 1–5% nanosilica to coatings can greatly improve print quality, without compromising brightness or opacity. Because of its hydrophilic surface, nanosilica also helps water-based ink adhesion, making it useful for both regular and special types of printing [46].

Nanostructured silica with a web-like structure, produced by carefully precipitating it from geothermal water, has been developed and used as a filler in newsprint paper. Similarly, nanostructured calcium silicate with a porous framework was created from sodium silicate solutions. Lab tests showed that incorporating these fillers into newsprint reduced ink absorption by approximately 40% and 53% for papers weighing 55 and 45 g/m^2, respectively [47]. The light scattering from precipitated calcium carbonate (PCC) fillers coated with novel silicate and zinc sulfide nanoparticles was studied, along with modifications where PCC was combined with aluminum-magnesium-silicate and zinc sulfide nanostructures. These modifications enhanced the paper's light scattering, which benefits pigment coating quality [49]. Eremeeva et al. [39] prepared a superhydrophobic paper surface by adding nano-TiO_2 to cellulose fibers. They mixed the nano-TiO_2 powder thoroughly using a high-speed homogenizer and then coated it with a coupling agent, 3-(trimethoxysilyl) propyl methacrylate. Results indicated water contact angles between about 127° and 154°, demonstrating superhydrophobicity. Precipitated calcium carbonate nanoparticles were also created for ultrathin, high-quality coated paper [70]. Several researchers prepared hydrophobic paper by spray-coating hydrophobic silica nanoparticles onto the paper surface [71, 83].

Silica nanoparticles and silica/calcium carbonate core–shell nanocomposites were produced by sol–gel process and incorporated as fillers in paper production applications. It was concluded that there was an increase in entrapment of silica nanoparticles, which incremented with increasing molar ratio of silica and calcium carbonate, because of the highly negatively charged surface that resulted in aggregation in the matrix, thereby enhancing the optical properties but slightly reducing the mechanical properties compared to the commercial PCC. On the other hand, [64] found that while mechanical properties did not significantly affect but hand sheet brightness, whiteness and opacity did improve compared to commercial PCC.

From a formulation standpoint, the use of nanosilica and inorganic fillers in paper coatings requires close attention to dispersion stability and binder compatibility. Also surface modification of nanosilica with coupling agent improves the compatibility with aqueous and polymeric binders like styrene-butadiene latex and polyvinyl alcohol. This step also plays a role in preventing filler agglomeration and sedimentation, which can compromise coating uniformity. Additionally, the use of inorganic fillers together, such as nanosilica with nanoclay, develops a synergetic effect, which in turn improves mechanical and barrier performance. Also, these multicomponent nanocoatings represent a very recent development in the field of sustainable and high-performance paper products.

2.3 Metal and Metal Oxide Nanoparticles

Metals and their oxide nanoparticles have appeared as functional additives in paper coatings, contributing improved barrier, optical, and antimicrobial properties. Their incorporation into paper coatings is typically achieved through dispersion in aqueous or solvent-based coating formulations, ensuring uniform distribution. The size and shape of these nanoparticles directly control their performance: smaller particles with higher surface area show more distinct effects due to better interaction with substrate. Zinc oxide nanoparticles (ZnO NPs) can be applied on paper surfaces without binders. Paper products coated with ZnO NPs exhibit antibacterial activity against *Escherichia coli*. ZnO NPs coating is done by initial ammonia application and heat treatment of zinc oxide and cellulose, followed by ultrasound treatment. The elimination of odor-causing bacteria and the neutralization of ammonia can be achieved by using nanosilver powder [17].

The latest work of researchers demonstrated that the brightness, whiteness, smoothness, print density and uniformity, picking velocity, oil absorption capacity, antifungal and UV-protecting properties of nano-ZnO-coated paper showed significant improvement in comparison to bulk ZnO-coated paper [77]. Researchers have also developed that the surface functionalization of paper with minimal concentrations of gold, silver, and titanium dioxide nanoparticles could yield devices with outstanding photocatalytic and antibacterial features [44]. Fe_2O_3 nanoparticles were synthesized and used as a filler to impart magnetic properties to paper made from cellulosic fibers [65]. A thin, narrow conductive coating ink made from nanosilver and organic polymer, polyaniline, has been developed for printing electronic applications [43, 102]. Cellulosic-TiO_2 nanocomposite was created by [4] and used as a protective coating for old manuscript papers.

2.4 Nanoclays

The Packaging Development Center at International Paper developed a technology for moisture and gas barriers in beverage packaging. This technology employs a nano clay blended coating for inkjet and digital printing paper [11]. Cellulose nanocrystals work similarly to nanoclays or carbon nanotubes, but they are readily available from renewable, recyclable natural resources. Fast-developing, biodegradable hydrophobic material film coatings made of cellulose latex nanocomposites have been used in packaging to enhance dimensional stability, hygro-expansivity, and toughness of sack paper, with potential applications in food packaging [35]. Yoon and Deng [95] employed a high dose of starch to modify clay and reported a 100 to 200% increase in the tensile strength of paper. Multiple studies support the proper application of nano pigments in papermaking [88]. These pigments are used in creating high-quality paper [53, 74]. Nanoclay and calcium carbonate-based coating color was developed to improve surface smoothness and air permeability of printing paper

[50]. Carbon nanotubes have also been used as fillers to produce electrically conductive paper [5, 6]. The concept of filler preflocculation using cationic and amphoteric starches in papermaking has been documented by the authors [16]. Additionally, the impact of particle size of fillers on the development of various paper properties has been explored [15, 18].

Despite the presence of nanopigments, only a few grades of paper currently incorporate nano–calcium carbonate, and a large-scale industrial process is yet to be developed [78]. The main challenges in applying nanopigments commercially are their poor dispersibility and the amount needed for use. Nano–calcium carbonate addresses these issues because of the availability of raw materials, along with its simple and low-cost production [89].

3 Nanotechnology in Printing Applications

Nanotechnology holds great significance across diverse industrial sectors focused on achieving superior quality, greater commercial prospects and better efficiency. An increase in the inclination of scientists in the nanometric domain is largely because nanomaterials exhibit better attributes in comparison to similar compositions with larger particles [18]. Over the past two decades, nanotechnology has been recognized as a crucial domain of development, facilitating the creation of novel nanomaterials with remarkable properties. Integrating these advancements with paper technology opens up new opportunities for the evolution of printing technologies. A wide variety of nanomaterials can be integrated into paper structures, enabling the development of paper products with innovative and enhanced properties.

The printing industry exemplifies how nanotechnology has driven progress to the forefront of technological capabilities, with the gradual integration of nanoparticles fostering industrial innovation and enhancing competitiveness [84]. Gumeniuk and Sherstyuk [34] stated that printing processes involve the transfer of material first onto the printing plate and subsequently from the plate to the substrates such as paper, plastics or metal. Both the surface of the printed material and the substances being transferred, i.e., printing inks, form a dispersed system. Depending on the technique, printing can occur either through direct contact or without contact between the printing elements and the substrate.

Nanotechnology has revolutionized paper printing by enhancing its performance, quality and functionality through the incorporation of various types of nanomaterials such as nanoinks, inorganic nanofillers, nano-pigments, cellulosic nanomaterials among others as discussed in the subsequent sections and shown in Fig. 3. Nanoparticles can significantly enhance paper surface properties, improve ink adhesion and minimize smudging for sharper prints. Nanopigments provide greater color intensity, brightness, and vibrancy, resulting in superior print quality. Nanotechnology also enables the development of conductive inks, paving the way for printed electronics. Additionally, nanoinks and coatings help reduce ink consumption and

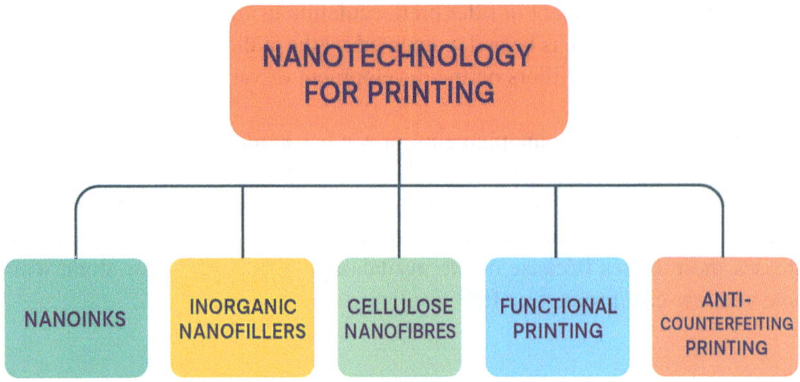

Fig. 3 Flowchart representing the various branches associated with nanotechnology-based printing

lower energy usage during printing, promoting sustainability. Overall, nanotechnology is transforming the printing industry by combining performance, innovation, and environmental responsibility.

3.1 Nanoinks

Nanotechnology-based inks have presented an entirely new dimension to printing, offering innovative possibilities that are reshaping traditional printing methods. The nanoinks amalgamate the appearance of both water-based and oil-based ink, thereby proving to be better than traditional inks in terms of their efficiency, array of substrates, print quality, and cost-effectiveness. They are fabricated from nanopigments, which exhibit physical characteristics comparable to those of traditional inks but offer a broader color spectrum due to the alignment of nanoscale pigments from 20 to 200 nm in size [23]. Nanoinks are characterized by their active hues and beyond color range, nanoinks offer enhanced durability and reliability compared to traditional press inks. Inorganic inks and nanomaterials offer significant advantages over organic semiconductors, particularly in terms of enhanced durability and superior performance. The development of inks that accumulate conductive or semiconductive nanomaterials—such as nanoparticles, nanowires, and nanotubes—is crucial, as these abstracts serve as the basal components for a deluge of electronic devices. Their affiliation enables the abundant potential of printing technologies to be acclimatized for the artifact of electronic accessories [8]. Three-dimensional (3D) printing is an emerging technology in electronic printing, having advantages such as low cost, high speed, and high precision. It presents a sustainable alternative to conventional printing methods, which are generally expensive and rely on organic inks that contribute to ecological pollution. Recently, there has been growing interest in accumulating nanomaterials into 3D printing, particularly through the development of nanostructured dispersions, known as nanoinks [12].

3.2 *Inorganic Nanofillers*

The mineral fillers enhance the paper properties such as brightness, opacity and print quality in addition to having lower costs than wood fibers. Nano-GCC, nano-PCC, nanoclay, nanokaolin, nano-SiO_2, nanotalc, nano-TiO_2, and nanosilicates are the frequently used inorganic nanofillers in papermaking. Various pigments affect light scattering differently, playing a crucial role in determining opacity, brightness and print quality. In a study carried out by [70], coating color made of nanoclay and calcium carbonate was used to lessen the permeability of air and to ameliorate the printing attributes and surface smoothness of paper. Clay nanopaper's production is considered environmentally conscious and remarkably simple, since it uses aqueous processes and mainly depends on the physical dispersion and arrangement of nanoparticles.

Nano-based calcium silicates were investigated as fillers in a study for 45–55 gsm newsprint, where its opacity increased [57, 62] by about 3.7 points and 1.3 points for the 45 gsm newsprint at 2 wt% loading and 55 gsm newsprint respectively [27]. This proved that nano–calcium silicate increased the opacity of the newsprint [57]. The nanostructured framework provides excellent light-scattering properties, leading to enhanced opacity and consequently, a remarkable improvement in print quality [92]. Nano–calcium silicate platelets stack together to build open mesh frameworks that allow rapid absorption of printing ink, effectively preventing show-through on the reverse face of paper [86]. This structure enhances the overall print quality by ensuring precise application of ink dots and minimizing unwanted ink spreading [87].

In recent decades, ZnO has emerged as one of the most important and versatile inorganic nanomaterials in the fields of electronics and photonics. Growing concern has also been geared at the usage of ZnO-based inks and their applications in advanced printing technologies [8]. Nano-ZnO and hybrid nano-ZnO are promising pigment materials for paper coating, known for enhancing brightness and overall printing performance. In the work by [45], the incorporation of nano-ZnO notably improved key paper properties, including whiteness, brightness, smoothness, print density and uniformity, picking velocity and oil sorption capacity, with optimal results achieved at specific nano-ZnO concentrations. Another breakthrough in the field of nanotechnology for paper printing is the introduction of nanosilica, which results in the improvement of optical characteristics and reduction of print-through of paper [25]. Gu et al. [33] reported that even less than 2 wt% addition of nanosilica reduced print-through by approximately 30% while increasing both opacity and brightness by about 1.7 points each, demonstrating a remarkable improvement. Additionally, the use of nanosilica resulted in a significant 40% reduction in print-through at an approximate 2 wt% addition for yellow directory papers [48]. Several studies have compared coated and uncoated nanosilica, affirming that coated nanosilica offers superior print quality, higher optical density, a broader color spectrum, improved water resistance, enhanced color, and better dimensional stability compared to its uncoated counterpart [9]. Nanostructured silica, synthesized through controlled precipitation from geothermal water, has been developed and utilized as a bleaching substance. Its application led to a significant reduction in print-through, achieving a 30% reduction for newsprint and a 40% reduction in full-scale paper mill production [67].

3.3 Cellulosic Nanofibers and Nanomaterials

The incorporation of nanocelluloses (NCs) into papermaking holds great potential for significantly enhancing paper quality. Nanocelluloses offer several key advantages, including greater surface area, exceptional optical attributes, impressive stiffness, low density, and superior mechanical strength, in addition to biodegradability, renewability, and environmental sustainability, which have sparked considerable interest among both researchers and industry stakeholders seeking greener solutions. Despite these promising benefits, employing cellulose nanofibrils (CNFs) and cellulose nanocrystals (CNCs) in the papermaking industry also presents several challenges. These include (1) poor retention of nanocelluloses within the fibrous matrix during paper formation; (2) relatively high production costs compared to conventional wet as well as dry strength additives; and (3) adverse effects on drainage, which slows down the paper manufacturing process by increasing consumption of energy and reducing machinery efficiency [72]. A balance between these advantages and limitations remains a key focus for advancing the practical application of nanocelluloses in the paper industry. The hierarchical and microscopic images of wood cellulosic fibers are depicted in Fig. 4.

Bacterial cellulose (BC), Cellulose nanofibers (CNFs), and cellulose nanocrystals (CNCs) represent three primary types of nanocelluloses (NCs), each distinguished by their unique dimensions, characteristics, functional roles, and methods of production [12]. The images of cellulose nanocrystals and nanoparticles are shown in Fig. 5. Compared to conventional wood-pulp-based papers, cellulose nanopapers possess significantly smoother surfaces, making them highly suitable substrates for inkjet printing applications [58]. The addition of clay or calcium carbonate pigments plays an important part in strengthening the distribution of CNF on the surface of the wet paper web, which in turn markedly improves both the physical strength and printability of the paper. For instance, [79] demonstrated that CNF could be effectively retained on the surface of copy paper by applying a porous precoating layer made of modified calcium carbonate, resulting in improved paper performance.

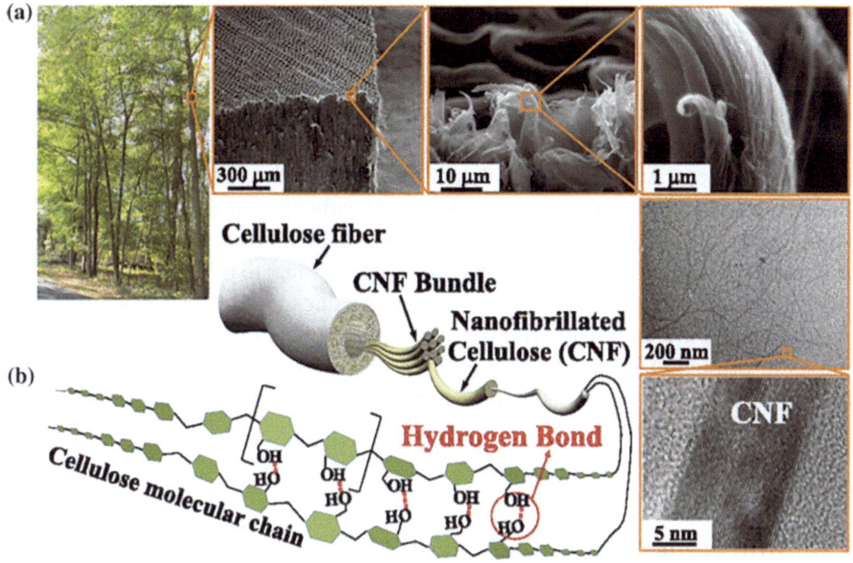

Fig. 4 Hierarchical and microscopic images of cellulose fibers (*Source* [82])

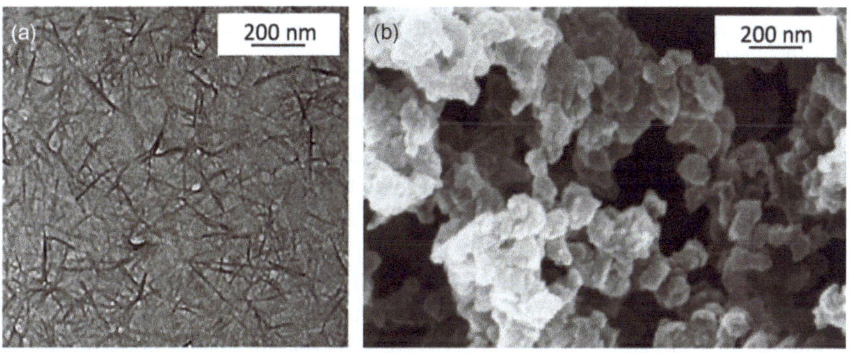

Fig. 5 Pictures of micrographs of cellulose nanocrystals (left) and cellulose nanoparticles (right) (*Source* [81])

3.4 Functional Printing: Conductive Inks for Printed Electronics

The advancement of accomplishment techniques that enable the transformation of functional nanomaterials into high-performance accessories in a scalable, absolute, and cost-effective manner presents significant analysis opportunities and challenges for the advancing decades. Additive manufacturing methods, including 2D and 3D printing, have emerged as effective strategies for architectural anatomic accessories from nanoscale apparatus attributable to their flexibility, accelerated prototyping capabilities, and ability to meet assorted artifact needs [97]. Printing technology offers a cost-effective way to apply conductive inks by dispersing metal nanomaterials in oil- or water-based solvents. These inks are advised to display an adorable backdrop, such as acceptable dispersibility, optimal fluidity, acceptable viscosity, and high solid content. An array of metal nanomaterials, such as gold (Au), silver (Ag), and copper (Cu), has been explored for bearing conductive inks. In a study by [101], direct ink writing technology based on nano-Ag conductive ink, which was primarily composed of Ag nanoparticles and adhesive, was activated to print circuit patterns on paper substrates. Cu is an alternative to Ag and Au due to its accomplished electrical application and low cost, authoritative it an ideal applicant for all-embracing applications in printed electronics. In an attempt, Ag–Cu conductive ink was prepared through a stepwise feeding method in order to disperse nano-Ag and nano-Cu in ethanol and water [101].

3.5 Anticounterfeiting Application Using Nano-Based Security Press Techniques

Security printing is based on patterns that are commonly accustomed yet are difficult to replicate, making them resistant to forgery. Since recognition is typically done through visual inspection, security printing often incorporates elements with distinct optical properties, the examples of which include holograms, colored fibers, and watermarks [19]. With the rise in global counterfeiting, a wide range of documents such as currency notes, passports, tickets, and academic authentication papers and certificates have become frequent targets [100]. To combat this, increasingly advanced techniques such as specialized security inks and papers are being developed to enhance document protection [68].

Crystalline nanocellulose, derived from one of the most abundant natural polymers, that is, cellulose, has emerged as a promising optical nanomaterial due to its unique structure and eco-friendly properties. Its polymer dispersions are widely being utilized as birefringent inks. When printed on dark paper, the text displays an interesting optical effect where it looks darker than the background under normal viewing, but appears brighter when viewed through crossed polarizers. This contrast inversion feature makes it a promising approach for enhancing security in printed materials [99]. This study explores the use of inkjet printing to fabricate cellulose nanocrystal configurations having adjustable optical characteristics, controlled by the caliper of the printed strata. One important aspect is developing patterns that can only be visible when using linearly polarized light. Eremeeva et al. [29] showed that inkjet printing can accurately create CNC patterns that are optically active and have adjustable optical properties. These findings are really important for applications where precise color visibility is crucial, such as in anticounterfeiting technology, decorative printing tasks, and color photo filters.

4 Environmental Dimensions of Printing and Coating Innovations

The use of nanotechnology in paper coating and printing offers both benefits and challenges for the environment. As the industry pushes for greener practices, nanomaterials can improve product performance while promising smaller ecological footprints. However, the long-term environmental effects of these materials, especially metal-based or synthetic nanoparticles, require careful and ongoing research. A coating with nanostructures can reduce the need for synthetic polymers, dyes, and large mineral fillers that have long dominated paper finishing. For example, adding nanocellulose or nanosilica can enhance barrier properties without relying on fluorocarbons or petroleum derivatives that persist in ecosystems [40]. In printing, nanopigments and nanoinks improve color brightness, increase contrast, and reduce ink usage—all while decreasing waste [52]. Their small size allows for more efficient dispersion, enabling higher brightness with less product. This also leads to lower ink consumption and less waste during printing [28]. Additionally, lighter ink coverage makes deinking easier, enhancing recycling efficiency and purity.

While nanotechnology offers many promising advantages, it also raises important environmental health and safety concerns. Materials such as nano-Ag, nano-TiO_2, and various carbon-based nanomaterials could pose risks to aquatic and terrestrial life if they are released into the environment during manufacturing, use, or disposal [69]. Studies indicate that nano-Ag can be harmful to freshwater organisms, including invertebrates and fish, potentially threatening aquatic biodiversity [30]. Similarly, nano-TiO_2, commonly used in coatings, can produce reactive oxygen species when exposed to sunlight, which may cause oxidative stress in microbes and other microorganisms [42]. Therefore, it is essential to thoroughly assess environmental impacts—particularly with regard to sustainability metrics highlighted by life cycle assessments (LCAs)—and to take proactive measures. Moving forward, as research progresses, it is crucial to focus on designing nanotechnologies that are not only effective but also safe, environmentally sustainable, and aligned with long-term ecological responsibility.

5 Conclusion and Future Outlook

In the pulp and paper industry, nanotechnology offers revolutionary possibilities, especially in the areas of coating and printing. Incorporating nanomaterials and nanoadditives into paper products provides clear benefits, including better adsorption, smoother surfaces, improved appearance, and customized functionalities. These advancements not only improve paper quality but also help the industry achieve goals of cost efficiency, energy savings, and sustainability. Recent progress in nanotechnology enables the creation of functionalized paper with qualities that are difficult to achieve with traditional materials, such as enhanced optical properties, increased barrier strength, and improved printability. Understanding how nanomaterials interact with the cellulose matrix, functional additives, and processing conditions is key to making these innovations successful. Factors like the synthesis method, structure, and shape of nanomaterials, along with their compatibility with the paper substrate and deposition techniques, significantly influence performance. While the use of inorganic and carbon-based nanomaterials has grown considerably, research into organic and bio-based nanomaterials is expanding rapidly, with great potential for developing versatile, eco-friendly, and biocompatible paper products. To overcome technical challenges and fully unlock nanotechnology's potential in papermaking, collaboration among researchers, industry leaders, and policymakers will be essential as the field progresses. The next wave of high-performance, environmentally friendly paper products will be driven by nanotechnology through ongoing innovation and interdisciplinary efforts.

Competing Interests The authors declare that they have no competing interest.

Declaration of Generative AI and AI-Assisted Technologies in the Writing Process During the preparation of this work, the book chapter author(s) have not taken any assistance from AI, and the book chapter author(s) reviewed and edited the content as needed and take(s) full responsibility for the content of the publication.

References

1. Abdollahi A, Alidaei-Sharif H, Roghani-Mamaqani H, Herizchi A (2020) Photoswitchable fluorescent polymer nanoparticles as high-security anticounterfeiting materials for authentication and optical patterning. J Mater Chem C 8(16):5476–5493. https://doi.org/10.1039/d0tc00937g
2. Afra E, Yousefi H, Lakani SA (2014) Properties of chemi-mechanical pulp filled with nanofibrillated and microcrystalline cellulose. J Biobased Mater Bioenergy 8(5):489–494. https://doi.org/10.1166/jbmb.2014.1462
3. Afra E, Yousefi H, Hadilam MM, Nishino T (2013) Comparative effect of mechanical beating and nanofibrillation of cellulose on paper properties made from bagasse and softwood pulps. Carbohyd Polym 97(2):725–730. https://doi.org/10.1016/j.carbpol.2013.05.032
4. Afsharpour M, Rad FT, Malekian H (2011) New cellulosic titanium dioxide nanocomposite as a protective coating for preserving paper-art-works. J Cult Herit 12(4):380–383. https://doi.org/10.1016/j.culher.2011.03.001
5. Agarwal M, Xing Q, Shim BS, Kotov N, Varahramyan K, Lvov Y (2009) Conductive paper from lignocellulose wood microfibers coated with a nanocomposite ofcarbon nanotubes and conductive polymers. Nanotechnology 20(21):215602. https://doi.org/10.1088/0957-4484/20/21/215602
6. Anderson RE, Guan J, Ricard M, Dubey G, Su J, Lopinski G, Dorris G, Bourne S, Simard B (2010) Multifunctional single-walled carbon nanotube–cellulose composite paper. J Mater Chem 20(12):2400–2407.https://doi.org/10.1039/B924260K
7. Andrade R, Skurtys O, Osorio F, Zuluaga R, Gañán P, Castro C (2014) Wettability of gelatin coating formulations containing cellulose nanofibers on banana and eggplant epicarps. LWT-Food Sci Technol 58(1):158–165. https://doi.org/10.1016/j.lwt.2014.02.034

8. Arrabito G, Aleeva Y, Pezzilli R, Ferrara V, Medaglia PG, Pignataro B, Prestopino G (2020) Printing ZnO inks: from principles to devices. Crystals 10(6):449. https://doi.org/10.3390/cryst10060449
9. Atik C, Ates S (2012) Mass balance of silica in straw from the perspective of silica reduction in straw pulp. BioResources 7(3):3274–3282
10. Aulin C, Gällstedt M, Lindström T (2010) Oxygen and oil barrier properties of microfibrillated cellulose films and coatings. Cellulose 17:559–574. https://doi.org/10.1007/s10570-009-9393-y
11. Azizi Samir MAS, Alloin F, Gorecki W, Sanchez JY, Dufresne A (2004) Nanocomposite polymer electrolytes based on poly (oxyethylene) and cellulose nanocrystals. J Phys Chem B 108(30):10845–10852. https://doi.org/10.1021/jp0494483
12. Balea A, Fuente E, Monte MC, Merayo N, Campano C, Negro C, Blanco A (2020) Industrial application of nanocelluloses in papermaking: a review of challenges, technical solutions, and market perspectives. Molecules 25(3):526. https://doi.org/10.3390/molecules25030526
13. Bayda S, Adeel M, Tuccinardi T, Cordani M, Rizzolio F (2019) The history of nanoscience and nanotechnology: from chemical–physical applications to nanomedicine. Molecules 25(1):112. https://doi.org/10.3390/molecules25010112
14. Bumbudsanpharoke N, Ko S (2019) Nanoclays in food and beverage packaging. J Nanomater 2019(1):8927167. https://doi.org/10.1155/2019/8927167
15. Chauhan VS, Bhardwaj NK (2012) Effect of particle size of talc filler on structural and optical properties of paper. Lignocellulose 1(3):241–259
16. Chauhan VS, Bhardwaj NK (2013) Effect of particle size and preflocculation of talc filler on sizing characteristics of paper. Appita: Technol Innov Manuf Environ 66(1):66–72. https://doi.org/10.3316/informit.083962724504325
17. Chauhan VS, Chakrabarti SK (2012) Use of nanotechnology for high performance cellulosic and papermaking products. Cellul Chem Technol 46(5):389–400
18. Chauhan VS, Bhardwaj NK, Chakrabarti SK (2013) Effect of particle size of magnesium silicate filler on physical properties of paper. Canadian J Chem Eng 91(5):855–861. https://doi.org/10.1002/cjce.21708
19. Chindawong C, Johannsmann D (2014) An anisotropic ink based on crystalline nanocellulose: potential applications in security printing. J Appl Polymer Sci 131(22). https://doi.org/10.1002/app.41063
20. Costa SV, Gonc¸alves AS, Zaguete MA, Mazonb T, Nogueira AF (2013) ZnO nanostructures directly grown on paper and bacterial cellulose substrates without any surface modification layer. Chem Commun 49(73):8096–8098. https://doi.org/10.1039/C3CC43152E
21. Despond S, Espuche E, Cartier N, Domard A (2005) Barrier properties of paper–chitosan and paper–chitosan–carnauba wax films. J Appl Polym Sci 98(2):704–710. https://doi.org/10.1002/app.21754
22. Dimic-Misic K, Ridgway C, Maloney T, Paltakari J, Gane P (2014) Influence on pore structure of micro/nanofibrillar cellulose in pigmented coating formulations. Transp Porous Media 103:155–179. https://doi.org/10.1007/s11242-014-0293-8
23. Dondi M, Blosi M, Gardini D, Zanelli C (2012) Ceramic pigments for digital decoration inks: an overview. In: Ceramic forum international (Vol 89, No 8–9, pp E59-E64). Castellon, Spain: Bauverlag GmbH
24. Dong C, Song D, Patterson T, Ragauskas A, Deng Y (2008) Energy saving in papermaking through filler addition. Ind Eng Chem Res 47(21):8430–8435. https://doi.org/10.1021/ie8011159
25. Duarte K, Justino CI, Pereira R, Panteleitchouk TS, Freitas AC, Rocha-Santos TA, Duarte AC (2013) Removal of the organic content from a bleached kraft pulp mill effluent by a treatment with silica-alginate-fungi biocomposites. J Environ Sci Health Part A 48(2):166–172. https://doi.org/10.1080/03601234.2012.716745
26. Dury-Brun C, Jury V, Guillard V, Desobry S, Voilley A, Chalier P (2006) Water barrier properties of treated-papers and application to sponge cake storage. Food Res Int 39(9):1002–1011. https://doi.org/10.1016/j.foodres.2006.07.003

27. El-Sherbiny S, El-Sheikh SM, Barhoum A (2015) Preparation and modification of nano calcium carbonate filler from waste marble dust and commercial limestone for papermaking wet end application. Powder Technol 279:290–300. https://doi.org/10.1016/j.powtec.2015.04.006
28. Enríquez E, Reinosa JJ, Fuertes V, Fernández JF (2022) Advances and challenges of ceramic pigments for inkjet printing. Ceram Int 48(21):31080–31101
29. Eremeeva E, Sergeeva E, Neterebskaia V, Morozova S, Kolchanov D, Morozov M,... Vinogradov A (2020) Printing of colorful cellulose nanocrystalline patterns visible in linearly polarized light. ACS Appl Mater Interfaces 12(40):45145–45154. https://doi.org/10.1021/acsami.0c11846
30. Fabrega J, Luoma SN, Tyler CR, Galloway TS, Lead JR (2011) Silver nanoparticles: behaviour and effects in the aquatic environment. Environ Int 37(2):517–531
31. Gaikwad KK, Ko S (2015) Overview on in polymer-nano clay composite paper coating for packaging application. J Mater Sci Eng 4(1):1000151. https://doi.org/10.4172/2169-0022.1000151
32. Gill RA (1995) Fillers for papermaking. In Applications of wet-end paper chemistry (pp 54–75). Dordrecht: Springer Netherlands. https://doi.org/10.1007/978-94-017-0756-5_4
33. Gu WJ, Li Y, Zhang XH (2012) Preparation of silica hollow spheres used as filler for paper. Appl Mech Mater 200:339–342. https://doi.org/10.4028/www.scientific.net/AMM.200.339
34. Gumeniuk O, Sherstyuk V (2010) Nanotechnologies in Printing and Packaging (Vol 3, pp 32–41). IC
35. Helbert W, Cavaille JY, Dufresne A (1996) Thermoplastic nanocomposites filled with wheat straw cellulose whiskers. Part I: processing and mechanical behavior. Polymer Compos 17(4):604–611. https://doi.org/10.1002/pc.10650
36. Kasmani JE, Samariha A (2025) Impact of nano-silica, cationic polyacrylamide, and cationic starch on long fiber utilization in recycled paper production. BioResources 20(3):5561–5574
37. Statista (2023) Leading pulp for paper producing countries worldwide in 2023. https://www.statista.com/statistics/1333386/pulp-for-paper-production-by-country/. Accessed 17 May 2025
38. Statista (2023) Market size of paper and pulp industry worldwide in 2021, with an estimated figure for 2022 and a forecast for 2029. https://www.statista.com/statistics/1073451/global-market-value-pulp-and-paper/. Accessed 7 June 2025
39. Huang L, Chen K, Lin C, Yang R, Gerhardt RA (2011) Fabrication and characterization of superhydrophobic high opacity paper with titanium dioxide nanoparticles. J Mater Sci 46:2600–2605. https://doi.org/10.1007/s10853-010-5112-1
40. Hubbe MA, Ferrer A, Tyagi P, Yin Y, Salas C, Pal L, Rojas OJ (2017) Nanocellulose in thin films, coatings, and plies for packaging applications: a review. BioResources 12(1):2143–2233
41. Hult EL, Iotti M, Lenes M (2010) Efficient approach to high barrier packaging using microfibrillar cellulose and shellac. Cellulose 17:575–586. https://doi.org/10.1007/s10570-010-9408-8
42. Hund-Rinke K, Simon M (2006) Ecotoxic effect of photocatalytic active nanoparticles (TiO_2) on algae and daphnids (8 pp). Environ Sci Pollut Res 13(4):225–232
43. Ihalainen P, Maattanen A, Jarnstrom J, Tobjork D, Osterbacka R, Peltonen J (2012) Influence of surface properties of coated papers on printed electronics. Ind Eng Chem Res 51(17):6025–6036. https://doi.org/10.1021/ie202807v
44. Imani R, Talaiepour M, Dutta J, Ghobadinezhad MR, Hemmasi AH, Nazhad MM (2011) Production of antibacterial filter paper from wood cellulose. BioResources 6(1):891–900. https://doi.org/10.15376/biores.6.1.891-900
45. Jang SD, Maniruzzaman M, Yun GY, Kim J (2013) Hybrid nanocomposites made with cellulose and ZnO nanoparticles and its biosensing application. In: Nanosensors, biosensors, and info-tech sensors and systems 2013 (Vol 8691, pp 169–173). SPIE. https://doi.org/10.1117/12.2010225
46. Jasmani L, Jami M, Ariffin KS, Nayan NHM, Hassan Z (2020) Application of nanotechnology in wood-based products industry: a review. Nanoscale Res Lett 15:207

47. Johnston JH, McFarlane AJ, Borrmann T, Moraes J (2004) Nano-structured silicas and silicates—new materials and their applications in paper. Curr Appl Phys 4(2–4):411–414. https://doi.org/10.1016/j.cap.2003.11.061
48. Julkapli NM, Bagheri S (2016) Developments in nano-additives for paper industry. J Wood Sci 62:117–130. https://doi.org/10.1007/s10086-015-1532-5
49. Juuti M, Koivunen K, Silvennoinen M, Paulapuro H, Peiponen KE (2009) Light scattering study from nanoparticle-coated pigments of paper. Colloids Surf A 352(1–3):94–98. https://doi.org/10.1016/j.colsurfa.2009.10.006
50. Kasmani JE, Mahdavi S, Alizadeh A, Nemati M, Samariha A (2013) Physical properties and printability characteristics of mechanical printing paper with LWC. BioResources 8(3):3646–3656
51. Kettle J, Lamminmäki T, Gane P (2010) A review of modified surfaces for high speed inkjet coating. Surf Coat Technol 204(12–13):2103–2109. https://doi.org/10.1016/j.surfcoat.2009.10.035
52. Khan HR, Khan FS, Syed AS, Akhtar J (2023) Nano-inks and their applications in packaging industries. In: Smart multifunctional nano-inks, pp 687–698. Elsevier
53. Koivunen K, Niskanen I, Peiponen KE, Paulapuro H (2009) Novel nanostructured PCC fillers. J Mater Sci 44:477–482. https://doi.org/10.1007/s10853-008-3095-y
54. Lee JY, Kim YH, Lee SR, Kim CH, Joo Sung Y, Lim GB, Kim SY, Kim JS, Park JH (2013) Effect of precipitated calcium carbonate on paper properties and drying energy reduction of duplex-board. J Korea Techn Assoc Pulp Paper Indus 45(6):24–29. https://doi.org/10.7584/ktappi.2013.45.6.024
55. Lei SG, Hoa SV, Ton-That MT (2006) Effect of clay types on the processing properties of polypropylene nanocomposites. Compos Sci Technol 66(10):1274–1279. https://doi.org/10.1016/j.compscitech.2005.09.012
56. Liu QX, Yin YN, Xu WC (2013) Study on application of hydrated calcium silicate in paper from wheat straw pulp. Adv Mater Res 774:1277–1280. https://doi.org/10.4028/www.scientific.net/AMR.774-776.1277
57. Liu QX, Yin YN, Xu WC (2013) Characteristics of hydrated calcium silicate and paper filler. Appl Mech Mater 395:577–581. https://doi.org/10.4028/www.scientific.net/AMM.395-396.577
58. Luu WT, Bousfield DW, Kettle J (2011) Application of nano-fibrillated cellulose as a paper surface treatment for inkjet printing. In: Paper conference and trade show, PaperCon 2011, pp 1152–1163. TAPPI Press
59. Martins NCT, Freire CSR, Pinto RJB, Fernandes SCM, Pascoal Neto C, Silvestre AJD, Trindade T (2012) Electrostatic assembly of Ag nanoparticles onto nanofibrillated cellulose for antibacterial paper products. Cellulose 19(4):1425–1436. https://doi.org/10.1007/s10570-012-9713-5
60. Mashkour M, Afra E, Resalati H, Mashkour M (2015) Moderate surface acetylation of nanofibrillated cellulose for the improvement of paper strength and barrier properties. RSC Adv 5(74):60179–60187. https://doi.org/10.1039/C5RA08161K
61. Mathiazhagan A, Joseph R (2011) Nanotechnology-a New prospective in organic coating-review. Int J Chem Eng Appl 2(4):225
62. Mathur V, Brennand B (2013) Novel silicate nano-fibers and super nano carbonates for dematerialization—basis weight reduction. In: Paper conference and trade show, PaperCon (Vol 2, pp 1092–1118)
63. Mazega A, Tarrés Q, Aguado R, Pèlach MÀ, Mutjé P, Ferreira PJ, Delgado-Aguilar M (2022) Improving the barrier properties of paper to moisture, air, and grease with nanocellulose-based coating suspensions. Nanomaterials 12(20):3675
64. Morsy FA, El-Sheikh SM, Barhoum A (2019) Nano-silica and $SiO_2/CaCO_3$ nanocomposite prepared from semi-burned rice straw ash as modified papermaking fillers. Arab J Chem 12(7):1186–1196. https://doi.org/10.1016/j.arabjc.2014.11.032
65. Munawar RF, Zakaria S, Radiman S, Hua CC, Abdullah M, Yamauchi T (2010) Properties of magnetic paper prepared via in situ synthesis method. Sains Malaysiana 39(4):593–598

66. Nayak L, Mohanty S, Nayak SK, Ramadoss A (2019) A review on inkjet printing of nanoparticle inks for flexible electronics. J Mater Chem C 7(29):8771–8795. https://doi.org/10.1039/C9TC01630A
67. Nelli LM (2007) Improving inkjet print performance with fumed silica at the size press. In: NIP & digital fabrication conference (Vol 23, pp 603–607). Society of Imaging Science and Technology. https://doi.org/10.2352/ISSN.2169-4451.2007.23.1.art00024_2
68. Ngo YH, Li D, Simon GP, Garnier G (2011) Paper surfaces functionalized by nanoparticles. Adv Coll Interface Sci 163(1):23–38. https://doi.org/10.1016/j.cis.2011.01.004
69. Nowack B, Bucheli TD (2007) Occurrence, behavior and effects of nanoparticles in the environment. Environ Pollut 150(1):5–22
70. Nypelo T, Osterberg M, Laine J (2011) Tailoring surface properties of paper using nanosized precipitated calcium carbonate particles. ACS Appl Mater Interfaces 3(9):3725–3731. https://doi.org/10.1021/am200913t
71. Ogihara H, Xie J, Okagaki J, Saji T (2012) Simple method for preparing superhydrophobic paper: spray-deposited hydrophobic silica nanoparticle coatings exhibit high water-repellency and transparency. Langmuir 28(10):4605–4608. https://doi.org/10.1021/la204492q
72. Oksman K, Moon RJ (2014) Characterization of nanocomposites structure. In: HANDBOOK OF GREEN MATERIALS: 2 Bionanocomposites: processing, characterization and properties (pp 89–105). https://doi.org/10.1142/9789814566469_0021
73. Ozcan A, Dogan T (2020) The effect of zeolite on inkjet coated paper surface properties and deinking. Nord Pulp Pap Res J 35(3):432–439. https://doi.org/10.1515/npprj-2020-0025
74. Park JK, Kim JK, Kim HK (2007) TiO_2–SiO_2 composite filler for thin paper. J Mater Process Technol 186(1–3):367–369. https://doi.org/10.1016/j.jmatprotec.2006.12.004
75. Pelton R, Geng X, Brook M (2006) Photocatalytic paper from colloidal TiO_2—fact or fantasy. Adv Coll Interface Sci 127(1):43–53. https://doi.org/10.1016/j.cis.2006.08.002
76. Petroudy SRD, Syverud K, Chinga-Carrasco G, Ghasemain A, Resalati H (2014) Effects of bagasse microfibrillated cellulose and cationic polyacrylamide on key properties of bagasse paper. Carbohyd Polym 99:311–318. https://doi.org/10.1016/j.carbpol.2013.07.073
77. Prasad V, Shaikh AJ, Kathe AA, Bisoyi DK, Verma AK, Vigneshwaran N (2010) Functional behaviour of paper coated with zinc oxide–soluble starch nanocomposites. J Mater Process Technol 210(14):1962–1967. https://doi.org/10.1016/j.jmatprotec.2010.07.009
78. Puurunen K, Vasara P (2007) Opportunities for utilising nanotechnology in reaching near-zero emissions in the paper industry. J Clean Prod 15(13–14):1287–1294. https://doi.org/10.1016/j.jclepro.2006.07.013
79. Ridgway CJ, Gane PA (2012) Constructing NFC-pigment composite surface treatment for enhanced paper stiffness and surface properties. Cellulose 19:547–560. https://doi.org/10.1007/s10570-011-9634-8
80. Salas C, Nypelö T, Rodriguez-Abreu C, Carrillo C, Rojas OJ (2014) Nanocellulose properties and applications in colloids and interfaces. Curr Opin Colloid Interface Sci 19(5):383–396. https://doi.org/10.1016/j.cocis.2014.10.003
81. Samyn P, Barhoum A (2018) Engineered nanomaterials for papermaking industry. In: Fundamentals of nanoparticles (pp 245–277). Elsevier. https://doi.org/10.1016/B978-0-323-51255-8.00009-4
82. Samyn P, Barhoum A, Öhlund T, Dufresne A (2018) Nanoparticles and nanostructured materials in papermaking. J Mater Sci 53(1):146–184. https://doi.org/10.1007/s10853-017-1525-4
83. Samyn P, Schoukens G, Kiekens P, Mast P, Abbeele HVD, Stanssens D, Vonck L (2010) Thermal resistance of organic nanoparticle coatings for hydrophobicity and water repellence of paper substrates. AUTEX Res J 10(4):100–109. https://doi.org/10.1515/aut-2010-100404
84. Sanchez C, Belleville P, Popall M, Nicole L (2011) Applications of advanced hybrid organic–inorganic nanomaterials: from laboratory to market. Chem Soc Rev 40(2):696–753. https://doi.org/10.1039/C0CS00136H
85. Saroha V, Dutt D, Bhowmick A (2019) PVOH modified nano-kaolin as barrier coating material for food packaging application. AIP Conf Proc 2162(1):020001. https://doi.org/10.1063/1.5130211

86. Shen J, Song Z, Qian X, Liu W (2009a) Modification of precipitated calcium carbonate filler using sodium silicate/zinc chloride based modifiers to improve acid-resistance and use of the modified filler in papermaking. Bioresources 4(4):1498–1519. https://doi.org/10.15376/biores.4.4.1498-1519
87. Shen J, Song Z, Qian X, Liu W (2009b) A preliminary investigation into the use of acid-tolerant precipitated calcium carbonate fillers in papermaking of deinked pulp derived from recycled newspaper. BioResources 4(3):1178–1189. https://doi.org/10.15376/biores.4.3.1178-1189
88. Shen J, Song Z, Qian X, Ni Y (2011) A review on use of fillers in cellulosic paper for functional applications. Ind Eng Chem Res 50(2):661–666. https://doi.org/10.1021/ie1021078
89. Shen J, Song Z, Qian X, Yang F, Kong F (2010) Nanofillers for papermaking wet end applications. BioResources 5(3):1328–1331
90. Vaseghi Z, Nematollahzadeh A (2020) Nanomaterials: types, synthesis, and characterization. Green synthesis of nanomaterials for bioenergy applications, 23–82.https://doi.org/10.1002/9781119576785.ch2
91. Wang X, Shu H, Hailan L, Xianxu Z, Mingjuan C, Zhenhua H, Maurizio M, Liping C, Kumar A, Nadda Q, Le V, Changlei X (2021) Photocatalytic degradation of surface-coated tourmaline-titanium dioxide for self-cleaning of formaldehyde emitted from furniture. J Hazard Mater 420:126565. https://doi.org/10.1016/j.jhazmat.2021.126565
92. Wu P, Zhang MY, Wang J, Liu F, Song SX (2013) Application of CPAM-bentonite microparticle system in papermaking with using fly ash based calcium silicate as filler. China Pulp Paper 32(9):5–10
93. Wu Y, Liang Y, Mei CT, Cai L, Nadda AK, Le QV, Peng Y, Lam SS, Sonne C, Xia C (2021) Advanced nanocellulose-based gas barrier materials: present status and prospects. Chemosphere 286:131891. https://doi.org/10.1016/j.chemosphere.2021.131891
94. Xiong R, Han Y, Wang Y, Zhang W, Zhang X, Lu C (2014) Flexible, highly transparent and iridescent all-cellulose hybrid nanopaper with enhanced mechanical strength and writable surface. Carbohyd Polym 113:264–271. https://doi.org/10.1016/j.carbpol.2014.06.069
95. Yoon SY, Deng Y (2007) Experimental and modeling study of the strength properties of clay– starch composite filled papers. Ind Eng Chem Res 46(14):4883–4890. https://doi.org/10.1021/ie0613523
96. Zaieda M, Bellakhal N (2009) Electrocoagulation treatment of black liquor from paper industry. J Hazard Mater 163(2):995–1000. https://doi.org/10.1016/j.jhazmat.2008.07.115
97. Zeng M, Zhang Y (2019) Colloidal nanoparticle inks for printing functional devices: emerging trends and future prospects. J Mater Chem A 7(41):23301–23336. https://doi.org/10.1039/C9TA07552F
98. Zhang X, Zhao Z, Ran G, Liu Y, Liu S, Zhou B, Wang Z (2013) Synthesis of lignin-modified silica nanoparticles from black liquor of rice straw pulping. Powder Technol 246:664–668. https://doi.org/10.1016/j.powtec.2013.06.034
99. Zhang YP (2013) Anti-counterfeiting method using synthesized Nanocrystalline Cellulose Taggants
100. Zhang Y, Aslan K, Previte MJ, Geddes CD (2008) Metal-enhanced fluorescence from paper substrates: Modified spectral properties of dyes for potential high-throughput surface analysis and assays and as an anti-counterfeiting technology. Dyes Pigm 77(3):545–549. https://doi.org/10.1016/j.dyepig.2007.08.007
101. Zhao C, Wang J, Lu L (2022) Preparation and application of water-based nano-silver conductive ink in paper-based 3D printing. Rapid prototyping journal 28(4):747–755. https://doi.org/10.1108/RPJ-05-2020-0112
102. Zhu H, Fang Z, Preston C, Li Y, Hu L (2014) Transparent paper: fabrications, properties, and device applications. Energy Environ Sci 7(1):269–287. https://doi.org/10.1039/C3EE43024C

Nanomaterials for Paper Recycling and Deinking

Dogan Tutak, Arif Ozcan, and Emine Arman Kandirmaz

Abstract Cellulose, which forms the raw material of paper, is the most abundant biopolymer in the universe. It can be easily produced by all plants, plankton, algae, and many microorganisms. In addition, it is a strong biopolymer that is renewable, environmentally friendly, and very easily biodegradable. There is no polymer other than cellulose that can mimic its functional mechanical and chemical properties. However, many problems such as breaking and shortening of paper fibers and removal of ink from fibers are encountered during the recycling of cellulose-based papers. In order to find solutions to these problems, many nanomaterials are used to prevent these problems. Various nanomaterials are used in the paper and printing industry. Nanomaterials used in the paper industry are nanocellulose, nanofillers, nanocoatings, and nano inks, which are applied to the paper. During the recycling process of printed papers, nanomaterials help in efficiently removing ink pigments easily without damaging the fibers. At the same time, it aims to improve the physical, mechanical, and optical properties of recycled papers obtained from recycling. Many studies have been done in this field and successful results have been obtained. In this study, the recycling of printed papers and the effects of nanotechnologies that will enable better paper recycling in these processes were examined, and this issue was highlighted.

Keywords Nanomaterials · Paper recycling · Deinking

1 Introduction

Recycling is the process of collecting waste materials (Paper, glass, Plastic, etc.) that have reached the end of their useful life, using them as raw materials, and converting them into new products. It is critically important for sustainability. It helps reduce environmental pollution and carbon footprint and protect natural resources [42].

D. Tutak (✉) · A. Ozcan · E. A. Kandirmaz
Department of Printing Technologies, Faculty of Applied Sciences, Marmara University, Istanbul, Turkey
e-mail: dtutak@marmara.edu.tr

R. K. Bachheti (ed.), *Nanotechnology in the Pulp and Paper Industry*, Smart Nanomaterials Technology, https://doi.org/10.1007/978-981-95-3473-9_7

Population growth, easy and popular online shopping opportunities, socioeconomic growth and development, and the increasing popularity of e-commerce are increasing the use of paper and cardboard for packaging worldwide. This has led to more paper and cardboard waste and increased disposal [36, 42].

Paper production processes, like other industries, have serious negative effects on the environment. These effects sometimes happen during the production and processing of raw materials. The use and processing of raw materials have various adverse effects on the environment. Different technologies and methods are used to reduce and make these negativities positive. Recycling is one of the most important of these methods. With recycling, both waste disposal can be done, and the consumption of primary raw materials decreases. With this reduction, the protection of natural resources is also ensured. The paper industry consumes a lot of energy and water. It is easier to obtain fibers in the recycled paper production process, which provides less energy and water consumption in paper production and reduces environmental pollution [10, 34].

However, recycling the recycled paper more than once can have negative effects on both the quality of the paper and the physical and optical properties of the paper [14] Each recycling process causes the fibers to shorten, which causes the strength of the paper to decrease [2, 20]. (In such cases, adding certain amounts of virgin fiber to the pulp during the paper recycling process increases the strength of the paper [12, 16]. Paper fibers cannot be recycled indefinitely. Recycling of a paper fiber is only possible seven times. Apart from this, one of the biggest problems is that the paper fibers deform after being wetted and dried in each paper recycling process. The swelling and shrinkage of the wetted and dried fibers also negatively affect the mechanical properties of the paper. At the same time, the color of the recycled paper produced is more grayish than that of paper produced from virgin fiber, and this is not a desirable situation for end users [42]. Various methods are used to improve paper properties, one of which is nanotechnological coatings. Nanotechnologies have shown different methods to improve the mechanical structures of fiber-based polymeric papers with a hydrophilic structure. Some studies have reported that nanomaterials can produce papers with stronger mechanical properties and smoother surface structures [21].

Nanoparticles have unique surface properties that accelerate the fundamental processes of the reaction phases in the matrix system. In recent times, many nanomaterials such as silver nanoparticles (AgNPs), nano Titanium dioxide (TiO_2), nano Titanium nitride (TiN), nano Zinc oxide (ZnO), Carbon nanotubes (CNTs), etc. are used as fillers in paper coatings and food packaging as functional additives in the development of fiber-based paper materials. Nanomaterials with different properties and structures impart different properties to these coatings due to their different properties [11]. Due to these properties, nanomaterials can be recycled from fiber-based wastes, facilitating the production of new fiber-based composites from these fibers and making them more efficient. This study aims to shed light on how to make the paper recycling process more efficient by using nanomaterials to produce higher-quality papers with better physical, chemical, and mechanical properties.

2 Paper Recycling

Since the late 1970s, recycling and reuse have been among the most important issues in the world. Both consumers and producers demand more environmentally friendly products. The primary input for paper production is wood-based fibers obtained from wood. Paper obtained from recycling is considered secondary fibers. Wood-based or secondary fibers obtained from recycling are processed using technologies with different material and energy requirements that produce products of different quality. The quality and technical characteristics of these papers produced may be different [9].

Recycled paper is the most important raw material for paper, cardboard, and paperboard production in the paper industry [15]. Paper and cardboard have become indispensable raw materials for the paper industry due to their low cost, sustainability, environmental compatibility, and easy processing for the recycling process [47] (Fig. 1).

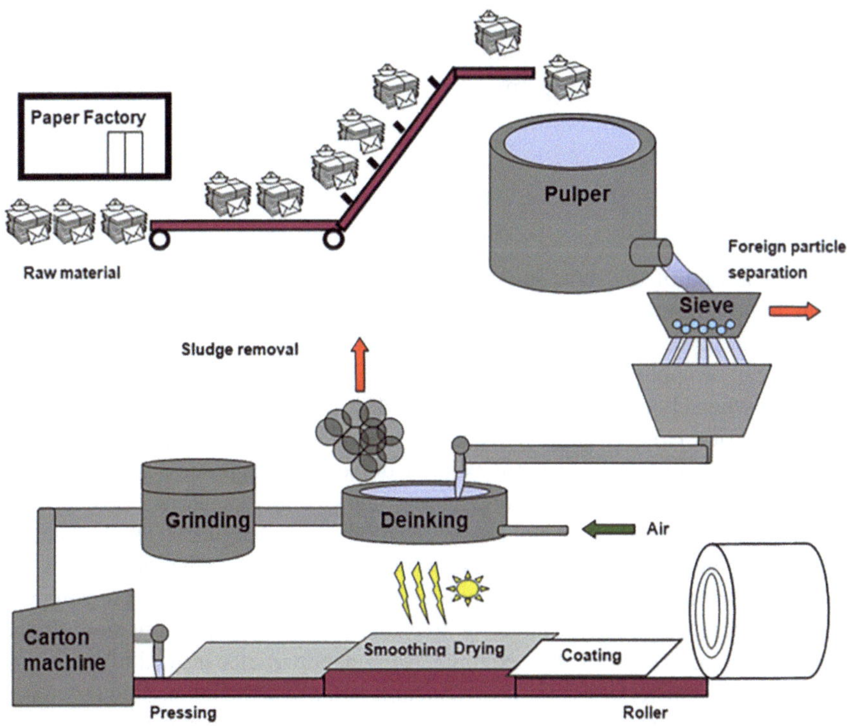

Fig. 1 Waste paper recycling process (www.ua-bw)

However, the majority of paper and cardboard to be recycled consists of printed papers that are difficult to remove. This makes it difficult to obtain whiter and higher-quality paper [47]. This process is called deinking. The deinking process plays an important role in the recycling process of waste paper, production cost, and quality.

2.1 Recycled Paper Production Processes

Recycled fibers pulp (RCF): Paper waste is collected for recycling and divided into bales [32]. The production of recycled pulp from waste paper is a simpler process than preparing wood-based wood pulp. The simplicity of the process also requires less energy consumption. The basic principle of recycled fibers pulp is to create the pulp by melting it in hot water to separate the fibers to be recycled without damaging them and without shortening their length [38]. Pulpers are mixers with special blades that are used to turn waste paper into pulp again and again and have an aqueous environment at a certain temperature. Dry fiber clusters are wetted and broken into pieces in these pulpers to obtain fibers that will be turned into paper again. Thus, the prepared fiber pulp can be easily sieved and cleaned. As a result, recycled fibers pulp (RCF) is prepared.

2.2 Deinking

Deinking is one of the most important stages in recycling printed papers [7]. It means removing ink particles and contaminants from pulped paper fibers during the deinking process. Thus, cleaner and brighter papers can be obtained. Different methods can be applied to deinking. The printing system and the type of ink used are important when choosing these methods [18, 45]. Nanomaterials are widely used in many industries. However, although it is also used in many areas for paper and pulp, it is quite challenging to give a serious example of the deinking process. Possible applications in the case of deinking using nanotechnology include surface modification of nano-sized molecules or nanoparticle-sized inks and nanofibers. Thus, easier-to-wash ink pigments and cleaner papers can be obtained. Since nano-air bubbles and nano-ink-fiber interaction have not yet been studied, it is quite difficult to comment on this subject. In summary, nanotechnology can improve deinking, but more studies need to be done to say this. For example, using nanoparticles instead of traditional inks will reduce material usage. This reduction will cause the ink pigment particles, contaminants in deinking processes, to be less on the final product, the paper surface. We can explain deinking processes simply with three basic methods.

2.2.1 Washing Method

The method of removing ink by washing has been used for many years. The system removes ink, filler and other contaminants from the pulp by washing with water. The effectiveness of this system is directly dependent on particle size. Although it is possible to remove particles smaller than 10–15 μm, the most effective particle size is 0.5–5 μm. These particles remain suspended during the washing phase and have no relation to the fibers. The chemicals used in the system weaken the print carrier network, release the retained pigment particles into the water, and help remove the printing ink (Fig. 2). It is the most suitable deinking method for nanotechnological inks. One of the biggest disadvantages of this method is excessive water consumption.

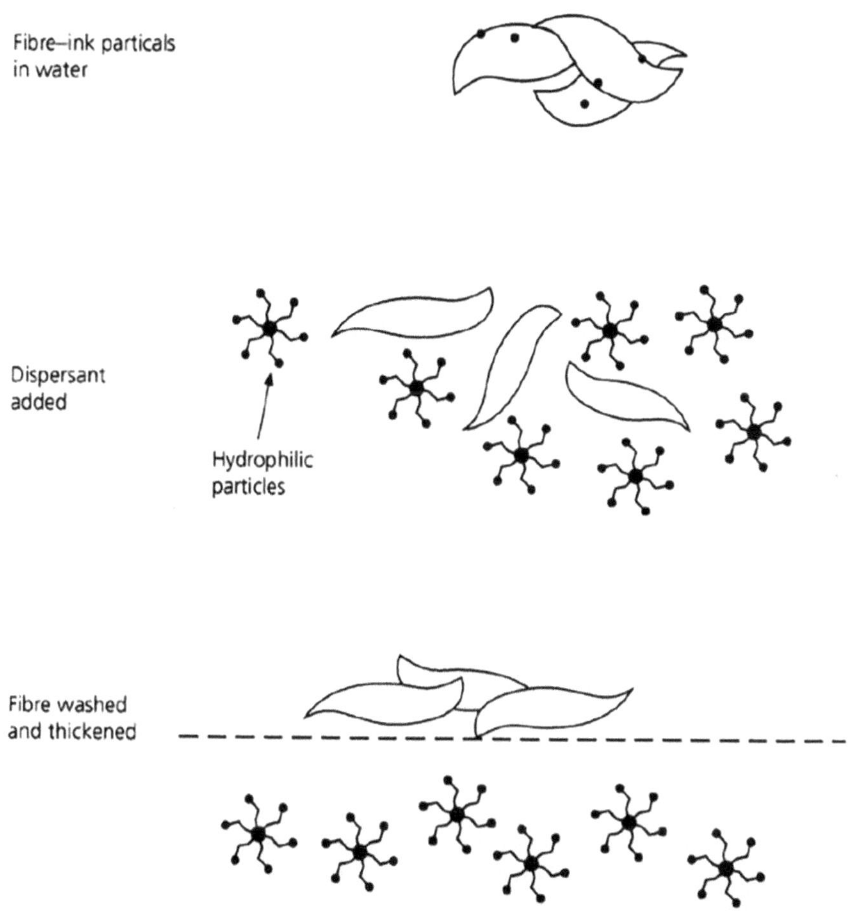

Fig. 2 Washing process in deinking [17]

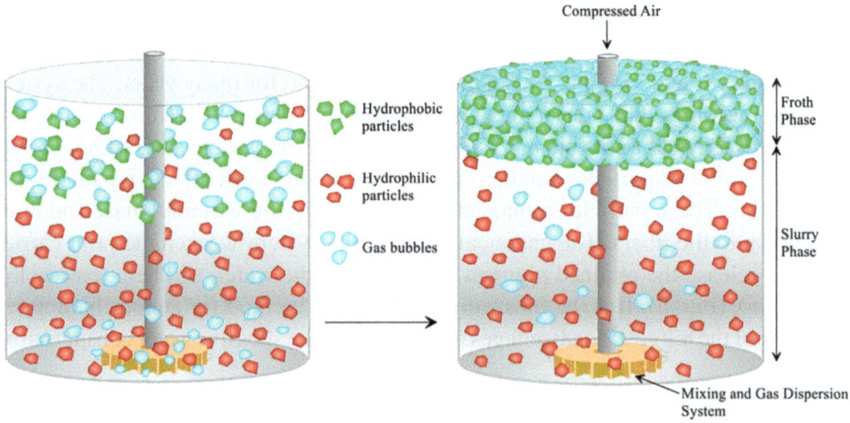

Fig. 3 Schematic illustration of separation of mineral particles by froth flotation (sizes not to scale) [35]

2.2.2 Flotation Method

The flotation method is a chemical–mechanical process method used to separate solid particles and liquids in waste paper pulp. In this method, the water-repellent properties of inks are eliminated with surface-active heteropolar chemical substances (alkaline salts of fatty acids, soap). These chemical substances added to the system surround the ink particles so that their hydrophilic ends are directed outward. Thus, the hydrophilic ink particles are separated from the fiber and dispersed into the water. The solids in the suspension are pushed upward with the air hoses created and are ensured to pass to the surface of the water. In this way, the solids are aimed to be removed from the system (Fig. 3).

2.2.3 Enzymatic Deinking

Due to the use of high amounts of sodium hydroxide, sodium carbonate, sodium silicate, magnesium sulfate, hydrogen peroxide, chelators, and surfactants (surface activators), the use of enzymes has emerged as an energy-saving, cost-reducing, and more environmentally friendly method [43]. The use of enzymes in paper production has a long history due to their use in the fiberization stage. In the deinking process, as stated above, while it is possible to easily remove hydrophobic structures in the flotation process, chemical, physical and enzymatic techniques must be applied to remove aged and hydrophilic structures. Especially on printed surfaces where thermoset toners are used, it is extremely difficult to separate these polymers from the fibers, and the use of enzymes stands out as an environmentally friendly solution, as stated above [41, 48]. The enzymes used in the enzyme method can be listed as cellulases, xylanases, laccases, amylases, lipase, pectinases. Enzymes take part

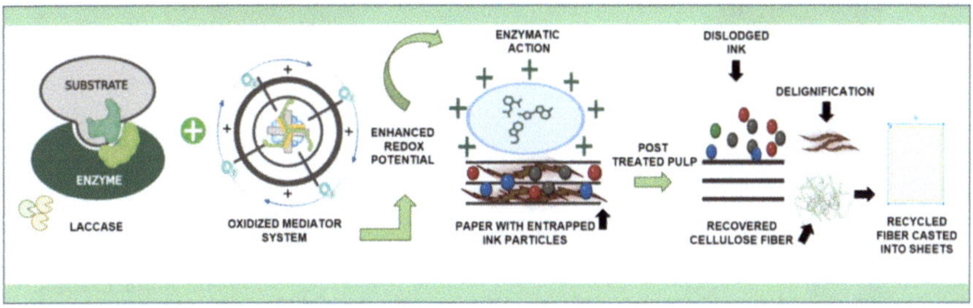

Fig. 4 An illustration of the efficacy of laccase in combination with an oxidized mediator system, enabling maximum deinking efficacy [33]

in the deinking process by removing the fibrils on the fiber surface and separating the ink particles from the fibers, making the particles separated from the fibrils more hydrophobic so that they can be more easily separated in the suspension, and changing the particle sizes to make them suitable for flotation and washing methods. Ambient temperature and pH are also important factors in the use of enzymes. In addition, it is known that mechanical processes are used together with enzymes, because they open the cellulose fibrils so that they can be accessed by enzymes [41, 48] (Fig. 4).

3 Nanotechnology

Nanotechnology deals with developing and applying materials smaller than 100 nm, which we call nanomaterials [4, 5, 26]. It is important in almost all modern industries targeting high quality and efficient market potential [22]. Nanomaterials (NMs), which are 1–100 nm in size in at least one dimension, are extremely popular in industrial applications: they give different properties to materials and restructure them. NMs are technological products [8]. The great interest in NMs is that NMs have much more properties than their non-nano counterparts, which have the same properties. Nanomaterial properties can be manipulated by precisely controlling the size, shape, synthesis conditions, and appropriate functionalization [6]. Therefore, it provides better changes in the material it is used in. In recent years, it has also been used extensively in pulp and paper coating applications [22]. NMs exhibit extremely superior physical and chemical properties. Therefore, they attract great attention from scientists working in different fields of environmental sciences [39].

Depending on their type and nature, many methods are used in the synthesis of nanomaterials. These methods can be classified as chemical, physical, or biological. The two most prominent methods are "bottom-up" and "top-down" (Fig. 5) [13]. NMs are synthesized from the basic level when the bottom-up method is used. In the top-down method, bulk materials are reduced to NMs. The methods used in the synthesis of NMs are

- Thermal decomposition

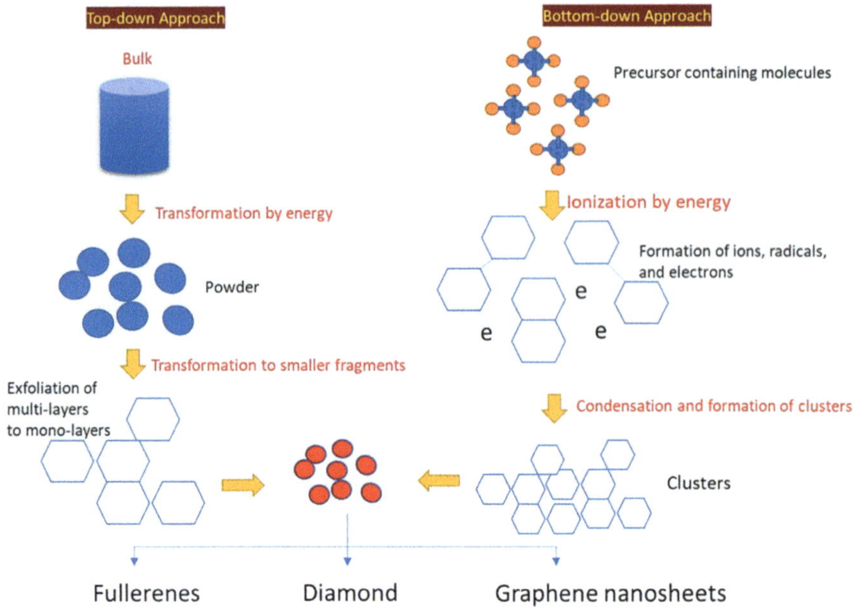

Fig. 5 Diagrammatic representation of top-down approach and bottom-down approach of making of nanomaterials [23]

- Chemical vapor deposition
- Gas phase method
- Combustion method
- Pulsed laser ablation
- Hydrothermal synthesis
- Solvothermal method
- Templating method
- Microwave synthesis
- Sol–Gel method [23]

Nanomaterials can be classified according to their structures as nanoparticles, layered or lamellar nanostructures, filamentary nanostructures and bulk nanostructured materials. Nanomaterials can be classified according to their size into the following categories:

- Nanorods and nanowires with dimensions <100 nm
- Tubes, fibers, and platelets with dimensions <100 nm
- 0 or 3 Dimensions <100 nm Particles, quantum dots, hollow spheres

According to their phase composition, we can classify nanomaterials based on their different phase structures as follows:

- Amorphous particles, crystalline, and layers are included in the single-phase solids class.
- Coated particles and matrix composites are included in multiphase solids.
- Colloids, ferrofluids, aerogels, etc. are included in multiphase nanomaterial systems [25].

3.1 Nanomaterials Used in the Paper Industry (NMs)

In recent years, studies have increased both in the paper industry and in universities to improve the existing physical and chemical properties of papers or to add new properties to the produced papers. Nanomaterials are generally used as nanocoatings applied to the paper surface or as nanofillers added to the pulp [40]. In addition, in the paper industry, nanocellulose, nano pigments, and nano inks are used to recycle waste papers more effectively. Due to their small size, these nanomaterials offer some advantages in deinking processes. The usage areas of nanomaterials in the paper industry are generally nanocellulose, nano inks, nanocoatings, and nanofillers. If we list the nanomaterials used in this context, we can name them as silver nanoparticles (AgNPs), nano Titanium dioxide (TiO_2), nano Zinc oxide (ZnO), nano Titanium nitride (TiN), carbon nanotube (CNT), etc.

3.1.1 Nanocellulose

Cellulose, the raw material of paper, is among the most depleted raw materials in the universe owing to its sustainable, environmentally friendly and renewable properties as a biopolymer. It is the basic building block of a plant cell wall formed by the condensation reaction of α-D glucose linked by 1–4 glycosidic bonds [19, 29]. Cellulose is obtained from nanocrystalline cellulose (Fig. 6), contains amorphous substances known as hemicellulose, and can be separated by mechanical grinding or hydrolysis, forming a highly crystalline or semicrystalline structure. The elasticity of the modulus can even reach 137 GPa [32]. Cellulose nanomaterials can be generally defined as cellulose consisting of crystalline and amorphous regions with diameters below 100 nm and lengths reaching several microns. A wide variety of cellulose sources have been used for the production of cellulose nanomaterials, including wood (softwood or hardwood), seed (cotton), bast (jute, flax, hemp), leaf (sisal), straw (rice, wheat), fruit (coconut fiber), tunics, algae, fungi, and bacteria [28].

Nanocellulose and its derivatives have been widely studied as additives in papermaking. The studies have determined that nanocellulose is used as an additive and that nanocellulose has superior performance. Nanocellulose additive strengthened the paper, ensured the retention of fine fibers, and increased in the barrier properties of the paper.

Natural hydroxyl groups help nanocellulose to combine appropriately with different fibers in pulp, thus enabling the crosslinking of fibers. This crosslinking

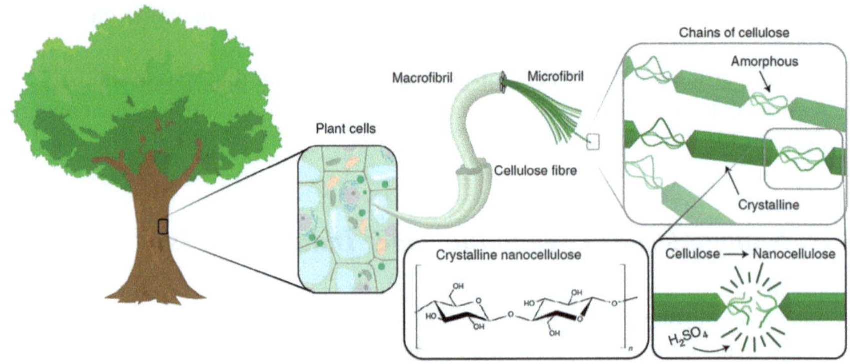

Fig. 6 Schematic of the hierarchical structure of crystalline nanocellulose [46]

gives some superior properties to paper. Natural hydroxyl groups help nanocellulose to combine appropriately with different fibers in the pulp, thus enabling the fibers to crosslink. This crosslinking gives paper some superior properties. It increases the tensile strength of paper, facilitates the adhesion of fine fibers, and increases the barrier properties of paper when coated on the surface of paper [24].

3.1.2 Nanocoatings

Coatings determine the surface properties of the paper and directly affect printability. In addition, coatings directly affect the recycling of paper. Scientific studies have proved that they give better results than uncoated papers. There is no contact between the ink and the fibers in coated paper, which is a nonissue. In uncoated paper, printing inks that adhere tightly to the fibers are more difficult to remove from the fibers and can significantly reduce the deinking of inks.

Pigments such as ultrafine calcium carbonate and ground calcium carbonate are widely used in paper coatings due to their high quality and low price. In addition, nanomaterials used in paper coating have been more successful in the recycling process after the printing process is completed. This is because nanomaterials reduce the amount of coating and are used instead of coatings that are difficult to recycle [44].

The porous structure of the paper and the hydrophilic structure of the fiber surface allow it to absorb nanoparticle suspensions with capillary forces and a high nanoparticle load is obtained during drying. Very small amounts of AgNP, TiO2, AuNP [1] nanocoatings applied to the paper, which are relatively cheap, provide excellent antibacterial and optical properties to the paper, while it can produce devices that show Surface Enhanced Raman Scattering (SERS) and Surface Plasmon Resonance (SPR) performances against counterfeiting. In addition, the paper can be easily degraded and is biodegradable, renewable, and biocompatible. In this respect, it has many advantages. Surface chemistry and morphology can be easily designed and made more easily recyclable using nanomaterials [30].

3.1.3 Nanoinks

Nanoinks are very small-sized materials that can be dispersed in various media owing to their sizes smaller than 100 nm. Although nanoinks are relatively new materials, there are a number of ways by which they can be synthesized. Some approaches are hot injection or noninjection organometallic, aqueous and biosynthetic methods [13]. Despite being used for many years, nanoinks have not yet become very popular. There are two reasons for this. The difficulty of bringing the pigment sizes of nanoinks to nanoscale and the dispersion of nanoscale dispersions. More studies should be carried out in this area to improve this situation. Nano-sized ink pigments provide important features to prints obtained with inks prepared with binders or solvents. For example, they provide higher optical densities and a wider color gamut volume [27]. Nanoinks have good surface wettability, which increases the printability of the ink and provides better adhesion to the surface. Especially in coated papers, nano-sized ink pigments that adhere to the surface coating can be more easily washed away during recycling.

4 Conclusion

Paper recycling is a necessity for all cellulose-based paper products at the end of their life cycle. It is directly linked to the printing method, inks, and coatings used. These stages (printing inks, noncelluloses, and nanocoatings) affect the recycling performance of the paper. According to Ozcan and Tutak, nano-zeolite coating contributed significantly to the recycling process of inkjet printed papers, which are very difficult to recycle. In addition, this coating also contributed to the brightness [31]. According to Ang et al., the original cellulose nanofiber sheets offer a groundbreaking solution for creating ultrahigh strength recycled cellulose nanofiber sheets in a fraction of the time it would take to produce equivalents [3]. Undoubtedly, cellulose nanofibers have enormous potential as they are sustainable and extremely recyclable. Therefore, efforts should be made to promote the widespread application of nanoscale building blocks. Nanoparticle applications during paper recycling help produce papers with higher brightness and lower steam production. In addition, the use of coating pigments such as nano-silica, nanoTiO_2, and nano zeolite can also reduce water usage during paper production and increase paper surface performance [37]. In summary, although nanomaterials are used directly in the recycling of printed or unprinted papers, it is seen in the studies that they are used in both pulp and coating processes of paper production. In addition, it is seen that the inks used in the printing processes are included in the process using nanotechnologies. It has been determined as a result of the studies that all of these have an indirect or direct effect on the recycling processes of paper.

Competing Interests The authors declare that they have no competing interest.

Declaration of Generative AI and AI-Assisted Technologies in the Writing Process During the preparation of this work, the book chapter author(s) have not taken any assistance from AI, and the book chapter author(s) reviewed and edited the content as needed and take(s) full responsibility for the content of the publication.

References

1. Adel AM, Al-Shemy MT, Diab MA, El-Sakhawy M, Toro RG, Montanari R,... Caschera D (2021) Fabrication of packaging paper sheets decorated with alginate/oxidized nanocellulose-silver nanoparticles bio-nanocomposite. Int J Biol Macromol 181:612–620
2. Amiandamhen SO, Osadolor SO (2020) Recycled waste paper–cement composite panels reinforced with kenaf fibres: durability and mechanical properties. J Mater Cycles Waste Manag 22:1492–1500
3. Ang S, Ghosh D, Haritos V, Batchelor W (2021) Recycling cellulose nanofibers from wood pulps provides drainage improvements for high strength sheets in papermaking. J Clean Prod 312:127731
4. Bachheti A (Joshi), Bachheti RK, Husen A (Eds) (2024) Carbon-based nanomaterials: synthesis, agricultural, biomedical, and environmental ınterventions. Springer Nature Singapore. https://doi.org/10.1007/978-981-97-0240-4
5. Bachheti RK, Bachheti A, Husen A (Eds) (2024a) Metal and metal-oxide based nanomaterials: Singapore: synthesis, agricultural, biomedical and environmental ınterventions. Springer Nature Singapore. https://doi.org/10.1007/978-981-99-7673-7
6. Baig N, Kammakakam I, Falath W (2021) Nanomaterials: a review of synthesis methods, properties, recent progress, and challenges. Mater Adv 2:1821–1871
7. Bajpai P (2024) Recycling and deinking of recovered paper. Elsevier
8. Behera A, Chatterjee S (2022) Industrial scale up applications of nanomaterials recycling. RNMs 341–361
9. Byström S, Lönnstedt L (2000) Paper recycling: a discussion of methodological approaches. Resources. Resour Conserv Recycl 28:55–65
10. Čabalová I, Kačík F, Geffert A, Kačíková D (2011) The effects of paper recycling and its environmental impact. Env Manage Pract 17:329–350
11. Chaudhary P, Fatima F, Kumar A (2020) Relevance of nanomaterials in food packaging and its advanced future prospects. J Inorg Organomet Polym Mater 30:5180–5192
12. Czene T, Koltai L (2020) The effect of the virgin fibers to the properties of different paper products. 10th GRID Serbia 147–151
13. de Souza FM, Gupta RK (2023) Nano-inks: fundamentals, synthesis, and energy applications. PS-IO-N ink 659–685 Elsevier
14. Egamberdiev EA, Norboyev SK (2022) Extraction of cellulose nanocrystals from secondary paper waste and their use in paper production. Techn Sci Innov 2022:215–222
15. Faul AM (2010) Quality requirements in graphic paper recycling. Cellul Chem Technol 44:451
16. Filipova I, Andze L, Skute M, Zoldners J, Irbe I, Dabolina I (2023) Improving recycled paper materials through the incorporation of hemp, wood virgin cellulose fibers, and nanofibers. Fibers 11:101
17. Fricker A, Thompson R, Manning A (2007) Novel solutions to new problems in paper deinking. Pigm Resin Techno 36:141–152
18. Ghanbarzadeh B, Ataeefard M, Etezad SM (2023) Enzymatic deinking of office waste printed paper: optimization via response surface methodology. Biomass Conv Bioref 13:215–223
19. Guan Y, Li W, Gao H, Zhang L, Zhou L, Peng F (2021) Preparation of cellulose nanocrystals from deinked waste newspaper and their usage for papermaking. Carbohydr Polym Technol Appl 2:100107

20. Han N, Zhang J, Hoang M, Gray S, Xie Z (2021) A review of process and wastewater reuse in the recycled paper industry. Environ Technol Innov 24:101860
21. Hasan KF, Horváth PG, Alpár T (2022) Nanotechnology for waste wood recycling. In: Nanotechnology in paper and wood engineering, pp 61–80. Elsevier
22. Julkapli NM, Bagheri S (2016) Developments in nano-additives for paper industry. J Wood Sci 62:117–130
23. Khan FA (2020) Synthesis of nanomaterials: methods & technology. Applications of Nanomaterials in Human Health. Springer, Singapore
24. Li A, Xu D, Luo L, Zhou Y, Yan W, Leng X,... Fan M (2021) Overview of nanocellulose as additives in paper processing and paper products. Nanotechnol Rev 10:264–281
25. Madkour LH (2019) Classification of Nanostructured Materials. In: Nanoelectronic materials. Advanced Structured Materials 116. Springer
26. Mazari SA, Ali E, Abro R, Khan FSA, Ahmed I, Ahmed M,... Shah A (2021) Nanomaterials: applications, waste-handling, environmental toxicities, and future challenges–a review. J Environ Chem Eng 9:105028
27. Mirković IB, Vukoje M, Bolanča Z, Vučković F (2021) Recycling of aged inkjet prints with nano technology based inks. Eur J Sustain Dev Res 1:5
28. Mokhena TC, John MJ (2020) Cellulose nanomaterials: new generation materials for solving global issues. Cellulose 27:1149–1194
29. Nagarajan KJ, Balaji AN, Rajan STK, Ramanujam NR (2020) Preparation of bio-eco based cellulose nanomaterials from used disposal paper cups through citric acid hydrolysis. Carbohydr Polym 235:115997
30. Ngo YH, Li D, Simon GP, Garnier G (2011) Paper surfaces functionalized by nanoparticles. Adv Colloid Interface Sci 163:23–38
31. Ozcan A, Tutak D (2020) The effect of zeolite on inkjet coated paper surface properties and deinking. NPPRJ 35:432–439
32. Ozola ZU, Vesere R, Kalnins SN, Blumberga D (2019) Paper waste recycling. circular economy aspects. Environ Climate Technol 23:260–273
33. Pandey S, Gupta S (2024) Exploring laccase: a sustainable enzymatic solution for the paper recycling domain. Arch Microbiol 206:211
34. Pathak P, Sharma C (2023) Processes and problems of pulp and paper industry: an overview. Phys Sci Rev 8:299–325
35. Pawlik M (2022) Fundamentals of froth flotation. ChemTexts 8:19
36. Peretz R, Mamane H, Wissotzky E, Sterenzon E, Gerchman Y (2021) Making cardboard and paper recycling more sustainable: recycled paper sludge for energy production and water-treatment applications. Waste Biomass Valorization 12:1599–1608
37. Porwal P, Taghiyari HR, Husen A (2023) Use of nanomaterials in the forest industry. In: Advances in smart nanomaterials and their applications, pp 69–487. Elsevier
38. Rahnama Mobarakeh M, Santos Silva M, Kienberger T (2021) Pulp and paper industry: decarbonisation technology assessment to reach CO2 neutral emissions—an Austrian case study. Energies 14:1161
39. Rizwan MD, Singh M, Mitra CK, Morve RK (2014) Ecofriendly application of nanomaterials: nanobioremediation. J Nanoparticles 1:431787
40. Samyn P, Barhoum A (2018) Engineered nanomaterials for papermaking industry. In: Fundamentals of nanoparticles, pp 245–277. Elsevier
41. Saxena A, Singh Chauhan P (2017) Role of various enzymes for deinking paper: a review. Crit Rev Biotechnol 37:598–612
42. Senarathna WGCM, Sulaksha LGT, Weerarathna DM, Jayathma WMVA, Gamage DGML, Thennakoon, TMTN,... Perera MDV (2023) Paper recycling for a sustainable future: global trends. J Res Technol Eng 4:164–186
43. Singh A, Kaur A, Dutt R, Ritu Y (2019) An efficient eco-friendly approach for recycling of newspaper waste. 3 Biotech 9:1–7
44. Steward F, Tsoi JC, Coles AM (2008) Innovation and sustainability in print-on-paper: a comparison of nanoparticle and deinking niches as emergent sociotechnical networks. J Clean Prod 16:957–966

45. Strickland KA (2024) Applications and methods for ıncreased liberation and separation in the deinking of recycled paper fiber. Doctoral Dissertation, Virginia Tech, USA
46. Williams NX, Bullard G, Brooke N, Therien MJ, Franklin AD (2021) Printable and recyclable carbon electronics using crystalline nanocellulose dielectrics. Nat Electron 4:261–268
47. Yang S, Shen J, He T (2022) Flotation de-inking for recycling paper: contrasting the effects of three mineral oil-free offset printing inks on its efficiency. Environ Sci Pollut Res 29:89283–89294
48. Yilgor N (2010) Mürekkep giderme sürecinde enzimlerin kullanilmasi using of enzymes in deinking process, 60:50–8. https://www.ua-bw.de/pub/beitrag_printversion.asp?subid=0&Thema_ID=3&ID=1767&Pdf=No&lang=EN

Nanotechnology for Paper Packaging Innovations

Arif Ozcan, Emine Arman Kandirmaz, and Dogan Tutak

Abstract Functional paper packaging has attracted a lot of attention lately because of its exceptional performance, adaptability, and promise to offer the packaging industry more sustainable alternatives. Paper-based packaging is characterized by its large surface area, adjustable porosity, and customizability, while functional paper-based packaging generally exhibits good mechanical strength and barrier properties. The production of many functional papers using nanotechnology is a rapidly advancing research topic for today's researchers and the packaging industry due to both sustainability and advanced features. Functional paper packages produced with nanotechnology exhibit advanced features, including high mechanical efficiency, improved gas barrier properties, and protection against moisture and temperature, as well as providing very good antibacterial properties. Additionally, functional paper packaging produced with nanotechnology enables us to provide smart, intelligent, or active packaging features. Packaging produced from paper is degradable/biodegradable materials. However, such packaging materials have poor barrier and mechanical properties compared to their petroleum-based counterparts. This study critically evaluates the improvement of disadvantages, such as product protection and shelf life extension, of paper-based packaging materials by using nanomaterials, as well as the limitations of applying nanosized materials. Nanosized materials used in functional paper packaging are examined in detail in the study. The findings will guide various researchers and manufacturers in the development of packaging for nanotechnology.

Keywords Nanotechnology · Functional paper packaging · Paper-based smart packaging · Paper based active packaging

A. Ozcan (✉) · E. Arman Kandirmaz · D. Tutak
Faculty of Applied Sciences, Department of Printing Technologies, University of Marmara, Istanbul, Turkey
e-mail: arifozcan@marmara.edu.tr

R. K. Bachheti (ed.), *Nanotechnology in the Pulp and Paper Industry*, Smart Nanomaterials Technology, https://doi.org/10.1007/978-981-95-3473-9_8

Abbreviations

AgNPs	Silver nanoparticles
BNC	Bacterial nanocelluloses
CNC	Cellulose nanocrystals
CNF	Cellulose nanofibers
CNTs	Carbon nanotubes
GRAS	Generally recognized as safe
MCP	Mechanochemical processing
PE	Polyethylene
PET	Polyethylene terephthalate
PP	Polypropylene
PS	Polistiren
PVS	Physical vapor synthesis
RFID	Radio-frequency identification
TiO_2 NPs	Titanium dioxide nanoparticles
TNPs	Titanium nanoparticles
ZnO NPs	Zinc oxide nanoparticles (ZnO NPs)
ZnO	Zinc oxide

1 Introduction

The primary component of paper, cellulose ($C_6H_{10}O_5$), is one of the safest, most environmentally friendly, and biodegradable materials utilized in the packaging business. It has a porous structure consisting of long-chain cellulose molecules in the crystalline state and microfibrils with amorphous regions that disrupt this crystalline structure [1]. Due to its hydrophilic structure and porosity, it is easily affected by environmental factors, has low resistance to moisture, and loses its physical and mechanical strength over time [2].

Paper is one of the most important packaging materials in the world, and paper consumption rates increase in proportion to the development level of countries. Packaging expenditures in developed countries continue to increase annually, and among these materials, paper and cardboard packaging account for more than one-third of all packaging materials. In other words, more paper is consumed than metal, plastic, and glass packaging materials [3].

In addition, the most commonly used materials, especially in the food and cleaning product packaging industries, are plastic-based polymers that are not biodegradable and pose a serious threat to humans and the environment [4], and studies on reducing plastics with increasing environmental awareness and the complete elimination of plastics expected in the near future have directed research to bionanocomposite materials. Therefore, in addition to the situation mentioned above, paper consumption and paper need will increase even more. Thanks to the use of biomaterials, in addition to the functionalization of packaging, the waste problem can also be solved to a significant extent [5, 6]. Currently, the use of biomaterials in packaging is quite limited [7].

The basis of paper production largely comes from forests, and the large amounts of water consumed in the pulp or paper-making process cannot be controlled or managed optimally. This situation highlights the importance of environmental approaches, particularly nanotechnology. In this way, environmentally friendly production methods will facilitate and promote the widespread adoption of wastewater treatment [3].

2 Nanotechnology

Nanotechnology stands out as a technology that offers significant advantages, closely followed by all industries today. It encompasses design, transformation, production, and applications that control the physical properties, reactive properties, and functionality of structures and systems with dimensions of 100 nm or less [8, 9]. This technology is a new discipline formed by the combination of different technological fields, focusing on producing or examining matter at the nanoscale [10]. The development of new nanomaterials and newly designed surfaces allows the production of better-performing products, which creates opportunities for all industries and increases product quality and safety [11], and there has been a significant increase in nanotechnology applications recently [12, 13]. It can be applied across all industries and bring significant innovations in various fields, including information technology, biotechnology, biomedical technology, and metrology. The most important applications of nanotechnology include the chemical industry, pharmacy, cosmetics, processed food, qualified materials, electronics, mechanics, optics, energy, and environmental sciences (Fig. 1). It is envisioned that some of today's serious problems, such as fatal diseases, energy shortages, and climate change, can be addressed with nanotechnology [14]. Some of the prominent features of the materials developed with nanotechnologies are strength, toxicity, optical, magnetic, and some physicochemical properties such as solubility, color, and high surface-to-volume ratio [15, 16]. In this way, nanotechnology has started a new industrial revolution by enabling the development of materials and systems with new or improved properties in agriculture, materials, food, medicine, medicine, and many other areas. The increase in the world population, the rise of epidemic diseases, and the threat of war all increase the need for food, and therefore, the importance of the agriculture and food industry is rapidly growing. New studies in this field focus on developing new materials and techniques for packaging, functionalizing, extending shelf life, and implementing and monitoring quality control of food [17]. Recently, interest in nanoparticle-based materials for novel and diverse applications in the packaging industry has increased significantly worldwide. Especially in food and medicine packaging, safety-enhancing elements are crucial and bionanotechnology comes to the forefront at this point.

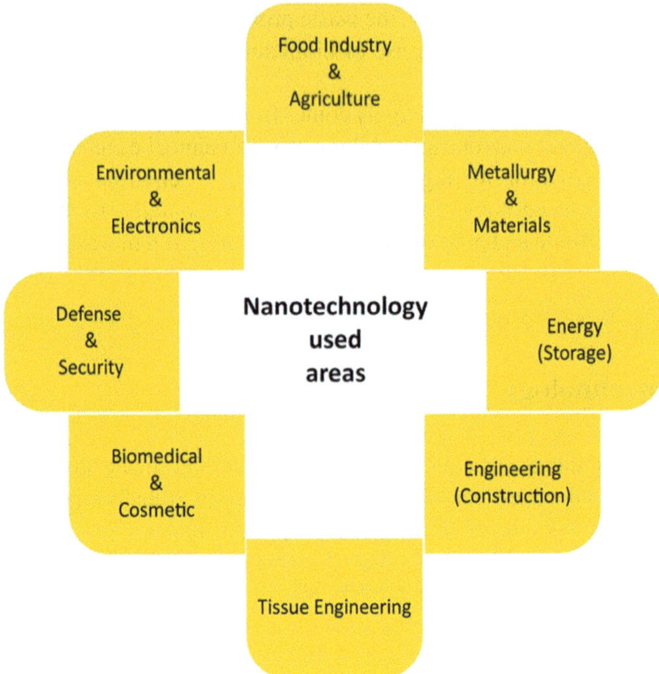

Fig. 1 Nanotechnology used areas

We can define bionanotechnology as the integration of biological molecules with existing nanomaterials [18].

Thanks to nanomaterials, technologies that extend the freshness and shelf life of products can be developed, methods and tools can be designed for instant monitoring of products, and most importantly, we can produce environmentally safe and degradable packaging. As the use of nanotechnology becomes widespread, significant improvements are seen in daily life. The use of nanomaterials contributes to the development and preservation of quality and the development of improved healthy products. Additionally, environmental impacts can be minimized due to their size [19]. Although nanotechnology is not new to the packaging industry, it is not very old either, and it is an area that is still open to development. Packaging materials can be developed using nanomaterials with various functions, thereby enhancing the shelf life and safety of packaged food, and driving innovations in the packaging industry [20].

3 Packaging

Packaging serves to safeguard the product, preserve its quality and safety, and cut down on waste. Packaging also aids in food preservation, informational content, and ease of use [21]. Nonbiocompatible petroleum-based products such as PET, PP, PS, and PE have long been the materials of choice for packaging due to their outstanding mechanical properties, affordability, long life, and resistance to degradation [22–25]. However, there are now significant environmental problems resulting from the widespread use of these nonbiodegradable, nonrenewable plastics [26, 27]. Disposable food packaging is also a significant factor here [28]. These plastic wastes pose a serious threat to the ecosystem and endanger wildlife [29].

A sustainable solution to the issues with conventional plastic packaging has been the emergence of biodegradable food packaging materials derived from renewable biological resources in recent years [30]. Microorganisms like bacteria, fungi, and algae can fully degrade biodegradable packaging into carbon dioxide, water, and biomass [31]. These materials can biodegrade more quickly when composted under regulated settings [32]. Concerns about the accumulation of plastic waste, growing consumer environmental consciousness, stringent government regulations regarding the disposal of packaging, and developments in technologies for the production and processing of biopolymer materials are the leading causes of the growing interest in biodegradable food packaging [33–35]. For these reasons, low environmental footprint and sustainable, smart packaging have recently gained prominence. Packaging waste constitutes a very large portion of all waste in the world, and plastic is at the forefront here. Half of all plastic waste originates from packaging [36, 37].

Global trends and technological developments in the food industry have triggered innovative approaches in the packaging industry to meet consumer expectations. Consumers have also increased their tendency to prefer products with more sustainable packaging materials over traditional packaging materials during shopping [38–40]. In all industries and the scientific world, the need for environmentally sensitive approaches, effective and efficient packaging, and the desire to work with recyclable, biocompatible materials have increased. Paper is also a material that fits these characteristics. Paper is a perfect material that can solve recyclable, bio-based, biodegradable, lightweight, and ecologically sustainable packaging problems in the form of a thin layer of cellulosic fibers [41, 42].

Paper is a material made of cellulose biofibers, usually composed of interlocking layers with different properties and functions [43]. Biodegradable packaging has many advantages, including sustainability, carbon neutrality, and waste management [44]. Nonbiodegradable packaging has a significant impact on the environment, including landfills and marine waste, and paper packaging can help address these issues. Due to its environmentally friendly nature, paper packaging has recently become more popular again [45]. Paper-based packaging has low barrier [46] properties for packaging applications due to its porous structure. The use of paper as packaging material without surface treatment brings about serious problems due to its weak barrier properties. Gas barrier properties, low moisture resistance, poor heat resistance, and mechanical strength, as well as oxygen permeability and diffusion of mineral oils, affect the quality of foods packaged in biopolymer-based packaging materials, such as paper [47]. Therefore, to impart specific properties to papers, surface coating is carried out using nanosized materials, thereby providing the necessary barrier or strength properties [48].

Nanomaterials and their applications have emerged as a key focus in both academic studies and commercial endeavors, as they can significantly enhance these expectations. The most common applications or trials include the development of lighter materials and the use of nanostructured composites or coatings to increase mechanical properties and improve barrier properties. It is also important to convert many packaging materials currently in use into bio-based and biodegradable materials, or to utilize nanotechnology to enhance the performance of some biodegradable materials. In this way, these materials can be utilized in a much wider range of food packaging applications, not just for specific foods. Several products are currently being developed as smart packaging materials: incorporating innovative processing techniques, active packaging that releases or absorbs chemicals, and nanosensors [49, 50]. The expectation is to revolutionize the packaging sector by developing sustainable, innovative, efficient, durable, lightweight, and more effective active and smart materials through nanotechnological products and applications [51–53]. With this transformation, traditional passive food packaging will disappear and be replaced by innovative systems commonly referred to as smart packaging and active packaging [54–56].

In addition, the development and growth of the food industry day by day provide an increase in the application of nanocomposites and similar hybrid packaging materials with desired packaging properties through the use of nanotechnology. Nanocomposites can be integrated with communication media, such as nanosensors [56], biosensors, barcodes [57], and radio-frequency identification (RFID) devices [58], and exhibit enhanced optical, electrical, mechanical, electronic, and barrier properties.

The new generation of smart packaging systems mainly aims to extend the shelf life of foods, improve their quality and ensure their safety [59], produce cheaper packaging materials [60], reduce processing times, reduce the use of preservatives in food formulas, and monitor the status of packaging and hence food quality in the supply chain [61, 62].

Based on their characteristics, nanoparticles can be classified into three categories: inorganic, mixed organic/inorganic, and surface modified. Silver, calcium carbonate, titanium dioxide, cerium dioxide hydroxides, quartz, zinc oxide, aluminum oxides, synthetic amorphous silica, and other metals and metal oxides are among the numerous metals and metal oxides that are found in inorganic nanoparticles. Some nanomaterials used in paper packaging are shown in Fig. 2. Organic nanoparticles are carbon-based nanoparticles such as nanostarch, nanocellulose, nanosaccharide, etc. The production of nanoparticles based on biomaterials has garnered renewed attention in recent times. Some studies have shown that polysaccharides, minerals, proteins, lipids, surfactants, and other food-based ingredients can be a good option to obtain organic and edible nanoparticles with very different properties. The significant role that nanotechnology plays in food packaging is yielding substantial commercial

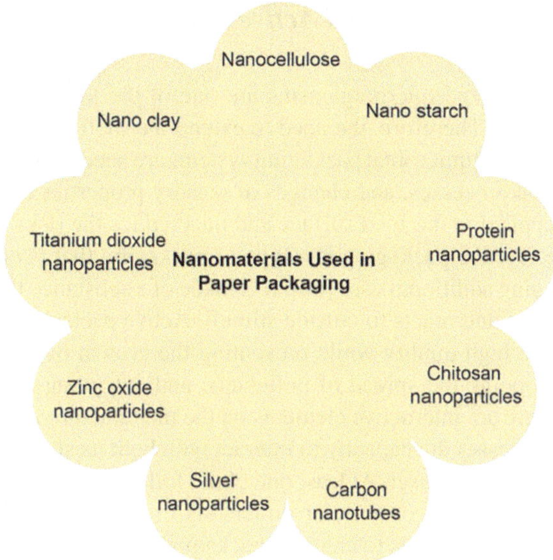

Fig. 2 Nanomaterials used in paper packaging

success for the food sector [63]. Carbon nanotubes, films, and various nanocoatings have recently garnered significant attention for use in food packaging materials. Nanoparticles are paving the way for the development of new-generation packaging materials that extend shelf life. Due to their antimicrobial properties, nanoparticles can act as solvents for enzymes, antioxidants, flavors, and other substances, extending the shelf life of products even under use and storage conditions [64]. Some organic and inorganic nanoparticles used in paper packaging are reviewed below.

3.1 Functional Packaging

One of the most important factors that determine our product preference today is packaging, and the striking element of packaging is its functionality. Functional packaging is a smart and innovative approach adopted by manufacturers that enhances the appearance and quality of your product. From the consumer's perspective, it is the driving force that convinces them to buy everything they visualize as attractive. We can utilize the concept of functional packaging for any type of packaging that ensures the product is safe and effective for the customer. Product safety is at the forefront in functional packaging, and at this point, smart packaging, which is evaluated in two categories as active packaging and smart packaging, comes into play.

3.2 Paper-Based Active Packaging

Foodborne microorganisms are one of the important health problems all over the world. Therefore, the need to extend the shelf life is increasing. Therefore, antioxidant/antimicrobial packaging systems are needed to control the contamination, oxidation processes, and changes in sensory properties that occur after the processes are applied to the food surface and thus reduce the risks of food poisoning [65].

Active packaging is a type of packaging that is created by giving passive packaging additional uses [66]. It is made of a substance that enhances fundamental functions and reacts to outside stimuli. Active packaging maintains the product's safety and high quality while preventing the growth of infections and harmful microbes, stopping the spread of pollutants, and extending its shelf life [67]. Put otherwise, there are interactive elements on the manufactured packaging. Intelligent packaging possesses the capacity to interact with both its surroundings within the supply chain and the end user. At least one of the following technologies—mechanical, chemical, electrical, electronic, or online—is used to accomplish this. The primary goal of these systems is to enhance packaging capabilities in response to growing customer demands, regulatory mandates, and safety awareness.

Active packaging is a type of packaging system that includes elements intended to release or absorb substances into the food being packaged or into the surrounding environment. Active packaging functions to preserve or extend the shelf life of a product, as well as to protect the quality, safety, and integrity of food products. Active packaging systems are commonly found in two groups: absorbers and sorbents. The former removes unwanted compounds (e.g., oxygen, moisture, ethylene, carbon dioxide, or odor) from the food or its environment, while the latter adds compounds such as antioxidants, antimicrobial complexes, flavors, carbon dioxide, ethylene, or ethanol to the packaged food or headspace [68]. Figure 3 shows active packaging systems for food applications and their overall effects.

3.3 Paper-Based Intelligent Packaging

Products in the food, beverage, and pharmaceutical industries commonly use smart packaging. Smart packaging is important to food companies; however, it has no immediate effect on the product. Materials and components that monitor the state of the food being packaged or the surrounding environment are referred to as smart packaging (Fig. 4). They do not engage with the product despite being able to convey the circumstances of the wrapped item. The goal is to keep an eye on the goods and guide customers. This information may be included in the packaging content, usage instructions, and storage conditions [69]. In order to notify consumers of any changes to the product or its existing state, smart packaging keeps an eye on the state of the packaged goods or the surroundings around the packaging [70]. A smart packaging system is one that uses scientific logic, sensing, detecting, recording, monitoring,

Active packaging type	Food type	Expectation
Active scavenging systems (absorber)		
Oxygen scavenger	Cooked meat products	Prevention of discoloration
	Grated cheese, bakery products	Prevention of mold growth
	Fruit and vegetable juices	Retention of vitamin C content, prevention of browning
	Seeds, nuts, and oils; fat-containing instant powders, fried snacks; dried meat products	Prevention of rancidity
Moisture scavenger	Mushrooms, tomatoes, strawberries, maize, grains, seeds, fresh fish, and meat	Extension of shelf life through maintaining moisture content, decrease in moisture condensation in the packaging, positive impact on the appearance, reduction in browning or discoloration
Ethylene absorber	Climacteric fruits and vegetables	Reduction in ripening and senescence, thereby enhancing quality and prolonging shelf-life
Active releasing systems (emitter)		
Antioxidant releaser	Fresh fatty fish and meat; fat-containing instant powders; seeds, nuts, and oils; fried products	Improvement of oxidative stability
Carbon dioxide emitter	Fresh fish and meat	Extension of microbiological shelf life, reduction in head space volume of modified atmosphere packaging
Antimicrobial packaging systems	Fresh and processed meat, fresh and smoked fish, fresh seafood, dairy products, fresh and processed fruits and vegetables, grain, cereals and bakery products, ready-to-eat meals	Inhibition or retardation of bacterial growth, extension of the shelf-life

Fig. 3 Forms of active packaging used in food applications [168]

and communication to enhance quality and safety, prolong shelf life, and aid in decision-making [71].

Smart packaging technologies have enhanced two purposes of packaging. Active packaging enhances protective features, and smart packaging enhances information functions. There are three primary components of smart packaging systems: indicators, sensors, and radio frequency identification (RFID) tags [72]. Color temperature, carbon dioxide and oxygen indicators, time–temperature, pathogen indicators, and breakage indicators are some of these systems.

4 Nanomaterials Used in Paper Packaging

4.1 Nanocellulose

Cellulose is a naturally abundant and biodegradable polymer found in the cell walls of plants and provides them with strength [73, 74]. Cellulose is the primary raw material for paper production, and its various derivatives are widely used as coating materials in packaging applications. Cellulose exhibits undesirable properties, including its hydrophilic structure [75], insolubility in water, high crystallinity, and poor film-forming ability. To turn these disadvantageous properties into advantages, cellulose derivatives have been produced and utilized as barrier materials [76]. The term “nanocellulose” is used to describe cellulose-based materials in the form of fibers

Fig. 4 Intelligent packaging functions

or crystals that have nano dimensions and can be extracted from lignocellulosic materials [77, 78].

Natural fibers, wood, food processing residue, agricultural waste, and other lignocellulosic materials are some of the sources of nanocellulose [79]. The global nanocellulose market size was 284.7 million US dollars in 2018, and this value is expected to increase to 661.3 million US dollars in 2023. Today, cellulose nanocrystals with a wide variety of structural properties and surface chemistry are sold commercially with a price tag of 4–10 US dollars per kilogram [80]. The high surface area and aspect ratio, high water retention capacity, crystallinity, tunable surface chemistry, and lack of cytotoxic properties of nanocellulose have increased their use in food applications [79, 81]. The hydroxyl groups on the surface of nanocellulose facilitate easier surface modification through various processes, such as carboxylation, esterification, amidation, and etherification, which increases its potential for use as a functional material [82]. Nanocellulose research, which has increased rapidly in recent years, shows that nanocellulose has excellent potential in terms of food applications. Nanocellulose is used in food applications, especially as food packaging [83], food additive [84], functional foods, coating material [85], encapsulation agent [86], stabilizer and cryoprotectant agent [87]. Additionally, it serves as a transport agent for bioactive substances, including probiotics, vitamins, minerals, amino acids, omega-3 fatty acids, curcumin, polyphenols, antioxidants, and essential oils, in biological systems [84, 88]. Nanocellulose, a plant-derived biopolymer, was first used as a food additive in 1983 [89], however, the product could not be commercialized due to the high cost of isolation methods [90]. The commercialization of nanocellulose is now possible due to advancements in its manufacture [79]. According to a research titled "Nanocellulose as a food additive" that was issued by the EU and the Swedish Institute for Food and Biotechnology, nanocellulose offers intriguing potential as an emulsion stabilizing agent [91].

Nanocelluloses exhibit different structural and morphological properties depending on the method of production. Under acid hydrolysis conditions, amorphous structures in nanocellulose are removed and cellulose nanocrystals (CNC) are formed in crystalline form. Cellulose nanofibers (CNF) are produced from lignocellulosic biomass with both amorphous and crystalline parts by mechanical processes (high-pressure homogenization, milling, etc.). Finally, bacterial nanocelluloses (BNC) are biologically synthesized by bacterial cells [92]. The chemical properties of CNC and CNF are generally comparable, although their physical characteristics and methods of extraction are different [93]. CNC is regarded as 90% pure, highly crystalline cellulose nanoparticles. The CNC is shaped like a short rod or needle, with a length of around 100–500 nm and a diameter of about 5–50 nm [94]. CNCs are generally synthesized by methods involving acid hydrolysis. They are usually obtained from the isolation of cellulose by dissolving the amorphous region and noncellulosic components, thus leaving the crystalline region behind [95]. CNF are elementary fibril bundles with a diameter of 5–50 nm and a length of several micrometers [96]. CNFs are mainly obtained from wood or plant sources such as cotton, flax, hemp, etc. CNF can be produced by chemical and biological processes or by mechanical methods and consists of alternation of crystalline (highly ordered structure) and amorphous region (disordered region). Bacteria from the genera *Agrobacteria*, *Acetobacter*, *Pseudomonas*, *Alkaligenes*, *Rhizobium*, and *Sarcina* produce BNC extracellularly in a matter of days to weeks [94]. Because it lacks lignin, hemicellulose, and pectin, BNC is a pure form of cellulosic material with high crystallinity when compared to sources of nanocellulose derived from plants [96]. BNC is a twisted ribbon structure with a diameter usually between 20 and 100 nm and a length measured in micrometers [97].

Nanocellulose-containing paper coatings or paper composites are used as preservatives in food product packaging during sales and transportation processes [98]. In addition, nanocelluloses limit microbial and oxidative spoilage of food due to their antimicrobial and antioxidant properties, thereby extending shelf life [99]. In studies conducted on nanocelluloses, in addition to their antibacterial properties, it has been determined that their use as a filler in paper or as a coating additive regulates oxygen and water vapor permeability, making them more effective in extending shelf life. It is also known that the addition of nanocelluloses improves the thermal and mechanical properties of paper packaging [100]. All forms of NC are generally regarded as biocompatible materials that enhance the growth, adhesion, migration, and differentiation of cells either on their own or in composites with other materials. They are also nontoxic, noninflammatory, nonimmunogenic, and nonallergic [101]. For this purpose, nanocelluloses, which are increasingly used both as fillers and coating materials in paper-derived packaging materials, are even more promising for the future.

4.2 Nanostarch

One of the most prevalent biopolymers in nature is starch. The worldwide industrial starch market research (2020–2028) projects that the starch market will reach a valuation of 97.85 billion US dollars in 2020 and will grow at an annual rate of 7% through 2028. In 2020, it is projected that the world will produce 88.1–97.7 million tons of starch. Of this, 75% is made up of corn starch, with the remaining 14% coming from cassava, 7% from wheat, and 4% from potato starches. Starch, whose chemical structure consists of amylose and amylopectins, is an edible biodegradable polymer. The first of these basic components found in the starch structure is amylose, which is a long chain with an alpha 1–4 glycosidic bond, while amylopectin contains an alpha 1–6 glycosidic bond in addition to this bond and is shorter in size but has a more meshwork structure. When starch is examined, it is stated in the literature that the average amylose content is 30% and the amylopectin content is 70% [102].

Various applications of starch are affected by its physicochemical, functional, and thermal properties. However, starch in its natural state has limited applications and, therefore, is subject to modifications for versatile use [103]. One such industrial application that has garnered significant interest is starch-based biodegradable food packaging, which reduces environmental pollution [104]. Starch is a polymer suitable for nanoparticle or nanocrystal production due to its natural degradability. Nanomaterials can be produced from starch using various methods, including enzymatic, chemical, or physical methods, and these materials are primarily utilized in sensor applications (such as smart packaging) or as reinforcing agents in food packaging [105].

Nanomaterials produced from starch can be grouped under two headings: starch nanocrystals and starch nanoparticles. Starch nanocrystals are generally produced by dual acid hydrolysis below the gelatinization temperature and can be used as fillers in food packaging to adjust mechanical properties [106].

Starch nanoparticles are amorphous starch nanogranules that are created at a nanoscale using methods like high-pressure homogenization, nanoprecipitation, and ultrasonication, and have less than 10% crystallinity [107]. Nanomaterials produced from starch offer significant advantages as a carrier phase due to their properties, including low viscosity and high surface area [108]. In the research conducted on food packaging containing starch nanomaterials, it was determined that these materials provide good flexibility, biodegradability and low molecular mass to the packaging material, as well as showing strength properties similar to low-density polyethylene, and also improve barrier properties against moisture and oxygen and provide protective properties against some microorganisms [109, 110].

When the literature is examined, it is seen that coatings with added starch nanoparticles protect food and reduce water vapor permeability and oxygen permeability [111]. It is known that coated papers produced with starch nanoparticles obtained from potatoes improve physicochemical character, increase tensile strength, and reduce water permeability [112]. In addition, it has been determined that starch nanoparticles are attracted to water and accelerate the diffusion of water into the sample when added to paper and coatings, thereby ensuring faster degradation of the packaging material [113]. Starch nanoparticles, or starch nanocrystals, are used as fillers to improve waterproofing, thermal and barrier properties in addition to their excellent properties such as strength, flexibility, and biodegradability [114, 115].

4.3 Protein Nanoparticles

The application of protein-based nanoparticles to paper-based packaging is increasing over time due to their properties such as biodegradability, easy metabolizability, suitability for surface modification, high nutritional value, and high storage stability. Nanoproteins can be easily produced from water-soluble (bovine or human albumin, etc.) and water-insoluble (zein, gliadin, etc.) proteins [116]. Protein nanoparticles are most commonly used to encapsulate different hydrophilic or lipophilic materials. Nanoproteins can be used in packaging both on their own and by combining with other molecules through electrostatic, van der Waals, or hydrophobic interactions [117]. Additionally, it is known that zein nanoparticles produced from whey protein and soybeans enhance the effectiveness and moisture barrier properties of paper packaging when added to it [118]. In addition, it has been determined that packaging materials made from nanoparticles obtained from whey facilitate sustainable packaging production by demonstrating biodegradability [119]. Protein-based nanoparticles have been used in food packaging, especially to improve moisture barrier properties [120]. In the literature, it has been observed that peanut protein nanoparticles improve strength, temperature resistance, and moisture barrier properties in protein-starch-based biocomposites [121], and zein nanoparticles improve the mechanical and moisture barrier properties of some films [118].

4.4 Chitosan Nanoparticles

Chitin is a type of polycarbonate found abundantly on earth and is used to produce chitosan by deacetylation [122]. Chitosan is a biocompatible and biodegradable polysaccharide [123, 124]). Some of the methods used for the production of chitosan nanoparticles are shown in Fig. 5. 1-Ionotropic gelation; 2-reverse micellar method; 3-emulsification; and 4-nanoprecipitation.

The qualities that make this material so crucial include its biodegradability, nontoxicity, antibacterial activity, biocompatibility, and high mechanical strength [125]. Chitin is transformed into chitosan by chemical and enzymatic processes [125]. Enzymatic and chemical methods are used to convert chitin into chitosan. Chitosan is a nontoxic, environmentally friendly, and antimicrobial material [126]. With these properties, chitosan can be utilized as a paper-coating material, offering significant advantages [127, 128]. It has been observed that chitosan enhances the moisture and gas barrier properties when added to a biodegradable polymer [129],

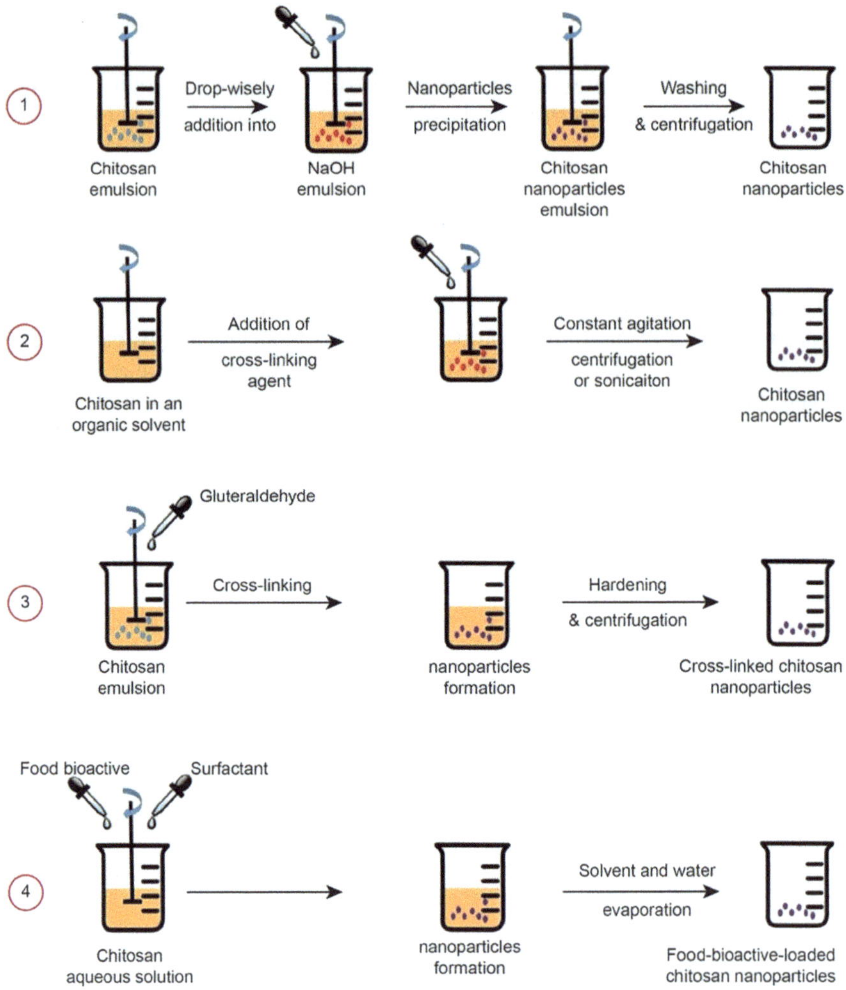

Fig. 5 Chitosan nanoparticle production methods [169]

and these nanoparticles improve the physical, mechanical, and barrier properties in biocomposite films. The life cycle of polylactic acid nanocomposites reinforced with chitosan nanoparticles is given in Fig. 6. The addition of chitosan and its derivatives to paper has been shown to enhance the strength properties of both unaged and aged paper sheets [130].

While the tear factor shows the opposite trend, the values of the breaking length obtained by adding nanochitosan and its derivatives to the pulp suspension are larger than those obtained by dipping the paper in nanochitosan and its derivatives solutions. More generally, it has been demonstrated that adding nanochitosan to paper manufacture improves its mechanical qualities. Additionally, the electrical, printing,

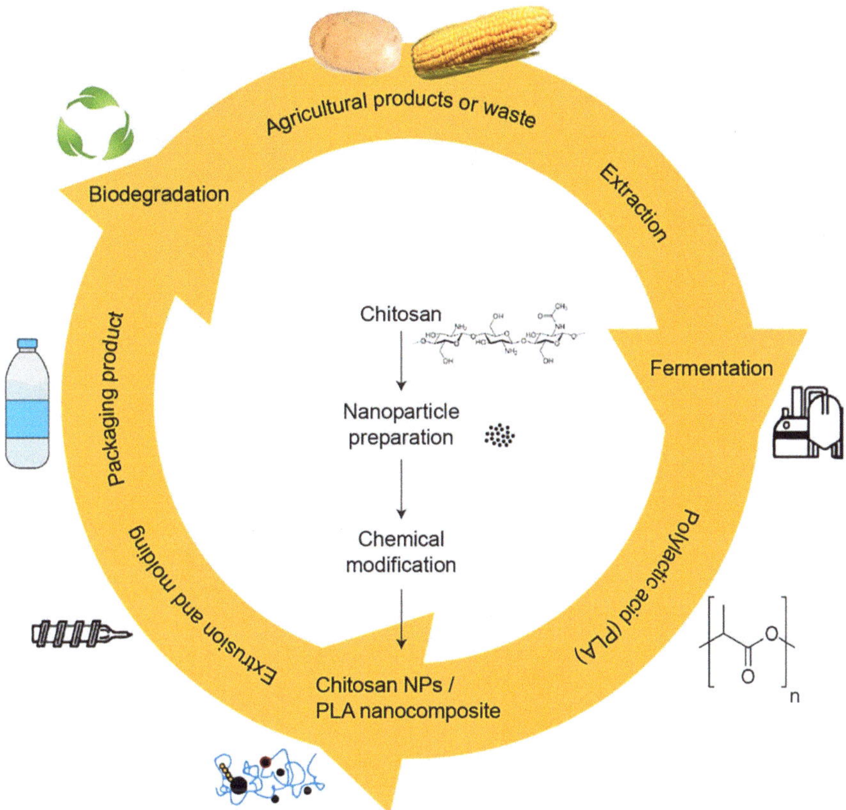

Fig. 6 Life cycle of polylactic acid nanocomposites reinforced with chitosan nanoparticles [129]

barrier, and antibacterial qualities of the paper are enhanced by nanochitosan and its derivatives.

4.5 *Carbon Nanotubes*

Carbon nanotubes are one-dimensional carbon-based materials. They come in two different types: single-walled and multiwalled, and both have an aspect ratio greater than 1000 (Fig. 7).

The use of CNTs in paper packaging is generally encountered in active packaging applications due to their antibacterial properties, which extend shelf life [131], and in integrated packaging applications, where they are utilized in conjunction with printed electronic systems [132]. There are many studies that demonstrate the improvement of thermal, mechanical, and tensile strength of paper by adding CNTs to coating materials [133]. The use of carbon nanotubes in the final product is currently limited

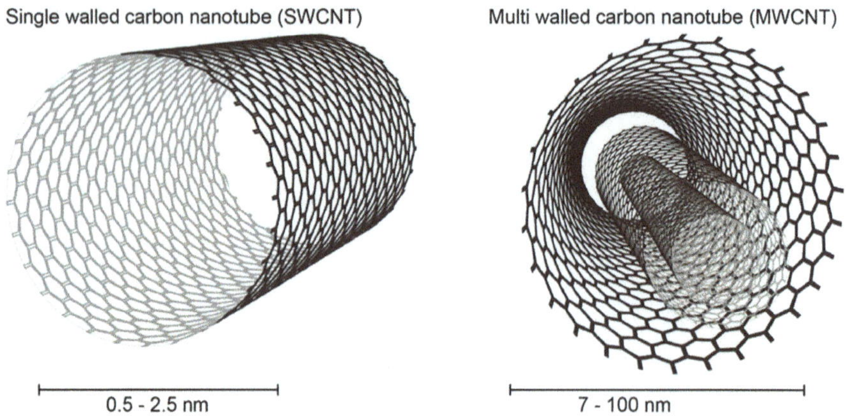

Fig. 7 Two different carbon nanotube structures

to luxury products due to the complicated and expensive production process, as well as the tendency for the paper to lose its white color [134, 135]. Different types of nano-biopolymers and the barrier properties they provide in the paper coating are shown in Fig. 8. In addition, the properties added by organic nanoparticles to coatings are shown in Fig. 9.

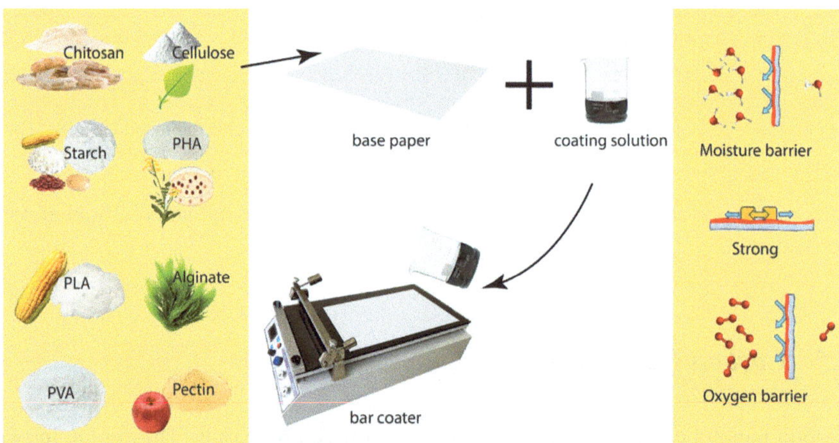

Fig. 8 Various biopolymer sources and how they affect coating paper substrates to improve barrier qualities [40]

Filler	Matrix	NP preparation method	NP modification	NC preparation method	Main findings
Cellulose NPs	PLA PLA, PHB PLA, PHB and plasticizer	Acid hydrolysis	Surfactant grafting L-lactide grafting	Solution casting or melt mixing	+Improvements in oxygen and water vapor permeabilities +Higher UV protection ±Similar thermal behavior -Decrease in transparency -Acceleration in disintegration -Higher stiffness with PLLA grafting
Starch NPs	PLA	Acid hydrolysis	Crosslinking with ECH PLA grafting	Solution casting	+Lower water vapor permeability ±Similar thermal behavior with PLA grafted SNPs -Lower glass transition temperature with epichlorohydrin crosslinked SNPs +Higher tensile strength and Young's modulus
Chitin NPs	PLA PLA, plasticizer PLA, PEG and plasticizer	Acid hydrolysis	L-lactide grafting	Solution casting or melt mixing	+Lower water vapor permeability +The use of PEG reduced the aggregation of NPs ±Increase in biodegradability of chitin NPs reinforced PLA film ±Similar thermal behavior -Lower elongation at break (%) values
Chitosan NPs	PLA	Radical polymerization SCASA Unknown	SMA grafting	Melt mixing	+Better miscibility between SMA-grafted CsNPs and PLA +Slight increase in elongation of PLA until 3% CsNPs content ±Similar thermal behavior

Fig. 9 Organic nanoparticles and their properties for coatings [129]

4.6 Silver Nanoparticles

Due to the disadvantages of organic materials, such as instability at high temperatures and reduced resistance to aggressive conditions, the use of inorganic or nanometals is more promising [136]. Among the inorganic nanomaterials currently in use, AgNPs are used in high amounts in food packaging technology due to their antibacterial properties compared to others.

Silver nanoparticles attract great attention due to their high and specific antibacterial properties against many different bacteria [137, 138]. The effect of silver nanoparticles on bacteria is that silver accumulated in the membrane of bacteria disrupts the membrane integrity, thus causing the death of bacteria. Another view is that the inhibition mechanism occurs through the formation of reactive oxygen species [139, 140].

Compared to other oils or organic nanomolecules used as antimicrobials, especially in food packaging, silver nanoparticles do not cause any organoleptic, taste, or sensory damage and differences, which is the reason for the preference for nanosilver [141]. In order to obtain the desired efficiency from silver nanoparticles in a coating or composite, the shape, size, and concentration of silver nanoparticles in the product are quite effective. In the study, it was determined that nanosilver-added composites were effective against Gram-positive and Gram-negative bacteria. At the same time, the most effective nanoparticles were those of 41 nm in size. The antibacterial effect decreased as the size increased, and it increased as the surface area increased in small-sized nanoparticles, with the interaction with microbial cells increasing accordingly [142]. A frequently encountered result in the literature is that the in vitro bacterial effect increases as the concentration of silver nanoparticles added to the coating or paper increases [143].

Apart from these properties, it has also been determined that nanosilver particles added to the coatings increase opacity, cause color changes, cause differences in mechanical properties depending on van der Waals interactions between silver particles in different binders and the binder polymer, decrease the contact angle with the addition of silver nanoparticles, significantly reduce the moisture absorption capacity of silver nanoparticles, silver nanoparticles cause a significant decrease in the water vapor permeability values of the biopolymer, etc. and provide properties to the paper [144–148]. Silver nanoparticles are made by a variety of methods, including physical, chemical, and biological synthesis. Every technique has its benefits and drawbacks. Biological approaches that utilize natural materials derived from microbes and plants have gained popularity recently due to their low cost, high efficiency, and low toxicity to both humans and the environment [149].

4.7 Zinc Oxide Nanoparticles (ZnO NPs)

Although the use of zinc oxide as an antibacterial substance was first reported in 1995, research on the topic dates back to the early 1950s [150]. As of right now, the US Food and Drug Administration has classified zinc oxide as generally recognized as safe (GRAS) [151]. Chemical reactions and synthesis techniques, such as precipitation, thermal decomposition, and hydrothermal synthesis, have also been employed in the literature, even though the two primary methods for producing ZnO nanoparticles for commercial use are mechanochemical processing (MCP) and physical vapor synthesis (PVS) [152]. Because of their antibacterial properties against both Gram-positive and Gram-negative bacteria, the generated zinc oxide nanoparticles are most frequently used in packaging [153]. The precise mechanism of ZnO nanoparticles' antibacterial action is still a mystery. Nonetheless, several processes are believed to be responsible for the antibacterial action of these nanoparticles. These mechanisms are given in Fig. 10.

Moreover, ZnO nanoparticles enable the improvement of packaging properties such as stability, barrier properties, and mechanical strength. Such characterization is essential in the design of a food package, because each property affects the physical integrity of the package and, therefore, ensures the protection of the packaged product [154–156].

4.8 Titanium Dioxide Nanoparticles (TiO_2 NPs)

Titanium dioxide nanoparticles offer exceptional biocompatibility, photocatalytic properties, and antibacterial capabilities, making them stable and reasonably priced metal oxide nanoparticles suitable for use in biodegradable packaging [157]. The properties that titanium nanoparticles add to packaging are given in Fig. 11.

Titanium nanoparticles are utilized in various industries, including medicine, food, cosmetics, and electronics [158]. TNPs have been utilized in food packaging to produce composite films with enhanced functional properties, resulting in active

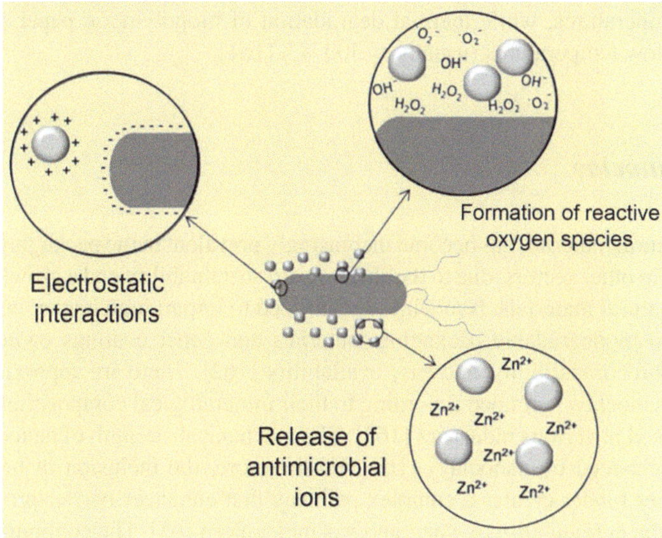

Fig. 10 Different mechanisms of antimicrobial activity of ZnO nanoparticles [153]

Fig. 11 Properties of titanium nanoparticles added package

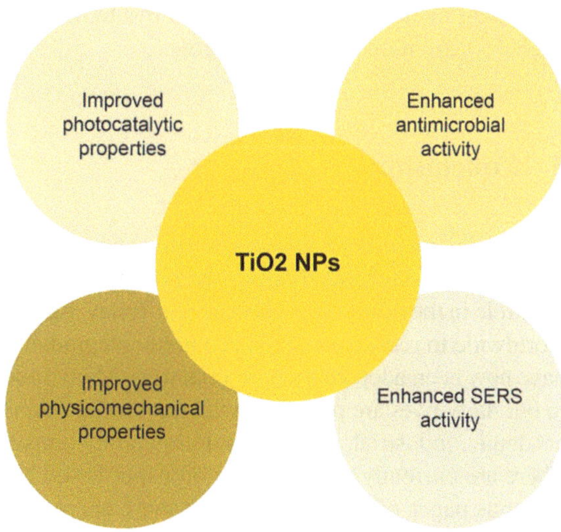

packaging films [159]. Furthermore, titanium nanoparticles interact with the film matrix to enhance its physical strength and gas barrier properties, in addition to its antibacterial properties. These qualities are crucial for active food packaging applications [157]. Furthermore, it has been established that opacity rises with the quantity of TiO_2 NPs added to the packing [160]. TiO_2 NPs added to the packaging material increase thermal stability, meaning that thermal decomposition occurs at higher temperatures, while thermal degradation of biopolymeric paper packaging occurs at low temperatures (usually $\leq$ 300 °C) [161].

4.9 Nanoclay

Clay, a natural material, has become increasingly prevalent in the packaging industry, as well as in other sectors, due to the shift toward sustainability and a growing preference for natural materials. Nanoclays are utilized to impart high barrier and thermal stability to biodegradable packaging materials and paper coatings owing to their low cost, biocompatibility, and easy availability [162]. There are approximately 30 different nanoclays, grouped according to their mineralogical compositions, that are actively used in various industries [163]. The mechanical strength of nanocomposite films is enhanced by nanoclays [164]. Furthermore, the inclusion of nanoclay in the polymer binder creates a complex pathway that enhances barrier properties by blocking the permeability of water and gas molecules [165]. The continuous release characteristic of active substances supported by nanoclays extends the shelf life and freshness of food products, as demonstrated by nanoclay-based nanocomposite packaging solutions [166]. Ethylene scavenging activity is demonstrated by active packaging methods that comprise halloysite nanoclay [167]. Figure 12 shows how nanoclays are used in food packaging.

5 Conclusion

The packaging sector continues to grow and progress every day. When we consider the need to package and transport everything in our daily lives, packaging also plays a vital role in the economies of countries. Today, the steps taken in developed countries worldwide to reduce and even ban nonbiodegradable plastic-like polymer materials have now been adopted by most countries. Many food products that we currently use in our daily lives are packaged with nonbiodegradable polymers. On the other hand, academic, industrial, and legal studies on biocompatible alternatives are promising. There are currently many studies on paper-based alternatives. As mentioned in the previous pages of this book section, there are very positive developments in eliminating the disadvantages arising from the structural features of paper-based products and turning them into advantages. At this point, the coating of the paper surface is of great importance. Nanotechnology applications enable us to utilize many biopolymers that have existed in our world for centuries in paper coating, thereby addressing the weaknesses of paper while incorporating the beneficial features of nanosized materials into the packaging, ultimately achieving superior performance. Coating with biopolymers, such as cellulose, starch, and chitosan, not only improves the properties of the base paper but also minimizes environmental impacts and adds extra functions to the packaging.

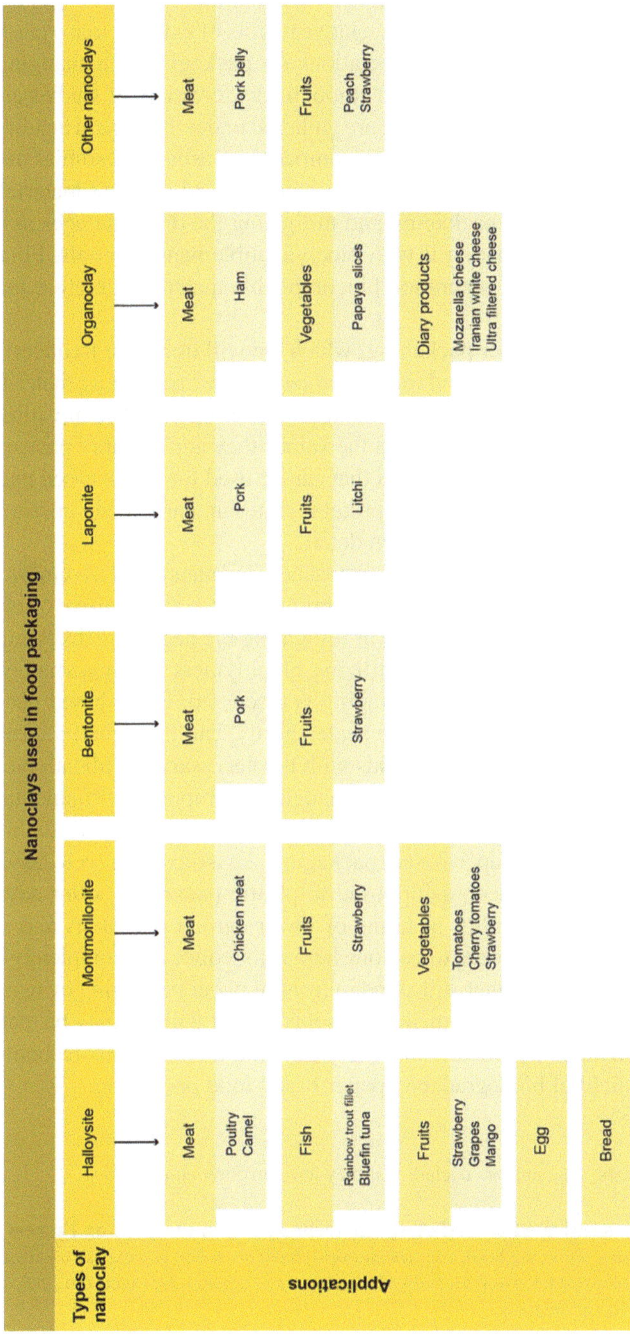

Fig. 12 Application of different nanoclay-based packaging materials in food system [170]

Although the coating of nanosized biopolymers on paper in packaging applications is not a new process, its development process continues. The emergence of new nanomaterials, the minimization of the negative effects of existing materials through nanotechnology, and the evolving expectations for packaging in a changing world continually drive this process forward. Although a wide variety of studies are being conducted, the focal points in packaging are gathered under three main headings. The first step is to reduce and eliminate nonbiocompatible plastics and similar materials. The second step is to use biocompatible materials instead of these materials. This involves switching to paper packaging and mitigating the disadvantages associated with paper packaging by coating it with biocompatible nanomaterials. Finally, the industrialization of the new generation biocompatible material, therefore making it cost-effective.

This study addresses paper packaging, which we will use more frequently in the coming period, and the coating and reinforcement of these packaging materials with nanomaterials to meet all requirements effectively. At this point, biocompatible nanomaterials used in the paper industry and the values they add to paper packaging are discussed. Biodegradable nanomaterials that can be used for paper-based packaging applications, such as cellulose, starch, protein, chitosan, carbon, silver, zinc oxide, titanium oxide and clay, are explained in detail.

Comprehensive research, innovative approaches, and nanomaterial reinforcement, which provide antimicrobial and antioxidant properties, are also focused on changing biopolymer properties. The importance of mitigating the negative effects caused by the structural properties of paper by utilizing biopolymers and nanomaterials and enhancing the properties of paper packaging is discussed. Paper-based packaging, which we will be using in a very large scale shortly, can be a good alternative to petroleum-derived plastic-based materials with the necessary modifications.

Dozens of studies have shown that the changing, disrupted environmental cycle, unsustainable carbon emissions, and the health risks associated with them can be reduced by switching to paper-based packaging. Existing life cycle assessment studies show that biopolymers, as opposed to plastic packaging, offer advantages for environmental sustainability in terms of lower carbon emissions and reduced usage of fossil fuels. What needs to be done here is to reduce and completely remove the commercial restrictions that still stand in front of paper packaging by using more nanotechnological materials and methods. In this way, it will not only be mandatory but also commercially sustainable. In general, developments in nanotechnology are exciting for the field of biodegradable, paper-based food packaging.

Competing Interests The authors declare that they have no competing interest.

Declaration of Generative AI and AI-Assisted Technologies in the Writing Process During the preparation of this work, the book chapter author(s) have not taken any assistance from AI, and the book chapter author(s) reviewed and edited the content as needed and take(s) full responsibility for the content of the publication.

References

1. Khwaldia K, Arab-Tehrany E, Desobry S (2010) Biopolymer coatings on paper packaging materials. Compr Rev Food Sci Food Saf 9(1):82–91
2. Bandyopadhyay A, Ramarao BV, Ramaswamy S (2002) Transient moisture diffusion through paperboard materials. Colloids Surf A 206(1–3):455–467. https://doi.org/10.1016/S0927-7757(02)00067-5
3. Huang J (2017) Sustainable development of green paper packaging. Environ Pollut 6(2):1–7
4. Thompson RC, Moore CJ, Vom Saal FS, Swan SH (2009) Plastics, the environment and human health: current consensus and future trends. Philos Trans R Soc B Biol Sci 364(1526):2153–2166. https://doi.org/10.1098/rstb.2009.0053
5. Calva-Estrada SJ, Jiménez-Fernández M, Lugo-Cervantes E (2019) Protein-based films: advances in the development of biomaterials applicable to food packaging. Food Eng Rev 11:78–92. https://doi.org/10.1007/s12393-019-09189-w
6. Kuswandi B (2017) Environmental friendly food nano-packaging. Environ Chem Lett 15(2):205–221. https://doi.org/10.1007/s10311-017-0613-7
7. Primožič M, Knez Ž, Leitgeb M (2021) (Bio) Nanotechnology in food science—food packaging. Nanomaterials 11(2):292. https://doi.org/10.3390/nano11020292
8. Bhushan B (ed) (2017) Springer handbook of nanotechnology. Springer.
9. Sattler KD (ed) (2010) Handbook of nanophysics: principles and methods, 1st edn. CRC Press. https://doi.org/10.1201/9781420075410
10. Luttge R (2011) Chapter 4-nanotechnology. Microfabrication for industrial applications. In: Luttge R (ed) Micro and nano technologies, pp 91–146
11. He X, Deng H, Hwang HM (2019) The current application of nanotechnology in food and agriculture. J Food Drug Anal 27(1):1–21
12. Bachheti RK, Bachheti A (eds) (2023) Secondary metabolites from medicinal plants: nanoparticles synthesis and their applications. CRC Press, Boca Raton, USA. https://doi.org/10.1201/9781003213727
13. Mlalila N, Kadam DM, Swai H, Hilonga A (2016) Transformation of food packaging from passive to innovative via nanotechnology: concepts and critiques. J Food Sci Technol 53:3395–3407
14. Logothetidis S (2012) Nanotechnology: principles and applications. In: Logothetidis S (ed) Nanostructured materials and their applications. NanoScience and technology. Springer, Berlin, Heidelberg. https://doi.org/10.1007/978-3-642-22227-6_1
15. Gupta A, Eral HB, Hatton TA, Doyle PS (2016) Nanoemulsions: formation, properties and applications. Soft Matter 12(11):2826–2841
16. Singh S, Gaikwad KK, Park SI, Lee YS (2017) Microwave-assisted step reduced extraction of seaweed (Gelidiella aceroso) cellulose nanocrystals. Int J Biol Macromol 99:506–510. https://doi.org/10.1016/J.IJBIOMAC.2017.03.004
17. Dasgupta N, Ranjan S, Mundekkad D, Ramalingam C, Shanker R, Kumar A (2015) Nanotechnology in agro-food: from field to plate. Food Res Int 69:381–400. https://doi.org/10.1016/j.foodres.2015.01.005
18. Ramsden JJ (2016) Chapter 11—bionanotechnology. In: Ramsden JJ (ed) Micro and nano technologies. Nanotechnology, 2nd edn, pp 263–278. William Andrew Publishing: Oxford, UK. ISBN 978-0-323-39311-9
19. Kuswandi B, Moradi M (2019) Improvement of food packaging based on functional nanomaterial. Nanotechnology: applications in energy. Drug Food 309–344. https://doi.org/10.1007/978-3-319-99602-8_16
20. Kuswandi B, Wicaksono Y, Jayus AA, Heng LY, Ahmad M (2011) Smart packaging: sensors for monitoring of food quality and safety. Sens Instrum Food Qual Saf 5:137–146. https://doi.org/10.1007/s11694-011-9120-x
21. Cheng J, Gao R, Zhu Y, Lin Q (2024) Applications of biodegradable materials in food packaging: a review. Alex Eng J 91:70–83. https://doi.org/10.1016/j.aej.2024.01.080

22. Asgher M, Qamar SA, Bilal M, Iqbal HM (2020) Bio-based active food packaging materials: sustainable alternative to conventional petrochemical-based packaging materials. Food Res Int 137:109625. https://doi.org/10.1016/j.foodres.2020.109625
23. Deshwal GK, Panjagari NR, Alam T (2019) An overview of paper and paper based food packaging materials: health safety and environmental concerns. J Food Sci Technol 56:4391–4403. https://doi.org/10.1007/s13197-019-03950-z
24. Fahmy HM, Eldin RES, Serea ESA, Gomaa NM, AboElmagd GM, Salem SA, Shalan AE (2020) Advances in nanotechnology and antibacterial properties of biodegradable food packaging materials. RSC Adv 10(35):20467–20484. https://doi.org/10.1039/D0RA02922J
25. Ncube LK, Ude AU, Ogunmuyiwa EN, Zulkifli R, Beas IN (2020) Environmental impact of food packaging materials: a review of contemporary development from conventional plastics to polylactic acid based materials. Materials 13(21):4994. https://doi.org/10.3390/ma13214994
26. Cazón P, Vázquez M (2021) Bacterial cellulose as a biodegradable food packaging material: a review. Food Hydrocoll 113:106530. https://doi.org/10.1016/j.foodhyd.2020.106530
27. Cheng H, Xu H, McClements DJ, Chen L, Jiao A, Tian Y, Jin Z (2022) Recent advances in intelligent food packaging materials: principles, preparation and applications. Food Chem 375:131738. https://doi.org/10.1016/j.foodchem.2021.131738
28. Jafarzadeh S, Jafari SM (2021) Impact of metal nanoparticles on the mechanical, barrier, optical and thermal properties of biodegradable food packaging materials. Crit Rev Food Sci Nutr 61(16):2640–2658. https://doi.org/10.1080/10408398.2020.1783200
29. Nilsen-Nygaard J, Fernández EN, Radusin T, Rotabakk BT, Sarfraz J, Sharmin N, Pettersen MK (2021) Current status of biobased and biodegradable food packaging materials: impact on food quality and effect of innovative processing technologies. Compr Rev Food Sci Food Saf 20(2):1333–1380. https://doi.org/10.1111/1541-4337.12715
30. Amin U, Khan MU, Majeed Y, Rebezov M, Khayrullin M, Bobkova E, Thiruvengadam M (2021) Potentials of polysaccharides, lipids and proteins in biodegradable food packaging applications. Int J Biol Macromol 183:2184–2198. https://doi.org/10.1016/j.ijbiomac.2021.05.182
31. Weligama Thuppahige VT, Karim MA (2022) A comprehensive review on the properties and functionalities of biodegradable and semibiodegradable food packaging materials. Comprehensive Reviews in Food Science and Food Safety 21(1):689–718. https://doi.org/10.1111/1541-4337.12873
32. Deng J, Zhu EQ, Xu GF, Naik N, Murugadoss V, Ma MG, Shi ZJ (2022) Overview of renewable polysaccharide-based composites for biodegradable food packaging applications. Green Chem 24(2):480–492. https://doi.org/10.1039/D1GC03898B
33. Eslamian Amiri M, Ahmady M, Ariaii P, Golestan L, Ghorbani-HasanSaraei A (2021) Use composite coating of chitosan-chia seed gum enriched with microliposomes of Bay laurel essential oil to increase the shelf life of quail fillets. Food Sci Nutr 9(12):6524–6537. https://doi.org/10.1002/fsn3.2578
34. Farajzadeh F, Motamedzadegan A, Shahidi SA, Hamzeh S (2016) The effect of chitosan-gelatin coating on the quality of shrimp (Litopenaeus vannamei) under refrigerated condition. Food Control 67:163–170. https://doi.org/10.1016/j.foodcont.2016.02.040
35. Soleimani-Rambod A, Zomorodi S, Naghizadeh Raeisi S, Khosrowshahi Asl A, Shahidi SA (2018) The effect of xanthan gum and flaxseed mucilage as edible coatings in cheddar cheese during ripening. Coatings 8(2):80. https://doi.org/10.3390/coatings8020080
36. Ahuja A, Samyn P, Rastogi VK (2024) Paper bottles: Potential to replace conventional packaging for liquid products. Biomass Convers Biorefinery 14(13):13779–13805. https://doi.org/10.1007/s13399-022-03642-3
37. Geyer R, Jambeck JR, Law KL (2017) Production, use, and fate of all plastics ever made. Sci Adv 3(7):e1700782. https://doi.org/10.1126/sciadv.1700782
38. Akhila K, Ramakanth D, Gaikwad KK (2022) Development of novel gallic acid-and cellulose acetate-coated paper as pH-responsive oxygen indicator for intelligent food packaging. J Coat Technol Res 19(5):1493–1506. https://doi.org/10.1007/s11998-022-00622-0

39. Escursell S, Llorach-Massana P, Roncero MB (2021) Sustainability in e-commerce packaging: a review. J Clean Prod 280:124314. https://doi.org/10.1016/J.JCLEPRO.2020.124314
40. Kunam PK, Ramakanth D, Akhila K, Gaikwad KK (2024) Bio-based materials for barrier coatings on paper packaging. Biomass Convers Biorefinery 14(12):12637–12652. https://doi.org/10.1007/s13399-022-03241-2
41. Kumar A, Deshmukh RK, Gaikwad KK (2024) Quality preservation in banana fruits packed in pine needle and halloysite nanotube-based ethylene gas scavenging paper during storage. Biomass Convers Biorefinery 14(5):6311–6320. https://doi.org/10.1007/s13399-022-02708-6
42. Lignou S, Oloyede OO (2021) Consumer acceptability and sensory profile of sustainable paper-based packaging. Foods 10(5):990. https://doi.org/10.3390/FOODS10050990
43. Basak S, Dangate MS, Samy S (2024) Oil-and water-resistant paper coatings: a review. Prog Org Coat 186:107938. https://doi.org/10.1016/j.porgcoat.2023.107938
44. Omerović N, Djisalov M, Živojević K, Mladenović M, Vunduk J, Milenković I, Vidić J (2021) Antimicrobial nanoparticles and biodegradable polymer composites for active food packaging applications. Compr Rev Food Sci Food Saf 20(3):2428–2454. https://doi.org/10.1111/1541-4337.12727
45. Oloyede OO, Lignou S (2021) Sustainable paper-based packaging: a consumer's perspective. Foods 10(5):1035. https://doi.org/10.3390/FOODS10051035
46. Jabeen N, Majid I, Nayik GA (2015) Bioplastics and food packaging: a review. Cogent Food Agri 1(1):1117749. https://doi.org/10.1080/23311932.2015.1117749
47. Motelica L, Ficai D, Ficai A, Oprea OC, Kaya DA, Andronescu E (2020) Biodegradable antimicrobial food packaging: trends and perspectives. Foods 9(10):1438. https://doi.org/10.3390/foods9101438
48. Kopacic S, Walzl A, Zankel A, Leitner E, Bauer W (2018) Alginate and chitosan as a functional barrier for paper-based packaging materials. Coatings 8(7):235. https://doi.org/10.3390/COATINGS8070235
49. Hejazian SA, Ghorbani-HasanSaraei A, Ahmadi M, Shahidi SA (2023) The effect of chia seed mucilage shell coating in combination with turmeric essential oil on egg shelf life and quality properties. J Food Meas Charact 17(4):4175–4190. https://doi.org/10.1007/s11694-023-01923-7
50. Salimbahrami SN, Ghorbani-HasanSaraei A, Tahermansouri H, Shahidi SA (2023) Synthesis, optimization via response surface methodology, and structural properties of carboxymethylcellulose/curcumin/graphene oxide biocomposite films/coatings for the shelf-life extension of shrimp. Int J Biol Macromol 253:126724. https://doi.org/10.1016/j.ijbiomac.2023.126724
51. Brizio APDR, Prentice C (2015) Development of an intelligent enzyme indicator for dynamic monitoring of the shelf-life of food products. Innov Food Sci Emerg Technol 30:208–217. https://doi.org/10.1016/j.ifset.2015.04.001
52. Mahieu A, Terrié C, Youssef BJIC (2015) Thermoplastic starch films and thermoplastic starch/polycaprolactone blends with oxygen-scavenging properties: influence of water content. Ind Crops Prod 72:192–199. https://doi.org/10.1016/j.indcrop.2014.11.037
53. Smolander M, Chaudhry Q (2010). Nanotechnologies in food packaging. https://doi.org/10.1039/9781847559883-00086
54. Meng X, Lee K, Kang TY, Ko S (2015) An irreversible ripeness indicator to monitor the CO_2 concentration in the headspace of packaged kimchi during storage. Food Sci Biotechnol 24:91–97. https://doi.org/10.1007/s10068-015-0014-2
55. Puligundla P, Jung J, Ko S (2012) Carbon dioxide sensors for intelligent food packaging applications. Food Control 25(1):328–333. https://doi.org/10.1016/j.foodcont.2011.10.043
56. Ranjan S, Dasgupta N, Chakraborty AR, Melvin Samuel S, Ramalingam C, Shanker R, Kumar A (2014) Nanoscience and nanotechnologies in food industries: opportunities and research trends. J Nanopart Res 16:1–23. https://doi.org/10.1007/s11051-014-2464-5
57. Sarapulova O, Sherstiuk V, Shvalagin V, Kukhta A (2015) Photonics and nanophotonics and information and communication technologies in modern food packaging. Nanoscale Res Lett 10:1–8. https://doi.org/10.1186/s11671-015-0939-7

58. Feng Y, Xie L, Chen Q, Zheng LR (2015) Low-cost printed chipless RFID humidity sensor tag for intelligent packaging. IEEE Sens J 15(6):3201–3208. https://doi.org/10.1109/JSEN.2014.2385154
59. Realini CE, Marcos B (2014) Active and intelligent packaging systems for a modern society. Meat Sci 98(3):404–419. https://doi.org/10.1016/j.meatsci.2014.06.031
60. Wu L, Yu Y, Zhi J (2015) Low cost and large-area fabrication of self-cleaning coating on polymeric surface based on electroless-plating-like solution deposition approach. RSC Adv 5(14):10159–10164. https://doi.org/10.1039/c4ra10513c
61. Bastarrachea L, Dhawan S, Sablani SS (2011) Engineering properties of polymeric-based antimicrobial films for food packaging: a review. Food Eng Rev 3:79–93. https://doi.org/10.1007/s12393-011-9034-8
62. De Jong AR, Boumans H, Slaghek T, Van Veen J, Rijk R, Van Zandvoort M (2005) Active and intelligent packaging for food: is it the future? Food Addit Contam 22(10):975–979. https://doi.org/10.1080/02652030500336254
63. Hassan NA, Darwesh OM, Smuda SS, Altemimi AB, Hu A, Cacciola F, Abedelmaksoud TG (2022) Recent trends in the preparation of nano-starch particles. Molecules 27(17):5497. https://doi.org/10.3390/molecules27175497
64. Weiss J, Takhistov P, McClements DJ (2006) Functional materials in food nanotechnology. J Food Sci 71(9):R107–R116. https://doi.org/10.1111/j.1750-3841.2006.00195.x
65. Rollini M, Mascheroni E, Capretti G, Coma V, Musatti A, Piergiovanni L (2017) Propolis and chitosan as antimicrobial and polyphenols retainer for the development of paper based active packaging materials. Food Packag Shelf Life 14:75–82. https://doi.org/10.1016/j.fpsl.2017.08.011
66. Brockgreitens J, Abbas A (2016) Responsive food packaging: recent progress and technological prospects. Compr Rev Food Sci Food Saf 15(1):3–15. https://doi.org/10.1111/1541-4337.12174
67. Ozdemir M, Floros JD (2004) Active food packaging technologies. Crit Rev Food Sci Nutr 44(3):185–193. https://doi.org/10.1080/10408690490441578
68. Ozcan A (2020) New approaches in smart packaging technologies. In Proceedings of the 10th international symposium on graphic engineering and design, GRID, pp 21–37. https://doi.org/10.24867/GRID-2020-p1
69. Müller P, Schmid M (2019) Intelligent packaging in the food sector: a brief overview. Foods 8(1):16. https://doi.org/10.3390/foods8010016
70. Biji KB, Ravishankar CN, Mohan CO, Srinivasa Gopal TK (2015) Smart packaging systems for food applications: a review. J Food Sci Technol 52:6125–6135. https://doi.org/10.1007/s13197-015-1766-7
71. Loucanova E, Kalamarova M, Parobek J (2017) The innovative approaches to packaging–comparison analysis of intelligent and active packaging perceptions in Slovakia. Studia Universitatis "Vasile Goldis" Arad Eco Ser 27(2):33–44. https://doi.org/10.1515/sues-2017-0007
72. Ghaani M, Cozzolino CA, Castelli G, Farris S (2016) An overview of the intelligent packaging technologies in the food sector. Trends Food Sci Technol 51:1–11. https://doi.org/10.1016/j.tifs.2016.02.008
73. Ashfaq A, Khursheed N, Fatima S, Anjum Z, Younis K (2022) Application of nanotechnology in food packaging: Pros and Cons. Journal of Agriculture and Food Research 7:100270. https://doi.org/10.1016/j.jafr.2022.100270
74. Singh T, Shukla S, Kumar P, Wahla V, Bajpai VK, Rather IA (2017) Application of nanotechnology in food science: perception and overview. Front Microbiol 8:1501. https://doi.org/10.3389/fmicb.2017.01501
75. Derluyn H, Janssen H, Diepens J, Derome D, Carmeliet J (2007) Hygroscopic behavior of paper and books. J Build Phys 31(1):9–34. https://doi.org/10.1177/1744259107079143
76. Nechita P, Roman M (2020) Review on polysaccharides used in coatings for food packaging papers. Coatings 10(6):566. https://doi.org/10.3390/COATINGS10060566

77. García A, Gandini A, Labidi J, Belgacem N, Bras J (2016) Industrial and crop wastes: a new source for nanocellulose biorefinery. Ind Crops Prod 93:26–38. https://doi.org/10.1016/j.indcrop.2016.06.004
78. Sharma A, Thakur M, Bhattacharya M, Mandal T, Goswami S (2019) Commercial application of cellulose nanocomposites–a review. Biotechnol Rep 21:e00316. https://doi.org/10.1016/j.btre.2019.e00316
79. Perumal AB, Nambiar RB, Moses JA, Anandharamakrishnan C (2022) Nanocellulose: recent trends and applications in the food industry. Food Hydrocoll 127:107484. https://doi.org/10.1016/j.foodhyd.2022.107484
80. Dhali K, Ghasemlou M, Daver F, Cass P, Adhikari B (2021) A review of nanocellulose as a new material towards environmental sustainability. Sci Total Environ 775:145871. https://doi.org/10.1016/j.scitotenv.2021.145871
81. Serpa A, Velásquez-Cock J, Gañán P, Castro C, Vélez L, Zuluaga R (2016) Vegetable nanocellulose in food science: a review. Food Hydrocoll 57:178–186. https://doi.org/10.1016/j.foodhyd.2016.01.023
82. Padhi S, Singh A, Routray W (2023) Nanocellulose from agro-waste: a comprehensive review of extraction methods and applications. Rev Environ Sci Bio/Technol 22(1):1–27. https://doi.org/10.1007/s11157-023-09643-6
83. Espinosa E, Rincón E, Morcillo-Martín R, Rabasco-Vílchez L, Rodríguez A (2022) Orange peel waste biorefinery in multi-component cascade approach: polyphenolic compounds and nanocellulose for food packaging. Ind Crops Prod 187:115413. https://doi.org/10.1016/j.indcrop.2022.115413
84. Alzate-Arbeláez AF, Dorta E, López-Alarcón C, Cortés FB, Rojano BA (2019) Immobilization of Andean berry (Vaccinium meridionale) polyphenols on nanocellulose isolated from banana residues: a natural food additive with antioxidant properties. Food Chem 294:503–517. https://doi.org/10.1016/j.foodchem.2019.05.085
85. Pacaphol K, Seraypheap K, Aht-Ong D (2019) Development and application of nanofibrillated cellulose coating for shelf life extension of fresh-cut vegetable during postharvest storage. Carbohyd Polym 224:115167. https://doi.org/10.1016/j.carbpol.2019.115167
86. Fitri IA, Mitbumrung W, Akanitkul P, Rungraung N, Kemsawasd V, Jain S, Winuprasith T (2022) Encapsulation of β-carotene in oil-in-water emulsions containing nanocellulose: impact on emulsion properties, in vitro digestion, and bioaccessibility. Polymers 14(7):1414. https://doi.org/10.3390/polym14071414
87. Li Z, Wang Q, Li S, Chang Y, Zheng X, Cao H, Zheng Y (2022) Usage of nanocrystalline cellulose as a novel cryoprotective substance for the Nemipterus virgatus surimi during frozen storage. Food Chem X 16:100506. https://doi.org/10.1016/j.fochx.2022.100506
88. Winuprasith T, Khomein P, Mitbumrung W, Suphantharika M, Nitithamyong A, McClements DJ (2018) Encapsulation of vitamin D3 in pickering emulsions stabilized by nanofibrillated mangosteen cellulose: impact on in vitro digestion and bioaccessibility. Food Hydrocoll 83:153–164. https://doi.org/10.1016/j.foodhyd.2018.04.047
89. Turbak AF, Snyder FW, Sandberg KR (1983) Microfibrillated cellulose, a new cellulose product: properties, uses, and commercial potential. J Appl Polym Sci Appl Polym Symp 37(9):815–827
90. Cervin NT, Aulin C, Larsson PT, Wågberg L (2012) Ultra porous nanocellulose aerogels as separation medium for mixtures of oil/water liquids. Cellulose 19:401–410. https://doi.org/10.1007/s10570-011-9629-5
91. Ström G, Öhgren C, Ankerfors M (2013) Nanocellulose as an additive in foodstuff. Innventia Rep 403:1–25
92. Mu R, Hong X, Ni Y, Li Y, Pang J, Wang Q, Zheng Y (2019) Recent trends and applications of cellulose nanocrystals in food industry. Trends Food Sci Technol 93:136–144. https://doi.org/10.1016/j.tifs.2019.09.013
93. Moriana R, Vilaplana F, Ek M (2016) Cellulose nanocrystals from forest residues as reinforcing agents for composites: a study from macro-to nano-dimensions. Carbohyd Polym 139:139–149. https://doi.org/10.1016/j.carbpol.2015.12.020

94. Ahmad Khorairi ANS, Sofian-Seng NS, Othaman R, Abdul Rahman H, Mohd Razali NS, Lim SJ, Wan Mustapha WA (2023) A review on agro-industrial waste as cellulose and nanocellulose source and their potentials in food applications. Food Rev Intl 39(2):663–688. https://doi.org/10.1080/87559129.2021.1926478
95. Lu P, Hsieh YL (2012) Preparation and characterization of cellulose nanocrystals from rice straw. Carbohyd Polym 87(1):564–573. https://doi.org/10.1016/j.carbpol.2011.08.022
96. Nechyporchuk O, Belgacem MN, Bras J (2016) Production of cellulose nanofibrils: a review of recent advances. Ind Crops Prod 93:2–25. https://doi.org/10.1016/j.indcrop.2016.02.016
97. Jozala AF, de Lencastre-Novaes LC, Lopes AM, de Carvalho Santos-Ebinuma V, Mazzola PG, Pessoa-Jr A, Chaud MV (2016) Bacterial nanocellulose production and application: a 10-year overview. Appl Microbiol Biotechnol 100:2063–2072. https://doi.org/10.1007/s00253-015-7243-4
98. Falguera V, Quintero JP, Jiménez A, Muñoz JA, Ibarz A (2011) Edible films and coatings: structures, active functions and trends in their use. Trends Food Sci Technol 22(6):292–303. https://doi.org/10.1016/j.tifs.2011.02.004
99. Appendini P, Hotchkiss JH (2002) Review of antimicrobial food packaging. Innov Food Sci Emerg Technol 3(2):113–126. https://doi.org/10.1016/S1466-8564(02)00012-7
100. Nambiar RB, Perumal AB, Sellamuthu PS, Sadiku ER (2021) Cellulose nanoparticle-based flexible. Adv Mater. https://doi.org/10.1039/9781788019545-00081
101. Eslahi N, Mahmoodi A, Mahmoudi N, Zandi N, Simchi A (2020) Processing and properties of nanofibrous bacterial cellulose-containing polymer composites: a review of recent advances for biomedical applications. Polym Rev 60(1):144–170. https://doi.org/10.1080/15583724.2019.1663210
102. Aldao DC, Šárka E, Ulbrich P, Menšíková E (2018) Starch nanoparticles–two ways of their preparation. Czech J Food Sci 36(2):133. https://doi.org/10.17221/371/2017-CJFS
103. Yazid NSM, Abdullah N, Muhammad N, Matias-Peralta HM (2018) Application of starch and starch-based products in food industry. J Sci Technol 10(2). https://doi.org/10.30880/jst.2018.10.02.023
104. Sozer N, Kokini JL (2009) Nanotechnology and its applications in the food sector. Trends Biotechnol 27(2):82–89. https://doi.org/10.1016/j.tibtech.2008.10.010
105. Le Corre D, Angellier-Coussy H (2014) Preparation and application of starch nanoparticles for nanocomposites: a review. React Funct Polym 85:97–120. https://doi.org/10.1016/j.reactfunctpolym.2014.09.020
106. González K, Retegi A, González A, Eceiza A, Gabilondo N (2015) Starch and cellulose nanocrystals together into thermoplastic starch bionanocomposites. Carbohyd Polym 117:83–90. https://doi.org/10.1016/j.carbpol.2014.09.055
107. Gardouh A, Srag El-Din AS, Moustafa Y, Gad S (2021) Formulation factors of starch-based nanosystems preparation and their pharmaceutical application. Rec Pharm Biomed Sci 5(Pharmacology-Pharmaceutics):28–39. https://doi.org/10.21608/RPBS.2020.51097.1080
108. Adeyeye SAO, Babu AS, Guruprasath N, Ganesh PS (2023) Starch nanocrystal and its food packaging applications. Curr Res Nutr Food Sci J 11(1):01–21. https://doi.org/10.12944/CRNFSJ.11.1.1
109. de Moraes JO, Müller CMO, Laurindo JB (2012) Influence of the simultaneous addition of bentonite and cellulose fibers on the mechanical and barrier properties of starch composite-films. Food Sci Technol Int 18(1):35–45. https://doi.org/10.1177/1082013211427622
110. Nascimento TA, Calado V, Carvalho CWP (2012) Development and characterization of flexible film based on starch and passion fruit mesocarp flour with nanoparticles. Food Res Int 49(1):588–595. https://doi.org/10.1016/j.foodres.2012.07.051
111. Sumit G (2012) Nanotechnology in food packaging a critical review. Russ J Agric Socio Econ Sci 10(10):14–24
112. Hasanin MS (2021) Simple, economic, ecofriendly method to extract starch nanoparticles from potato peel waste for biological applications. Starch-Stärke 73(9–10):2100055. https://doi.org/10.1002/star.202100055

113. Chavan P, Sinhmar A, Sharma S, Dufresne A, Thory R, Kaur M, Nain V (2022) Nanocomposite starch films: a new approach for biodegradable packaging materials. Starch-Stärke 74(5–6):2100302. https://doi.org/10.1002/star.202100302
114. Campelo PH, Sant'Ana AS, Clerici MTPS (2020) Starch nanoparticles: production methods, structure, and properties for food applications. Curr Opin Food Sci 33:136–140. https://doi.org/10.1016/j.cofs.2020.04.007
115. Dularia C, Sinhmar A, Thory R, Pathera AK, Nain V (2019) Development of starch nanoparticles based composite films from non-conventional source-water chestnut (Trapa bispinosa). Int J Biol Macromol 136:1161–1168. https://doi.org/10.1016/j.ijbiomac.2019.06.169
116. Verma D, Gulati N, Kaul S, Mukherjee S, Nagaich U (2018) Protein based nanostructures for drug delivery. J Pharm 2018(1):9285854. https://doi.org/10.1155/2018/9285854
117. Bourbon AI, Pereira RN, Pastrana LM, Vicente AA, Cerqueira MA (2019) Protein-based nanostructures for food applications. Gels 5(1):9. https://doi.org/10.3390/gels5010009
118. Oymaci P, Altinkaya SA (2016) Improvement of barrier and mechanical properties of whey protein isolate based food packaging films by incorporation of zein nanoparticles as a novel bionanocomposite. Food Hydrocolloids 54:1–9. https://doi.org/10.1016/j.foodhyd.2015.08.030
119. Rana S, Fangueiro R, Thakur VK, Joshi M, Thomas S, Fiedler B (2017) Nanomaterials from natural products for industrial applications. J Nanomater 2017. https://doi.org/10.1155/2017/4817653
120. Zubair M, Ullah A (2020) Recent advances in protein derived bionanocomposites for food packaging applications. Crit Rev Food Sci Nutr 60(3):406–434. https://doi.org/10.1080/10408398.2018.1534800
121. Li X, Ji N, Qiu C, Xia M, Xiong L, Sun Q (2015) The effect of peanut protein nanoparticles on characteristics of protein-and starch-based nanocomposite films: a comparative study. Ind Crops Prod 77:565–574. https://doi.org/10.1016/j.indcrop.2015.09.026
122. Muzzarelli RA, Boudrant J, Meyer D, Manno N, DeMarchis M, Paoletti MG (2012) Current views on fungal chitin/chitosan, human chitinases, food preservation, glucans, pectins and inulin: a tribute to Henri Braconnot, precursor of the carbohydrate polymers science, on the chitin bicentennial. Carbohyd Polym 87(2):995–1012. https://doi.org/10.1016/j.carbpol.2011.09.063
123. Bachheti RK, Bachheti A, Husen A (eds) (2025) Chitin-based nanoparticles for the agriculture sectors. Springer. https://doi.org/10.1007/978-981-96-0920-8
124. Baldrick P (2010) The safety of chitosan as a pharmaceutical excipient. Regul Toxicol Pharmacol 56(3):290–299. https://doi.org/10.1016/J.YRTPH.2009.09.015
125. Aider M (2010) Chitosan application for active bio-based films production and potential in the food industry. LWT Food Sci Technol 43(6):837–842. https://doi.org/10.1016/j.lwt.2010.01.021
126. Divya K, Jisha MS (2018) Chitosan nanoparticles preparation and applications. Environ Chem Lett 16:101–112. https://doi.org/10.1007/s10311-017-0670-y
127. Leceta I, Guerrero P, Ibarburu I, Dueñas MT, De la Caba K (2013) Characterization and antimicrobial analysis of chitosan-based films. J Food Eng 116(4):889–899. https://doi.org/10.1016/j.jfoodeng.2013.01.022
128. Ozcan A, Kandirmaz EA, Buyukpehlivan GA (2023) Chitosan-titanium nanoparticle coated papers for active packaging. J Food Eng 356:111584. https://doi.org/10.1016/j.jfoodeng.2023.111584
129. Yanat M, Schroën K (2021) Preparation methods and applications of chitosan nanoparticles; with an outlook toward reinforcement of biodegradable packaging. React Funct Polym 161:104849. https://doi.org/10.1016/j.reactfunctpolym.2021.104849
130. Nada AMA, El-Sakhawy M, Kamel S, Eid MAM, Adel AM (2005) Effect of chitosan and its derivatives on the mechanical and electrical properties of paper sheets. Egypt J Solids 28(2):359–377. https://doi.org/10.21608/EJS.2005.149335
131. Salimi F, Moradi M, Tajik H, Molaei R (2021) Optimization and characterization of eco-friendly antimicrobial nanocellulose sheet prepared using carbon dots of white mulberry (Morus alba L.). J Sci Food Agri 101(8):3439–3447. https://doi.org/10.1002/jsfa.10974

132. Boschetto G, Carapezzi S, Todri-Sanial A (2021) Graphene and carbon nanotubes for electronics nanopackaging. IEEE Open J Nanotechnol 2:120–128. https://doi.org/10.1109/OJNANO.2021.3127652
133. Yu HY, Qin ZY, Sun B, Yang XG, Yao JM (2014) Reinforcement of transparent poly (3-hydroxybutyrate-co-3-hydroxyvalerate) by incorporation of functionalized carbon nanotubes as a novel bionanocomposite for food packaging. Compos Sci Technol 94:96–104. https://doi.org/10.1016/j.compscitech.2014.01.018
134. Liu H, Nishide D, Tanaka T, Kataura H (2011) Large-scale single-chirality separation of single-wall carbon nanotubes by simple gel chromatography. Nat Commun 2(1):309. https://doi.org/10.1038/ncomms1313
135. Rao R, Pint CL, Islam AE, Weatherup RS, Hofmann S, Meshot ER, Hart AJ (2018) Carbon nanotubes and related nanomaterials: critical advances and challenges for synthesis toward mainstream commercial applications. ACS Nano 12(12):11756–11784. https://doi.org/10.1021/acsnano.8b06511
136. Metak AM (2015) Effects of nanocomposite based nanosilver and nano-titanium dioxide on food packaging materials. Int J Appl Sci Technol 5(2):26–40
137. Kalantari K, Mostafavi E, Afifi AM, Izadiyan Z, Jahangirian H, Rafiee-Moghaddam R, Webster TJ (2020) Wound dressings functionalized with silver nanoparticles: promises and pitfalls. Nanoscale 12(4):2268–2291. https://doi.org/10.1039/C9NR08234D
138. Leyu AM, Debebe SE, Bachheti A, Rawat YS, Bachheti RK (2023) Green synthesis of gold and silver nanoparticles using invasive alien plant Parthenium hysterophorus and their antimicrobial and antioxidant activities. Sustainability 15(12):9456
139. Flores-López LZ, Espinoza-Gómez H, Somanathan R (2019) Silver nanoparticles: electron transfer, reactive oxygen species, oxidative stress, beneficial and toxicological effects. Mini review. J Appl Toxicol 39(1):16–26. https://doi.org/10.1002/jat.3654
140. Song Z, Wu Y, Wang H, Han H (2019) Synergistic antibacterial effects of curcumin modified silver nanoparticles through ROS-mediated pathways. Mater Sci Eng C 99:255–263. https://doi.org/10.1016/j.msec.2018.12.053
141. Dhall RK (2013) Advances in edible coatings for fresh fruits and vegetables: a review. Crit Rev Food Sci Nutr 53(5):435–450. https://doi.org/10.1080/10408398.2010.541568
142. De Moura MR, Mattoso LH, Zucolotto V (2012) Development of cellulose-based bactericidal nanocomposites containing silver nanoparticles and their use as active food packaging. J Food Eng 109(3):520–524. https://doi.org/10.1016/j.jfoodeng.2011.10.030
143. Bahrami A, Mokarram RR, Khiabani MS, Ghanbarzadeh B, Salehi R (2019) Physico-mechanical and antimicrobial properties of tragacanth/hydroxypropyl methylcellulose/beeswax edible films reinforced with silver nanoparticles. Int J Biol Macromol 129:1103–1112. https://doi.org/10.1016/j.ijbiomac.2018.09.045
144. Arman Kandirmaz E, Ozcan A (2019) Antibacterial effect of Ag nanoparticles into the paper coatings. Nord Pulp Pap Res J 34(4):507–515. https://doi.org/10.1515/npprj-2019-0034
145. Chang PR, Jian R, Yu J, Ma X (2010) Fabrication and characterisation of chitosan nanoparticles/plasticised-starch composites. Food Chem 120(3):736–740. https://doi.org/10.1016/j.foodchem.2009.11.002
146. Ji N, Liu C, Zhang S, Xiong L, Sun Q (2016) Elaboration and characterization of corn starch films incorporating silver nanoparticles obtained using short glucan chains. LWT 74:311–318. https://doi.org/10.1016/j.lwt.2016.07.065
147. Ortega F, Giannuzzi L, Arce VB, García MA (2017) Active composite starch films containing green synthetized silver nanoparticles. Food Hydrocolloids 70:152–162. https://doi.org/10.1016/j.foodhyd.2017.03.036
148. Shankar S, Tanomrod N, Rawdkuen S, Rhim JW (2016) Preparation of pectin/silver nanoparticles composite films with UV-light barrier and properties. Int J Biol Macromol 92:842–849. https://doi.org/10.1016/j.ijbiomac.2016.07.107
149. Almatroudi A (2020) Silver nanoparticles: Synthesis, characterisation and biomedical applications. Open life Sci 15(1):819–839. https://doi.org/10.1515/biol-2020-0094

150. Sawai J (2003) Quantitative evaluation of antibacterial activities of metallic oxide powders (ZnO, MgO and CaO) by conductimetric assay. J Microbiol Methods 54(2):177–182. https://doi.org/10.1016/S0167-7012(03)00037-X
151. FDA (2011) Part 182—substances generally recognized as safe. Food and drug administration, Washington DC, USA. https://www.ecfr.gov/current/title-21/chapter-I/subchapter-B/part-182. Accessed 11 Oct 2024
152. Casey P (2006) Nanoparticle technologies and applications. In: Nanostructure control of materials, pp 1–31. Woodhead Publishing. https://doi.org/10.1533/9781845691189.1
153. Espitia PJP, Soares NDFF, Coimbra JSDR, de Andrade NJ, Cruz RS, Medeiros EAA (2012) Zinc oxide nanoparticles: synthesis, antimicrobial activity and food packaging applications. Food Bioprocess Technol 5:1447–1464. https://doi.org/10.1007/s11947-012-0797-6
154. Arman Kandirmaz E (2020) Printing of UV-curable transparent conductive polymer composite. J Macromol Sci Part A 57(2):139–144. https://doi.org/10.1080/10601325.2019.1674665
155. Cho JW, Paul DR (2001) Nylon 6 nanocomposites by melt compounding. Polymer 42(3):1083–1094. https://doi.org/10.1016/S0032-3861(00)00380-3
156. Ghule K, Ghule AV, Chen BJ, Ling YC (2006) Preparation and characterization of ZnO nanoparticles coated paper and its antibacterial activity study. Green Chem 8(12):1034–1041. https://doi.org/10.1039/B605623G
157. Alizadeh-Sani M, Khezerlou A, Ehsani A (2018) Fabrication and characterization of the bionanocomposite film based on whey protein biopolymer loaded with TiO_2 nanoparticles, cellulose nanofibers and rosemary essential oil. Ind Crops Prod 124:300–315. https://doi.org/10.1016/j.indcrop.2018.08.001
158. Sani MA, Maleki M, Eghbaljoo-Gharehgheshlaghi H, Khezerlou A, Mohammadian E, Liu Q, Jafari SM (2022) Titanium dioxide nanoparticles as multifunctional surface-active materials for smart/active nanocomposite packaging films. Adv Coll Interface Sci 300:102593. https://doi.org/10.1016/j.cis.2021.102593
159. Sani MA, Ehsani A, Hashemi M (2017) Whey protein isolate/cellulose nanofibre/TiO_2 nanoparticle/rosemary essential oil nanocomposite film: its effect on microbial and sensory quality of lamb meat and growth of common foodborne pathogenic bacteria during refrigeration. Int J Food Microbiol 251:8–14. https://doi.org/10.1016/j.ijfoodmicro.2017.03.018
160. Arezoo E, Mohammadreza E, Maryam M, Abdorreza MN (2020) The synergistic effects of cinnamon essential oil and nano TiO_2 on antimicrobial and functional properties of sago starch films. Int J Biol Macromol 157:743–751. https://doi.org/10.1016/j.ijbiomac.2019.11.244
161. Shao X, Sun H, Zhou R, Zhao B, Shi J, Jiang R, Dong Y (2020) Effect of bovine bone collagen and nano-TiO_2 on the properties of hydroxypropyl methylcellulose films. Int J Biol Macromol 158:937–944. https://doi.org/10.1016/j.ijbiomac.2020.04.107
162. Müller K, Bugnicourt E, Latorre M, Jorda M, Echegoyen Sanz Y, Lagaron JM, Schmid M (2017) Review on the processing and properties of polymer nanocomposites and nanocoatings and their applications in the packaging, automotive and solar energy fields. Nanomaterials 7(4):74. https://doi.org/10.3390/nano7040074
163. Guo F, Aryana S, Han Y, Jiao Y (2018) A review of the synthesis and applications of polymer–nanoclay composites. Appl Sci 8(9):1696. https://doi.org/10.3390/app8091696
164. Risyon NP, Othman SH, Basha RK, Talib RA (2020) Characterization of polylactic acid/halloysite nanotubes bionanocomposite films for food packaging. Food Packag Shelf Life 23:100450. https://doi.org/10.1016/j.fpsl.2019.100450
165. Lee MH, Kim SY, Park HJ (2018) Effect of halloysite nanoclay on the physical, mechanical, and antioxidant properties of chitosan films incorporated with clove essential oil. Food Hydrocoll 84:58–67. https://doi.org/10.1016/j.foodhyd.2018.05.048
166. Bumbudsanpharoke N, Ko S (2015) Nano-food packaging: an overview of market, migration research, and safety regulations. J Food Sci 80(5):R910–R923. https://doi.org/10.1111/1750-3841.12861

167. Boonsiriwit A, Xiao Y, Joung J, Kim M, Singh S, Lee YS (2020) Alkaline halloysite nanotubes/low density polyethylene nanocomposite films with increased ethylene absorption capacity: applications in cherry tomato packaging. Food Packag Shelf Life 25:100533. https://doi.org/10.1016/j.fpsl.2020.100533
168. Yildirim S, Röcker B, Pettersen MK, Nilsen-Nygaard J, Ayhan Z, Rutkaite R, Coma V (2018) Active packaging applications for food. Compr Rev Food Sci Food Saf 17(1):165–199. https://doi.org/10.1111/1541-4337.12322
169. Garavand F, Cacciotti I, Vahedikia N, Rehman A, Tarhan Ö, Akbari-Alavijeh S, Jafari SM (2022) A comprehensive review on the nanocomposites loaded with chitosan nanoparticles for food packaging. Crit Rev Food Sci Nutr 62(5):1383–1416. https://doi.org/10.1080/10408398.2020.1843133
170. Nath D, Santhosh R, Pal K, Sarkar P (2022) Nanoclay-based active food packaging systems: a review. Food Packag Shelf Life 31:100803. https://doi.org/10.1016/j.fpsl.2021.100803

Paper-Based Nanocomposites: Preparation, Application, and Challenges

Shrutikona Das and Kumar Anupam

Abstract The incorporation of nanomaterials in paper matrices has received considerable attention for diversifying the use of traditional paper-based materials. The incorporation of nanomaterials into a paper matrix results in paper-based nanocomposites that enhance the mechanical, barrier, optical, and functional qualities of the base paper. Various nanomaterials, such as those made from wood components, inorganic, organic, carbon-based, and composite nanoparticles, can be added to the paper. Interaction among nanomaterials, paper matrix, and other additives is crucial for understanding paper functionality while preserving the basic properties of the paper matrix. Different types of nanomaterials, such as nanoclay, nano-silica, cellulose nanofiber, nano-titanium dioxide, silver nanoparticles, carbon nanotubes, etc., give rise to various types of paper nanocomposites. This chapter discusses the aforementioned types of nanocomposites, along with their preparation strategies, including dip coating, in situ synthesis, vacuum-assisted filtration, blending, drop casting, layer-by-layer assembly, surface treatment, wet-end addition, size press treatment, direct assembly, and the sol–gel method. By using these techniques, nanomaterials can be successfully incorporated into the paper matrix, improving its qualities and capabilities. The application of these paper-based nanocomposites in flexible electronics, antibacterial packaging, water purification, energy storage, and smart packaging has also been discussed in this chapter. This chapter also highlights the challenges of cost, compatibility, scale-up, and performance stability associated with paper-based nanocomposites. Environmental impact, regulatory compliance, and health and safety issues are also featured.

Keywords Nanomaterial · Nanocomposite · Synthesis · Pulp and paper · Sustainability

S. Das · K. Anupam (✉)
Chemical Recovery and Biorefinery Division, Central Pulp and Paper Research Institute, Saharanpur, Uttar Pradesh, India
e-mail: kumaranupam@live.com

R. K. Bachheti (ed.), *Nanotechnology in the Pulp and Paper Industry*, Smart Nanomaterials Technology, https://doi.org/10.1007/978-981-95-3473-9_9

1 Introduction

Nanocomposites represent a groundbreaking advancement in materials science, combining nanoscale fillers with bulk matrices to create materials with unprecedented properties and performance characteristics. These engineered materials typically consist of a host matrix—which may be polymeric, metallic, or ceramic—reinforced with nanoscale constituents such as carbon nanotubes (CNTs), graphene, nanoclay, metallic nanoparticles, or quantum dots, with at least one dimension in the 1–100 nm range [1, 2]. The extraordinary properties of nanocomposites stem from their exceptionally high surface-area-to-volume ratios and quantum effects that emerge at the nanoscale, which enable dramatic improvements in mechanical strength, thermal stability, electrical conductivity, and optical properties compared to conventional composites [3–5]. For instance, the incorporation of just 0.5–5 wt% of well-dispersed carbon nanotubes into a polymer matrix can enhance tensile strength by 200–300%. At the same time, nanoclay reinforcements can improve gas barrier properties by up to 100-fold [6]. When properly dispersed, nanoparticles create an enormous interfacial area—up to 1000 m^2/g for some carbon nanomaterials—that dramatically alters the matrix's molecular mobility and crystallization behavior [7]. Three primary mechanisms govern property enhancements in nanocomposites: (1) restriction of polymer chain mobility near nanoparticle surfaces, (2) formation of percolation networks for electrical/thermal conductivity, and (3) crack deflection and energy dissipation at nanoparticle–matrix interfaces [8–11]. For example, in epoxy–CNT nanocomposites, the nanotubes' high aspect ratio (typically 100–1000) enables efficient stress transfer while their intrinsic conductivity ($\approx 10^6$ S/m) creates conductive pathways at low loadings (0.1–1 vol%) [12]. Similarly, layered silicates like montmorillonite (MMT, nanoclay can improve flame retardancy by creating tortuous diffusion paths for oxygen and volatile degradation products [13]. However, achieving optimal performance requires overcoming significant challenges in nanoparticle dispersion, interfacial bonding, and scalable manufacturing—issues that have driven extensive research in surface functionalization, compatibilizers, and advanced processing techniques.

Polymer nanocomposites constitute the largest and most commercially successful category, with global markets projected to exceed $15 billion by 2025 [14]. Polypropylene (PP)–nanoclay composites, for instance, are widely used in automotive parts (e.g., bumpers and dashboards), where they provide 40–50% weight reduction compared to metals while meeting stringent mechanical requirements [15]. In aerospace applications, epoxy–CNT nanocomposites enable up to 30% weight savings in aircraft components while providing electromagnetic interference (EMI) shielding—critical for avionics protection [16]. The electronics industry extensively employs polyimide–silver (Ag) nanowire nanocomposites for flexible printed circuits, where they combine high conductivity ($\approx 10^5$ S/m) with exceptional bendability (>100,000 cycles at 5-mm radius) [17, 18]. Biomedical applications leverage chitosan–silver nanoparticle (AgNP) nanocomposites for wound dressings that provide sustained antimicrobial activity while maintaining biocompatibility [19]. Recent advances in stimuli-responsive nanocomposites—such as pH-sensitive hydrogels with graphene oxide (GO) or temperature-responsive polymers with magnetic nanoparticles—are enabling smart drug delivery systems with precise spatiotemporal control [12]. Alumina–silicon carbide nanocomposites (5–10 vol% SiC) exhibit fracture toughness values (8–10 MPa·m½) nearly double those of monolithic alumina, making them ideal for cutting tools and ballistic armor [20]. In the energy sector, yttria-stabilized zirconia (YSZ)–graphene nanocomposites demonstrate 300% higher ionic conductivity than conventional YSZ electrolytes, potentially revolutionizing solid oxide fuel cells [21]. Metal matrix nanocomposites, such as aluminum–carbon nanotube systems, combine the lightweight nature of aluminum (2.7 g/cm^3) with strength rivaling that of some steels (yield strength >500 MPa), enabling the development of next-generation automotive and aerospace components [22].

The manufacturing of nanocomposites presents unique challenges that have spurred innovations in processing technologies. Solution mixing remains the most common method for polymer nanocomposites, where ultrasonication (typically 20–100 W for 10–60 min) and high-shear mixing (5000–20,000 rpm) help achieve nanoparticle dispersion [23]. Melt compounding using twin-screw extruders (barrel temperatures of 150–300 °C, depending on the polymer) offers more industrially scalable production, but it risks nanoparticle degradation at high shear rates [24]. In situ polymerization—where nanoparticles are dispersed in monomer before polymerization—provides exceptional dispersion but requires careful control of reaction kinetics [25]. For ceramic and metal nanocomposites, colloidal processing routes (e.g., slip casting, tape casting) combined with pressure-assisted sintering help maintain nanoscale microstructures [26]. Emerging techniques, such as 3D printing of nanocomposites, enable the creation of complex geometries with locally tuned properties. For instance, graphene-reinforced polylactic acid (PLA) filaments printed with 0.5–2 wt% loading show an 80% improvement in flexural strength while maintaining printability [27].

Future directions in nanocomposite research focus on multifunctional smart materials, including self-healing nanocomposites (e.g., microcapsule-embedded systems that autonomously repair cracks), energy-harvesting composites (piezoelectric nanofillers in polymer matrices), and bioinspired nanocomposites that mimic natural materials like nacre [8–11]. Advances in computational materials science, particularly machine learning-assisted design of nanocomposite architectures, are accelerating the discovery of optimal nanoparticle–matrix combinations [28–32]. As manufacturing scalability improves and costs decrease, nanocomposites are poised to transform industries ranging from flexible electronics to sustainable construction, ushering in a new era of high-performance, multifunctional materials. In Table 1, the types of nanocomposites, their key properties, and possible matrix compositions are presented.

Table 1 Key properties of nanocomposites along with possible matrix composition

Serial no	Nanocomposite type	Key properties	Possible matrix materials	References
[1]	Polymer–clay	Enhanced strength, thermal stability, barrier properties	Polypropylene (PP), epoxy, nylon	[33]
[2]	CNT–polymer	High electrical conductivity, mechanical reinforcement	Polyethylene (PE), polystyrene (PS)	[34]
[3]	Graphene–polymer	Superior electrical/thermal conductivity, flexibility	Epoxy, polyurethane (PU)	[35]
[4]	SiO_2–polymer	Scratch resistance, UV protection	PMMA, polycarbonate (PC)	[36]
[5]	TiO_2–polymer	Photocatalytic, UV blocking	Polyester, PVC	[36]
[6]	ZnO–polymer	Antimicrobial, UV shielding	Polyethylene (PE), latex	[36]
[7]	Al_2O_3–polymer	High hardness, thermal resistance	Epoxy, polyimide	[37, 38]
[8]	Fe_3O_4–polymer	Magnetic properties, biocompatibility	PLA, chitosan	[39]
[9]	Ag–polymer	Antibacterial, conductive	PVA, polyethylene (PE)	[36]
[10]	Cellulose nanofiber–polymer	Biodegradable, high strength	PLA, PVA, starch	[39]
[11]	Halloysite–polymer	Improved flame retardancy, mechanical strength	Epoxy, rubber	[33]
[12]	Boron nitride–polymer	High thermal conductivity, electrically insulating	Silicone, epoxy	[37, 38]
[13]	Quantum dot–polymer	Tunable optical properties	PDMS, acrylic	[40]
[14]	MoS_2–polymer	Lubrication, semiconductor properties	Nylon, epoxy	[37, 38]
[15]	LDH–polymer	Flame retardant, corrosion inhibition	Polyolefins, PVC	[33]

(continued)

Table 1 (continued)

Serial no	Nanocomposite type	Key properties	Possible matrix materials	References
[16]	Carbon black–polymer	Electrical conductivity, reinforcement	Rubber, polyethylene (PE)	[34]
[17]	Nanodiamond–polymer	Extreme hardness, thermal conductivity	PEEK, epoxy	[37, 38]
[18]	Chitosan–nanoparticle	Biocompatible, antimicrobial	Alginate, collagen	[39]
[19]	POSS–polymer	Enhanced thermal stability, low density	Epoxy, silicone	[37, 38]
[20]	MXene–polymer	High conductivity, mechanical strength	PVDF, hydrogels	[40]

Other than polymer matrix mentioned in Table 1, paper is also one of the possible matrices for incorporation of nanomaterials. These are called paper-based nanomaterials. Paper-based nanocomposites offer several advantages over other nanocomposites mentioned above. This chapter provides a detailed overview of paper-based nanocomposites, including their classification, preparation pathways, application routes, and the challenges associated with their use. This chapter is categorized into five sections. In Section One, there is a brief introduction to nanocomposite, whereas Section Two focuses on paper-based nanocomposites and their impact on base paper matrix. The usefulness of paper matrices in composition with different nanomaterials, such as nano-silica, nanoclay, nano-titanium dioxide (TiO_2), silver nanoparticles, cellulose nanofibers (CNFs), and carbon nanotubes, has been overviewed in this section. Section Three provides various preparation routes for nanocomposites, including dip coating, vacuum-assisted filtration, layer-by-layer (LBL) assembly, blending, in situ synthesis, drop casting, wet-end addition, surface treatment, size reduction treatment, sol–gel method, and direct assembly. Applications of paper-based nanocomposites in flexible electronics, antibacterial packaging-grade paper, energy storage devices, water purification, and smart packaging materials have been discussed in Section Four. Section Five highlights several challenges like, cost, compatibility, environmental impact, health and safety, regulatory compliance, scale-up, and performance stability associated with paper-based nanocomposites.

2 Paper-Based Nanocomposites

The paper industry uses nanomaterials such as nanoclay, nano-silica, cellulose nanofibers, nano-titanium dioxide, silver nanoparticles, graphene, or carbon nanotubes (CNTs) in a paper matrix to formulate new-generation nanocomposites for industrial use. Figure 1 shows different types of paper-based nanocomposites. The impact of paper-based nanocomposites on paper properties is provided in Table 2. Details of various types of paper-based nanocomposites are provided in the following subsections.

2.1 Nanoclay-Based Paper Nanocomposites

Nanoclay is widely used in paper applications to enhance performance, functionality, and sustainability. In barrier coatings, it reduces the permeability of oxygen, water vapor, and oil, improving the protective qualities of paper and board. When used in packaging paper, especially for food products, it serves as an eco-friendly alternative to plastic or aluminum laminates. In printing paper, nanoclay improves

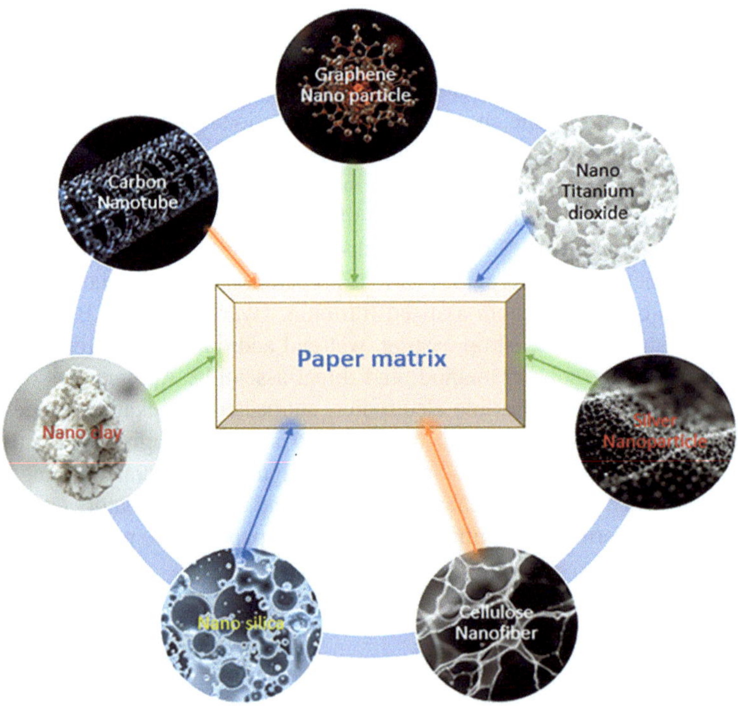

Fig. 1 Different compositions of paper-based nanocomposites

Table 2 Impact of paper-based nanocomposites on paper properties

Property	Description	Enhancement due to nanomaterials	References
Mechanical strength	Tensile strength, flexibility, and durability of paper composites	Incorporation of nanocellulose or carbon nanotubes improves tensile strength and elasticity	[41]
Barrier properties	Resistance to moisture, oxygen, and grease penetration	Nanoclay or graphene oxide coatings enhance hydrophobicity and gas barrier performance	[42]
Thermal stability	Ability to withstand high temperatures without degradation	Cellulose nanofibers (CNFs) and silica nanoparticles improve thermal resistance	[43–46]
Electrical conductivity	Conductivity for applications in flexible electronics	Addition of silver nanowires or carbon nanotubes enables conductive paper	[47]
Biodegradability	Eco-friendly decomposition compared to synthetic composites	Nanocellulose retains biodegradability while enhancing strength	[48, 49]
Optical transparency	Light transmittance for applications like flexible displays	Thin films of cellulose nanofibers maintain transparency with improved mechanical properties	[50]

surface smoothness, printability, and ink absorption control. It also contributes to flame-retardant paper by increasing thermal stability and fire resistance. Additionally, when mixed with cellulose nanofibers or added to the wet-end slurry, nanoclay enhances the mechanical strength and stiffness of paper. In greaseproof paper, it is used in coatings to block the penetration of grease and oil, offering a biodegradable alternative to traditional fluorochemical-based coatings. Chauhan and Chakrabarti [51] reported that the potential use of nanoclay composites includes high-density polyethylene (HDPE) containers, multilayer flexible films for ketchup and potato crisps, and barrier layers in multilayer polyethylene terephthalate (PET) bottles, as well as meat and cheese containers in food packaging. The product created a waterproof barrier when it was applied to a paper surface [51]. Nanoclay-based paper nanocomposites represent a significant advancement in sustainable material engineering, combining the renewable nature of cellulose fibers with the exceptional reinforcing properties of nanoclays. These composites are typically fabricated by incorporating layered silicates, such as montmorillonite (MMT) or kaolinite, into paper matrices through techniques like solution casting, layer-by-layer assembly, or in situ polymerization [52, 53]. The high aspect ratio (100–1000) and large surface area (750–800 m^2/g) of exfoliated nanoclay platelets significantly enhance paper properties, including tensile strength (40–60% improvement at 3–5 wt% loading), thermal stability (decomposition temperature increased by 50–80 °C), and barrier performance (oxygen permeability reduced by 90–95%) [54]. These enhancements stem from the nanoclay's ability to create tortuous diffusion paths for gases and restrict the mobility of polymer chains within the cellulose matrix [55, 56].

Nanoclay–paper composites find extensive applications in active food packaging, where their improved barrier properties extend shelf life while maintaining biodegradability [57]. For instance, MMT–chitosan nanocomposite coatings on paper demonstrate exceptional antimicrobial activity (log 4 reduction against *Escherichia coli*) due to the synergistic effects of chitosan's biocidal properties and nanoclay's ion-exchange capacity [58]. The materials also show promise in flame-retardant packaging, where nanoclay's char-forming ability reduces peak heat release rates by 40–60% [59]. Despite these advantages, challenges persist in achieving complete nanoclay exfoliation and uniform dispersion within the porous paper structure, often requiring surface modification with quaternary ammonium salts or silanes [60]. Recent advances include hybrid systems combining nanoclay with cellulose nanofibers (CNFs) to create hierarchical structures with tunable mechanical and barrier properties [8–11]. With growing emphasis on sustainable packaging, nanoclay–paper nanocomposites offer an eco-friendly alternative to plastic films while maintaining compatibility with conventional paper recycling processes [61].

2.2 *Nano-Silica-Based Paper Nanocomposites*

Nano-silica-based paper nanocomposites are advanced materials that incorporate silica nanoparticles (SiO_2) into paper matrices to enhance mechanical, thermal, barrier, and functional properties. These composites leverage the high surface area, reactivity, and tunable chemistry of nano-silica to improve traditional paper performance for specialized applications. Jin et al. [62] strengthened and extended the lifespan of the paper fibers in their work by using nano-silica based on polyurethane. Researchers investigated the relationship between the mass concentration of polyurethane-based nano-silica and the specific characteristics of the papers. The findings demonstrated that nano-silica-based polyurethane could significantly enhance the mechanical characteristics (such as tearing strength, tensile strength, and folding endurance) and age resistance of papers [62]. Nano-silica-based paper nanocomposites have garnered significant attention due to their enhanced mechanical, thermal, and barrier properties, making them suitable for various applications, including packaging, filtration, and biomedical uses. The incorporation of nano-silica into paper matrices improves tensile strength, water resistance, and thermal stability by forming a strong interfacial bond with cellulose fibers [43–46]. The high surface area and reactivity of nano-silica facilitate uniform dispersion within the paper matrix, leading to improved optical properties and reduced porosity [63, 64]. Additionally, these nanocomposites exhibit antimicrobial and flame-retardant properties, expanding their use in food packaging and safety applications [65, 66]. The sol–gel method and in situ precipitation are common techniques for synthesizing nano-silica paper composites, offering cost-effectiveness and scalability [67]. Environmental benefits, such as biodegradability and reduced chemical usage, further enhance their appeal as sustainable alternatives to conventional paper coatings [68]. However, challenges such as nanoparticle agglomeration and high production costs necessitate further research to optimize processing techniques and ensure commercial viability [48, 49]. Recent advancements in functionalized nano-silica, including hydrophobic and hydrophilic modifications, have further tailored these composites for specialized applications [69–72]. Overall, nano-silica-based paper nanocomposites represent a promising innovation in materials science, combining performance enhancements with eco-friendly attributes.

2.3 Cellulose Nanofiber-Based Paper Nanocomposites

Cellulose nanofiber (CNF)-based paper nanocomposites have emerged as a revolutionary class of sustainable materials due to their exceptional mechanical strength, biodegradability, and tunable surface chemistry, making them ideal for advanced packaging, flexible electronics, and biomedical applications. CNFs, derived from renewable biomass sources such as wood, agricultural residues, or bacterial cellulose, exhibit high crystallinity, large surface area, and excellent reinforcement capabilities when incorporated into paper matrices [73]. The hierarchical structure of CNFs allows for strong hydrogen bonding with cellulose fibers, significantly enhancing tensile strength (up to 200–300 MPa) and Young's modulus (10–20 GPa) compared to conventional paper [74]. Additionally, the nanoscale dimensions of CNFs (3–20 nm in width and several micrometers in length) enable transparent and lightweight nanocomposites with improved barrier properties against oxygen, water vapor, and ultraviolet (UV) radiation, which are critical for food packaging and protective coatings [75]. Functional modifications, such as carboxylation, TEMPO-mediated oxidation, or polymer grafting, further enhance CNF dispersibility and interfacial adhesion with hydrophobic matrices like polylactic acid (PLA) or polyethylene (PE), expanding their compatibility with industrial processing techniques [76]. Moreover, CNF-based nanocomposites exhibit remarkable thermal stability (decomposition temperatures above 250 °C) and flame-retardant properties when combined with additives like clay or boron nitride, making them suitable for high-performance applications [77]. In the biomedical field, CNF papers are explored for wound dressings, drug delivery carriers, and tissue engineering scaffolds due to their biocompatibility, high porosity, and ability to immobilize bioactive molecules [78]. Environmental advantages, including low carbon footprint, recyclability, and compostability, align CNF nanocomposites with circular economy principles, addressing global plastic pollution challenges [79]. However, challenges such as high production costs, scalability limitations in CNF isolation (e.g., mechanical fibrillation, enzymatic pretreatment), and moisture sensitivity under high-humidity conditions require further research to optimize cost-effective manufacturing and chemical cross-linking strategies [55, 56]. Recent advancements in hybrid nanocomposites—integrating CNFs with nanoparticles like silver, graphene, or nano-silica—have enabled multifunctional materials with antimicrobial, conductive, or self-cleaning properties [80].

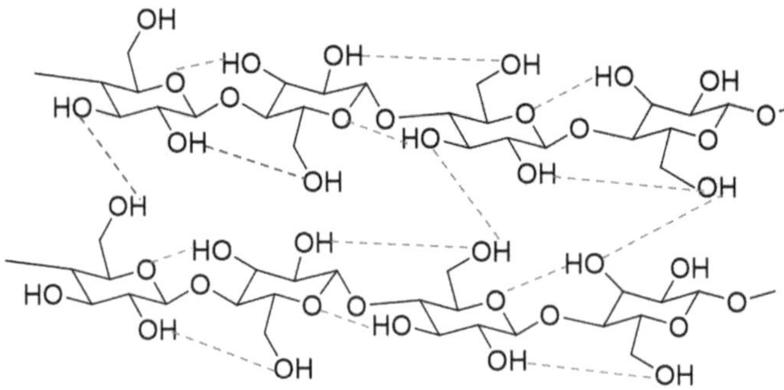

Fig. 2 Structure of cellulose nanofiber

For instance, CNF–silver nanocomposites exhibit potent antibacterial activity for medical packaging, while CNF–graphene hybrids show promise in flexible supercapacitors and sensors [81, 82]. Industry adoption is growing, with pilot-scale production of CNF-reinforced papers for automotive interiors, electronic substrates, and biodegradable films, supported by policies promoting sustainable materials [83]. Future research should focus on life cycle assessment (LCA) to validate environmental benefits, standardize CNF grades, and develop high-throughput processing methods to bridge the lab-to-industry gap [84]. In summary, CNF-based paper nanocomposites represent a paradigm shift in materials science, offering a unique combination of performance, sustainability, and versatility to meet the demands of next-generation applications while reducing ecological impact. In Fig. 2, organic skeleton of cellulose nanofiber is presented.

2.4 Nano-Titanium Dioxide-Based Paper Nanocomposites

Lysenkov et al. added modified nano-TiO_2 to cellulosic fibers to create a new kind of paper with a superhydrophobic surface. After using a high-speed homogenizer to disperse the nano-TiO_2 powder, the coupling agent 3-(trimethoxysilyl)propyl methacrylate (MPS) was used to modify the powder's surface. According to the findings, the water contact angles of the modified paper varied between 126.5° and 154.2°. Furthermore, a large number of evenly spaced nano-TiO_2 protuberances were seen on the paper's surface, further confirming that the nanostructure of the resulting paper made it superhydrophobic. Comparative optical analyses on the paper hand sheets showed that the sample containing the MPS-modified TiO_2 nanoparticles had significantly higher opacity [85]. Nano-titanium dioxide (TiO_2)-based paper nanocomposites have gained significant attention in advanced material applications due to their unique photocatalytic, UV-blocking, antimicrobial, and self-cleaning properties, making them highly valuable for packaging, construction, and environmental remediation sectors. The incorporation of TiO_2 nanoparticles (typically 10–100 nm in size) into paper matrices enhances mechanical strength, thermal stability, and optical properties while maintaining flexibility and biodegradability [86]. TiO_2 exists in three primary crystalline phases (anatase, rutile, and brookite), with anatase being the most photocatalytically active, enabling self-cleaning paper surfaces that decompose organic pollutants and stains under UV or visible light irradiation [87]. This photocatalytic activity is further enhanced through doping with nitrogen, carbon, or metals (e.g., silver or copper), which extends light absorption into the visible spectrum, improving efficiency in indoor applications [88]. In packaging, TiO_2 nanocomposites offer exceptional UV shielding, preventing the photodegradation of perishable goods while exhibiting potent antibacterial effects against pathogens such as *E. coli* and *Staphylococcus aureus*, which are crucial for food safety [89]. Additionally, these composites demonstrate superior barrier properties against moisture and oxygen, extending shelf life without synthetic polymer coatings [90]. The mechanical performance of TiO_2-reinforced paper, including tensile strength (up to 40% improvement) and abrasion resistance, arises from the nanoparticles' high surface area and strong interfacial bonding with cellulose fibers [91, 92]. Environmental applications include air and water purification, where TiO_2-coated paper filters degrade volatile organic compounds (VOCs) and dyes under solar irradiation, offering a low-cost solution for pollution control [93]. However, challenges such as nanoparticle agglomeration, potential cytotoxicity, and high-energy synthesis methods (e.g., sol–gel, hydrothermal) necessitate surface modifications with silanes or polymers to ensure dispersion and biocompatibility [94]. Recent advances include hybrid systems combining TiO_2 with cellulose nanofibers (CNFs) or graphene oxide (GO) to create multifunctional papers with enhanced conductivity or mechanical robustness for flexible electronics and sensors [3–5]. For instance, TiO_2–GO nanocomposites exhibit improved photocatalytic efficiency and mechanical strength, suitable for wearable UV sensors or photocatalytic reactors [95–100]. Despite these advantages, large-scale production faces hurdles such as high TiO_2 production costs (~$50–100/kg) and the need for UV activation, prompting research into visible-light-active dopants and scalable deposition techniques, including spray coating and roll-to-roll processing [101]. Regulatory concerns regarding the release of nanoparticles during recycling or composting also necessitate life cycle assessments to ensure environmental safety [102]. Future directions focus on bio-based TiO_2 synthesis using plant extracts, integration with smart sensors for real-time monitoring, and development of recyclable photocatalytic papers for circular economy applications [103]. In summary, nano-TiO_2 paper nanocomposites represent a transformative technology merging sustainability with high functionality, though commercialization demands further optimization in cost, safety, and performance under real-world conditions.

2.5 *Silver Nanoparticles-Based Paper Nanocomposites*

Silver nanoparticle (AgNP)-based paper nanocomposites have emerged as a highly promising class of functional materials due to their exceptional antimicrobial, conductive, and optical properties, making them invaluable for applications in food packaging, medical products, flexible electronics, and sensing technologies. The incorporation of AgNPs (typically 5–100 nm in diameter) into paper matrices imparts strong antibacterial and antifungal activity against pathogens such as *Escherichia coli*, *Staphylococcus aureus*, and *Candida albicans*, significantly extending the shelf life of packaged food and reducing hospital-acquired infections [104]. The antimicrobial mechanism involves the release of silver ions (Ag^+), which disrupts microbial cell membranes, inhibits respiratory enzymes, and damages DNA, with studies showing a greater than 99% reduction in bacterial load at AgNP concentrations as low as 0.1–1.0 wt% [105]. Beyond biocidal properties, AgNP-based paper composites exhibit enhanced mechanical strength (up to 35% increase in tensile modulus) due to nanoparticle reinforcement and interfacial bonding with cellulose fibers while maintaining flexibility and printability [106]. The high electrical conductivity of AgNPs (6.3×10^7 S/m) enables applications in flexible electronics, including radio frequency identification (RFID) tags, biosensors, and conductive inks, where paper substrates functionalized with AgNP networks achieve sheet resistances below 1 Ω/sq [107]. Optical properties, such as localized surface plasmon resonance (LSPR), enable these nanocomposites to function as colorimetric sensors for humidity, pH, or pathogens, exhibiting visible color shifts upon target interaction [108]. Synthesis methods, such as chemical reduction (using $NaBH_4$ or citrate), green synthesis (utilizing plant extracts), or in situ photoreduction, offer tunable control over AgNP size, distribution, and loading (10–500 ppm, typical for antimicrobial efficacy) [109]. However, challenges include nanoparticle aggregation, oxidative dissolution (reducing long-term activity, and potential cytotoxicity, necessitating surface modifications with polymers (e.g., polyvinylpyrrolidone [PVP], chitosan or silica coatings to improve stability and biocompatibility [97–100]. Environmental concerns over AgNP leaching during recycling have spurred the development of immobilized AgNP systems, such as layer-by-layer assemblies or covalent bonding to cellulose, to minimize the ecological impact [110]. Recent advances include hybrid nanocomposites that combine AgNPs with graphene oxide (GO) or cellulose nanofibers (CNFs) to achieve synergistic mechanical and conductive performance, as well as stimuli-responsive systems where Ag^+ release is triggered by moisture or microbial enzymes [81, 82]. Cost remains a barrier (AgNPs ~ \$50–200/g), driving research into waste-derived silver sources (e.g., photographic waste) and low-loading formulations (<50 ppm) without compromising functionality [111, 112]. Regulatory frameworks (e.g., Environmental Protection Agency [EPA], Food and Drug Administration [FDA]) increasingly address AgNP safety, requiring life cycle assessments to validate environmental and health risks [113]. Future directions focus on scalable manufacturing methods (roll-to-roll coating, 3D printing), biodegradable AgNP alternatives (e.g., CuNPs), and integration with Internet of Things (IoT) devices for smart packaging [114]. In summary, AgNP-based paper nanocomposites represent a versatile platform that merges sustainability with multifunctionality, however, widespread adoption hinges on overcoming cost, stability, and regulatory challenges.

2.6 Graphene-Based Paper Nanocomposites

Graphene-based paper nanocomposites represent a groundbreaking advancement in multifunctional materials, combining the exceptional mechanical, electrical, thermal, and barrier properties of graphene with the flexibility, sustainability, and cost-effectiveness of paper substrates, enabling transformative applications in flexible electronics, energy storage, sensors, and advanced packaging. The incorporation of graphene derivatives—such as graphene oxide (GO), reduced graphene oxide (rGO), or functionalized graphene nanosheets—into paper matrices through techniques like vacuum filtration, spray coating, or in situ polymerization dramatically enhances tensile strength (up to 400% improvement) and Young's modulus (exceeding 50 GPa) due to graphene's intrinsic strength (130 GPa) and robust interfacial bonding with cellulose fibers [115]. These composites exhibit unparalleled electrical conductivity ($\sim 10^3$–10^6 S/m), making them ideal for flexible circuits, antennas, and transparent electrodes in wearable electronics, where they outperform conventional indium tin oxide (ITO) [116]. The impermeability of graphene to gases and liquids (e.g., oxygen transmission rates < 10^{-3} $cm^3/m^2/day$) revolutionizes packaging by extending food shelf life and protecting sensitive devices from moisture and corrosion [117]. Additionally, graphene's high thermal conductivity (~5000 W/mK) enables the development of heat-dissipating papers for electronic cooling, while its plasmonic and electrochemical properties facilitate the creation of ultrasensitive biosensors for detecting pathogens, toxins, or biomarkers at femtomolar concentrations [118, 119]. Functionalization with polymers (e.g., polydopamine) or nanoparticles (e.g., Ag, TiO_2) further tailors these composites for antimicrobial, photocatalytic, or self-cleaning applications [120]. Challenges include achieving uniform dispersion at industrial scales, mitigating graphene's hydrophobicity for aqueous processing, and reducing production costs (graphene prices range from $50 to $500/g), although advances in laser scribing and roll-to-roll manufacturing are addressing these barriers [121]. Environmental and health concerns related to graphene toxicity have prompted the development of bio-based graphene analogues (e.g., graphene derived from lignin or chitosan) and rigorous life cycle assessments [122]. Recent innovations include graphene-based paper batteries with energy densities exceeding 300 Wh/kg, foldable supercapacitors that retain 95% of their capacitance after 10,000 cycles, and smart packaging featuring embedded RFID or freshness indicators [123, 124]. The integration of graphene with cellulose nanofibers (CNFs) or MXenes creates hybrid nanocomposites with synergistic mechanical–electrical properties for aerospace or automotive applications [125]. Regulatory frameworks are evolving to address the novel risks associated with graphene, necessitating the development of standardized protocols for safety and biodegradability [126]. Future research focuses on scalable green synthesis, multifunctional graphene inks for printed electronics, and bioresorbable composites for medical implants [127]. In summary, graphene-based paper nanocomposites bridge the gap between nanotechnology and sustainable materials, offering unparalleled performance metrics, though widespread commercialization requires overcoming scalability, cost, and regulatory hurdles.

2.7 Carbon Nanotube (CNT)-Based Paper Nanocomposites

Carbon nanotube (CNT)-based paper nanocomposites have emerged as a revolutionary class of advanced materials that combine the extraordinary mechanical, electrical, and thermal properties of CNTs with the flexibility, sustainability, and cost-effectiveness of paper substrates, enabling breakthrough applications in flexible electronics, energy storage, aerospace, and smart packaging. The incorporation of single-walled carbon nanotubes (SWCNTs) or multi-walled carbon nanotubes (MWCNTs) into paper matrices through techniques such as vacuum filtration, inkjet printing, or solution casting dramatically enhances tensile strength (up to 500% improvement) and Young's modulus (exceeding 60 GPa) due to CNTs' exceptional strength ($\approx$100 GPa) and stiffness ($\approx$1 TPa), along with their high aspect ratio (typically 100–10,000) that facilitates effective stress transfer to cellulose fibers [34]. These nanocomposites exhibit unparalleled electrical conductivity (ranging from 10^2 to 10^5 S/m), making them ideal for applications such as flexible circuits, electromagnetic shielding (achieving >60 dB attenuation), and transparent conductive electrodes with sheet resistances below 100 Ω/sq at 80% transparency [128]. The anisotropic thermal conductivity of CNTs ($\approx$3000 W/mK axially) enables the development of thermal management papers for electronics cooling, while their large surface area ($\approx$250–1500 m^2/g) and porous structure make them excellent candidates for supercapacitor electrodes with specific capacitances exceeding 300 F/g [97–100]. Functionalization of CNTs with carboxyl, amine, or polymer groups improves their dispersion in aqueous solutions and enhances interfacial bonding with cellulose, while the addition of nanoparticles like silver or titanium dioxide creates multifunctional composites with antimicrobial or photocatalytic properties [129]. Major challenges include achieving uniform CNT dispersion at industrial scales, reducing production costs (high-quality CNTs cost $50–500/g), and addressing potential toxicity concerns, though recent advances in bio-based CNT synthesis and scalable manufacturing techniques like roll-to-roll processing are mitigating these limitations [130]. Environmental applications include water filtration membranes with >99% removal efficiency for heavy metals and organic contaminants, while biomedical uses encompass biosensors with attomolar detection limits and neural tissue scaffolds that promote cell growth [28–32]. Recent innovations include CNT–paper batteries with energy densities >400 Wh/kg, strain sensors with gauge factors >1000, and smart packaging with integrated freshness indicators [118, 119]. The combination of CNTs with other nanomaterials like graphene oxide or cellulose nanofibers produces hybrid nanocomposites with synergistic properties for aerospace applications, where weight reduction and electromagnetic interference shielding are critical [131]. Future research focuses on sustainable carbon nanotube (CNT) production from biomass, development of self-healing CNT–paper composites, and integration with Internet of Things (IoT) devices for real-time monitoring applications [132]. Regulatory frameworks are evolving to address concerns about CNT safety, emphasizing the need for standardized characterization protocols and life cycle assessments [133]. In summary, CNT-based paper nanocomposites represent a paradigm shift in materials science, offering unprecedented performance characteristics that bridge the gap between nanotechnology and sustainable manufacturing. However, widespread commercialization requires further progress in cost reduction, scalable processing, and comprehensive safety evaluations.

3 Preparation of Nanocomposites

For preparing nanocomposites, nanomaterial, such as carbon-based (carbon nanotubes, graphene, graphene oxide), metal/metal oxides (silver [Ag], gold [Au], titanium dioxide [TiO_2], zinc oxide [ZnO]), polymeric (nanoclays, chitosan nanoparticles), cellulose-based (cellulose nanocrystals [CNCs], cellulose nanofibers [CNFs]), etc., is selected first. Then, the nanomaterial is incorporated into or onto paper, adopting different integration methods. These methods are represented in Fig. 3, which include dip coating, in situ synthesis, vacuum-assisted filtration, blending, drop casting, and layer-by-layer assembly. These methods are described in detail in the following subsections.

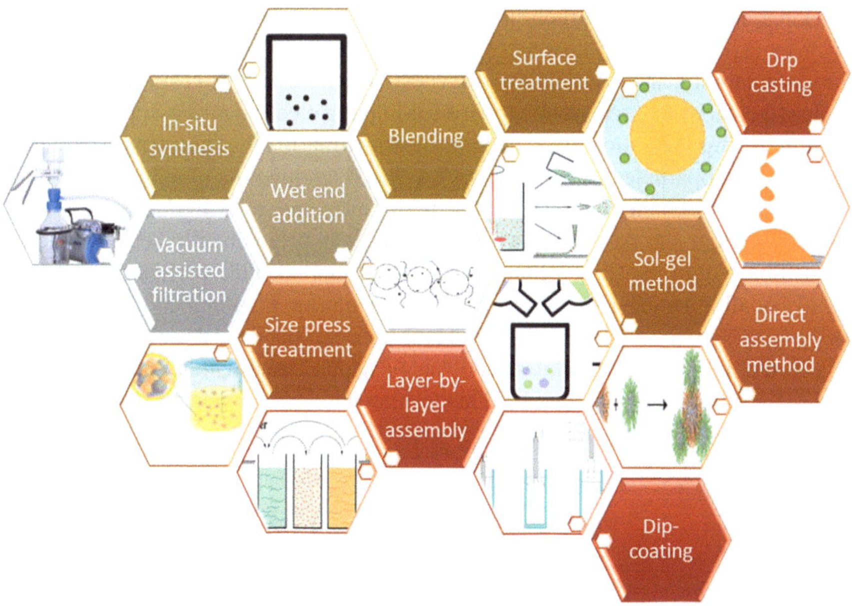

Fig. 3 Synthesis routes of paper-based nanocomposites

3.1 Dip Coating

Dip coating is a cost-effective and scalable technique used to apply nanocomposite coatings onto paper substrates, enhancing functionalities such as water resistance, mechanical strength, antimicrobial properties, and gas barrier performance [134]. In Fig. 4, the dip-coating process of nanomaterial synthesis is presented.

The process involves preparing a nanocomposite solution by dispersing nanoparticles (e.g., nanocellulose, nanoclays, Ag, ZnO, or graphene) in a polymer matrix (e.g., chitosan, polyvinyl alcohol [PVA], or starch) using ultrasonication or high-shear mixing to prevent agglomeration [135]. The paper substrate is immersed in the solution and withdrawn at a controlled speed (1–10 mm/s), where a higher withdrawal speed yields thicker coatings, followed by drying under ambient or heated conditions to form a uniform film [136]. Key parameters affecting coating quality include solution viscosity, nanoparticle concentration, drying conditions, and paper porosity. Multiple dips or posttreatments (e.g., thermal annealing or chemical cross-linking) further improve performance [137]. Applications range from food packaging (oxygen and moisture barriers, antimicrobial surfaces) to flexible electronics (conductive coatings) and medical products (antibacterial wound dressings) [138]. Despite advantages such as simplicity and uniformity, challenges including nanoparticle agglomeration, cracking, and the need for multiple coatings persist [139]. Recent advances include layer-by-layer (LbL) dip coating for enhanced barriers, bio-based nanocomposites for sustainability, and smart coatings for responsive packaging [140, 141].

Jebali et al. [142] used atomic force and scanning electron microscopy to examine the dip-coating deposition of nanocomposite thin films made of polyvinylpyrrolidone and silica nanoparticles. Particles up to 5 wt% did not alter the film thickness, which was determined by the classical additive law of an evaporative regime and a draining regime. Neither the particles nor the polymer was preferentially retrieved during deposition within the film volume. This was in line with the particles being sufficiently small to avoid becoming trapped in the meniscus. The number of particles seen on the surface of the thicker films made in the pure capillary and draining regimes was unaffected by the film's thickness and was fully consistent with a uniform distribution of the particles within the film. On the other hand, the number of particles varied

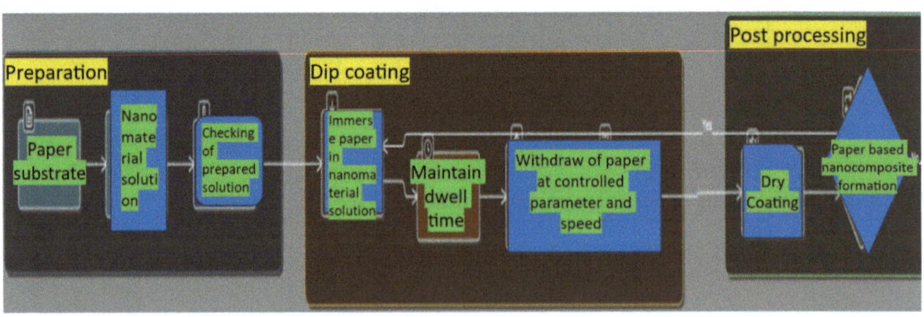

Fig. 4 Synthesis of nanomaterials via dip coating

between the two regimes for the thinner films (close to the critical withdrawal speed). Due to the existence of an exclusion zone at the interfaces, the number in the draining branch increased as the thickness decreased, as predicted theoretically. A surprising particle depletion was seen in the capillary branch, which may have been caused by competition and interaction between time-dependent processes that occurred during the dip coating. When creating nanocomposite thin films, this behavior must be carefully taken into account when surface characteristics like wettability or roughness are taken into account [142].

3.2 *In Situ Synthesis*

The in situ synthesis of paper-based nanocomposites involves the direct formation and incorporation of nanoparticles within the cellulose matrix of paper, creating a homogeneous composite with enhanced functional properties. This approach provides superior nanoparticle dispersion and stronger interfacial bonding compared to conventional ex situ methods, resulting in enhanced mechanical strength, improved barrier properties, and multifunctionality [52, 53]. For example, silver nanoparticles (AgNPs) can be synthesized in situ on paper fibers by reducing $AgNO_3$ with natural reducing agents such as polyphenols present in lignin, resulting in highly antimicrobial paper suitable for food packaging and medical applications [43–46]. Similarly, zinc oxide (ZnO) and titanium dioxide (TiO_2) nanoparticles can be precipitated within the paper's porous structure through sol–gel techniques, imparting UV-blocking and photocatalytic self-cleaning properties [143]. The process typically involves immersing paper in precursor solutions (e.g., metal salts) followed by chemical, thermal, or photochemical reduction to form nanoparticles directly on the cellulose fibers [144]. This method minimizes nanoparticle aggregation, enhances durability, and reduces material waste, making it an eco-friendly and scalable approach for producing functional paper [145]. However, challenges such as controlling nanoparticle size distribution, ensuring uniform deposition, and maintaining paper porosity must be addressed to optimize performance [146]. Recent advancements have included the use of biopolymers, such as chitosan and starch, as stabilizing agents to enhance nanoparticle adhesion and biocompatibility [147–151].

3.3 *Vacuum-Assisted Filtration*

Vacuum-assisted filtration is an efficient technique for depositing nanocomposite coatings onto paper substrates, enabling the formation of thin, uniform, and high-performance functional layers. This method involves dispersing nanomaterials such as cellulose nanofibers (CNFs), graphene oxide (GO), or montmorillonite (MMT) in an aqueous suspension, which is then filtered through a porous paper substrate under vacuum pressure [152–154]. The vacuum pressure facilitates the rapid removal of water while ensuring the homogeneous deposition of nanoparticles onto the paper fibers, resulting in dense and well-packed nanostructured coatings [155]. For instance, vacuum filtration of GO suspensions onto paper creates conductive and mechanically robust coatings suitable for flexible electronics and sensors [156]. Similarly, CNF-based nanocomposites deposited via this method exhibit enhanced barrier properties against oxygen and water vapor, making them ideal for sustainable food packaging [157]. The process parameters, including nanoparticle concentration, filtration time, and vacuum pressure, significantly influence the coating thickness and uniformity [158, 159]. Compared to conventional coating techniques, vacuum-assisted filtration offers advantages such as scalability, minimal material waste, and ability to produce multilayered structures with tailored functionalities [97–100]. However, challenges such as clogging of the filter membrane and the need for posttreatment drying must be addressed to optimize the process [160, 161]. Recent advancements include combining vacuum filtration with cross-linking agents or thermal treatments to further enhance the mechanical and barrier properties of nanocomposite-coated paper [28–32].

3.4 Blending

The blending process is a widely used technique for incorporating nanomaterials into paper substrates to create functional nanocomposites with enhanced properties. This method involves the homogeneous mixing of nanoparticles such as nanocellulose, carbon nanotubes, or metal oxides with paper pulp or coating formulations before sheet formation [162]. During blending, nanoparticles are dispersed in aqueous or solvent-based systems using high-shear mixing or ultrasonication to prevent aggregation and ensure uniform distribution within the cellulose matrix [163]. In Fig. 5, a schematic of paper-based nano-composite synthesis via the blending process is provided. The combination of TEMPO-Oxidized cellulose nanofibres (CNFs) and wood pulp fibres which enhanced the barrier qualities and mechanical strength of the final paper [164]. In a similar research, blending of antimicrobial silver nanoparticles or photocatalytic TiO_2 resulted in antibacterial activity and self-cleaning capabilities of paper products [165]. Nanocomposite paper sheets from combined suspension can be processed via standard papermaking methods [166]. Blending offers compatibility and scalability with recent paper production strategies. It modifies nanocomposite properties depending on blending parameters, concentration and nanoparticle type [167]. During processing to overcome retention problems and nanoparticle sedimentation, optimization in dispersion stability and retention aids was needed [168]. Some innovations based on bio-based dispersants and coupling chemicals were made to get enhanced nanoparticle-pulp interactions and improvement in final product performance [169].

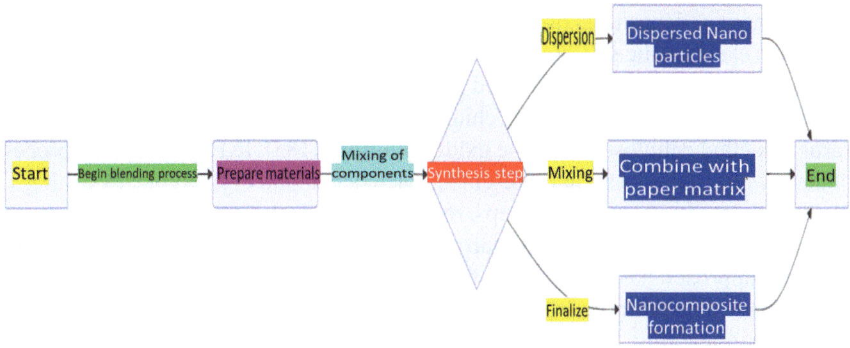

Fig. 5 Paper-based nanocomposite synthesis by blending process

3.5 *Drop Casting*

The drop-casting method offers a simple and versatile approach for depositing nanocomposite coatings onto paper substrates, making it particularly suitable for laboratory-scale fabrication of functional materials. In this process, a controlled volume of nanoparticle dispersion (typically containing cellulose nanocrystals, graphene oxide, or metallic nanoparticles) is dispensed onto the paper surface using a micropipette or syringe, allowing the solvent to evaporate under ambient or controlled conditions [140]. This technique enables precise localization of nanomaterials, making it ideal for creating patterned coatings or sensor elements on paper-based devices [170]. For instance, drop-casting silver nanowire suspensions have been employed to fabricate conductive tracks for flexible paper electronics with sheet resistances as low as 5 Ω/sq [171, 172]. The self-assembly of nanoparticles during solvent evaporation leads to the formation of thin, uniform films whose thickness and morphology depend on the concentration of the dispersion, surface wettability, and drying kinetics [173]. While drop casting provides excellent control over coating geometry and composition, challenges include coffee-ring effects from uneven drying and limited scalability for industrial applications [174]. Recent advancements combine drop casting with posttreatment methods, such as plasma activation or chemical cross-linking, to improve coating adhesion and durability [175–179]. The technique's simplicity and low equipment requirements make it particularly valuable for prototyping antimicrobial papers, electrochemical sensors, and energy storage devices [180–184].

3.6 *Layer-By-Layer Assembly*

The layer-by-layer (LbL) assembly technique has emerged as a powerful method for constructing multifunctional nanocomposite coatings on paper substrates through the alternating deposition of oppositely charged polyelectrolytes and nanoparticles. This electrostatic self-assembly process enables precise nanoscale control over coating thickness, composition, and functionality by sequentially immersing the paper substrate in cationic (e.g., chitosan, polyethyleneimine) and anionic (e.g., carboxymethyl cellulose, graphene oxide) solutions [185]. In Fig. 6, the LbL process of nanocomposite formation is presented. Each deposition cycle, typically lasting 1–5 min with intermediate rinsing steps, results in the adsorption of a molecular monolayer, allowing for the gradual buildup of tailored nanocomposite architectures [186]. The LbL approach has been successfully employed to create paper-based materials with enhanced barrier properties (5–10 bilayers of chitosan/montmorillonite reduced oxygen permeability by 98%), antimicrobial functionality (silver nanoparticle-embedded coatings), and electrical conductivity (carbon nanotube/polyaniline multilayers) [175–179]. Key advantages include ability to incorporate diverse nanomaterials at room temperature, water-based processing, and potential for industrial-scale implementation through automated dip-coating or spray-LbL systems [8–11]. Recent innovations focus on stimuli-responsive layer-by-layer (LbL) systems, where properties change in response to pH, humidity, or temperature, enabling smart packaging applications [180–184]. However, challenges remain in optimizing deposition conditions for porous paper substrates and improving coating durability under mechanical stress [123, 124]. The technique's versatility is further demonstrated in advanced applications such as paper-based biosensors, where 20–30 LbL cycles can create highly selective molecular recognition surfaces [187–190].

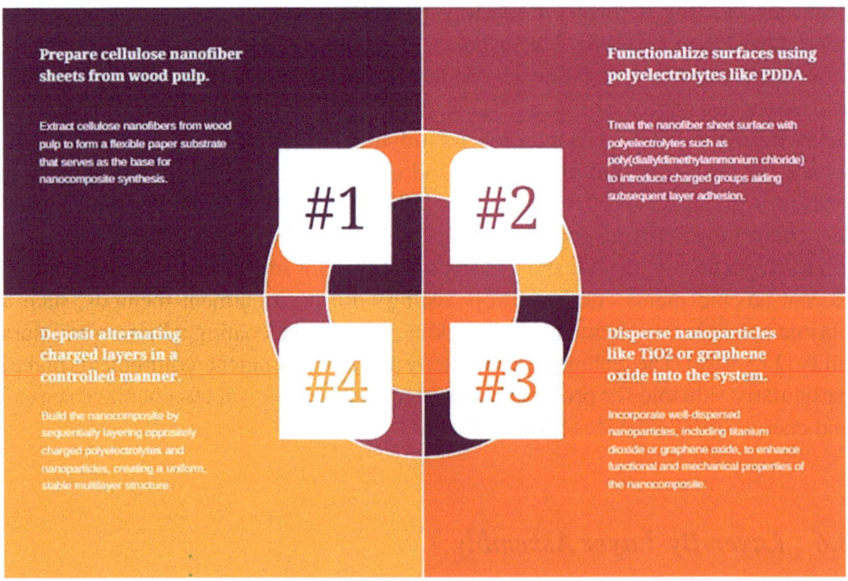

Fig. 6 Layer by layer process of nanocomposite formation

3.7 *Other Methods*

Adel [191], in his study, reported wet-end addition, surface treatment, size press treatment, sol–gel method, and direct assembly method for paper-based nanocomposite formation. In wet-end addition, prior to the formation of paper sheets, the nano-metal particles were integrated into distinct fibers. In order to improve smoothness, gloss, thermal stability, optical properties, and other paper matrix characteristics, commercial nanopowders were used as fillers in the papermaking process. Weak bonds like hydrogen bonds and van der Waals forces were used to incorporate these fillers into cellulose fibers. Therefore, adding binders, linkers, and retention aids during the papermaking process can improve retention in the finished product. Since the crystalline characteristics of the coated substrate varied depending on the shape and type of the coated surface, the surface treatment technique was crucial when applying nano-metal particles as a coat to a cellulosic fiber substrate. The surface treatment method involves coating or impregnating the surface of paper sheets with nano-metal particles via a chemical bath.

Photocatalytic papers were created using the size press process, which involved coating paper sheets with a combination of TiO_2 and various binder types, including silicon binders, resins, and colloidal silica. These binders can prevent photodegradation of paper by incorporating TiO_2 into paper matrices. Paper sheets were coated with a mixture of colloidal silica and TiO_2. When TiO_2 was deposited onto cellulose fibers using the sol–gel method, negatively charged functional groups were added, which caused the TiO_2 to become anchored on the cellulose matrix. Self-cleaning, antimicrobial, and photochromic surfaces can be created using the simple and repeatable anchoring technique.

The citrate group served as a stabilizing and reducing agent in the direct assembly process, which created Au and Ag nano-metal particles by citrate reduction of their aqueous salt solutions. The particles were then collected directly onto the paper matrices. In order to prevent agglomeration in solution, the weakly bound citrate ions transferred negative surface charges to the nano-metal particles, which then incorporated these negative charges onto the substrates through electrostatic self-assembly [191].

4 Applications of Paper-Based Nanocomposites

Paper-based nanocomposites find applications in smart packaging, flexible electronics, antibacterial packaging, water treatment, and energy storage devices. In Figure 7, a pictorial representation of the application of paper-based nanocomposites is provided.

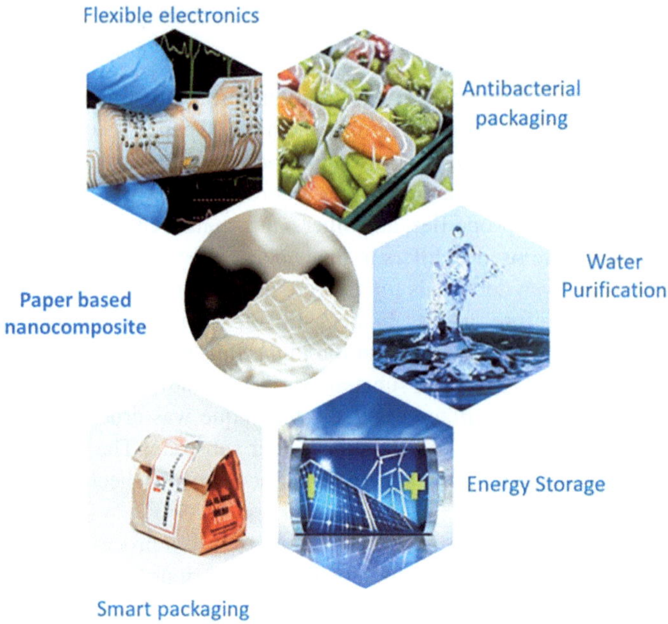

Fig. 7 Application of paper-based nanocomposites

4.1 Flexible Electronics

Flexible electronics fabricated from paper-based nanocomposites represent a transformative advancement in sustainable and high-performance electronic devices, combining the mechanical flexibility, biodegradability, and low-cost processing of paper with the enhanced electrical, thermal, and mechanical properties of nanomaterials. These nanocomposites typically incorporate conductive nanomaterials such as graphene, carbon nanotubes (CNTs), silver nanoparticles (AgNPs), or conductive polymers (e.g., PEDOT:PSS) into cellulose-based paper matrices through techniques like inkjet printing, screen printing, or vacuum filtration, enabling the development of lightweight, foldable, and eco-friendly electronic components [47]. The porous structure of paper facilitates excellent adsorption and uniform dispersion of nanomaterials, while its natural flexibility allows for bending radii as small as 1 mm without significant loss of functionality, making it ideal for wearable sensors, foldable displays, and flexible energy storage devices [192–194]. Similarly, CNT-based paper circuits demonstrate exceptional conductivity ($\approx 10^3$ S/m) and electromagnetic interference (EMI) shielding efficiency (>40 dB), crucial for portable electronics and 5G communication devices [97–100]. The integration of functional nanoparticles, such as zinc oxide (ZnO) or titanium dioxide (TiO_2), further enables multifunctional applications, including UV sensors, photocatalytic systems, and self-cleaning surfaces [175–179]. Paper-based nanocomposites also excel in energy storage, with supercapacitors achieving specific capacitances >300 F/g and batteries demonstrating energy densities > 200 Wh/kg, all while maintaining flexibility and recyclability [69–72]. Challenges include improving environmental stability (e.g., moisture resistance) through hydrophobic coatings (e.g., polydimethylsiloxane [PDMS]), scaling up production via roll-to-roll manufacturing, and ensuring biocompatibility for medical applications [195]. Recent innovations include paper-based triboelectric nanogenerators (TENGs) for energy harvesting, achieving power densities >3 W/m^2, and biodegradable sensors for real-time health monitoring [196]. The use of natural binders (e.g., chitosan, alginate) and low-temperature processing (<150 °C) further enhances sustainability, aligning with circular economy principles [197]. Future directions focus on integrating Internet of Things (IoT) capabilities, developing self-healing materials, and exploring large-area electronics for smart packaging and disposable diagnostics [198]. With their unique combination of performance, sustainability, and versatility, paper-based nanocomposites are poised to revolutionize flexible electronics, offering a viable alternative to traditional petrochemical-derived substrates while addressing global e-waste challenges.

4.2 *Antibacterial Packaging*

The growing demand for sustainable and functional packaging has driven the development of paper-based nanocomposites with potent antibacterial properties, offering an eco-friendly alternative to conventional plastic packaging while addressing microbial contamination risks in food and medical applications. These advanced materials integrate antimicrobial nanoparticles—such as silver (Ag), zinc oxide (ZnO), copper oxide (CuO), titanium dioxide (TiO_2), or chitosan—into cellulose matrices through techniques like in situ synthesis, dip coating, or layer-by-layer assembly, creating active packaging that inhibits pathogenic bacteria (e.g., *E. coli*, *S. aureus*) and extends product shelf life [199]. Silver nanoparticles (AgNPs) are particularly effective, demonstrating >99% bacterial reduction at low concentrations (0.1–1 wt%) by releasing Ag^+ ions that disrupt microbial cell membranes and DNA replication, while ZnO and CuO nanoparticles provide broad-spectrum antimicrobial activity through reactive oxygen species (ROS) generation and metal ion leaching [89]. The porous structure of paper enhances nanoparticle dispersion and contact efficiency with microbes, while cellulose's biodegradability ensures environmental compatibility [35]. Functionalization with natural polymers like chitosan or alginate further improves antibacterial efficacy and biocompatibility, reducing potential toxicity concerns associated with synthetic additives [200].

Key advantages of these nanocomposites include their mechanical robustness (tensile strength improvements up to 40%), barrier properties against moisture and oxygen (water vapor transmission rates <10 g/m^2/day), and scalability via roll-to-roll processing [175–179]. For instance, AgNP-coated papers exhibit log 5–6 reductions in bacterial load within 24 h, while TiO_2-integrated composites offer additional UV-blocking and self-cleaning capabilities under light exposure [87]. Challenges include nanoparticle migration, which is mitigated by embedding nanoparticles within cellulose nanofiber (CNF) networks or using silica coatings [114]. Regulatory compliance (e.g., FDA, EFSA) requires rigorous leaching tests to ensure safety, with recent studies confirming minimal metal migration (<0.1 ppm) below permissible limits [102].

Emerging innovations include stimuli-responsive systems that trigger antimicrobial activity in response to pH changes or humidity, as well as hybrid nanocomposites combining Ag/ZnO for synergistic effects [187–190]. Economic viability is enhanced by using waste-derived cellulose or biosynthesized nanoparticles (e.g., plant-extracted AgNPs), aligning with circular economy goals [147–151]. Future research focuses on optimizing nanoparticle–cellulose interactions for long-term stability, developing edible coatings for direct food contact, and integrating smart indicators (e.g., colorimetric sensors for real-time spoilage detection) [201]. Ling et al. [202] quaternized carboxymethyl chitosan/organic montmorillonite (QAOM) nanocomposite loaded with AgNP. By employing quaternized carboxymethyl chitosan (QCMC) and organic montmorillonite (OMMT) as stabilizing and reducing agents for AgNP synthesis, a novel QAOM nanocomposite was produced in an eco-friendly manner. QAOM nanocomposites on incorporation into paper matrix enhanced tear, tensile, and burst strengths of final paper. Surface-coating also imparts antibacterial properties to paper [202]. Zinc oxide nanoparticles contribute to natural defence of paper against the harmful effects of bacteria, mould, and pollutants, as well as to heat and light accelerated aging [203]. In Table 3, the application of paper-based nanocomposites in antibacterial packaging is provided.

4.3 Water Purification

Access to clean drinking water remains a critical global challenge due to increasing pollution and contamination from industrial, agricultural, and domestic sources. Conventional water purification methods, such as chlorination, reverse osmosis, and activated carbon filtration, are often energy-intensive, costly, or ineffective against certain pollutants. In recent years, nanotechnology has emerged as a promising solution for water treatment, with paper-based nanocomposites gaining attention due to their affordability, biodegradability, and high efficiency. Paper-based nanocomposites incorporate nanomaterials, such as carbon nanotubes, graphene oxide, silver nanoparticles, or metal–organic frameworks (MOFs), into cellulose matrices, thereby enhancing their adsorption, catalytic, and antimicrobial properties. These composites can remove heavy metals, organic dyes, pathogens, and even microplastics from contaminated water through mechanisms like electrostatic attraction, photocatalytic degradation, and size exclusion [63, 64]. One significant advantage of paper-based nanocomposites is their scalability and ease of use. Unlike complex filtration systems, these materials can be designed as portable filters, making them suitable for decentralized water treatment in rural or disaster-stricken areas [211]. For instance, a study by Melchor-Durán et al. [212] demonstrated that a graphene oxide-coated paper

Table 3 Use of paper-based nanocomposites in antibacterial packaging

Nanocomposite type	Antibacterial agent	Target microorganisms	Mechanism of action	Key advantages	Reference
Silver nanoparticle (AgNP)-coated paper	Silver ions (Ag^+)	*Escherichia coli*, *Staphylococcus aureus*, *Listeria monocytogenes*	Release of Ag^+ ions disrupts cell membranes and DNA	Broad-spectrum, long-lasting activity	[204]
Chitosan–zinc oxide (ZnO) nanocomposite paper	ZnO nanoparticles	*Salmonella* spp., *Pseudomonas* spp.	Reactive oxygen species (ROS) generation and membrane damage	Biodegradable, enhances mechanical strength	[205]
Graphene oxide (GO)-modified paper	GO nanosheets	*Bacillus subtilis*, *E. coli*	Physical disruption of cell walls by sharp edges	Thermal stability, no metal leaching	[206–208]
Cellulose nanofiber (CNF)–titanium dioxide (TiO_2) paper	TiO_2 nanoparticles	Mold, yeast, Gram ± bacteria	Photocatalytic ROS generation under UV/visible light	Self-cleaning, UV protection	[209]
Essential oil-embedded nanocellulose paper	Thymol, cinnamaldehyde	*Listeria*, *E. coli*	Disruption of lipid membranes and metabolic pathways	Natural, food-safe, aroma-active	[210]

filter could remove over 95% of lead (Pb^{2+}) and arsenic (As^{3-}) ions from water due to the high surface area and functional groups of graphene oxide that bind metal ions effectively. Similarly, silver nanoparticle-embedded paper composites exhibit potent antibacterial activity against *Escherichia coli* and *Vibrio cholerae*, providing a dual function of physical filtration and microbial disinfection [147–151]. Furthermore, photocatalytic nanocomposites utilizing titanium dioxide (TiO_2) or zinc oxide (ZnO) nanoparticles can degrade organic pollutants, such as pesticides and dyes, when exposed to sunlight, offering a sustainable purification method [175–179]. Despite their potential, challenges remain in the large-scale application of paper-based nanocomposites, including nanoparticle leaching, long-term stability, and cost-effectiveness. Research is ongoing to enhance the mechanical strength and reusability of these materials while minimizing environmental risks associated with the release of nanomaterials [37, 38]. Future advancements in biodegradable nanocomposites and green synthesis methods could further enhance their sustainability. In Table 4, the application of paper-based nanocomposites in wastewater treatment, along with their purification efficiency, is provided.

Table 4 Paper-based nanocomposite in wastewater treatment

Nanocomposite type	Target contaminant	Purification mechanism	Efficiency	References
Graphene oxide (GO)-coated paper	Heavy metals (Pb^{2+}, Cd^{2+}), dyes	Adsorption via oxygen functional groups; electrostatic interactions	>90% removal for Pb^{2+}, ~85% for methylene blue	[171, 172]
Cellulose nanofiber (CNF)–silver nanoparticle paper	Bacteria (*E. coli, S. aureus*)	Antimicrobial action of Ag^+ ions; physical filtration by nanofibers	>99.9% bacterial reduction	[152–154]
TiO_2 nanoparticle-embedded paper	Organic pollutants (phenols, pesticides)	Photocatalytic degradation under UV light	~95% degradation of bisphenol A in 2 h	[213]
Carbon nanotube (CNT)-modified paper	Oils, organic solvents	Super hydrophobicity; selective absorption	>98% separation efficiency for oil–water mixtures	[91, 92]
Chitosan–nanocellulose composite paper	Dyes, heavy metals	Chelation and adsorption via amino ($-NH_2$) and hydroxyl (–OH) groups	~90% removal of Cr(VI)	[214]

4.4 Energy Storage

The growing demand for sustainable and efficient energy storage systems has driven research into innovative materials, including paper-based nanocomposites. These composites integrate nanomaterials such as graphene, carbon nanotubes, conductive polymers, or metal oxides into cellulose-based paper, enhancing energy storage capabilities while maintaining flexibility, low cost, and eco-friendliness [69–72]. Paper-based nanocomposites are particularly promising for applications in supercapacitors, batteries, and flexible electronics due to their high surface area, porosity, and tunable electrical properties [97–100]. One key advantage of paper-based nanocomposites in energy storage is their lightweight and mechanically robust structure, which allows for flexible and wearable energy devices. For instance, graphene-coated paper electrodes have demonstrated high capacitance and cycling stability in supercapacitors, making them suitable for portable electronics [158, 159]. Similarly, lithium-ion batteries incorporating cellulose nanofiber-supported silicon nanoparticles exhibit improved charge–discharge rates and reduced electrode degradation compared to traditional materials [8–11]. Additionally, conductive polymers like polyaniline (PANI) embedded in paper matrices enhance charge storage through pseudocapacitive mechanisms, offering a cost-effective alternative to metal-based electrodes [147–151]. Another significant benefit is the sustainability of paper-based nanocomposites. Unlike conventional energy storage materials that rely on toxic or nonrenewable components, cellulose-based composites are biodegradable and can be produced using eco-friendly processes [111, 112]. Recent studies have also explored the use of recycled paper waste as a substrate for energy storage, further reducing environmental impact [28–32]. However, challenges such as long-term stability, scalability, and integration with existing energy systems remain areas of active research. In Table 5, the application of paper-based nanocomposites in energy storage is provided.

4.5 Smart Packaging

The packaging industry is undergoing a transformation with the integration of smart technologies to enhance food safety, shelf life, and sustainability. Paper-based nanocomposites have emerged as a promising solution, combining the biodegradability of cellulose with the advanced functionalities of nanomaterials such as nanosensors, antimicrobial agents, and gas barriers [224]. These composites enable real-time monitoring of food freshness, prevent microbial contamination, and improve mechanical strength while remaining environmentally friendly [225]. A key application of paper-based nanocomposites in smart packaging is the detection of spoilage indicators. For instance, pH-sensitive nanoparticles embedded in paper can change color in response to ammonia or carbon dioxide released by decaying meat or fish, providing consumers with a visual freshness indicator [180–184]. Similarly,

Table 5 Paper-based nanocomposites in energy storage applications

No	Application	Nanocomposite used	Key findings	References
1	Flexible supercapacitors	Carbon nanotube (CNT)-coated paper	High conductivity (~1000 S/m) and capacitance retention after bending cycles	[215]
2	Lithium-ion battery electrodes	Cellulose paper with $LiFePO_4$ nanoparticles	Improved ion diffusion and mechanical flexibility	[216]
3	Wearable energy storage devices	Graphene oxide (GO)-modified paper	Lightweight, foldable, and biocompatible for wearable tech	[217]
4	Paper-based microbial fuel cells	Conductive polymer-coated paper	Sustainable energy generation from organic waste	[218]
5	Zinc–air batteries	MnO_2 nanowire-embedded paper	Enhanced oxygen reduction reaction (ORR) efficiency	[219]
6	Triboelectric nanogenerators (TENGs)	Cellulose nanofiber–TiO_2 composite	Self-powered sensors with high energy conversion efficiency	[220]
7	Printed paper batteries	Silver nanowire–ink on paper	Low-cost, disposable power sources for medical diagnostics	[221]
8	Supercapacitor–textile hybrids	Carbonized paper with polypyrene	Washable and stretchable energy storage for smart textiles	[222]
9	Sodium-ion battery anodes	SnO_2 nanosheets on cellulose paper	High capacity (~600 mAh/g) and eco-friendly alternative to Li-ion	[43–46]
10	Solar energy storage systems	Paper-based MoS_2/PANI nanocomposite	Photochargeable supercapacitors with 85% efficiency retention	[223]

graphene oxide and silver nanoparticle-infused paper exhibit antimicrobial properties, inhibiting the growth of *E. coli* and *Salmonella* while extending product shelf life [192–194]. Additionally, nanocellulose films reinforced with clay or chitosan enhance their oxygen and moisture barrier properties, thereby reducing food waste by maintaining optimal storage conditions [65, 66]. Another advantage of paper-based smart packaging is its sustainability. Unlike conventional plastic packaging, these composites are biodegradable and can be produced using renewable resources [226]. Recent innovations include edible coatings made from starch and cellulose nanocrystals that further minimize environmental impact [17, 18]. However, challenges such as scalability, cost-effectiveness, and regulatory approval for food contact nanomaterials remain areas of ongoing research.

5 Challenges Associated with the Use of Nano-Additives

Some of the key challenges associated with the efficient use of nanomaterials in the pulp and paper industry at a commercial scale are shown in Fig. 8 and described in detail below.

5.1 *Cost*

The cost of producing nanomaterials can be high, and incorporating them into large-scale papermaking operations could raise costs considerably. One major challenge in adopting paper-based nanocomposites is their high production cost. The integration of nanomaterials, such as carbon nanotubes, graphene, or metal nanoparticles, into paper matrices increases expenses due to the complex synthesis and functionalization processes [152–154]. Additionally, specialized equipment and controlled environmental conditions are required, further escalating costs [227]. Scalability remains an issue, as mass-producing these composites while maintaining consistency is difficult and expensive [228]. The cost of raw materials, particularly high-purity nanomaterials, also contributes to financial constraints [229, 230]. Moreover, functionalizing paper with nanomaterials often involves multiple steps, increasing labor and time expenditures [231, 232].

Despite their advantages in flexibility, sustainability, and sensing applications, the high cost limits widespread industrial adoption. Researchers are exploring cost-effective alternatives, such as waste-derived nanomaterials, to mitigate expenses [3–5]. Until more affordable production methods are developed, the high cost remains a significant barrier.

Fig. 8 Challenges associated with usage of paper-based nanocomposites

5.2 Compatibility

It can be not easy to guarantee that nanomaterials are compatible with the machinery and procedures used in papermaking today. Equipment and procedure changes might be required. A key challenge in paper-based nanocomposites is material compatibility. Integrating nanomaterials (e.g., graphene, metal oxides) with cellulose fibers often results in poor dispersion, which weakens their mechanical and electrical properties [206–208]. Hydrophobic nanomaterials may resist binding to hydrophilic paper, thereby reducing their effectiveness [233]. Additionally, chemical treatments to enhance compatibility can degrade paper's flexibility [234]. Variability in paper porosity further complicates the uniform distribution of nanomaterials [235]. Overcoming these issues requires optimized surface modifications, but this increases production complexity and cost.

5.3 Environmental Impact

Nanomaterials' effects on the environment, including any potential emissions during manufacture and use, must be carefully considered and controlled. Despite being perceived as eco-friendly, paper-based nanocomposites pose significant environmental challenges. The incorporation of nanomaterials (e.g., carbon nanotubes, silver nanoparticles, or graphene oxide) into paper matrices raises concerns about toxicity, biodegradability, and long-term ecological effects [147–151]. Many nanomaterials are not inherently biodegradable, which can lead to potential accumulation in ecosystems if improperly disposed of [236]. Studies indicate that nanoparticles can leach into soil and water, posing risks to microorganisms, plants, and aquatic life [237].

Additionally, the production of nanocomposites often involves energy-intensive processes and the use of hazardous chemicals, contributing to pollution [180–184]. Recycling process of paper-based nanocomposites is quite challenging. Separation of nanomaterials from cellulose fibres is very complicated and also it contaminates paper recycling streams [3–5]. Renewable nature of paper on incorporation of synthetic nanomaterials reduces its sustainabilty [238]. Certain nanoparticles, including silver and titanium oxide, have antimicrobial properties that can hamper natural microbial ecosystems upon release [239]. The regulatory frameworks governing nanoparticle recycling and diposal are inadequately developed amplifying the risk of unrestricted environmental exposure [240]. Researchers are examining evironmentally friendly synthesis methods, enhanced waste mangement strategies, and biodegradable nanomaterials to address these challenges [241]. Until these solutions attain scalability, environmental risks related with paper-based nanocomposites will continue to pose a weighty obstacle to their sustainable implementation.

5.4 Health and Safety

To safeguard employees and customers, the health and safety risks associated with working with nanomaterials must be fully recognized and controlled. The integration of nanomaterials into paper-based composites presents significant health and safety concerns for workers and end users. During manufacturing, airborne nanoparticles may be released, posing inhalation risks that could lead to pulmonary inflammation or systemic toxicity [242]. Many engineered nanomaterials, such as carbon nanotubes and metal oxides, have shown potential carcinogenic effects in long-term exposure scenarios [243]. Worker protection is particularly challenging because conventional personal protective equipment (PPE) may not effectively filter nanosized particles [244]. Additionally, the lack of standardized safety protocols for handling nanocomposites increases occupational hazards [245]. Dermal exposure during processing is another concern, as some nanoparticles can penetrate skin barriers and cause cellular damage [246]. End-user safety is equally problematic. Wearable or food-packaging nanocomposites may release nanoparticles through abrasion or moisture exposure [247]. The potential for nanoparticle migration raises concerns about chronic exposure, particularly for immunocompromised individuals [248]. Furthermore, combustion of nanocomposites could release toxic fumes, creating additional hazards during disposal [249]. Current regulations struggle to keep pace with these emerging risks, leaving gaps in safety guidelines [250]. While engineering controls and improved PPE can mitigate some risks, comprehensive toxicological studies and stricter safety standards are needed to ensure safe adoption of these materials.

5.5 Regulatory Compliance

It can be difficult and time-consuming to comply with safety and environmental regulations when using nanomaterials in paper products. Paper-based nanocomposites pose substantial reguatory challenges for development and commercialisation due to their novel nature. Current regulatory frameworks find it challenging to appropriately classify nanocomposites which combine traditional paper with engineered nanomaterials that may have unique properties [251]. This makes it challenging to identify which safety standards are applicable, especially for applications in food packaging or medical devices where material migration may occur [252]. One major compliance issue is the lack of standardised testing protocols for toxicity assessment and nanoparticle release in composite materials [253]. Regulatory agencies like the FDA and EPA have different requirements for nanomaterials which leads to complicated approval processes [254]. The European Union's Registration, Evaluation, Authorisation and Restriction of Chemicals (REACH) regulation requires extensive safety data, but the high cost of compliance studies often discourages small manufacturers [255]. Furthermore, the vast differences in international regulations create trade barriers for products made of nanocomposite materials [256]. For instance, while some countries treat nanocomposites as standard materials, others impose strict labeling and safety assessments that are specific to nanotechnology [257]. Because nanotechnology is dynamic, regulations need to be regularly reviewed and modified [258]. These compliance concerns increase development and time-to-market costs, which could impede innovation in sustainable packaging and other fascinating areas [259]. Until standardised international standards and more transparent regulatory pathways are established, manufacturers must navigate a complex and dynamic compliance environment.

5.6 Scale-Up

Maintaining the desired characteristics of nanomaterials can be challenging when transitioning from laboratory-scale production to large-scale manufacturing. The transition from laboratory-scale production to industrial manufacturing of paper-based nanocomposites presents several significant challenges. One major obstacle is maintaining uniform dispersion of nanomaterials at larger production volumes, as agglomeration becomes more prevalent with increased batch sizes [229, 230]. The high viscosity of nanomaterial suspensions often leads to uneven distribution within the paper matrix during continuous processing, compromising product consistency [231, 232]. Process control becomes increasingly challenging during scale-up, as parameters such as drying temperature, pressure, and coating thickness exhibit greater variability in industrial settings [206–208]. The specialized equipment required for nanocomposite production, such as precision coating machines and controlled drying systems, represents a substantial capital investment that many manufacturers are reluctant to make [187–190]. Additionally, the current batch-to-batch variation in nanomaterial properties makes quality control challenging at commercial scales [260]. The extended processing times needed for proper nanomaterial integration often conflict with the high-speed requirements of industrial paper production lines [192–194]. Supply chain limitations for consistently producing high-quality nanomaterials in industrial quantities further complicate scale-up efforts [261]. Many promising nanocomposite formulations that perform well in the lab cannot be economically produced at tonnage scales due to these combined technical and logistical challenges [262, 263].

5.7 Performance Stability

Commercial viability depends on the stability and steady performance of paper products enhanced with nanomaterials over time. Temperature and humidity induced degradation [180–184], disruption of nanomaterial network due to swelling of cellulosic fibres, reduction in electrical conductivity [264], decrement in antimicrobial efficacy due to oxidation [69–72], thermal instability, strctural deformation above

80 °C [187–190], etc. are measure stability issues associated with PBN. This thermal sensitivity severely limits applications in packaging and electronics. Long-term UV exposure causes photodegradation of both the paper matrix and organic nanomaterials, with studies showing 40% reduction in mechanical strength after 500 h of sunlight exposure [262, 263]. Mechanical fatigue through bending or folding creates microcracks that disrupt the percolation network of conductive nanomaterials [160, 161]. Repeated stress cycles can increase electrical resistance by orders of magnitude, making these materials unreliable for flexible electronics. The inherent porosity of paper also allows gradual nanoparticle migration, leading to performance drift over time [265]. Chemical stability is equally problematic. Many nanocomposites show pH-dependent behavior, with acidic or alkaline conditions causing nanoparticle leaching [266]. This raises concerns for food packaging applications where pH variations occur. Biological degradation by microorganisms further compromises long-term stability in humid environments [267]. Current stabilization strategies, including polymer encapsulation and cross-linking, often compromise the eco-friendly advantages of paper substrates [28–32]. Developing stabilization methods that maintain both performance and sustainability remains a critical research challenge.

6 Conclusion

Paper-based nanocomposites represent a transformative innovation in materials science, offering unparalleled opportunities for enhancing material performance across diverse industries. While significant progress has been made in understanding and optimizing these materials, ongoing research is essential to overcome existing challenges and unlock their full potential. The innovation of these composites in water purification combines the benefits of nanotechnology with low-cost, eco-friendly materials to address global water scarcity and contamination issues. The revolution of paper-based nanocomposites in energy storage methods provides high performance with sustainability. Their versatility makes them ideal for next-generation flexible and green energy technologies, supporting the transition toward renewable energy solutions. These nanocomposites represent a breakthrough in smart packaging, offering enhanced functionality while aligning with global sustainability goals. Their ability to monitor, protect, and preserve products makes them ideal for future eco-conscious packaging solutions. By addressing issues related to the application of paper-based nanocomposites, evolving demands for the next generation of high-performance paper-based nanocomposites in the pulp and paper sector can be successfully incorporated on a commercial scale.

7 Competing Interests

The authors declare that they have no competing interest.

8 Declaration of Generative AI and AI-Assisted Technologies in the Writing Process

During the preparation of this work, the book chapter author(s) did not take any assistance from AI, and the book chapter author(s) reviewed and edited the content as needed and take(s) full responsibility for the content of the publication.

References

1. Gonfa YH, Bachheti A, Husen A, Bachheti RK (2024) Carbon-based smart nanomaterials: an overview. Carbon-based nanomaterials: synthesis, agricultural, biomedical, and environmental interventions. Springer Nature Singapore, Singapore, pp 1–77
2. Sanchez C, JuliánB BP, Popall M (2005) Applications of hybrid organic-1028 inorganic nanocomposites. J Mater Chem 15:3559–3592. https://doi.org/10.1039/B509097K
3. Senthil MS, Raju N, Sampath PS, Jayakumari LS (2014). Effects of nanomaterials on polymer composites—an expatiate view. Rev Adv Mater Sci 38:40–54
4. Razi Z, Mozammel M, Abbasloo S, Mohammadzadeh E (2024) Recent innovations and challenges in composite recycling: toward sustainable circular economies
5. Thippeswamy BH, Maligi AS, Hegde G (2021) Roadmap of effects of biowaste-synthesized carbon nanomaterials on carbon nano-reinforced composites. Catalysts 11(12):1485. https://doi.org/10.3390/catal11121485
6. Zare Y, Rhee KY, Park SJ (2022) Two-stage modeling of tensile strength for a carbon-nanotube-based system applicable in the biomedical field. JOM 74. https://doi.org/10.1007/s11837-022-05270-8
7. Nnamani Akinniyi J, Abimbola Atanda S, Ariyo DO, AhmedT, Solomon Ayanda I, Omolola Badmos F, Oiza Jimoh M, Queen Adegbola R (2024) Perspective chapter: nanocomposites—unlocking the potentials for diverse applications. IntechOpen. https://doi.org/10.5772/intechopen.114914
8. Mirmehdi S, Hein PRG, de Luca Sarantópoulos CIG, Dias MV, Tonoli GHD (2018) Cellulose nanofibrils/nanoclay hybrid composite as a paper coating: effects of spray time, nanoclay content and corona discharge on barrier and mechanical properties of the coated papers. Food Packag Shelf Life 15:87–94. https://doi.org/10.1016/j.fpsl.2017.11.007
9. Díez-Pascual AM, Rahdar A (2022) LbL Nano-Assemblies: a versatile tool for biomedical and healthcare applications. Nanomaterials 12(6):949. https://doi.org/10.3390/nano12060949
10. Hwang J, Jung M, Park J-J, Kim E-K, Lee G, Lee KJ, Choi J-H, Song W-J (2022) Preparation and electrochemical characterization of Si@C nanoparticles as an anode material for lithium-ion batteries via solvent-assisted wet coating process. Nanomaterials 12(10):1649. https://doi.org/10.3390/nano12101649
11. Kausar A, Ahmad I, Maaza M, Bocchetta P (2023) Self-Healing nanocomposites—advancements and aerospace applications. J Compos Sci 7:148. https://doi.org/10.3390/jcs7040148
12. Das SS, Bharadwaj P, Bilal M, Barani M, Rahdar A, Taboada P, Bungau S, Kyzas GZ (2020) Stimuli-responsive polymeric nanocarriers for drug delivery, imaging, and theragnosis. Polymers 12(6):1397. https://doi.org/10.3390/polym12061397
13. Xue Y, Guo YH, Rafailovich M (2019) Flame retardant polymer nanocomposites and interfaces. IntechOpen. https://doi.org/10.5772/intechopen.79548
14. Grand View Research (2023) Polymer nanocomposites market size, share & trends analysis report by product (carbon nanotubes, metal oxide), by application (packaging, automotive, electronics), by region, and segment forecasts, 2023–2030 (Report ID: GVR-1–68038–991–0). https://www.grandviewresearch.com/industry-analysis/polymer-nanocomposites-market

15. Naskar A, Keum J, Boeman R (2016) Polymer matrix nanocomposites for automotive structural components. Nat Nanotechnol 11:1026–1030. https://doi.org/10.1038/nnano.2016.262
16. Suresh SS, George K, Mohanty S, Nayak SK (2021) Architect of polymer nanocomposites for aerospace applications. https://doi.org/10.4018/978-1-7998-8591-7.ch055.
17. Nair NM, Pakkathillam JK, Kumar K, Arunachalam K, Ray D, Swaminathan P (2020) Printable silver nanowire and PEDOT:PSS nanocomposite ink for flexible transparent conducting applications. ACS Appl Electron Mater XXXX. https://doi.org/10.1021/acsaelm.0c00061.
18. Pirozzi A, Ferrari G, Donsì F (2021) The use of nanocellulose in edible coatings for the preservation of perishable fruits and vegetables. Coatings 11(8):990. https://doi.org/10.3390/coatings11080990
19. Alven S, Aderibigbe BA (2024) Chitosan-based scaffolds incorporated with silver nanoparticles for the treatment of infected wounds. Pharmaceutics 16(3):327. https://doi.org/10.3390/pharmaceutics16030327
20. Thompson GS, Harmer MP (2001) Nanoscale ceramic. Composites. https://doi.org/10.1016/B0-08-043152-6/01032-9
21. Kausar A, Ahmad I, Zhao T, Aldaghri O, Ibnaouf KH, Eisa MH (2023) Graphene nanocomposites as innovative materials for energy storage and conversion—design and headways. Int J Mol Sci 24(14):11593. https://doi.org/10.3390/ijms241411593
22. Bakshi S, Lahiri D, Agarwal A (2013) Carbon nanotube reinforced metal matrix composites—a review. Int Mater Rev. https://doi.org/10.1179/095066009X12572530170543
23. Joseph K (2006) Polymer nanocomposites: processing, characterization, and applications
24. Pegel S, Villmow T, Kasaliwal G, Pötschke P (2012) Polymer-carbon nanotube composites: melt processing, properties and applications. https://doi.org/10.3139/9781569905258.005
25. Xu H, Hao Z, Wang C, Deng J, Wang T, Zhang J (2023) In situ emulsion polymerization to multifunctional polymer nanocomposites: a review. Macromolecular Chem Phys 224. https://doi.org/10.1002/macp.202300185
26. Palmero P (2015) Structural ceramic nanocomposites: a review of properties and powders' synthesis methods. Nanomaterials 5(2):656–696. https://doi.org/10.3390/nano5020656
27. Tang B (2025) Research progress of 3D printing polymer nanocomposites. Appl Comput Eng 200:41–49. https://doi.org/10.54254/2755-2721/2026.KA28777.
28. Keneshbekova A, Smagulova G, Kaidar B, Imash A, Ilyanov A, Kazhdanbekov R, Yensep E, Lesbayev A (2024) MXene/Carbon nanocomposites for water treatment. Membranes 14(9):184. https://doi.org/10.3390/membranes14090184
29. Torun I, Celik N, Ruzi M, Onses M (2020) Transferring the structure of paper for mechanically durable superhydrophobic surfaces. Surf Coat Technol 405:126543. https://doi.org/10.1016/j.surfcoat.2020.126543
30. Agboola O, Fayomi OSI, Ayodeji A, Ayeni AO, Alagbe EE, Sanni SE, Okoro EE, Moropeng L, Sadiku R, Kupolati KW, Oni BA (2021) A review on polymer nanocomposites and their effective applications in membranes and adsorbents for water treatment and gas separation. Membranes 11(2):139. https://doi.org/10.3390/membranes11020139
31. Qasim M, Kareem A, El-Kadri OM, Alnaser AS (2025) Femtosecond laser-structured wastepaper as a biodegradable flexible current collector for supercapacitor applications. RSC Adv 15(44):37152–37165. https://doi.org/10.1039/d5ra06223c. PMID: 41064143; PMCID: PMC12501962
32. Yang Z, Ren J, Zhang Z, Chen X, Guan G, Qiu L, Zhang Y, Peng H (2015) Recent advancement of nanostructured carbon for energy applications. Chem Rev 115(11):5159–5223. https://doi.org/10.1021/cr5006217. Epub 2015 May 19 PMID: 25985835
33. Azeez A, Rhee KY, Park S-J, Hui D (2013) Epoxy clay nanocomposites—processing, properties and applications: a review. Compos B Eng 45:308–320. https://doi.org/10.1016/j.compositesb.2012.04.012
34. Baughman RH, Zakhidov AA, De Heer WA (2002) Carbon nanotubes-the route toward applications. Science (New York, N.Y.) 297:787–792. https://doi.org/10.1126/science.1060928.

35. Fortunati E, Rinaldi S, Peltzer M, Bloise N, Visai L, Armentano I, Jimenez A, Latterini L, Kenny J (2014) Nano-biocomposite films with modified cellulose nanocrystals and synthesized silver nanoparticles. Carbohyd Polym 101:1122–1133. https://doi.org/10.1016/j.carbpol.2013.10.055
36. Hassan T, Salam A, Khan A, Khan SU, Khanzada H, Wasim M, Kim IS(2021) Functional nanocomposites and their potential applications: a review. J Polym Res 28. https://doi.org/10.1007/s10965-021-02408-1
37. Singh G, Thakur N, Kumar R (2024) Nanoparticles in drinking water: assessing health risks and regulatory challenges. Sci Total Environ 949:174940. https://doi.org/10.1016/j.scitotenv.2024.174940
38. Pleşa I, Noţingher PV, Schlögl S, Sumereder C, Muhr M (2016) Properties of polymer composites used in high-voltage applications. Polymers (Basel) 8(5):173. https://doi.org/10.3390/polym8050173. PMID: 30979265; PMCID: PMC6432377
39. Norizan MN, Shazleen SS, Alias AH, Sabaruddin FA, Asyraf MRM, Zainudin ES, Norli A, Samsudin MS, Kamarudin SH, Norrahim MNF (2022) Nanocellulose-based nanocomposites for sustainable applications: a review. Nanomaterials 12:1–53. https://doi.org/10.3390/nano12193483.
40. Iravani S, Varma RS (2022) Smart MXene quantum dot-based nanosystems for biomedical applications. Nanomaterials 12(7):1200. https://doi.org/10.3390/nano12071200
41. Dufresne A (2012) Nanocellulose: from nature to high performance tailored materials. De Gruyter. https://doi.org/10.1515/9783110254600
42. Xu Y, Wu Z, Li A, Chen N, Rao J, Zeng Q (2024) Nanocellulose composite films in food packaging materials: a review. Polym 16(3):423. https://doi.org/10.3390/polym16030423
43. Li T, Chen C, Brozena AH et al (2021) Developing fibrillated cellulose as a sustainable technological material. Nature 590:47–56. https://doi.org/10.1038/s41586-020-03167-7
44. Xu D, Chen C, Xie J, Zhang B, Miao L, Cai J, Yunhui H, Zhang L (2016) A hierarchical N/S-Codoped carbon anode fabricated facilely from cellulose/polyaniline microspheres for high-performance sodium-ion batteries. Adv Energy Mater 6. https://doi.org/10.1002/aenm.201501929
45. Khan RA, Beck S, Dussault D, Salmieri S, Bouchard J, Lacroix M (2013) Mechanical and barrier properties of nanocrystalline cellulose reinforced poly(caprolactone) composites: effect of gamma radiation. J Appl Polymer Sci 129. https://doi.org/10.1002/app.38896
46. Maršík D, Thoresen PP, Maťátková O, Masák J, Sialini P, Rova U, Tsikourkitoudi V, Christakopoulos P, Matsakas L, Jarošová Kolouchová I (2024) Synthesis and characterization of lignin-silver nanoparticles. Molecules 29(10):2360. https://doi.org/10.3390/molecules29102360
47. Jansson E, Lyytikäinen J, Tanninen P, Eiroma K, Leminen V, Immonen K, Hakola L (2022) Suitability of paper-based substrates for printed electronics. Materials 15(3):957. https://doi.org/10.3390/ma15030957
48. Shange MG, Khumalo NL, Mohomane SM, Motaung TE (2024) Factors affecting silica/cellulose nanocomposite prepared via the sol-gel technique: a review. Materials 17(9):1937. https://doi.org/10.3390/ma17091937
49. Thomas B, Raj MC, Athira BK, Joy J, Moores A, Drisko GLD, Sanchez, C (2018) Nanocellulose, a versatile green platform: from biosources to materials and their applications. Chem Rev 118(24):11575–11625.https://doi.org/10.1021/acs.chemrev.7b00627
50. Nogi M, Karakawa M, Komoda N et al (2015) Transparent conductive nanofiber paper for foldable solar cells. Sci Rep 5:17254. https://doi.org/10.1038/srep17254
51. Chauhan V, Chakrabarti S (2012) Use of nanotechnology for high performance cellulosic and papermaking products. Cellul Chem Technol 46:389
52. Handayani W, Listyarini A, Yasman Y, Imawan C (2020) A simple synthesis of silver nanoparticles on cellulose paper for antimicrobial applications. Mater Sci Forum 990:191–196. https://doi.org/10.4028/www.scientific.net/MSF.990.191
53. Gu F, Yang W, Song J, Xiao H, Wang W, Cai Z (2022) Crosslinked PVA/nanoclay hydrogel coating for improving water vapor barrier of cellulose-based packaging at high temperature and humidity. Coatings 12(10):1562. https://doi.org/10.3390/coatings12101562

54. Nath D, Santhosh R, Pal K, Sarkar P (2022) Nanoclay-based active food packaging systems: a review. Food Packag Shelf Life 31:100803. https://doi.org/10.1016/j.fpsl.2021.100803.
55. Lavoine N, Desloges I, Dufresne A, Bras J (2012) Microfibrillated cellulose—its barrier properties and applications in cellulosic materials: a review. Carbohyd Polym 90:735–764. https://doi.org/10.1016/j.carbpol.2012.05.026
56. Sun Q, Deng Y(2006) Polymer/nanoclay nanocomposites for paper barrier coating. In: 2006 TAPPI international conference on nanotechnology.
57. Llanos JH, Tadini CC (2017) Preparation and characterization of bio-nanocomposite films based on cassava starch or chitosan, reinforced with montmorillonite or bamboo nanofibers. Int J Biol Macromol 107. https://doi.org/10.1016/j.ijbiomac.2017.09.001.
58. Brandelli A (2024) Nanocomposites and their application in antimicrobial packaging. Front Chem 26(12):1356304. https://doi.org/10.3389/fchem.2024.1356304. PMID: 38469428; PMCID: PMC10925673
59. Mishra K, Vaidyanathan R (2019) The influence of nanoclay on the flame retardancy and mechanical performance of recycled carpet composites. Recycling 4(2):22. https://doi.org/10.3390/recycling4020022
60. Shen Z, Rajabi-Abhari A, Oh K, Yang G, Youn HJ, Lee HL (2021) Improving the barrier properties of packaging paper by polyvinyl alcohol based polymer coating—effect of the base paper and nanoclay. Polymers 13(8):1334. https://doi.org/10.3390/polym13081334
61. Nath D, Santhosh R, Pal K, Sarkar P (2022) Nanoclay-based active food packaging systems: a review. Food Packag Shelf Life 31:100803. https://doi.org/10.1016/j.fpsl.2021.100803.
62. Jin S, Qi Y, Shen Y, Li H(2020) A transparent polyurethane based on nanosilica in reinforcing papers. Nordic Pulp Paper Res J 1. https://doi.org/10.1515/npprj-2019-0099.
63. Suthabanditpong W, Buntem R, Takai C, Fuji M, Shirai T (2015) The quantitative effect of silica nanoparticles on optical properties of thin solid silica UV-cured films. Surf Coat Technol 279. https://doi.org/10.1016/j.surfcoat.2015.07.068
64. Voisin H, Bergström L, Liu P, Mathew AP (2017) Nanocellulose-based materials for water purification. Nanomaterials 7:57. https://doi.org/10.3390/nano7030057.
65. Karimi Dehbakri S, Ehtesham Nia A, Mumivand H et al (2025) Effect of nanocellulose-based edible coatings enriched with α-pinene and myrtle essential oil on the postharvest quality of strawberry. BMC Plant Biol 25:812. https://doi.org/10.1186/s12870-025-06757-7
66. Abu-Thabit N, Hakeem AS, Mezghani K, Ratemi E, Elzagheid M, Umar Y, Primartomo A, Al Batty S, Azad AK, Al Anazi S, Ahmad A (2020) Preparation of pH-indicative and flame-retardant nanocomposite films for smart packaging applications. Sensors 20(19):5462. https://doi.org/10.3390/s20195462
67. Abdul Khalil HPS, Adnan AS, Yahya EB, Olaiya NG, Safrida S, Hossain MS, Balakrishnan V, Gopakumar DA, Abdullah CK, Oyekanmi AA, Pasquini D (2020) A review on plant cellulose nanofibre-based aerogels for biomedical applications. Polymers (Basel) 12(8):1759. https://doi.org/10.3390/polym12081759. PMID: 32781602; PMCID: PMC7465206
68. Dias V, Kuznetsova A, Tedim J, Yaremchenko A, Zheludkevich M, Portugal I, Evtuguin D (2015) Silica-based nanocoating doped by layered double hydroxides to enhance the paper-board barrier properties. World J Nano Sci Eng 5:126–139. https://doi.org/10.4236/wjnse.2015.54015
69. Shao Y, Wang H, Zhang Q, Li Y (2013) High-performance flexible asymmetric supercapacitors based on 3D porous graphene/MnO_2 nanorod and graphene/Ag hybrid thin-film electrodes. J Mater Chem C 1:1245–1251. https://doi.org/10.1039/C2TC00235C
70. Fan X, Liu B, Ding J, Deng Y, Han X, Hu W Zhong C (2020) Flexible and wearable power sources for next-generation wearable electronics. Batteries Supercaps 3. https://doi.org/10.1002/batt.202000115
71. Song H, Zheng L (2013) Nanocomposite films based on cellulose reinforced with nano-SiO_2: microstructure, hydrophilicity, thermal stability, and mechanical properties. Cellulose 20. https://doi.org/10.1007/s10570-013-9941-3
72. Pushparaj VL, Shaijumon MM, Kumar A, Murugesan S, Ci L, Vajtai R, Linhardt RJ, Nalamasu O, Ajayan PM. Flexible energy storage devices based on nanocomposite paper. Proc Natl

Acad Sci U S A 104(34):13574–13577 (2007). https://doi.org/10.1073/pnas.0706508104. Epub 2007 Aug 15. PMID: 17699622; PMCID: PMC1959422
73. Klemm D, Kramer F, Moritz S, Lindström T, Ankerfors M, Gray D, Dorris A (2011) Nanocelluloses: a new family of nature-based materials. Angew Chem Int Ed Engl 50(24):5438–5466. https://doi.org/10.1002/anie.201001273. Epub 2011 May 20 PMID: 21598362
74. Isogai A, Saito T, Fukuzumi H (2011) TEMPO-oxidized cellulose nano fibers. Nanoscale 3(1):71–85.https://doi.org/10.1039/C0NR00583E
75. Nogi M, Iwamoto S, Nakagaito AN, Yano H (2019) Optically transparent nanofiber paper. Adv Mater 21(16):1595–1598. https://doi.org/10.1002/adma.200803174
76. Habibi Y, Lucia LA, Rojas OJ (2021) Cellulose nanocrystals: chemistry, self-assembly, and applications. Chem Rev 110(6):3479–3500. https://doi.org/10.1021/cr900339w
77. Wu Y, Wang X, Yao L, Chang S, Wang X (2023) Thermal insulation mechanism, preparation, and modification of nanocellulose aerogels: a review. Molecules 28(15):5836. https://doi.org/10.3390/molecules28155836
78. Lin N, Huang J, Dufresne A (2012) Preparation, properties and applications of polysaccharide nanocrystals in advanced functional nanomaterials: a review. Nanoscale 4:3274–3294. https://doi.org/10.1039/c2nr30260h
79. Moon R, Martini A, Simonsen J, Youngblood J (2011) ChemInform abstract: cellulose nanomaterials review: structure, properties and nanocomposites. Chem Soc Rev 40:3941–3994. https://doi.org/10.1039/c0cs00108b
80. Dufresne A (2017) Nanocellulose: from nature to high performance tailored materials, 2nd edn. https://doi.org/10.1515/9783110480412
81. Wu CN, Saito T, Fujisawa S, Fukuzumi H, Isogai A (2012) Ultrastrong and high gas-barrier nanocellulose/clay-layered composites. Biomacromol 13(6):1927–1932. https://doi.org/10.1021/bm300465d. Epub 2012 May 15. PMID: 22568705
82. Wu J, Zheng Y, Wen X, Lin Q, Chen X, Wu Z (2014) Silver nanoparticle/bacterial cellulose gel membranes for antibacterial wound dressing: investigation in vitro and in vivo. Biomed Mater 9(3):035005. https://doi.org/10.1088/1748-6041/9/3/035005. Epub 2014 Apr 16. PMID: 24739469.
83. Bras J, Hassan M, Bruzesse C, Hassan E, El-Wakil N, Dufresne A (2010) Mechanical, barrier, and biodegradability properties of bagasse cellulose whiskers reinforced natural rubber nanocomposites. Ind Crops Prod 32:627–633. https://doi.org/10.1016/j.indcrop.2010.07.018
84. Siró I, Plackett D (2010) Microfibrillated cellulose and new nanocomposite materials: a review. Cellulose 17(3):459–494. https://doi.org/10.1007/s10570-010-9405-y
85. Lysenkov E, Sytnyk, I, Lysenkova I, Demchenko V (2025) Microstructure and mechanical properties of polymer nanocomposites based on polylactic acid and carbon nanotubes. Physicochem Mech Mater 61:58–65. https://doi.org/10.15407/pcmm2025.05.058.
86. Fujishima A, Zhang X, Tryk DA (2008) TiO_2 photocatalysis and related surface phenomena. Surf Sci Rep 63(12):515–582. https://doi.org/10.1016/j.surfrep.2008.10.001
87. Pelaez M, Nolan NT, Pillai SC, Seery MK, Falaras P, Kontos AG, Dunlop PSM, Hamilton JWJ, Byrne JA, O'Shea K, Entezari MH, Dionysiou DD (2012) A review on the visible light active titanium dioxide photocatalysts for environmental applications. Appl Catal B: Environ 125:331–349. https://doi.org/10.1016/j.apcatb.2012.05.036
88. Asahi R, Morikawa T, Ohwaki T, Aoki K, Taga Y (2001) Visible-light photocatalysis in nitrogen-doped titanium oxides. Science 293(5528):269–271. https://doi.org/10.1126/science.1061051. PMID: 11452117
89. Diez-Pascual A, Díez-Vicente A (2014) Poly(3-hydroxybutyrate)/ZnO bionanocomposites with improved mechanical, barrier and antibacterial properties. Intern J Mol Sci (IJMS) 15:10950–10973. https://doi.org/10.3390/ijms150610950
90. Camarillo R, Toston S, Martinez F, Jimenez C, Rincon J (2016) Preparation of TiO_2-based catalysts with supercritical fluid technology: characterization and photocatalytic activity in CO_2 reduction. J Chem Technol Biotechnol 92. https://doi.org/10.1002/jctb.5169
91. He Q, Guo Z, Ma S, He Z (2022) Recent advances in superhydrophobic papers for oil/water separation: a mini-review. ACS Omega 7(48):43330–43336. https://doi.org/10.1021/acsomega.2c05886.PMID:36506134;PMCID:PMC9730453

92. Wang Y, He Y, Lai Q, Fan M (2014) Review of the progress in preparing nano TiO_2: an important environmental engineering material. J Environ Sci (China) 26(11):2139–2177. https://doi.org/10.1016/j.jes.2014.09.023. Epub 2014 Oct 12. PMID: 25458670
93. Qu X, Brame J, Li Q, Alvarez PJ (2013) Nanotechnology for a safe and sustainable water supply: enabling integrated water treatment and reuse. Acc Chem Res 46(3):834–843. https://doi.org/10.1021/ar300029v. Epub 2012 Jun 27. PMID: 22738389
94. Bavykin D, Friedrich J, Walsh F (2006) Protonated titanates and TiO_2 nanostructured materials: synthesis, properties, and applications. Adv Mater 18:2807–2824. https://doi.org/10.1002/adma.200502696
95. Agresti A, Di Giacomo F, Pescetelli S, Di Carlo A (2024) Scalable deposition techniques for large-area perovskite photovoltaic technology: a multi-perspective review. Nano Energy 122. https://doi.org/10.1016/j.nanoen.2024.109317.
96. Zhang L, Dillert R, Bahnemann D, Vormoor M (2012) Photo-induced hydrophilicity and selfcleaning: models and reality. Energy Environ Sci 5:7491–7507. https://doi.org/10.1039/C2EE03390A
97. Zhang Q, Huang J-Q, Zhao M-Q, Qian WZ (2011) Carbon nanotube mass production: principles and processes. ChemSusChem 4:864–889. https://doi.org/10.1002/cssc.201100177
98. Maity CK, De S, Verma K, Moniruzzaman M, Sahoo S (2023) Nanocellulose: a versatile nanostructure for energy storage applications. Ind Crops Prod 204:117218. https://doi.org/10.1016/j.indcrop.2023.117218.
99. Zhang X-F, Liu Z-G, Shen W, Gurunathan S (2016) Silver nanoparticles: synthesis, characterization, properties, applications, and therapeutic approaches. Int J Mol Sci 17(9):1534. https://doi.org/10.3390/ijms17091534
100. Zhang Y, Liu H, Cui K, Yu J, Yan M, Ge S, Cheng X, Zhang L (2018) Flexible electronics based on micro/nanostructured paper. Adv Mater 30(51):1801588. https://doi.org/10.1002/adma.201801588
101. Schneider J, Matsuoka M, Takeuchi M, Zhang J, Horiuchi Y, Anpo M, Bahnemann DW (2014) Understanding TiO_2 photocatalysis: mechanisms and materials. Chem Rev 114(19):9919–9986. https://doi.org/10.1021/cr5001892. Epub 2014 Sep 19 PMID: 25234429
102. Bondarenko O, Juganson K, Ivask A, Kasemets K, Mortimer M, Kahru A (2013) Toxicity of Ag, CuO and ZnO nanoparticles to selected environmentally relevant test organisms and mammalian cells in vitro: a critical review. Arch Toxicol 87(7):1181–200. https://doi.org/10.1007/s00204-013-1079-4. Epub 2013 Jun 1. PMID: 23728526; PMCID: PMC3677982.
103. Khan MM, Adil SF, Al-Mayouf A (2015) Metal oxides as photocatalysts. J Saudi Chem Soc 19(5):462–464. https://doi.org/10.1016/j.jscs.2015.04.003
104. Rai M, Yadav A, Gade A (2021) Silver nanoparticles as a new generation of antimicrobials. Biotechnol Adv 27(1):76–83. https://doi.org/10.1016/j.biotechadv.2008.09.002
105. Durán N, Durán M, de Jesus MB, Seabra AB, Fávaro WJ, Nakazato G (2016) Silver nanoparticles: a new view on mechanistic aspects on antimicrobial activity. Nanomedicine 12(3):789–799. https://doi.org/10.1016/j.nano.2015.11.016. Epub 2015 Dec 24. PMID: 26724539
106. Fortunati E, Rinaldi S, Peltzer M, Bloise N, Visai L, Armentano I, Jimenez A, Latterini L, Kenny J (2014) Nano-biocomposite films with modified cellulose nanocrystals and synthesized silver nanoparticles. Carbohyd Polym 101:1122–1133. https://doi.org/10.1016/j.carbpol.2013.10.055
107. Magdassi S, Grouchko M, Kamyshny A (2010) Copper nanoparticles for printed electronics: routes towards achieving oxidation stability. Materials (Basel) 3(9):4626–4638. https://doi.org/10.3390/ma3094626.PMID:28883344;PMCID:PMC5445770
108. El-Naggar ME, Hassabo AG, Mohamed AL, Shaheen TI (2017) Surface modification of SiO_2 coated ZnO nanoparticles for multifunctional cotton fabrics. J Colloid Interface Sci 15(498):413–422. https://doi.org/10.1016/j.jcis.2017.03.080. Epub 2017 Mar 21. PMID: 28349884

109. Ahmed S, Ahmad M, Swami BL, Ikram S (2016) A review on plants extract mediated synthesis of silver nanoparticles for antimicrobial applications: a green expertise. J Adv Res 7(1):17–28. https://doi.org/10.1016/j.jare.2015.02.007. Epub 2015 Mar 9. PMID: 26843966; PMCID: PMC4703479.
110. Siripattanakul-Ratpukdi S, Fürhacker M (2014) Review: issues of silver nanoparticles in engineered environmental treatment systems. Water Air Soil Pollut 225. https://doi.org/10.1007/s11270-014-1939-4
111. Zolin L, Destro M, Curtil D, Chaussy D, Penazzi N, Beneventi D, Gerbaldi C (2014) Flexible cellulose-based electrodes: towards eco-friendly all-paper batteries. Chem Eng Trans 41:361–366. https://doi.org/10.3303/CET1441061
112. Soman V, Vishwakarma K, Poddar MK (2023) Ultrasound assisted synthesis of polymer nanocomposites: a review. J Polymer Res 30. https://doi.org/10.1007/s10965-023-03786-4
113. Nowack B, Krug HF, Height M (2011) 120 Years of nanosilver history: implications for policy makers. Environ Sci Technol 45. https://doi.org/10.1021/es200435m
114. Llorens A, Lloret E, Picouet PA, Trbojevich R, Fernandez A (2012) Metallic-based micro and nanocomposites in food contact materials and active food packaging. Trends Food Sci Technol 24. https://doi.org/10.1016/j.tifs.2011.10.001
115. Novoselov KS, Geim AK, Morozov SV, Jiang D, Zhang Y, Dubonos SV, Grigorieva IV, Firsov AA (2004) Electric field effect in atomically thin carbon films. Science 306(5696):666–669. https://doi.org/10.1126/science.1102896. PMID: 15499015
116. Geim AK, Grigorieva IV (2013) Van der waals heterostructures. Nature 499(7459):419–425. https://doi.org/10.1038/nature12385. PMID: 23887427
117. Compton OC, Nguyen ST (2010) Graphene oxide, highly reduced graphene oxide, and graphene: versatile building blocks for carbon-based materials. Small 6(6):711–723. https://doi.org/10.1002/smll.200901934
118. Liu J, Tang J, Gooding JJ (2012) Strategies for chemical modification of graphene and applications of chemically modified graphene. J Mater Chem 22(25):12435–12452. https://doi.org/10.1039/C2JM31218B
119. Liu Z, Xu Z, Hu X, Gao C (2013) Lyotropic liquid crystal of polyacrylonitrile-grafted graphene oxide and its assembled continuous strong nacre-mimetic fibers. Macromolecules 46(17):6931–6941
120. Zhu Y, Murali S, Cai W, Li X, Suk JW, Potts JR, Ruoff RS (2010) Graphene and graphene oxide: synthesis, properties, and applications. Adv Mater 22(35):3906–3924
121. Park S, Ruoff RS (2009) Chemical methods for the production of graphenes. Nat Nanotechnol 4(4):217–24. https://doi.org/10.1038/nnano.2009.58. Epub 2009 Mar 29. Erratum in: Nat Nanotechnol. 2010 Apr;5(4):309. PMID: 19350030
122. Eigler S, Hirsch A (2014) Chemistry with graphene and graphene oxide-challenges for synthetic chemists. Angew Chem Int Ed Engl 53(30):7720–7738. https://doi.org/10.1002/anie.201402780. Epub 2014 Jun 24 PMID: 24962439
123. Shao Y, El-Kady MF, Wang LJ, Zhang Q, Li Y, Wang H, Mousavi MF, Kaner RB (2015) Graphene-based materials for flexible supercapacitors. Chem Soc Rev 44(11):3639–3665. https://doi.org/10.1039/c4cs00316k. PMID: 25898904
124. Zheng Z, McDonald J, Khillan R, Su Y, Shutava T, Grozdits G, Lvov YM (2006) Layer-by-layer nanocoating of lignocellulose fibers for enhanced paper properties. J Nanosci Nanotechnol 6:624–632. https://doi.org/10.1166/jnn.2006.081
125. Naguib M, Mochalin VN, Barsoum MW, Gogotsi Y (2014) 25th anniversary article: MXenes: a new family of two-dimensional materials. Adv Mater 26(7):992–1005. https://doi.org/10.1002/adma.201304138. Epub 2013 Dec 19. PMID: 24357390
126. Fadeel B, Bussy C, Merino S, Vázquez E, Flahaut E, Mouchet F, Evariste L, Gauthier L, Koivisto AJ, Vogel U, Martín C, Delogu LG, Buerki-Thurnherr T, Wick P, Beloin-Saint-Pierre D, Hischier R, Pelin M, Candotto Carniel F, Tretiach M, Cesca F, Benfenati F, Scaini D, Ballerini L, Kostarelos K, Prato M, Bianco A (2018) Safety assessment of graphene-based materials: focus on human health and the environment. ACS Nano 12(11):10582–10620. https://doi.org/10.1021/acsnano.8b04758. Epub 2018 Nov 12. PMID: 30387986

127. Yoon HJ, Yang JH, Zhou Z, Yang SS, Cheng MMC (2011) Carbon dioxide gas sensor using a graphene sheet. Sens Actuator B-Chem 157:310–313. https://doi.org/10.1016/j.snb.2011.03.035
128. De Volder MF, Tawfick SH, Baughman RH, Hart AJ (2013) Carbon nanotubes: present and future commercial applications. Science 339(6119):535–539. https://doi.org/10.1126/science.1222453. PMID: 23372006
129. Tasis D, Tagmatarchis N, Bianco A, Prato M (2006) Chemistry of carbon nanotubes. Chem Rev 106(3):1105–1136. https://doi.org/10.1021/cr050569o. PMID: 16522018
130. Jorio A, Dresselhaus G, Dresselhaus M (2008) Carbon nanotubes: advanced topics in the synthesis. structure, properties and applications. https://doi.org/10.1007/978-3-540-72865-8
131. Coleman JN, Khan U, Blau WJ, Gun'ko YK (2006) Small but strong: a review of the mechanical properties of carbon nanotube–polymer composites. Carbon 44(9):1624–1652. https://doi.org/10.1016/j.carbon.2006.02.038.
132. Terrones M, Botello-Méndez AR, Campos-Delgado J, López-Urías F, Vega-Cantú YI, Rodríguez-Macías FJ, Elías AL, Muñoz-Sandoval E, Cano-Márquez AG, Charlier JC, Terrones H (2010) Graphene and graphite nanoribbons: morphology, properties, synthesis, defects and applications. Nano Today 5(4):351–372
133. Lam CW, James JT, McCluskey R, Arepalli S, Hunter RL (2006) A review of carbon nanotube toxicity and assessment of potential occupational and environmental health risks. Crit Rev Toxicol 36(3):189–217. https://doi.org/10.1080/10408440600570233. PMID: 16686422
134. Xie F (2023) Biopolymer-based multilayer films and coatings for food preservation: an update of the recent development. Curr Food Sci Technol Rep 1(1):1–12
135. De Azeredo HM (2009) Nanocomposites for food packaging applications. Food Res Int 42(9):1240–1253
136. Brinker CJ, Scherer GW (1990) Sol-gel science: the physics and chemistry of sol-gel processing. Gulf Professional Publishing
137. Kirkland C, Laufer G, Grunlan J (2012, August) Clay-chitosan nanobrick walls: renewable thin films with exceptional gas barrier. In: Abstract of papers of the American Chemical Society, vol 244. American Chemical Society, 1155 16TH ST, NW, Washington, DC 20036 USA
138. Duncan TV (2011) Applications of nanotechnology in food packaging and food safety: barrier materials, antimicrobials and sensors. J Colloid Interface Sci 363(1):1–24
139. Plackett D, Siró I (2012) Nanocomposites for food and beverage packaging. In: Emerging food packaging technologies, pp 239–273. Woodhead Publishing
140. Koppolu R (2024) High-throughput processing of nanocelluloses into barrier coatings: a focus on nanocellulose rheology and multilayer barrier properties
141. Baughman RH, Zakhidov AA, de Heer WA (2002) Fuzzy nanoassemblies: toward layered polymeric multicomposites. Science 297
142. Jebali S, Vayer M, Khaled B, Mahut F, Sinturel C (2023) Dip-coating deposition of nanocomposite thin films based on water-soluble polymer and silica nanoparticles. Colloids Surf A 680:132688. https://doi.org/10.1016/j.colsurfa.2023.132688
143. Afsharpour M, Imani S (2017) Preventive protection of paper works by using nanocomposite coating of zinc oxide. J Cult Herit 25:142–148
144. Salama A, Abouzeid RE, Owda ME, Cruz-Maya I, Guarino V (2021) Cellulose–silver composites materials: preparation and applications. Biomolecules 11(11):1684
145. Xuan S, Guo J, Li M, Qin X, Hou Y, Zhan Q, Jin P, Zheng Y, Wu Z (2025) Long-acting antibacterial nanochitin/nanocellulose bioplastic packaging film. Intern J Biol Macromol 309:142862
146. Noorian SA, Hemmatinejad N, Navarro JA (2020) Ligand modified cellulose fabrics as support of zinc oxide nanoparticles for UV protection and antimicrobial activities. Int J Biol Macromol 154:1215–1226
147. Zhang H, Cai J, Su J, Zhu M, Gong Y, Xu B (2024) Calcium Alginate/Silver nanosheet microencapsulated phase change materials with controlled morphology for thermal energy storage and antibacterial application. ACS Appl Nano Mater 7(14):16360–16371

148. Xiong C, Wang T, Han J, Zhang Z, Ni Y (2024) Recent research progress of paper-based supercapacitors based on cellulose. Energy Environ Mater 7(3):e12651
149. Muthamma K, Sunil D (2022) Cellulose as an eco-friendly and sustainable material for optical anticounterfeiting applications: an up-to-date appraisal. ACS Omega 7(47):42681–42699
150. Dube E, Okuthe GE (2025) Silver nanoparticle-based antimicrobial coatings: sustainable strategies for microbial contamination control. Microbiol Res 16(6):110. https://doi.org/10.3390/microbiolres16060110
151. Sharma VK, Yngard RA, Lin Y (2009) Silver nanoparticles: green synthesis and their antimicrobial activities. Adv Colloid Interface Sci 145(1–2):83–96. https://doi.org/10.1016/j.cis.2008.09.002. Epub 2008 Sep 17. PMID: 18945421
152. Verma RK, Kesarwani S, Xu J, Davim JP (2023) Polymer nanocomposites: fabrication to applications. https://doi.org/10.1201/9781003343912.
153. Pirozzi A, Ferrari G, Donsì F (2021) The use of nanocellulose in edible coatings for the preservation of perishable fruits and vegetables. Coatings 11(8):990. https://doi.org/10.3390/coatings11080990
154. Raheem SA, Alfatlawi AH (2021) Surface-water purification using cellulose paper impregnated with silver nanoparticles. Drink Water Eng Sci 14:95–102. https://doi.org/10.5194/dwes-14-95-2021
155. He X, Chen R, Cai X, Hong X, Yang Z, Cheng L, Shen Z (2026) Facile fabrication of graphene film with improved orientation and electrical conductivity by high hydraulic pressure. Carbon 246:120917. https://doi.org/10.1016/j.carbon.2025.120917
156. Hu L, Choi JW, Yang Y, Jeong S, La Mantia F, Cui L, Cui Y (2009) Highly conductive paper for energy-storage devices. Proc Natl Acad Sci USA 106(51):21490–21494. https://doi.org/10.1073/pnas.0908858106
157. Cai J, Chen Q, Chang C (2023) Spray-assisted LBL assembly of chitosan/nanocellulose as coatings of commercial membranes for oil-in-water emulsion separation. Int J Biol Macromol 242:124852
158. Ye L, Hong Y, Liao M, Wang B, Wei D, Peng H, Ye L, Hong Y, Liao M, Wang B, Wei D, Peng H (2020) Recent advances in flexible fiber-shaped metal-air batteries. Energy Storage Mater 28:364–374
159. Shanmugam K, Doosthosseini H, Varanasi S, Garnier G, Batchelor W (2018) Flexible spray coating process for smooth nanocellulose film production. Cellulose 25. https://doi.org/10.1007/s10570-018-1677-7
160. Wang C, Liu Q, Song H, Jiang Q (2023) Vacuum filtration method towards flexible thermoelectric films. OAE Publishing Inc
161. Li H, Lv S, Fang Y (2020) Bio-inspired micro/nanostructures for flexible and stretchable electronics. Nano Res 13:1–9. https://doi.org/10.1007/s12274-020-2628-9
162. Roy R, Eldhose M, George C, Joseph A (2023) Nanocellulose-based (Bio)composites for optoelectronic applications. https://doi.org/10.1007/978-981-19-0710-4_39.
163. Lavoine N, Desloges I, Bras J (2014) Microfibrillated cellulose coatings as new release systems for active packaging. Carbohyd Polym 103:528–537
164. Fukuzumi H, Saito T, Iwata T, Kumamoto Y, Isogai A (2009) Transparent and high gas barrier films of cellulose nanofibers prepared by TEMPO-mediated oxidation. Biomacromol 10(1):162–165. https://doi.org/10.1021/bm801065u. PMID: 19055320
165. Dankovich TA, Gray DG (2011) Bactericidal paper impregnated with silver nanoparticles for point-of-use water treatment. Environ Sci Technol 45(5):1992–1998
166. Calvo V, Martínez-Barón C, Fuentes L, Maser WK, Benito AM, González-Domínguez JM (2024) Nanocellulose: the ultimate green aqueous dispersant for nanomaterials. Polymers 16(12):1664. https://doi.org/10.3390/polym16121664
167. Mucci VL, Aranguren MI (2025) Plant-based cellulose nanopapers with applications for packaging, protective films and energy devices. J Renew Mater 13(8)
168. Merayo N, Balea A, de la Fuente E, Blanco Á, Negro C (2017) Interactions between cellulose nanofibers and retention systems in flocculation of recycled fibers. Cellulose 24. https://doi.org/10.1007/s10570-016-1138-0.

169. Chakrabarty A, Teramoto Y (2018) Recent advances in nanocellulose composites with polymers: a guide for choosing partners and how to incorporate them. Polymers 10(5):517. https://doi.org/10.3390/polym10050517
170. Singh AT, Lantigua D, Meka A, Taing S, Pandher M, Camci-Unal G (2018) Based sensors: emerging themes and applications. Sensors 18(9):2838
171. Scandurra G, Arena A, Ciofi C (2023) A brief review on flexible electronics for IoT: solutions for sustainability and new perspectives for designers. Sensors 23(11):5264. https://doi.org/10.3390/s23115264
172. Joshi NC, Prateek G (2022) Advances of graphene oxide based nanocomposite materials in the treatment of wastewater containing heavy metal ions and dyes. Curr Res Green Sustain Chem 5. https://doi.org/10.1016/j.crgsc.2022.100306
173. Jambhulkar S, Ravichandran D, Zhu Y, Thippanna V, Ramanathan A, Patil D, Fonseca N, Thummalapalli SV, Sundaravadivelan B, Sun A, Xu W, Yang S, Kannan AM, Golan Y, Lancaster J, Chen L, Joyee EB, Song K (2024) Nanoparticle assembly: from self-organization to controlled micropatterning for enhanced functionalities. Small 20(6):2306394
174. Ooi Y, Hanasaki I, Mizumura D, Matsuda Y (2017) Suppressing the coffee-ring effect of colloidal droplets by dispersed cellulose nanofibers. Sci Technol Adv Mater 18(1):316–324
175. Chen D, Cheng Y, Zhou N, Chen P, Wang Y, Li K, Huo S, Cheng P, Peng P, Zhang R, Wang L, Liu H, Liu Y, Ruan R (2020) Photocatalytic degradation of organic pollutants using TiO_2-based photocatalysts: a review. J Cleaner Prod 268. https://doi.org/10.1016/j.jclepro.2020.121725
176. Min T, Zhou L, Sun X, Du H, Zhu Z, Wen Y (2022) Electrospun functional polymeric nanofibers for active food packaging: a review. Food Chem 391:133239. https://doi.org/10.1016/j.foodchem.2022.133239
177. Wang X, Yao C, Wang F, Li Z (2017) Cellulose-based nanomaterials for energy applications. Small 13(42):1702240
178. Spagnuolo L, D'Orsi R, Operamolla A (2022) Nanocellulose for paper and textile coating: the importance of surface chemistry. ChemPlusChem 87(8):e202200204
179. Kalnins D (2019) Preparation of conductive coatings. Solution modification and CVD furnished coatings. Doctoral dissertation
180. Ha M, Lim S, Park J, Um DS, Lee Y, Ko H (2015) Electronic skin: bioinspired interlocked and hierarchical design of ZnO nanowire arrays for static and dynamic pressure-sensitive electronic skins. Adv Funct Mater 25. https://doi.org/10.1002/adfm.201500453.
181. Jamwal V, Mittal A (2024) Recent progresses in nanocomposite films for food-packaging applications: synthesis strategies, technological advancements, potential risks and challenges. Food Rev Intl 40(10):3634–3665
182. Rossa V, Ferreira LEM, da Costa Vasconcelos S, Shimabukuro ETT, da Costa Madriaga VG, Carvalho AP, Pergher SBC, da Silva FDC, Ferreira VF, Conte Junior CA, de Melo Lima T (2022) Nanocomposites based on the graphene family for food packaging: historical perspective, preparation methods, and properties. RSC Adv 12(22):14084–14111
183. Caratelli V, Di Meo E, Colozza N, Fabiani L, Fiore L, Moscone D, Arduini F (2022) Nanomaterials and paperbased electrochemical devices: merging strategies for fostering sustainable detection of biomarkers. J Mater Chem B 10(44):9021–9039
184. Hervy M, Evangelisti S, Lettieri P, Lee K-Y (2015) Life cycle assessment of nanocellulose-reinforced advanced fibre composites. Compos Sci Technol 118:154–162. https://doi.org/10.1016/j.compscitech.2015.08.024
185. Richardson JJ, Cui J, Bjornmalm M, Braunger JA, Ejima H, Caruso F (2016) Innovation in layer-by-layer assembly. Chem Rev 116(23):14828–14867
186. Mishra AK, Maurya R, Jagtap CY, Singh R, Kumar V (2025) Current status and future prospects of application of engineered nanomaterials. https://doi.org/10.1201/9781003484189-20.
187. Casati R, Vedani M (2014) Metal matrix composites reinforced by nano-particles: a review. Metals 4(1):65–83. https://doi.org/10.3390/met4010065

188. Iost R, Crespilho F (2011) Layer-by-layer self-assembly and electrochemistry: applications in biosensing and bioelectronics. Biosens Bioelectron 31:1–10. https://doi.org/10.1016/j.bios.2011.10.040
189. Kumar S, Shukla A, Baul PP, Mitra A, Halder D (2018) Biodegradable hybrid nanocomposites of chitosan/gelatin and silver nanoparticles for active food packaging applications. Food Packag Shelf Life 16:178–184
190. Szczyglewska P, Feliczak-Guzik A, Nowak I (2023) Nanotechnology-general aspects: a chemical reduction approach to the synthesis of nanoparticles. Molecules 28(13):4932. https://doi.org/10.3390/molecules28134932
191. Adel AM (2016) Incorporation of nano-metal particles with paper matrices. Interdisc J Chem 1(2):36–46
192. Sharma S, Sharma N, Kaushal N (2023) Utilization of novel bacteriocin synthesized silver nanoparticles (AgNPs) for their application in antimicrobial packaging for preservation of tomato fruit. Front Sustain Food System 7:1072738
193. Lim J-V, Bee S-T, Tin Sin L, Ratnam CT, Abdul Hamid ZA (2021) A review on the synthesis, properties, and utilities of functionalized carbon nanoparticles for polymer nanocomposites. Polymers 13(20):3547. https://doi.org/10.3390/polym13203547
194. Li R, Zhang K, Chen G (2019) Highly transparent, flexible and conductive CNF/AgNW paper for paper electronics. Materials 12(2):322. https://doi.org/10.3390/ma12020322
195. Kim J, Kumar R, Bandodkar AJ, Wang J (2017) Advanced materials for printed wearable electrochemical devices: a review. Adv Electron Mater 3(1):1600260
196. Yao S, Swetha P, Zhu Y (2018) Nanomaterial-enabled wearable sensors for healthcare. Adv Healthcare Mater 7(1):1700889
197. Torres FG, Commeaux S, Troncoso OP (2012) Biocompatibility of bacterial cellulose based biomaterials. J Funct Biomater 3(4):864–878. https://doi.org/10.3390/jfb3040864
198. Kumar N, Lee SY, Park SJ (2024) Recent progress and challenges in paper-based microsupercapacitors for flexible electronics: a comprehensive review. ACS Appl Mater Interfaces 16(17):21367–21382
199. Ahmed J, Arfat YA, Bher A, Mulla M, Jacob H, Auras R (2018) Active chicken meat packaging based on polylactide films and bimetallic Ag–Cu nanoparticles and essential oil. J Food Sci 83(5):1299–1310
200. Yadav S, Mehrotra GK, Dutta PK (2021) Chitosan based ZnO nanoparticles loaded gallic-acid films for active food packaging. Food Chem 334:127605
201. Ghaani M, Cozzolino CA, Castelli G, Farris S (2016) An overview of the intelligent packaging technologies in the food sector. Trends Food Sci Technol 51:1–11
202. Ling Y, Luo Y, Luo J, Wang X, Sun R (2013) Novel antibacterial paper based on quaternized carboxymethyl chitosan/organic montmorillonite/Ag NP nanocomposites. Ind Crops Prod 51:470–479
203. Afsharpour M, Imani S (2017) Preventive protection of paper works by using nanocomposite coating of zinc oxide. J Cult Herit 25:142–148
204. More PR, Pandit S, Filippis AD, Franci G, Mijakovic I, Galdiero M (2023) Silver nanoparticles: Bactericidal and mechanistic approach against drug resistant pathogens. Microorganisms 11(2):369. https://doi.org/10.3390/microorganisms11020369
205. Al-Naamani L, Dobretsov S, Dutta J (2016) Chitosan-zinc oxide nanoparticle composite coating for active food packaging applications. Innov Food Sci Emerg Technol 38. https://doi.org/10.1016/j.ifset.2016.10.010
206. Rossa V, Monteiro Ferreira LE, da Costa Vasconcelos S, Tai Shimabukuro ET, Gomes da Costa Madriaga V, Carvalho AP, Castellã Pergher SB, de Carvalho da Silva F, Ferreira VF, Conte Junior CA, de Melo Lima T (2022) Nanocomposites based on the graphene family for food packaging: historical perspective, preparation methods, and properties. RSC Adv 12(22):14084–14111. https://doi.org/10.1039/d2ra00912a. PMID: 35558848; PMCID: PMC9094098
207. Zhao W, Wen J, Zhao Y, Wang Z, Shi Y, Zhao Y (2020) Hierarchically porous carbon derived from biomass reed flowers as highly stable Li-Ion battery anode. Nanomaterials 10(2):346. https://doi.org/10.3390/nano10020346

208. Gareev KG, Grouzdev DS, Koziaeva VV, Sitkov NO, Gao H, Zimina TM, Shevtsov M (2022) Biomimetic Nanomaterials: diversity, technology, and biomedical applications. Nanomaterials 12(14):2485. https://doi.org/10.3390/nano12142485
209. Anaya-Esparza LM, Villagrán-de la Mora Z, Ruvalcaba-Gómez JM, Romero-Toledo R, Sandoval-Contreras T, Aguilera-Aguirre S, Montalvo-González E, Pérez-Larios A (2020) Use of titanium dioxide (TiO_2) nanoparticles as reinforcement agent of polysaccharide-based materials. Processes 8(11):1395. https://doi.org/10.3390/pr8111395
210. Firmanda A, Fahma F, Warsiki E, Syamsu K, Arnata I, Sartika D, Suryanegara L, Mayanullah Q, Suyanto A (2023) Antimicrobial mechanism of nanocellulose composite packaging incorporated with essential oils. Food Control 147:109617. https://doi.org/10.1016/j.foodcont.2023.109617
211. Ghaffari SB, Sarrafzadeh MH (2023) Cationic cellulose filter papers modified with ZnO/Ag/GO nanocomposite as point of use gravity-driven filters for bacterial removal from water. Sci Rep 13(1):22604. https://doi.org/10.1038/s41598-023-50114-3. PMID: 38114628; PMCID: PMC10730911
212. Melchor-Durán AP, Moreno-Virgen MR, Bonilla-Petriciolet A, Reynel-Ávila HE, Lucio Ortiz E, González-Vázquez OF (2024) Heavy metal removal from water using graphene oxide in magnetic-assisted adsorption systems: characterization, adsorption properties, and modelling. Separations 11(10):294. https://doi.org/10.3390/separations11100294
213. Zhu XD, Liu Y, Li Z (2017) Surface coating performance of TiO_2 nanoparticle-modified veneered panels and their influence on formaldehyde emission. J Coat Technol Res 14(6):1271–1278
214. Kumalo FI, Mugudamani I, Atangana E, Gumede TP (2025) Recent developments in cellulose/chitosan biopolymer composites for winery wastewater treatment and reuse: a review. Materials 18(21):5028. https://doi.org/10.3390/ma18215028
215. Hu L, Choi JW, Yang Y, Jeong S, La Mantia F, Cui LF, Cui Y (2009) Highly conductive paper for energystorage devices. Proc Natl Acad Sci 106(51):21490–21494
216. Leijonmarck S, Cornell A, Lindbergh G, Wågberg L (2013) Single-paper flexible Li-ion battery cells through a paper-making process based on nano-fibrillated cellulose. J Mater Chem A 1(15):4671–4677
217. Say MG, Brett CJ, Edberg J, Roth SV, Soderberg LD, Engquist I, Berggren M (2022) Scalable paper supercapacitors for printed wearable electronics. ACS Appl Mater Interfaces 14(50):55850–55863
218. Fraiwan A, Mukherjee S, Sundermier S, Lee H-S, Choi S (2013) A paper-based microbial fuel cell: instant battery for disposable diagnostic devices. Biosens Bioelectron 49C:410–414. https://doi.org/10.1016/j.bios.2013.06.001
219. Zhong Y, Dai J, Xu X, Su C, Shao Z (2020) Facilitating oxygen redox on manganese oxide nanosheets by tuning active species and oxygen defects for zinc-air batteries. ChemElectroChem 7:4949–4955. https://doi.org/10.1002/celc.202001419
220. Han J, Xu N, Liang Y, Ding M, Zhai J, Sun Q, Wang ZL (2021) Paper-based triboelectric nanogenerators and their applications: a review. Beilstein J Nanotechnol 1(12):151–171. https://doi.org/10.3762/bjnano.12.12. PMID: 33614382; PMCID: PMC7871030
221. Ramya J, Hemalatha J (2015) Paper battery-a promising energy solution in new IT era. Int J Eng Res Tech (IJERT) NCICCT 3(12)
222. Aishwarya N, Charan P (2023) Paper batteries: a flexible and multifunctional energy storage technology. Intern J Res Appl Sci Eng Tech 11:1015–1017. https://doi.org/10.22214/ijraset.2023.49155.
223. Chen Q, Xie F, Wang G, Ge K, Ren H, Yan M, Qiyang W, Bi H (2021) Hybrid MoS2@PANI materials for high-performance supercapacitor electrode. Ionics 27:1–14. https://doi.org/10.1007/s11581-021-04147-1
224. Sukhavattanakul P, Srikhao N, Ummartyotin S, Narain R (2025) Sensing food spoilage with nanotechnology: a review of current research and challenges. Measurement: Food 18:100221. https://doi.org/10.1016/j.meafoo.2025.100221

225. Tavassoli M, Alizadeh Sani M, Khezerlou A, Ehsani A, Jahed-Khaniki G, McClements DJ (2022) Smart biopolymer-based nanocomposite materials containing pH-sensing colorimetric indicators for food freshness monitoring. Molecules 27(10):3168. https://doi.org/10.3390/molecules27103168
226. El-Sayed SM, Youssef AM (2023) Eco-friendly biodegradable nanocomposite materials and their recent use in food packaging applications: a review. Sustain Food Technol 1(2):215–227
227. Divya BT, Anjali G (2024) Nanocomposites as sustainable smart materials: a review. J Reinf Plastic Compos. https://doi.org/10.1177/07316844241233162
228. Sanchez C, Julián B, Belleville P, Popall M (2005) Applications of hybrid organic–inorganic nanocomposites. J Mater Chem 15(35–36):3559–3592
229. Jieming Q, Rui X (2025) Bio-inspired multifunctional nanocellulose materials. https://doi.org/10.1002/9783527849857.ch08
230. Ali AMM, Jabir SM, Kadhim AA, Almagtome A (2022) Nanotechnology practices and cost restructure for effective cost management under industry 4.0 based manufacturing systems. TEM J 11:1193–1199. https://doi.org/10.18421/TEM113-26
231. Lee SJ, Yoon SJ, Jeon I-Y (2022) Graphene/polymer nanocomposites: preparation, mechanical properties, and application. Polymers 14(21):4733. https://doi.org/10.3390/polym14214733
232. Abdallah K, Mayar A, Almoataz A, Ammar E (2021) Fabrication techniques of polymeric nanocomposites: a comprehensive review. ARCHIVE Proc Inst Mech Eng Part C J Mech Eng Sci 1989–1996 203–210(236):1–19. https://doi.org/10.1177/09544062211055662
233. Antony Jose S, Cowan N, Davidson M, Godina G, Smith I, Xin J, Menezes PL (2025) A comprehensive review on cellulose nanofibers, nanomaterials, and composites: manufacturing, properties, and applications. Nanomaterials 15(5):356. https://doi.org/10.3390/nano15050356
234. Wang H, Zhang X, Wang N, Li Y, Feng X, Huang Y, Chunsong Z, Zhenglian L, Minghao F, Gang O, Huajian G, Xiaoyan L, Wu H (2017) Ultralight, scalable, and high-temperature–resilient ceramic nanofiber sponges. Sci Adv 3. https://doi.org/10.1126/sciadv.1603170
235. Gupta D, Sharma S, Bharti A, Kumar R (2025) Two-dimensional nanomaterials for energy storage. https://doi.org/10.1007/978-981-96-7952-2_9
236. Batley GE, Kirby JK, McLaughlin MJ (2012) Fate and risks of nanomaterials in aquatic and terrestrial environments. Acc Chem Res 46. https://doi.org/10.1021/ar2003368
237. Nel A, Xia T, Mädler L, Li N (2006) Toxic potential of materials at the nanolevel. Science 311(5761):622–627. https://doi.org/10.1126/science.1114397. PMID: 16456071
238. Jeffery S (2009) Nanomaterials: risks and benefits. https://doi.org/10.1007/978-1-4020-9491-0.
239. Nowack B, Bucheli TD (2007) Occurrence, behavior and effects of nanoparticles in the environment. Environ Pollut (Barking, Essex: 1987) 150:5–22. https://doi.org/10.1016/j.envpol.2007.06.006
240. Corsi I, Bellingeri A, Eliso MC, Grassi G, Liberatori G, Murano C, Sturba L, Vannuccini ML, Bergami E (2021) Eco-interactions of engineered nanomaterials in the marine environment: towards an eco-design framework. Nanomaterials 11(8):1903. https://doi.org/10.3390/nano11081903
241. Ahmed SF, Mofijur M, Rafa N, Chowdhury AT, Chowdhury S, Nahrin M, Saiful Islam ABM Ong HC (2022) Green approaches in synthesising nanomaterials for environmental nanobioremediation: technological advancements, applications, benefits and challenges. Environ Res 204:111967. https://doi.org/10.1016/j.envres.2021.111967
242. Oberdörster G, Oberdörster E, Oberdörster J (2005) Nanotoxicology: an emerging discipline evolving from studies of ultrafine particles. Environ Health Perspect 113(7):823–839. https://doi.org/10.1289/ehp.7339. Erratum in: Environ Health Perspect 2010Sep;118(9):A380. PMID: 16002369; PMCID: PMC1257642
243. Martínez G, Merinero M, Pérez-Aranda M, Pérez-Soriano EM, Ortiz T, Villamor E, Begines B, Alcudia A (2021) Environmental impact of nanoparticles application as an emerging technology: a review. Materials 14(1):166. https://doi.org/10.3390/ma14010166

244. Dolez PI (2015) Progress in personal protective equipment for nanomaterials. In: Nanoengineering: global approaches to health and safety issues, pp 607–635. https://doi.org/10.1016/B978-0-444-62747-6.00019-1
245. Schulte PA, Leso V, Niang M, Iavicoli I (2019) Current state of knowledge on the health effects of engineered nanomaterials in workers: a systematic review of human studies and epidemiological investigations. Scand J Work Environ Health 45(3):217–238. https://doi.org/10.5271/sjweh.3800. Epub 2019 Jan 17. PMID: 30653633; PMCID: PMC6494687
246. Kumar L, Ramakanth D, Konala A, Gaikwad K (2021) Edible films and coatings for food packaging applications: a review. Environ Chem Lett 20:26. https://doi.org/10.1007/s10311-021-01339-z
247. Noonan GO, Whelton AJ, Carlander D, Duncan TV (2014) Measurement methods to evaluate engineered nanomaterial release from food contact materials. Compr Rev Food Sci Food Saf 13(4):679–692. https://doi.org/10.1111/1541-4337.12079. PMID: 33412706
248. Honarvar Z, Hadian Z, Mashayekh M (2016) Nanocomposites in food packaging applications and their risk assessment for health. Electron Phys 8(6):2531–2538. https://doi.org/10.19082/2531.PMID:27504168;PMCID:PMC4965203
249. Pirela SV, Miousse IR, Lu X, Castranova V, Thomas T, Qian Y, Bello D, Kobzik L, Koturbash I, Demokritou P (2016) Effects of laser printer–emitted engineered nanoparticles on cytotoxicity, chemokine expression, reactive oxygen species, DNA methylation, and DNA damage: a comprehensive in vitro analysis in human small airway epithelial cells, macrophages, and lymphoblasts. Environ Health Perspect 124(2):210–219
250. Mackevica A, Hansen SF (2016) Release of nanomaterials from solid nanocomposites and consumer exposure assessment—a forward-looking review. Nanotoxicology 10(6):641–653. https://doi.org/10.3109/17435390.2015.1132346
251. Xuan L, Ju Z, Skonieczna M, Zhou PK, Huang R (2023) Nanoparticles-induced potential toxicity on human health: applications, toxicity mechanisms, and evaluation models. MedComm 4(4):e327. https://doi.org/10.1002/mco2.327
252. Allan J, Belz S, Hoeveler A, Hugas M, Okuda H, Patri A, Rauscher H, Silva P, Slikker W, Sokull-Kluettgen B, Tong W, Anklam E (2021) Regulatory landscape of nanotechnology and nanoplastics from a global perspective. Regul Toxicol Pharmacol 122:104885. https://doi.org/10.1016/j.yrtph.2021.104885
253. Gupta RK, Guha P, Srivastav PP (2024) Investigating the toxicological effects of nanomaterials in food packaging associated with human health and the environment. J Hazard Mater Lett 5:100125 https://doi.org/10.1016/j.hazl.2024.100125
254. Laux P, Tentschert J, Riebeling C, Braeuning A, Creutzenberg O, Epp A, Fessard V, Haas KH, Haase A, Hund-Rinke K, Jakubowski N, Kearns P, Lampen A, Rauscher H, Schoonjans R, Störmer A, Thielmann A, Mühle U, Luch A (2018) Nanomaterials: certain aspects of application, risk assessment and risk communication. Arch Toxicol 92(1):121–141. https://doi.org/10.1007/s00204-017-2144-1. Epub 2017 Dec 22. PMID: 29273819; PMCID: PMC5773666
255. Peter L, Jutta T, Christian R, Albert B, Otto C, Astrid E, Valerie F, Karl-Heinz H, Andrea H, Hund-Rinke K, Norbert J, Peter K, Alfonso L, Hubert R, Reinhilde S, Angela S, Axel T, Uwe M, Andreas L (2018). Nanomaterials: certain aspects of application, risk assessment and risk communication. Arch Toxicol 92. https://doi.org/10.1007/s00204-017-2144-1
256. Froggett S, Clancy S, Boverhof D, Canady R (2014) A review and perspective of existing research on the release of nanomaterials from solid nanocomposites. Part Fibre Toxicol 11:17. https://doi.org/10.1186/1743-8977-11-17
257. Oomen AG, Steinhäuser KG, Bleeker EA, van Broekhuizen F, Sips A, Dekkers S, Wijnhoven SWP, Sayre PG (2018) Risk assessment frameworks for nanomaterials: scope, link to regulations, applicability, and outline for future directions in view of needed increase in efficiency. NanoImpact 9:1–13. https://doi.org/10.1016/j.impact.2017.09.001
258. Gupta RK, Abd El Gawad F, Ali EA, Karunanithi S, Yugiani P, Srivastav PP (2024) Nanotechnology: current applications and future scope in food packaging systems. Measurement: Food 13:100131. https://doi.org/10.1016/j.meafoo.2023.100131

259. Dhingra R, Naidu S, Upreti G, Sawhney R (2010) Sustainable nanotechnology: through green methods and lifecycle thinking. Sustainability 2(10):3323–3338
260. Sanchez C, Rozes L, Ribot F, Laberty C, Grosso D, Sassoye C, Boissiere C, Nicole L (2010) "Chimie douce": a land of opportunities for the designed construction of functional inorganic and hybrid organic-inorganic nanomaterials. Comptes Rendus Chimie - C R CHIM 13:3–39. https://doi.org/10.1016/j.crci.2009.06.001
261. Otilia C, Maria T (2021) Multifunctional ZnO nanoparticle: based coatings for cultural heritage preventive conservation. https://doi.org/10.5772/intechopen.94070
262. Kumar R, Rai B, Kumar G (2019) A simple approach for the synthesis of cellulose nanofiber reinforced chitosan/PVP bio nanocomposite film for packaging. J Polym Environ 27(12):2963–2973
263. Saleh TA (2022) Nanomaterials and hybrid nanocomposites for CO_2 capture and utilization: environmental and energy sustainability. RSC Adv 12(37):23869–23888
264. Han M, Zhang X, Zhang H (2019) Flexible and stretchable triboelectric nanogenerator devices. https://doi.org/10.1002/9783527820153
265. Sanchez C, Boissiere C, Coupé A, Goettmann F, Grosso D, Julian-Lopez B, Llusar M, Nicole L (2005) Design of functional nano-structured inorganic and hybrid materials. Stud Surf Sci Catal 156:19–36. https://doi.org/10.1016/S0167-2991(05)80186-7
266. Gao Z, Li Y, Shang X, Hu W, Gao G, Duan L (2019) Bio-inspired adhesive and self-healing hydrogels as flexible strain sensors for monitoring human activities. Mater Sci Eng C 106:110168. https://doi.org/10.1016/j.msec.2019.110168
267. Dong Y, Xie Y, Ma X, Yan L, Yu HY, Yang M, Abdalkarim SYM, Jia B (2023) Multi-functional nanocellulose based nanocomposites for biodegradable food packaging: Hybridization, fabrication, key properties and application. Carbohydr Polym 321:121325. https://doi.org/10.1016/j.carbpol.2023.121325
268. Kumar S, Devi L (2011) Review on modified TiO2 photocatalysis under UV/visible light: selected results and related mechanisms on interfacial charge carrier transfer dynamics. J Phys Chem A 115:13211–13241. https://doi.org/10.1021/jp204364a
269. Novoselov KS, Mishchenko A, Carvalho A, Castro Neto AH (2020) Graphene-polymer nanocomposites for flexible electronics. Science 353(6298):aac9439. https://doi.org/10.1126/science.aac9439

Recent Advancements in Nanotechnology and Biotechnology in Pulp Bleaching

Shiva Shaddel Khalifelu, Aziz Babapoor, and Seyyed Mojtaba Mousavi

Abstract Pulp and paper manufacturing has advanced, but environmental issues persist. Efforts are being made to expand the opportunity for raw substances and cleanser production techniques to reduce strength intake. Biotechnology, particularly enzymatic processes, is being utilized to enhance pulp bleaching, with xylanases and laccases likely providing environmentally friendly bleaching for wood and agro-based pulps. Xylanase treatment eliminates white pulp flicker, reducing flicker in white light. A chlorine-free white method has been developed using an enzymatic treatment with manganese peroxidase (MnP), thereby reducing chemical intake and increasing the bioavailability of organic halides in the effluent. This technique, followed by hydrogen peroxide separation, is being investigated for increasing pulp and paper manufacturing. Advancements in nanotechnology and biotechnology, including the use of enzymes for biobleaching, can significantly reduce the environmental footprint of pulp and paper production while enhancing product quality and reducing energy consumption.

Keywords Nanotechnology · Wood industry · Pulp · Paper · Nanofibers · Xylanase · Biotechnology · Laccases

1 Introduction

Since its discovery, wood has been the primary raw fabric for the pulp and paper (PP) business. Within nature, lignocellulosic cloth made of hemicellulose, lignin, and cellulose is the most commonplace for renewable biopolymer. However, lignin's particular shape and features as a resistant plant coat restrict the entry of cellulose and hemicellulose, saving lignocellulolytic enzymes from penetrating [1]. The pulping

S. S. Khalifelu · A. Babapoor (✉)
Department of Chemical Engineering, University of Mohaghegh Ardabil, Ardabil, Iran
e-mail: babapoor@uma.ac.ir

S. M. Mousavi
Department of Chemical Engineering, National Taiwan University of Science and Technology, Taipei, Taiwan

R. K. Bachheti (ed.), *Nanotechnology in the Pulp and Paper Industry*, Smart Nanomaterials Technology, https://doi.org/10.1007/978-981-95-3473-9_10

process separates lignocellulose from cellulose, the primary component of paper [2]. The bleaching method, which is the most polluting level because it releases mutagenic and ecologically detrimental by-products such as dioxins, absorbable organic halides (AOXs), chlorinated lignin derivatives, and furans, removes the residual lignin to reduce the brown hue. Therefore, new ecologically friendly bleaching strategies are required [3].

Numerous enzymes released by particular bacterial and fungal species can catalyze a wide range of oxidative and hydroxylation processes. Researchers examined the biodegradation of ligocellulosic biomass (LCBM) in microorganisms known as basidiomycetes, which are capable of breaking down wood. Figure 1 lists the five distinct extracellular enzymes involved in the enzyme-mediated breakdown of lignin [4].

When carried out in total chlorine-free (TCF) bleaching sequences, the use of lignin-degrading microorganisms or their enzymes is a unique, environmentally

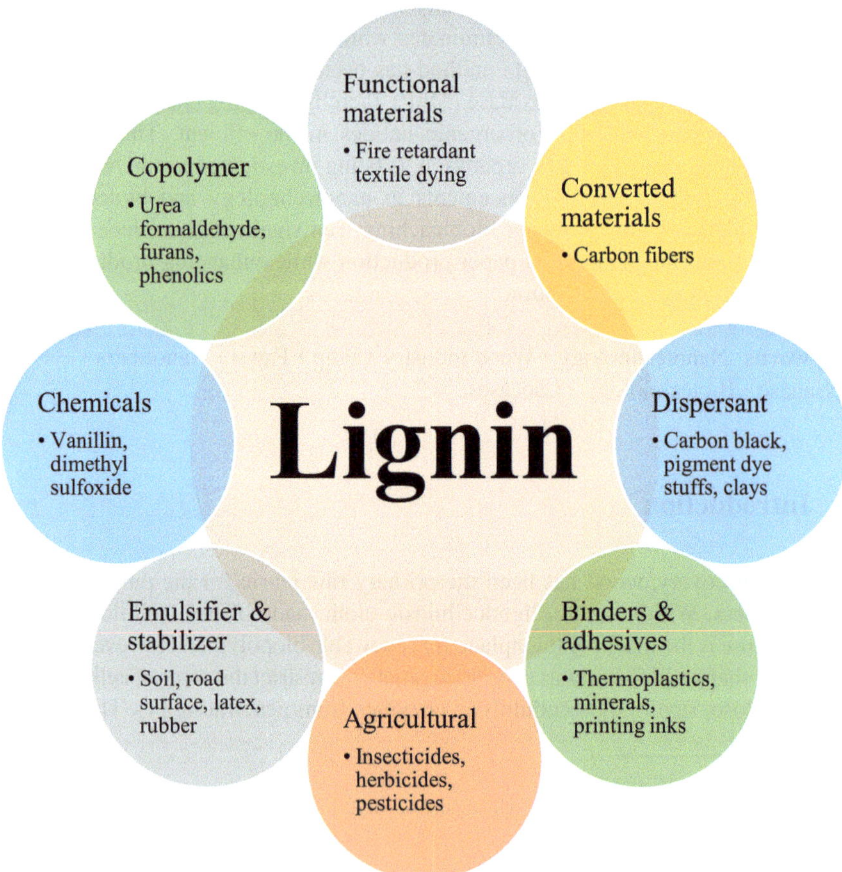

Fig. 1 Different types of extracellular enzymes involved in lignin degradation [4]

friendly method that has shown excellent outcomes in enhancing the final pulp brightness and reducing the cost of the bleaching system [5]. Relying on the implemented hardwoods and softwoods, xylanase remedy in mill-scale assessments caused a 12–25% reduction in AOX formation and a 10–25% reduction in the consumption of total elemental chlorine [3]. There are numerous categories of microorganisms that break down lignin, including fungi, Gram-positive and Gram-negative bacteria, and other microorganisms. Environmental rules limit the use of genetically engineered fungi, despite their potential, because they produce more laccase enzymes [6]. However, due to their greater metabolic versatility and adaptability, microorganisms may also thrive in various environmental niches, including animal stomachs, rhizosphere soil, lake and sea sediments, and decaying plant matter [7]. Among the most important phyla of bacteria are the Actinobacteria, which can be filamentous and incorporate DNA with a high G + C mol% composition. By breaking down complex polymers and lifeless natural components and returning them to the soil, they play an essential component in the composition of the soil [8]. To be able to perceive Actinobacteria with the ability to biobleach, this study screened a variety of lignocellulose-rich soil samples. It then tested the effect of realistic elements at the bleaching hobby of the most hit isolate. At some point during the investigation, the mechanical pulp—which was made from 25 and 75% robotically ground Fagus orientalis and Carpinus beetles wooden—turned utilized [9]. With numerous enzymatic applications reported in the literature, the pulp and paper industry is integrating environmentally friendly production methods into its processes [10]. Xylanases in prebleaching kraft pulps and extraordinary enzymes in paper recycling are examples of business utilization. Enzyme strategies may be the foundation for future cost-added gadgets [11].

However, as new uses do not align with present-day strategies, the enzyme or device should be modified. Economic approach and technological viability are characteristics of the device [12]. This study explores novel enzyme programs in the pulp and paper industries, offering insights into their use in enhancing unit operations, pulp product efficiency, or both [11].

Table 1 shows the houses of paper sheets made from bleached softwood pulp dealt with tremendous quantities of potassium permanganate. The bleached softwood pulp undergoes oxidation and removal of some noncellulosic components because of the softness of KMnO4. It also oxidizes the very last hemicellulose and in situ lignin. This explains the high-quality profits in terms of water retention rate (WRV), brightness, and breaking length that have been achieved. Treating paper sheets with 2% KMnO4 increases their breaking strength by 25% due to the oxidation of hemicellulose and the decrease in degree of polymerization (DP) cellulose, which forms strength boosters. However, it shows that along the facet KMnO4, bleached moderate timber pulp can increase its brightness and shorten its bleaching period, making it a viable bleached pulp development. Potassium permanganate is a potential treatment for the problematic results observed in shipping and storage facilities prior to papermaking, as it also serves as a purifier and deodorizer [13].

Table 1 Properties of paper made from bleached softwood pulp treated with varying amounts of potassium permanganate [13]

$KMnO_4$ quantity/ 100 g pulp	Breaking length in meters	Percent increase in breaking length	Brightness %	WRV%
None (untreated)	3606	–	71.13	74.60
0.50	3775	4.69	75.75	62.90
0.25	3749	3.97	72.88	73.50
2.00	4526	25.51	72.63	63.30
1.00	3809	5.63	75.18	53.30
1.50	4047	12.23	72.90	59.20

Paper recycling entails numerous fundamental processes, such as pulping, washing, screening, flotation, bleaching, and wastewater remedy. It provides a considerable fiber supply [14, 15]. Enzymes can enhance fiber production, decrease electricity consumption at some stage in refining, or modify the fiber primarily during these processes [16]. Cellulases have embellished the kraft pulping of sycamore chips and other pulp assets viably, as they provide extra efficient delignification and chemical savings through the bleaching method without sacrificing the paper's energy [17]. However, there is difficulty in using enzymes due to their high cost and uncertainties regarding how well great enzymes support low-molecular-weight pulping chemicals [18].

2 What is Paper Pulp?

Fibers from flora, minerals, or artificial assets may be observed in paper pulp, a raw material used to make paper. Remove the moisture, and it becomes a felted or disheveled sheet. Cellulosic fibers, which are most commonly observed in forests, are the primary source of most papers [19]. Extra resources encompass flax, agricultural waste, jute, cotton, Manila hemp, sisal, recycled fibers from cardboard containers, and old and scrap paper. The pleasant woods for sulfite paste and mechanical paste are hemlock, fir, and balsam fir. Other hardwoods work well for sulfate paste. Wood pulp can be categorized into two types: mechanical and chemical [20]. By eliminating contaminants and reducing shade from wood pulp, paper pulp bleaching is crucial in producing high-quality paper [21]. It produces strong, high-quality white paper that is captivating. Elemental chlorine-free (ECF), totally chlorine-based, and total chlorine-free (TCF) are popular techniques. Although chlorine-based bleaching works properly, it raises environmental issues [22, 23]. Options that are more environmentally friendly are TCF and ECF. Bleaching paper pulp has the benefits of better paper fine, brighter whiteness, and more extended sturdiness. However, the process can also have an unfavorable impact on the sustainability of paper production and the environment [24].

Pulp is a fibrous lignocellulosic substance combined with water and different chemical substances or plant-based components used in the production of paper-making and other paper items. It can be made from timber, fiber crops, waste paper, or rags [25].

Earlier than Cai Lun's discovery of papermaking in China in 105 AD, prehistoric societies used raw plant sources to make writing materials equivalent to paper, which included papyrus and amate [26]. A display screen extracts cellulose fibers from a solution and dries them into sheets or rolls, creating a finer, more consistent slurry due to the maceration system [27]. Hemp rag, leftover Internet, and bast fibers from the paper mulberry plant have been used to make the primary paper in China. W3, four the domestication of the mulberry tree for pulp manufacture commenced in the sixth century [28]. The extra pulp materials covered straw, cotton, hibiscus bark, bamboo, and blue sandalwood. In the thirteenth century, the artwork of producing paper from hemp and linen fibers from ragged clothes, fishing nets, and fabric baggage made its way to Europe [29, 30]. The manufacturing and affordability of rag paper, which aided in the development of printing, depended closely on the usage of rags. Over 95% of the pulp produced globally is made from pulpwood and other tree merchandise [25, 31, 32]. Pulping is the procedure of changing wooden or different agricultural sources into bendy fibers to make paper. One famous approach is kraft pulping, which eliminates lignin and produces a sturdy, bendy cellulose fiber. The printing grades on paper involve pulping acid sulfite. While pulping by mechanical, thermo-mechanical, and chemo-thermomechanical means produces fibers appropriate for printing or magazine stock, they have a lower strength and are optically exceptional. Making white paper requires pulping, cleansing, and bleaching the fibers. In addition, manipulations, such as adding filler and coloration and converting the dimensions, can sooner or later result in the very last paper output.

Paper pulp is a fibrous cloth made from plant fibers, including hemicellulose, cellulose, lignin, pigment, pectin, ash, and resin: chemical or mechanical separation from uncooked substances, including waste paper, straw, wood, and sugarcane. Paper pulp is one of the most extensively used hand-raw resources in the world for the production of paint, plastic, paper, synthetic fiber, and other merchandise [33]. Turning plant fibers—wood fiber being the most common type—into paper pulp involves crushing, screening, boiling, washing, bleaching, cleansing, and drying. The industries that use paper pulp the most frequently include paint, interlayer, plastic, artificial fiber, and papermaking [25].

3 How to Whiten Paper Pulp?

Pulp bleaching is crucial in producing pulp that ensures brightness, softness, and cleanliness. Adsorbable organic halides (AOXs) are the pollutants in pulp and paper enterprise effluent that need to worry you the maximum, considering they may be quite dangerous to fish and zooplankton. The pulp and paper (PP) sector is running to limit AOX emissions, so take a look at rigorous rules. The Fenton approach is

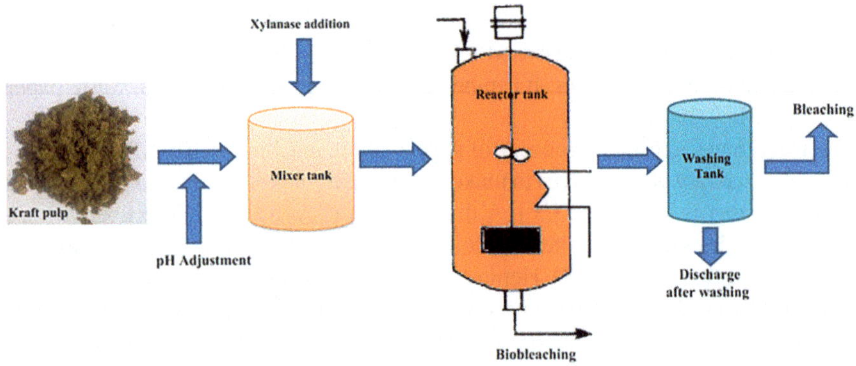

Fig. 2 How to whiten paper and paper pulp [127]

a complex oxidation technique for reducing refractory materials, along with AOX [34, 35]. Recent research aims to enhance the overall performance of the Fenton method in the pulp and paper industry by developing heterogeneous catalysts that can be separated from treated wastewater, reused, and operated at milder pH conditions, thereby controlling the leaching of iron and other ions. This approach should overcome barriers and enhance the effectiveness of the Fenton method [36, 37]. The conventional technique of manufacturing paper on an industrial scale includes heating pulp to an excessive temperature of as much as 170 °C in an alkaline way and then using a chemical bleaching approach to do away with lignin. Several natural derivatives of chlorine are produced during the chemical bleaching process, which is often accomplished via sellers based on chlorine, including hypochlorite, C_lO_2, and C_{l2} [38] Most of them are considered as mighty mutagens and viable cancer agents. As an environmentally friendly replacement for whitening pulps, chemicals are used to apply microbial enzymes, such as xylanases and laccases [39]. The use of biological shops in the bleaching of pulp for paper, which includes enzymes or entire microorganisms, is known as biobleaching (Fig. 2).

The types of raw materials used, the process technologies employed, internal recirculation of the effluent, and management techniques all impact the characteristics of the wastewater produced by the pulp and paper industry. A thorough understanding of the specific industrial effluent's properties is required to choose and create an appropriate treatment method for effluent reuse in the process application. Figure 3 lists the main features of wastewater generated at different phases of the process as well as the sources of pollution from many companies in and around the Erode district of Tamil Nadu, India, where samples were gathered [40].

Chlorine-based bleaching is a broadly used approach for paper pulp bleaching. It uses chlorine fuel, chlorine dioxide, or sodium hypochlorite to remove impurities such as lignin. However, it also releases poisonous pollutants, such as dioxins, into the environment [25].

Regardless of the type, it offers efficiency and a low production rate, producing shiny white paper with actual optical properties. Elemental chlorine-free (ECF)

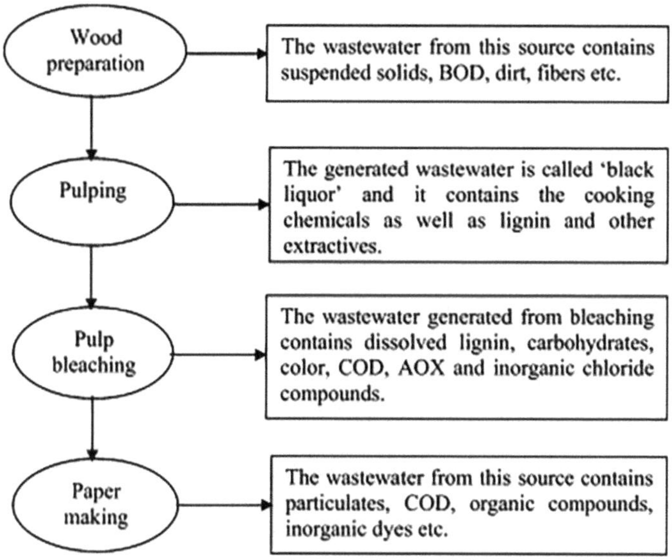

Fig. 3 Pulp and paper industry effluent sources [40]

bleaching is an environmentally friendly alternative to chlorine-based absolute bleaching, involving the use of chlorine dioxide in an elemental chlorine process to remove impurities from wood pulp [38]. It produces color and white paper with suitable optical properties and is notably low in price (TCF) bleaching. This is the most environmentally friendly choice, as the usage of non-chlorine bleaches, such as hydrogen peroxide or ozone, helps to remove impurities [38]. TCF bleaching is the most costly but less powerful technique for producing vibrant, white paper, with minimal environmental impact and the ability for advanced energy efficiency and durability, alongside other non-chlorine bleaching methods [35] Those strategies are less commonly used, but they are nonetheless powerful in generating brilliant papers [41]. Bleaching is a crucial step in paper production, extending the lifespan of paper by removing impurities and reducing the risk of mold development and deterioration. This approach is vital for developing exceptional, expert-grade papers that are visually appealing and sturdy [42]. Bleaching reduces the chance of paper becoming yellow or brittle, making it first class for files and materials that need to be preserved for prolonged durations [43].

The selection of the bleaching approach is based on environmental impact, value, and paper quality, with paper pulp bleaching improving paper quality by creating a uniform, clean surface [35]. Bleaching strengthens paper fibers, making them long-lasting and resistant to tearing while improving whiteness, which makes them visually appealing and suitable for industries such as publishing, printing, and advertising and marketing [44–47].

Additionally, xylanases have reduced the enterprise's chemical costs without affecting contemporary techniques [46, 48, 49]. De-bottlenecking in the generator has increased throughput in chlorine dioxide bleaching, thereby enhancing the bleaching rate in ECF and TCF tactics [11].

Xylanases are enzymes used in the pulp and paper industry to bleach kraft pulp and address environmental concerns related to chlorine in bleaching. Their mechanism of action has been extensively studied [50]. As ultraviolet (UV) soaking up hexenuronic acid happens throughout the kraft pulping procedure, xylanase remedy improves lignin extraction, modifies carbohydrate-lignin interactions, and cleaves redeposited xylan [51]. At the same time as a moderate acid remedy selectively removes colored components, restricting brightness reversion in handled kraft pulps, the xylan cleavage approach reduces the need for bleaching chemical substances [11, 52].

Moreover, xylanase treatment improves the alkali's capacity to extract lignin, which enhances the bleaching efficacy of the fiber's residual lignin [50, 53]. As it blocks lignin from accessing the fiber floor, xylan removal throughout kraft pulping may also restrict lignin extraction [54]. Studies suggest that not all pulps contain xylan that continuously reprecipitates on the floor, and our understanding of the xylanase bleaching mechanism does not indicate whether a selected xylanase could enhance bleaching [55].

Researchers are investigating the viable applications of xylanases and the need for enzymes that can bleach pulp in various settings and species [56].

Isolating and cloning novel xylanases from all sources have been the focus of recent research, which emphasizes how well-suited those enzymes are for the kraft process without the need for pH or temperature adjustments [57].

Bleaching sulfite pulps with xylanase has shown various levels of success; the efficacy of the remedy increases with the addition of xylanase and alkaline oxygen treatments. However, it may not be financially viable to perform additional remedies without recycling the enzymes [58]. Using xylanase, it became viable to lower the final pentose content of dissolving pulps, however, the method required a low amount of pentosans to be present in the pulp, as well as repeated treatment and oxygen treatment [59].

Nagar et al. [53], in keeping with the research of Jimenez et al., states that biobleaching wheat straw pulp resulted in 3.0 points for peroxidase enzyme-activated chlorine bleach and a 2.4-factor growth in brightness for peroxide enzyme bleaching. Even though xylanase treatment reduced yellowness in bleached hardwood kraft pulp, enzyme remedy may additionally nonetheless affect the output. Pulpzyme HC, an industrial enzyme, showed a 1.5% brightness improvement, up to 15% discount in laptop number, and a 10 to 15% decrease in HexA. These results were probable because of the hydrolysis of xylan on the fiber floor and extraction of stabilized quinine chromophoric structures. Bajpai and Bajpai [39] also reported the effects of diverse cellulase-loose commercial xylanases on pulp brightness.

Pulps produced with the aid of pulping approaches are too dark for many paper products without bleaching. Bleaching dissolves and removes residual lignin to enhance the pulp's brightness. It also aims to purify the pulp, enhance color balance, and reduce environmental impact [60, 61]. Increasing the pulp's brightness is the number one purpose of pulp bleaching. The dark, greater firmly bonded residual lignin corporations that make up the chromophoric components of pulps are useful organizations of degraded and modified lignin, rather than the unique type. Bleaching techniques are categorized based on the use of chlorine-containing retailers [62]. Traditional bleaching, which makes use of all forms of bleaching reagents, consists of chlorine gasoline, while ECF bleaching makes use of molecular chlorine, at the same time, TCF bleaching uses no chlorinated agents [63].

Laccase is a versatile biocatalyst with a vast capacity in papermaking. It can be used for pulping, delignification, biografting, decolorizing, and stabilizing paper mill effluent, as well as biotransforming excessive molecular weight lignins into lower molecular weight, fragrant compounds [64]. It additionally has programs for deinking vintage newsprints and pitch removal from numerous pulps. However, the commercialization of laccases is restricted due to their significantly lower redox capacity and the high cost of in-depth, large-scale manufacturing. Despite its versatility, laccases have been limited in commercialization due to their low redox capacity and the need for mediators for non-phenolic substrate oxidation [64].

4 An Introduction to the Science of Nanotechnology

Biotechnology can enhance the manufacture of pulp and paper by increasing product prices, reducing production costs, and generating high-value goods. Enzyme technologies can trade fiber traits and lessen environmental problems [25]. The sector desires new technology to integrate into cutting-edge strategies or reduce costs, as it is highly capital-intensive. Despite this, the enterprise has recognized the use of enzymes in generating paper, as it can meet market expectations universally [12].

Higher information on enzyme mechanisms and activities on fibers, the advent of better methods for their utilization, and the modification of enzymes to perform better under modern operating conditions are all necessary for the innovation and implementation of these enzyme applications [12]. Performance is critical because the end result needs to be strong and consistent enough to support a shift in the technique era [65]. Although economies of scale can decrease the fee of manufacturing enzymes, a paper producer will not use the generation, except it offers a considerable advantage. Technical boundaries can also save you the expense of employing an enzyme in a procedure step, as doing so may also require adjusting operational settings, introducing chemicals, or incurring capital costs [66].

Combining modern enzyme methods with contemporary techniques is the most effective way to ensure their adoption [18].

Enzymes have significantly improved fiber recycling in the production of paper, lowering costs, and enhancing the final product [16]. They can remove impurities from the recycling process, decrease the buildup of pitch during paper production from pulps with excessive resin content, and make it easier for recycled paper to deink [67]. Additionally, they improve pulp drainage, allowing paper machines to operate more efficiently and reduce capital costs [25].

The paper and pulp industries generate substantial amounts of effluent, contaminating land and water [68]. The effluent substantially affects the ecology because it contains dangerous organochlorinated chemical compounds released at various stages of the papermaking system [69]. Some of the low-priced and environmentally friendly methods made feasible via biotechnology include using enzyme cocktails, which significantly reduce the amount of waste-filled water resources [70]. Due to the fact that enzyme combinations, including manganese peroxidase, laccase, lignin peroxidase, lipase, cellulase, protease, xylanase, and amylase, are moderate, inexpensive, nonpolluting, extraordinarily powerful, and selective, they have widespread promise as biobleaching agents [71]. They also embellish the quality of the paper. Even though they are within the studies and development ranges, debarking, enzymatic deinking, and vessel picking reduction using enzymes have a significant capacity to reduce energy utilization [70, 72].

5 The Position of Nanotechnology in the Pulp and Paper Industry

The rapidly growing area of nanotechnology is predicted to significantly transform how materials are used in the twenty-first century [73]. Delivering effective and eco-friendly technology makes it feasible to discover novel substances, strategies, and phenomena on the nanoscale [74]. New directions in theoretical and practical studies provide new possibilities for developing modern nanosystems and nanostructured substances. Nanotechnology is crucial due to its numerous benefits, including enhanced production performance, the combination of living and nonliving gadgets; decreased size, speed, power, and material consumption; and specific phenomena and properties [73]. Malaysia has focused on nanotechnology development since the early twenty-first century, having commenced its own microtechnology and nanotechnology research in 2000 [74]. Due to launching its own studies and improvement application in 2001, the government has been closely monitoring the development of nanotechnology [75]. The countrywide nanotechnology tasks were initiated in 2006, and the third business master plan (2005–2020) acknowledges nanotechnology as an advanced field [76].

By supplying an asset control system for papermaking and beyond the mill, nanotechnology is revolutionizing papermaking [77, 78]. Papermakers can comprise shape and order into sheet designs because of traits in fiber research, minerals, and additives. In situ synthesis and lumen loading strategies create magnetic sheets [79]. Due to the tiny particles produced by chemical coprecipitation, which show superb paramagnetic pastime, lumin loading sheets have greater first-rate magnetic characteristics [80]. During showering and papermaking, these particles agglomerate and bind firmly, proscribing floor energy and generating low detachment quotes [81, 82].

However, magnetic debris deposited on fiber surfaces can adversely affect surface paper electricity. To create novel paper materials, Agarwal has effectively blended nano-, micro-, and macroscale substances [83–85].

However, the features of intact cellular partitions, interactions among cell wall layers, and the traits of nanoscale domain names consisting of hemicellulose, cellulose, and lignin—which function as the muse for the system's houses—continue to be largely unknown [86].

Utilizing nanotechnology enables the creation of timber-based substances and goods, leveraging the considerable capacity of trees [77]. By correctly harnessing the benefits that wood-based substances offer, this technology aims to meet the demands of both current and future technologies [68, 77]. Novel substances, wet-end chemistry, coating, nanofiltration, calendar rolls, sensors, microscopes, and nanoscale assemblers are some of the applications [87–89]. Wood fiber-based items using nanoparticles can be moisture-resistant mobile phone elements, sophisticated membranes and filters, high-quality loudspeaker cones, and paint, coating, and adhesive additives. Academia, industry, and society must collaborate to overcome obstacles and provide expertise and methods for advancing nanotechnology [90].

6 Recent Problems in Pulp Bleaching

Paper pulp bleaching is crucial in producing high-quality paper, and various strategies are available to achieve this goal. Chlorine-based bleaching, for example, has been observed to release poisonous pollutants such as dioxins into the environment that can damage aquatic lifestyles and contaminate soil. Nonchlorine bleaching methods, consisting of ECF and TCF bleaching, have notably reduced these pollution levels; nonetheless, they pose health risks to humans and ecosystems [69, 91].

Sustainability concerns are another issue related to paper pulp bleaching. The system requires enormous quantities of water and electricity, and the chemical substances used can be expensive and potentially hazardous [63]. The use of virgin wooden pulp, which is frequently bleached, can contribute to deforestation and habitat destruction. To address these sustainability issues, many paper manufacturers have adopted additional sustainable practices, including the use of recycled paper and non-wood fibers, and have implemented closed-loop structures that reduce the amount of water and chemical substances used in the bleaching process [92].

Regulatory requirements additionally play a function in paper pulp bleaching. In the USA, the Environmental Safety Employer (EPA) units limit the amount of pollutants released through bleaching. In Europe, chlorine-based bleaching has been banned totally, and paper producers must use ECF or TCF strategies instead [93].

In the end, while paper pulp bleaching is important for producing remarkable paper, it is far more important to balance the demand for remarkable paper with the need to protect the environment and promote sustainability. By doing so, we will continue to revel in the advantages of excellent paper while minimizing the adverse effects on our planet.

Xylanase is an enzyme that can embellish the dissolving pulp's brightness, particularly when dealing with fungal pretreatment. The enzyme treatment stage may vary, even though examples include pre- or post-pulping, alkaline treatment, bleaching treatment, and so on [94]. Combining alkaline leaching and enzyme therapy is not beneficial, as they operate in distinct, noncomplementary methods. The enterprise must improve bleaching with peracetic acid, ozone, chlorates, and peroxide because bleaching techniques vary [95].

Because xylanase gets rid of reactants that combine with the bleaching agent or obstructions to the bleaching activity, its effect at the very last pulp brightness appears indirect. While there are occasional cases of xylanase treatments leading to direct brightening, most enzyme bleaching that occurs immediately uses oxidative enzymes that concentrate on the lignin's color-producing components [96].

In kraft pulp, the fungi Phanerochaete chrysosporium and Trametes versicolor break down the lignin and boom its brightness [97]. They generate lignin peroxidase (Lip) and manganese peroxidase (MnP). Each fungus produces cellobiose dehydrogenase, which may be discovered to be used inside the breakdown of wood [97].

Fungal growth at the pulp produced a more exquisite brightness than remedy with fungal enzymes, which only barely improved brightness. One way to overcome the difficulty of cellulose hydrolysis is to use oxidative enzymes rather than whole organisms [98].

Complicated combinations of components eliminated from the pulp during bleaching, by and large, residual lignin and fragrant extractives, partly oxidized polymeric molecules that may be difficult to biodegrade, are accountable for the robust hue of kraft pulp bleaching filtrates. These materials, including phytosterols, are related to estrogenic hobby and filtrate toxicity in eucalypt kraft pulp-bleaching filtrates [99].

Complicated combinations of additives removed from the pulp throughout bleaching, often including residual lignin and aromatic extractives, as well as partly oxidized polymeric molecules that may be hard to biodegrade, are responsible for the intense hue of kraft pulp bleaching filtrates. These materials, including phytosterols, are linked to estrogenic pastime and filtrate toxicity in eucalypt kraft pulp-bleaching filtrates [100].

The pulp and paper enterprise produces much fresh water, which produces dangerous effluent with an excessive chemical oxygen demand (COD), a massive biochemical oxygen demand (BOD), and occasional biodegradability. The wastewater includes more than 700 natural and inorganic additives, consisting of color, phenolic compounds, absorbable natural halogens, saturated solids (SS), tannins, fatty acids, resin acids, lignin and its derivatives, excessive amounts of BOD and COD, and suspended solids [101]. The uncooked materials and specific manner of chemical compounds affect the properties of the bleaching wastewater. Due to their unique qualities, which make them incompatible with conventional municipal wastewater treatment, dedicated wastewater treatment is required. Pollutant removal may also require more than just organic strategies, particularly if there is a significant resistant component in the influent stream and the extreme best limitations are in the region. An extensively used approach for removing SS from pulp and paper effluents is chemical coagulation–flocculation, as observed through sedimentation [102, 103].

7 An Evaluation of Pulp Bleaching and the Latest Developments

The growing environmental focus has caused a developing need for eco-friendly industrial approaches, especially within the pulp and paper industries. The traditional bleaching technique in papermaking involves chlorination–alkali extraction and chlorine dioxide treatment to remove lignin from the pulp. However, this procedure may be toxic and dangerous to biological systems. Biobleaching of pulps using microbial enzymes is a appropriate opportunity [104]. Lignolytic enzymes, such as laccases, help eliminate lignin from the pulp, while hemicellulolytic enzymes, like xylanases and mannanases, depolymerize hemicellulose, increasing the accessibility of bleaching chemicals to the lignin. However, enzymes ought to fulfill certain uniqueness, such as being lively at excessive temperatures and pH, cellulose-loose, and no longer sufficient for powerful biobleaching. A cocktail of enzymes can be prepared using distinct techniques, which include developing unique microorganisms separately or co-culturing them to produce extraordinary enzymes. However, those strategies may be economically unviable and inappropriate for all industries [104, 105].

Pulp bleaching is the method of altering or disposing of colored additives from pulp to make it whiter. Enzymes that break down lignin were the first concept to gain widespread recognition. Then again, xylanases were a helpful bleaching agent for pulps as early as 1986. Current studies have examined the capacity of laccases or manganese peroxidases for pulp bleaching [98, 106]. Within the pulp and paper industry, xylanases have been one of the most significant success stories of enzymes, supporting the resolution of environmental problems associated with the use of chlorine for bleaching. Researchers are now investigating how xylanase aids in bleaching; it can cleave redeposited xylan, trade the connection among carbohydrates and lignin, and decorate lignin extraction [96, 107, 108].

Treating high cellulose pulps to create soluble reactive carbohydrate chains enables them to dissolve [109]. Endoglucanase will increase reactivity in eucalyptus and acacia pulp using decreasing viscosity and chain duration. At the identical protein dose, cellulases hydrolyze pulp more efficiently than cellobiohydrolases [110]. Using cellulases and alkali extract in steps produces an extra soluble dissolving pulp than treating it first with enzymes and then alkali [111]. However, the 2-forestall technique produces fibers with an incredibly reduced solubility. Reductions in hemicellulose content, appropriate viscosity, and alkali solubility arise for the duration of the alkali extraction and subsequent alkali extraction of dissolving pulps comprised of bleached hardwood kraft fiber and recycled paper treated with xylanase. Pulp bleaching is the act of altering or disposing of colored additives from pulp to make it whiter. Before everything, lignin-degrading enzymes were the biggest, but in 1986, xylanases were proven to be efficient as a bleaching agent. Latest studies have investigated the potential use of laccases or manganese peroxidases in pulp bleaching [60, 106].

Recent advancements have demonstrated the potential for strengthening paper by incorporating nanocellulose or nanofibrillated cellulose into its composition. Advanced profits in energy qualities have resulted from the combination of poly(amideamine) epichlorohydrin (PAE) and cellulose nanofibrils used as a strength-reinforcing agent [112, 113]. A unique type of paper composed of cellulose nanofibers has been created, utilizing nanofibrils with varying molar loads and nanopapers with various porosities [114]. The high mechanical performance of nanofibrils and the nanofibrillar community shape of extremely porous nanopaper are related to its exceptional strength. Better fibril strength, extra tremendous interfibril adhesion properties, and fewer, more uniformly dispersed flaws are perhaps the reasons behind nanopaper's superiority [115, 116].

Xylanases are enzymes used to bleach various varieties of pulp fibers, including hardwoods, eucalyptus, softwoods, bamboo, bagasse, ramie fibers, wheat straw, and khar grass. They are powerful in helping to bleach pulp. However, not all xylanases are effective with all pulp resources, and some enzymes work better with different xylanases or carbohydrases. Maximum xylanases utilized in bleaching are smaller than the ones in family F (or 10) and are contributors of own family G (or 11) of enzymes. A few xylanases interact extra favorably with carbohydrates or other xylanases [50].

Table 2 presents an expansion of research that examines how adding xylanase influences the bleaching procedure. Xylanases promote the discharge of lignin and, in the end, improve the bleaching impact of chemical reagents by doing away with the xylan move-linking layer of internal pulp fibers. Using xylanases in biobleaching dramatically reduces the quantity of chlorine wanted, from 10 to 50% (Table 2). Besides lowering total paper production charges, biobleaching improves paper best. 4-O-Methylglucuronic acid is one organization that replaces xylan polymers; after alkaline cooking, it transforms into hexenuronic acid. According to Roncero Blanca et al. [117], this substance increases the kappa variety and causes pulp yellowing. Therefore, a discounted launch of hexenuronic acid is besides an outcome of xylan removal through xylanase pre-remedy. There are numerous regulations for the use of xylanase in biobleaching. One is that because the cellulase hobby extensively reduces paper satisfaction, enzyme preparations must consist of as little of it as viable. The use of positive microorganisms that produce large quantities of xylanases with little to no cellulase activity can quickly overcome this limitation. Using heterolothat lackproduced xyactivitygenes cadditional in host microorganisms missing cellulase pastime is an extra method. The types of raw materials used have similar restrictions on their softness as pre-bleaching retailers. In agreement with Ghatora et al. [118], isolates from thermophilic fungi such as Chaetomium thermophilum, Melanocarpus sp., Humicola insolens, Malbranchea sp., and Thermoascus aurantiacus produced alkaline-active xylanases that are more suitable for pulp biobleaching.

The type of pulp being produced, including mechanical, sulfite, dissolving, kraft, and nonmechanical pulps, also plays a role in determining the use of xylanases in

Table 2 Enhancement of the brightness of pulses using cellulase-free commercial xylanases in a CDEHD bleaching sequence [60]

Sample tested	Reaction temp. (°C)	Reaction pH	Enzyme dose (IU/g)	Reaction time (h)	Increase in brightness (% ISO)	Brightness (% ISO)
Control	–	–	–		–	87.1
Bleachzyme B	40 to 50	7 to 7.5	10.0		0.8	87.9
Bleachzyme F	45 to 50	6 to 6.5	6.0	3.0	1.5	88.6
VAI-xylanase	50 to 60	5 to 7	3.5		1.3	88.4
Ecopulp-X200	50 to 55	5 to 6	10.0		1.4	88.5
Cartazyme HS-10	40 to 55	4 to 5	13.0		1.5	88.6
Pulpzyme HB	45 to 55	7 to 8	14.0		1.1	88.2
Pulpzyme HC	60 to 70	8 to 9	14.0		0.9	88.0
Irgazyme 40S	50 to 60	7 to 8	7.5		1.5	88.6
Irgazyme 10	40 to 50	5.5 to 6	12.0		1.4	88.5

CD, chlorination; D, chlorine dioxide; H, hypochlorite; E, extraction

[a] Based on data from Bajpai and Bajpai

bleaching. Xylanase treatment of kraft pulps is generally applicable, but it complicates the usage of xylanase due to the need for empirical adjustment of enzyme doses. Oxygen delignification makes extra xylan available, and the enzyme doses for bleaching are lower than those for kraft pulp [11].

Fungal cultures have yielded herbal mediators for pulp bleaching; a few have proposed that MnP is a more significant herbal mediator than laccase [119]. Although hydrogen peroxide is essential for the action of MnP and LiP, pretty massive concentrations of the substance render them inactive. The inferior output of those enzymes produced using fungal cultures, along with their price, hydrogen peroxide requirements, and susceptibility to hydrogen peroxide, limited their initial application. The desired hydrogen peroxide may be produced or added continuously at a low level using a device [120].

A TCF method for pulp bleaching involves using xylanase and MnP in conjunction with caustic extraction and hydrogen peroxide bleaching. After MnP treatment, hydrogen peroxide bleaching is green, enhancing the final brightness by utilizing 10 factors [121]. During prolonged bleaching with regular hydrogen peroxide addition, the concurrent use of xylanase and MnP resulted in a synergistic effect that enhanced lignin elimination. However, using laccase at the same time was not a success [122].

MnP and laccase can bleach at comparable levels, although laccase is more readily accessible and easier to produce. This results in an initial darkening of the pulp, which is later eliminated through washing and bleaching [72]. Researchers have discovered new laccases and mediators from various sources, including microorganisms such as Bacillus subtilis and Azospirillum lipoferum. Researchers have compared the results of not unusual laccase mediators, 1-hydroxybenzotriazole (1-HBT), and a couple of 2'-azinobis-(3-ethylbenzthiazoline-6-sulfonate) (ABTS) [123].

The use of oxidative enzymes in pulp bleaching is promising. However, the system is not economically feasible due to the price of the enzyme and the need for a cheaper, realistic, dependable mediator for laccase treatment. In fiber recycling, enzymes can enhance dewatering rates, increase contaminant removal, and strengthen the bond in recovered fibers. The susceptibility of MnP to hydrogen peroxide and the requirement for an extremely low-value, nontoxic, and green mediator for laccase therapy remain significant troubles [106].

Moreover, enzyme remedies can enhance recycled fiber drainage rates over and above what polymer addition can do. To regulate the fiber floor structure, enzyme treatments require longer retention times and sufficient mixing. With proper mixing, enzyme addition can be completed in as little as 15 min at the suitable pH, temperature, and consistency [124]. The formula and interest of the enzyme determine the dosage and length of therapy. The power required to meet power requirements may be reduced, or energy attributes can be improved at a given refining level by adding enzymes earlier in the refining process. After refining, enzymes promote freedom and allow quicker device operation. For higher paper formation, decreased foundation weight, more suitable strength, and increased utilization of recycled fibers, enzyme-assisted refining can permit decreased head box uniformity [125].

Oil palm empty fruit bunches are a precious feedstock with biotechnological potential. Biopulping, which utilizes fungi-producing xylanases and ligninases, can enhance pulp quality. Aspergillus sp. LPB-5 produces an enzymatic cocktail in solid-state biopulping. Thermal, chemical, and biological pretreatments were compared, and enzymatic biobleaching was implemented in the pretreated pulps [126]. Biopulping and biobleaching mixture ended in 36.8% lignin loss, 26.27% hemicellulose discount, 74.36% pulp yield, and 36.56% digestibility. Alkaline and biobleaching mixture eliminated 81.90% hemicellulose and 93.89% lignin, with 73.59% digestibility. Enzymatic biobleaching elevated pulp digestibility in all pretreatments. This tender technique has eco-friendly potential [9, 126].

8 Conclusions

Within the pulp and paper industry, the use of biotechnology and nanotechnology in pulp bleaching represents a significant step toward adopting more environmentally friendly practices. Producers can also minimize their impact on the surroundings without sacrificing the capability and quality of their products by utilizing this technology. Sufficient research and development efforts are crucial to fully actualize the potential benefits and address any barriers that may arise during the commercialization of these innovative methodologies.

Competing Interests The authors declare that they have no competing interest.

Declaration of Generative AI and AI-Assisted Technologies in the Writing Process During the preparation of this work, the book chapter author(s) have not taken any assistance from AI, and the book chapter author(s) reviewed and edited the content as needed and take(s) full responsibility for the content of the publication.

References

1. Zamani A (2015) Introduction to lignocellulose-based products. In: Lignocellulose-based bioproducts. pp 1–36
2. Eugenio ME, Ibarra D, Martín-Sampedro R, Espinosa E, Bascón I, Rodríguez A (2019) Alternative raw materials for pulp and paper production in the concept of a lignocellulosic biorefinery. Cellulose 12(4):78
3. Shi L, Ge J, Nie S, Qin C, Yao S (2019) Effect of lignin structure on adsorbable organic halogens formation in chlorine dioxide bleaching. R Soc Open Sci 6(2):182024. https://doi.org/10.1098/rsos.182024
4. Yadav VK, Gupta N, Kumar P, Dashti MG, Tirth V, Khan SH, Yadav KK et al (2022) Recent advances in synthesis and degradation of lignin and lignin nanoparticles and their emerging applications in nanotechnology. Materials. https://doi.org/10.3390/ma15030953
5. Viikari L, Suurnäkki A, Grönqvist S, Raaska L, Ragauskas A (2009) Forest products: biotechnology in pulp and paper processing. In: Encyclopedia of microbiology, 3rd edn. Academic Press, pp 80–94. https://doi.org/10.1016/B978-012373944-5.00123-1
6. Brink DP, Ravi K, Lidén G, Gorwa-Grauslund MF (2019) Mapping the diversity of microbial lignin catabolism: experiences from the ELignin database. Appl Microbiol Biotechnol 103:3979–4002
7. Sharma J, Sharma D, Sharma A, Vishwakarma V, Dubey A, Namdeo H (2020) The contribution of microbial biotechnology for achieving sustainable development. Microb Biotechnol Basic Res Appl 1–18. https://doi.org/10.1007/978-981-15-2817-0_1
8. Kaviya N, Upadhayay VK, Singh J, Khan A, Panwar M, Singh AV (2019) Enzyme processes for pulp and paper: a review of recent developments. In: ACS symposium mycorrhizosphere and pedogenesis. pp 25–52. https://doi.org/10.1007/978-981-13-6480-8_2
9. Hamedi J, Vaez Fakhri A, Mahdavi S (2020) Biobleaching of mechanical paper pulp using streptomyces rutgersensis UTMC 2445 isolated from a lignocellulose-rich soil. J Appl Microbiol 128(1):161–170. https://doi.org/10.1111/jam.14489
10. Singh G, Kaur S, Khatri M, Arya SK (2019) Biobleaching for pulp and paper industry in india: emerging enzyme technology. Biocatal Agric Biotechnol 17:558–565. https://doi.org/10.1016/j.bcab.2019.01.019
11. Matos JMS, Evtuguin DV, António P, de Sousa M, Graça M, Carvalho VS (2024) Xylanase treatment of eucalypt kraft pulps: effect of carryover. Appl Microbiol Biotechnol 108(1):210. https://doi.org/10.1007/s00253-024-13027-3
12. Wiltschi B, Cernava T, Dennig A, Casas MG, Geier M, Gruber S, Haberbauer M et al (2020) Enzymes revolutionize the bioproduction of value-added compounds: from enzyme discovery to special applications. Biotechnol Adv 40:107520. https://doi.org/10.1016/j.biotechadv.2020.107520
13. Fahmy TYA, Mobarak F, Kassem N, Abdel-Kader AH (2008) New approach for upgrading pulp & paper quality: mild potassium permanganate treatment of already bleached pulps. Carbohyd Polym 74(4):892–894
14. Akbarpour I (2023) Surface characterization of pulp fiber from mixed waste newspaper and magazine deinked-pulp with combined cellulase and laccase-violuric acid system (LVS). Bioresour Technol Rep 23:101551
15. Okayama T (2002) The effects of recycling on pulp and paper properties. Kami Pa Gikyoshi/Japan Tappi J 56(7):62–68. https://doi.org/10.2524/jtappij.56.986

16. Barrios N, Smith MM, Venditti RA, Pal L (2024) Enzyme-assisted dewatering and strength enhancement of cellulosic fibers for sustainable papermaking: a bench and pilot study. J Clean Prod 434:140094
17. Cui L, Meddeb-Mouelhi F, Laframboise F, Beauregard M (2015) Effect of commercial cellulases and refining on kraft pulp properties: correlations between treatment impacts and enzymatic activity components. Carbohyd Polym 115:193–199. https://doi.org/10.1016/j.carbpol.2014.08.076
18. Robinson PK (2015) Enzymes: principles and biotechnological applications. Essays Biochem 59:1–41. https://doi.org/10.1042/BSE0590001
19. Robertson GL (2011) Packaging materials for biscuits and their influence on shelf life. In: Manley's technology of biscuits, crackers and cookies, 4th edn. Elsevier, pp 247–67. https://doi.org/10.1533/9780857093646.2.247
20. Britannica T (2024) Editors of encyclopaedia lead. Encyclopedia Britannica. https://www.britannica.com/science/lead-chemical-element
21. Quintana E, Valls C, Blanca Roncero M (2024) Dissolving-grade pulp: a sustainable source for fiber production. Wood Sci Technol 58(1):23–85. https://doi.org/10.1007/s00226-023-01519-w
22. Bajpai P (2001) Microbial degradation of pollutants in pulp mill effluents. Adv Appl Microbiol 48:79–134. https://doi.org/10.1016/s0065-2164(01)48001-4
23. Hashemi S (2024) Optimization of washing steps prior to bleaching of a pulp produced by cooking with ionic liquid
24. Bajpai P (2005) ECF and TCF bleaching. In: Environmentally benign approaches for pulp bleaching. pp 177–192. https://doi.org/10.1016/b978-044451724-1/50005-8
25. Bajpai P (2018) Brief description of the pulp and papermaking process. In: Biotechnology for pulp and paper processing, pp 9–26. https://doi.org/10.1007/978-981-10-7853-8_2
26. Bainbridge A (2023) Materials and identification. In: Conservation of books. Routledge, pp 254–364
27. Patterson G (2021) Cellulose before cell: historical themes. Carbohyd Polym 252:117182. https://doi.org/10.1016/j.carbpol.2020.117182
28. Helman-Wazny A (2021) A tale of papermaking along the silk road. In: Exploring written artefacts. pp 423–40. https://doi.org/10.1515/9783110753301-022
29. Li P, Yanpeng X, Yin L, Liang X, Wang R, Liu K (2023) Development of raw materials and technology for pulping—a brief review. Polymers 15(22):4465. https://doi.org/10.3390/polym15224465
30. Tsien T-H (2020) Collected writings on chinese culture. The Chinese University of Hong Kong Press. https://doi.org/10.2307/j.ctt1pb61wm
31. El-Sayed A, Essam S, El-Sakhawy M, El-Sakhawy MA (2020) Non-wood fibers as raw material for pulp and paper industry. Nord Pulp Pap Res J 35(2):215–230. https://doi.org/10.1515/Npprj-201
32. Särkkä T (2018) The quest for raw materials in the british paper trade: the development of the bamboo pulp and paper industry in British India up to 1939. In: Technological transformation in the global pulp and paper industry 1800–2018: comparative perspectives. pp 237–59. https://doi.org/10.1007/978-3-319-94962-8_11
33. Samanta KK, Basak S, Chattopadhyay SK (2015) Recycled fibrous and nonfibrous biomass for value-added textile and nontextile applications. Environ Footprints Eco-Des Prod Proc 167–212 https://doi.org/10.1007/978-981-287-643-0_8
34. Ramos MDN, Santana CS, Velloso CCV, da Silva AHM, Magalhães F, Aguiar A (2021) A review on the treatment of textile industry effluents through fenton processes. Process Saf Environ Prot 155:366–386. https://doi.org/10.1016/j.psep.2021.09.029
35. Sharma N, Bhardwaj NK, Singh RBP (2020) Environmental issues of pulp bleaching and prospects of peracetic acid pulp bleaching: a review. J Clean Prod 256:120338
36. Ganiyu SO, Zhou M, Martínez-Huitle CA (2018) Heterogeneous electro-fenton and photoelectro-fenton processes: a critical review of fundamental principles and application for water/wastewater treatment. Appl Catal B 235:103–129. https://doi.org/10.1016/j.apcatb.2018.04.044

37. Ribeiro JP, Sarinho L, Neves MC, Nunes MI (2022) Valorisation of residual iron dust as fenton catalyst for pulp and paper wastewater treatment. Environ Pollut 310:119850. https://doi.org/10.1016/j.envpol.2022.119850
38. Bajpai P (2012b) Environmentally benign approaches for pulp bleaching, 2nd Edn. Elsevier. https://doi.org/10.1016/B978-0-444-51724-1.X5000-X
39. Maalej-Achouri I, Guerfali M, Belhaj I, Belghith H (2012) The effect of talaromyces thermophilus cellulase-free xylanase and commercial laccase on lignocellulosic components during the bleaching of kraft pulp. Int Biodeterior Biodegradation 75(November):43–48. https://doi.org/10.1016/j.ibiod.2012.04.015
40. Sridhar R, Sivakumar V, Prince Immanuel V, Prakash Maran J (2011) Treatment of pulp and paper industry bleaching effluent by electrocoagulant process. J Hazard Mater 186(2):1495–1502. https://doi.org/10.1016/j.jhazmat.2010.12.028
41. Snyder H (2019) Literature review as a research methodology: an overview and guidelines. J Bus Res 104:333–339. https://doi.org/10.1016/j.jbusres.2019.07.039
42. Brückle I (2009) Bleaching in paper production versus conservation. Restaurator 30(4):280–293. https://doi.org/10.1515/rest.016
43. Zervos S, Alexopoulou I (2015) Paper conservation methods: a literature review. Cellulose 22(5):2859–2897. https://doi.org/10.1007/s10570-015-0699-7
44. Blechschmidt J, Heinemann S, Putz HJ, Laufmann M, Kogler W, Gliese T, Auhorn WJ (2006) Raw materials for paper and board manufacture. In: Handbook of paper and board. pp 20–61. https://doi.org/10.1002/3527608257.ch2
45. Dhiman J, Anupam K, Kumar V, Saruchi (2024) Varieties, characteristics, and applications of nanofillers in pulp and paper industry. In: Handbook of nanofillers. Springer, pp 1–32. https://doi.org/10.1007/978-981-99-3516-1_43-1
46. Jani SM, Rushdan I, Saad MJ, Ibrahim R (2014) 51 Effect of bleaching on coir fibre pulp and paper properties (Kesan Pelunturan Terhadap Sifat-Sifat Pulpa Dan Kertas Sabut Kelapa). J Trop Agric Fd Sc 42(1):51–61
47. Tillmann O (2006) Paper and board grades and their properties. In: Handbook of paper and board. pp 446–66. https://doi.org/10.1002/3527608257.ch11
48. Twede D, Selke SEM, Kamdem D-P, Shires D (2014) Cartons, crates and corrugated board: handbook of paper and wood packaging technology. DEStech Publications Inc.
49. Kogler W, Tietz M, Auhorn WJ (2012) Paper and board, 7. coating of paper and board. In: Ullmann's encyclopedia of industrial chemistry. https://doi.org/10.1002/14356007.o18_o15.
50. Almeida N, Meyer V, Burnet A, Boucher J, Talens-Perales D, Pereira S, Ihalainen P et al (2022) Use of a novel extremophilic xylanase for an environmentally friendly industrial bleaching of kraft pulps. Int J Mol Sci 23(21):13423. https://doi.org/10.3390/ijms232113423
51. Shatalov AA, Pereira H (2009) Impact of hexenuronic acids on xylanase-aided bio-bleaching of chemical pulps. Biores Technol 100(12):3069–3075
52. Jiang ZH, Van Lierop B, Berry R (2000) Hexenuronic acid groups in pulping and bleaching chemistry. Tappi J 83(1)
53. Nagar S, Jain RK, Thakur VV, Gupta VK (2013) Biobleaching application of cellulase poor and alkali stable xylanase from bacillus pumilus SV-85S. 3 Biotech 3:277–285
54. Liu Q, Fan H, Mou H, Liu J, Huang J, Dong X, Song H (2021) Preparation and characterization of xylan by an efficient approach with mechanical pretreatments. Ind Crops Prod 165:113420
55. Shatalov AA, Pereira H (2008) Effect of xylanases on peroxide bleachability of eucalypt (E. Globulus) kraft pulp. Biochem Eng J 40(1):19–26
56. Liab K, Azadi P, Collins R, Tolan J, Kim JS, Eriksson K-E (2000) Relationships between activities of xylanases and xylan structures. Enzyme Microb Technol 27(1–2):89–94
57. Ellatif SA, Abdel Razik ES, Al-Surhanee AA, Al-Sarraj F, Daigham GE, Mahfouz AY (2022) Enhanced production, cloning, and expression of a xylanase gene from endophytic fungal strain trichoderma harzianum Kj831197.1: unveiling the in vitro anti-fungal activity against phytopathogenic fungi. J Fungi 8(5). https://doi.org/10.3390/jof8050447
58. Manji AH (2006) Extended usage ofxylanase enzyme to enhance the bleaching of softwood kraft pulp. Tappi J 5(1):23–26

59. Bajpai P, Bajpai P (2018) Production of dissolving-grade pulp. In: Biotechnology for pulp and paper processing. pp 291–311
60. Gangwar AK, Prakash NT, Prakash R (2014) Applicability of microbial xylanases in paper pulp bleaching: a review. BioResources 9(2)
61. Lehr M, Miltner M, Friedl A (2021) Removal of wood extractives as pulp (Pre-)treatment: a technological review. SN Appl Sci 3(12):886. https://doi.org/10.1007/s42452-021-04873-1
62. Bajpai P (2021) Nonwood plant fibers for pulp and paper. Elsevier
63. Sousa AM, Pinto ISS, Machado L, Gando-Ferreira L, Quina MJ (2023) Sustainability of Kraft pulp mills: bleaching technologies and sequences with reduced water use. J Ind Eng Chem 125:58–70
64. Singh G, Arya SK (2019) Utility of laccase in pulp and paper industry: a progressive step towards the green technology. Int J Biol Macromol 134:1070–1084. https://doi.org/10.1016/j.ijbiomac.2019.05.168
65. Gruetzemacher R, Whittlestone J (2022) The transformative potential of artificial intelligence. Futures 135:102884
66. Jegannathan KR, Nielsen PH (2013) Environmental assessment of enzyme use in industrial production–a literature review. J Clean Prod 42:228–240
67. Gutiérrez A, del Río JC, Martínez MJ, Martínez AT (2001) The biotechnological control of pitch in paper pulp manufacturing. Trends Biotechnol 19(9):340–348
68. Singh AK, Kumar A, Bilal M, Chandra R (2022) Environmental pollutants of paper industry wastewater and their toxic effects on human health and ecosystem. Bioresour Technol Rep 20:101250. https://doi.org/10.1016/j.biteb.2022.101250
69. Singh AK, Chandra R (2019) Pollutants released from the pulp paper industry: aquatic toxicity and their health hazards. Aquat Toxicol 211:202–216
70. Kumar A, Bilal M, Singh AK, Ratna S, Rameshwari KT, Ahmed I, Iqbal HM (2022) Enzyme cocktail: a greener approach for biobleaching in paper and pulp industry. In: Nanotechnology in paper and wood engineering. Elsevier, pp 303–28
71. Boruah P, Sarmah P, Das PK, Goswami T (2019) Exploring the lignolytic potential of a new laccase producing strain Kocuria Sp. PBS-1 and its application in bamboo pulp bleaching. Int Biodeteriorat 143:104726
72. Saleem R, Khurshid M, Ahmed S (2018) Laccases, manganese peroxidases and xylanases used for the bio-bleaching of paper pulp: an environmental friendly approach. Protein Pept Lett 25(2):180–186
73. Malik S, Muhammad K, Waheed Y (2023) Nanotechnology: a revolution in modern industry. Molecules 28(2):661. https://doi.org/10.3390/molecules28020661
74. Nasrollahzadeh M, Sajadi SM, Sajjadi M, Issaabadi Z (2019) An introduction to nanotechnology. In: Interface science and technology, vol 28. Elsevier, pp 1–27
75. Rajalingam K (2024) Nanotechnology education and workforce development through higher education: the case of Malaysia. Asian J Res Educ Soc Sci 6(1):215–227
76. Roco MC (2016) Building foundational knowledge and infrastructure for nanotechnology: 2000–2030. In: Nanotechnology: delivering on the promise, vol 1. ACS Publications, pp 39–52
77. Jasmani L, Rusli R, Khadiran T, Jalil R, Adnan S (2020) Application of nanotechnology in wood-based products industry: a review. Nanoscale Res Lett 15:1–31
78. Julkapli NM, Bagheri S (2016) Developments in nano-additives for paper industry
79. Zakaria S, Ong BH, Van de Ven TGM (2004) Lumen loading magnetic paper II: mechanism and kinetics. Coll Surf A 251(1–3):31–36
80. Chia CH, Zakaria S, Ahamd S, Abdullah M, Mohd Jani S (2006) Preparation of magnetic paper from kenaf: lumen loading and in situ synthesis method. Am J Appl Sci 3(3):1750–1754
81. Garrett TR, Bhakoo M, Zhang Z (2008) Bacterial adhesion and biofilms on surfaces. Prog Nat Sci 18(9):1049–1056
82. Lee KH, Youn HJ, Lee HL (2020) Analysis of spotty deposits on fine paper and investigation of key factors affecting alkyl ketene dimer spot formation. ACS Omega 5(25):15529–15536

83. De Cicco D, Asaee Z, Taheri F (2017) Use of nanoparticles for enhancing the interlaminar properties of fiber-reinforced composites and adhesively bonded joints—a review. Nanomaterials 7(11):360
84. Rahman M, Islam KS, Dip TM, Chowdhury MFM, Debnath SR, Maruf SM, Sadman HM, Sakib TS, Padhye R, Houshyar S (2024) A review on nanomaterial-based additive manufacturing: dynamics in properties, prospects, and challenges. Prog Addit Manuf 9(4):1197–1224
85. Raj RS, Prince Arulraj G, Anand N, Kanagaraj B, Lubloy E, Naser MZ (2023) Nanomaterials in geopolymer composites: a review. Dev Built Environ 13:100114
86. Moon RJ, Frihart CR, Wegner T (2006) Nanotechnology applications in the forest products industry. For Prod J 56(5):4–10
87. Aydin D, Gübbük İH, Ersöz M (2024) Recent advances and applications of nanostructured membranes in water purification. Turk J Chem 48(1):1–20
88. Chai Z, Childress A, Busnaina AA (2022) Directed assembly of nanomaterials for making nanoscale devices and structures: mechanisms and applications. ACS Nano 16(11):17641–17686
89. Shen J, Song Z, Qian X, Yang F, Kong F (2010) Nanofillers for papermaking wet end applications. BioResources 5(3):1328–1331
90. Bajpai P (2016) Pulp and paper industry: nanotechnology in forest industry. Elsevier
91. Erhardt CS, Basegio TM, Capela I, Rodríguez AL, Machado ÊL, López DAR, Tarelho L, Bergmann CP (2021) AOX degradation of the pulp and paper industry bleaching wastewater using NZVI in two different agitation processes. Environ Technol Innov 22:101420
92. Del Rio F, Dylan D, Sovacool BK, Griffiths S, Bazilian M, Kim J, Foley AM, Rooney D (2022) Decarbonizing the pulp and paper industry: a critical and systematic review of sociotechnical developments and policy options. Renew Sustain Energy Rev 167:112706. https://doi.org/10.1016/j.rser.2022.112706
93. Söderholm K, Bergquist AK, Söderholm P (2017) The transition to chlorine free pulp revisited: nordic heterogeneity in environmental regulation and R&D collaboration. J Clean Prod 165. https://doi.org/10.1016/j.jclepro.2017.07.190
94. Walia A, Guleria S, Mehta P, Chauhan A, Parkash J (2017) Microbial xylanases and their industrial application in pulp and paper biobleaching: a review. 3 Biotech 7(1):11. https://doi.org/10.1007/s13205-016-0584-6
95. Tian D, Chen Y, Shen F, Luo M, Huang M, Jinguang H, Zhang Y, Deng S, Zhao L (2021) Self-Generated peroxyacetic acid in phosphoric acid plus hydrogen peroxide pretreatment mediated lignocellulose deconstruction and delignification. Biotechnol Biofuels 14(1):224. https://doi.org/10.1186/s13068-021-02075-w
96. Roncero MB, Torres AL, Colom JF, Vidal T (2005) The effect of xylanase on lignocellulosic components during the bleaching of wood pulps. Biores Technol 96(1):21–30. https://doi.org/10.1016/j.biortech.2004.03.003
97. Asemoloye MD, Bello TS, Oladoye PO, Gbadamosi MR, Babarinde SO, Adebami GE, Olowe OM, Temporiti MEE, Wanek W, Marchisio MA (2023) Engineered yeasts and lignocellulosic biomaterials: shaping a new dimension for biorefinery and global bioeconomy. Bioengineered 14(1):2269328
98. Bajpai P (2012a) Biotechnology for pulp and paper processing. https://doi.org/10.1007/978-1-4614-1409-4.
99. Coimbra E, Mounteer A, Carmo A, Michielsen M, Totola L, Guerino J, Gonçalves J, Silva P (2021) Electrocoagulation of kraft pulp bleaching filtrates to improve biotreatability. Process Saf Environ Prot 147(March):346–355. https://doi.org/10.1016/j.psep.2020.09.039
100. Patel K, Patel N, Vaghamshi N, Shah K, Dugdirala SM, Dudhagara P (2021) Trends and strategies in the effluent treatment of pulp and paper industries: a review highlighting reactor options. Curr Res Microb Sci 2:100077. https://doi.org/10.1016/j.crmicr.2021.100077
101. Covinich L, Bengoechea D, Fenoglio R, Area M (2014) Advanced oxidation processes for wastewater treatment in the pulp and paper industry: a review. Am J Environ Eng. https://doi.org/10.5923/j.ajee.20140403.03

102. Mainardis M, Buttazzoni M, De Bortoli N, Mion M, Goi D (2020) Evaluation of ozonation applicability to pulp and paper streams for a sustainable wastewater treatment. J Clean Prod 258. https://doi.org/10.1016/j.jclepro.2020.120781
103. Panda S (2020) Magnetic separation of ferrous fractions linked to improved bioleaching of metals from waste-to-energy incinerator bottom ash (IBA): a green approach. Environ Sci Pollut Res Int 27(9):9475–9489. https://doi.org/10.1007/s11356-020-07615-9
104. Angural S, Kumar A, Kumar D, Warmoota R, Sondhi S, Gupta N (2020) Lignolytic and hemicellulolytic enzyme cocktail production from bacillus tequilensis LXM 55 and its application in pulp biobleaching. Bioprocess Biosyst Eng 43(12):2219–2229. https://doi.org/10.1007/s00449-020-02407-4
105. Woldesenbet F, Virk AP, Gupta N, Sharma P (2013) Biobleaching of mixed wood kraft pulp with alkalophilic bacterial xylanase, mannanase and laccase-mediator system. J Microbiol Biotechnol 3
106. Bajpai P (2004) Biological bleaching of chemical pulps. Crit Rev Biotechnol 24(1):1–58. https://doi.org/10.1080/07388550490465817
107. Roncero MB, Torres AL, Colom JF, Vidal T (2000) Effects of xylanase treatment on fibre morphology in totally chlorine free bleaching (TCF) of eucalyptus pulp. Process Biochem 36(1–2):45–50. https://doi.org/10.1016/S0032-9592(00)00178-3
108. Gupta GK, Dixit M, Kapoor RK, Shukla P (2022) Xylanolytic enzymes in pulp and paper industry: new technologies and perspectives. Mol Biotechnol 64(2):130–143. https://doi.org/10.1007/s12033-021-00396-7
109. Bernardi M, Tangorra RR, Palmisano P, Bogliano A (2016) Chemicals from renewable sources. In: Reference module in chemistry, molecular sciences and chemical engineering. https://doi.org/10.1016/B978-0-12-409547-2.11722-6
110. Sango C, Kaur P, Bhardwaj NK, Sharma J (2018) Bacterial cellulase treatment for enhancing reactivity of pre-hydrolysed kraft dissolving pulp for viscose. 3 Biotech 8(6):271. https://doi.org/10.1007/s13205-018-1293-0
111. Kaur P, Bhardwaj NK, Sharma J (2021) A study elucidating the relation between cellulose dissolution and crystallinity after cellulase treatment at different doses. 3 Biotech 11(8):371. https://doi.org/10.1007/s13205-021-02920-7
112. Abdul Khalil HPS, Davoudpour Y, Islam MN, Asniza Mustapha K, Sudesh RD, Jawaid M (2014) Production and modification of nanofibrillated cellulose using various mechanical processes: a review. Carbohyd Polym 99:649–665. https://doi.org/10.1016/j.carbpol.2013.08.069
113. Hollertz R, Durán VL, Larsson PA, Wågberg L (2017) Chemically modified cellulose micro- and nanofibrils as paper-strength additives. Cellulose 24. https://doi.org/10.1007/s10570-017-1387-6
114. Sethi J, Liimatainen H, Sirviö JA (2021) Fast and filtration-free method to prepare lactic acid-modified cellulose nanopaper. ACS Omega 6(29):19038–19044. https://doi.org/10.1021/acsomega.1c02328
115. Henriksson M, Berglund LA, Isaksson P, Lindström T, Nishino T (2008) Cellulose nanopaper structures of high toughness. Biomacromol 9(6):1579–1585. https://doi.org/10.1021/bm800038n
116. Jakob M, Mahendran AR, Gindl-Altmutter W, Bliem P, Konnerth J, Müller U, Veigel S (2021) The strength and stiffness of oriented wood and cellulose-fibre materials: a review. Prog Mater Sci 125(December):100916. https://doi.org/10.1016/j.pmatsci.2021.100916
117. Roncero MB, Torres A, Colom J, Vidal T (2003) Effect of xylanase on ozone bleaching kinetics and properties of eucalyptus kraft pulp. J Chem Technol Biotechnol 78(August):1023–1031. https://doi.org/10.1002/jctb.893
118. Ghatora SK, Chadha BS, Badhan AK, Saini HS, Bhat MK (2006) Identification and characterization of diverse xylanases from thermophilic and thermotolerant fungi. BioResources 1. https://doi.org/10.15376/biores.1.1.18-33
119. Fillat A, Colom JF, Vidal T (2010) A new approach to the biobleaching of flax pulp with laccase using natural mediators. Biores Technol 101(11):4104–4110. https://doi.org/10.1016/j.biortech.2010.01.057

120. Falade AO, Nwodo UU, Iweriebor BC, Green E, Mabinya LV, Okoh AI (2017) Lignin peroxidase functionalities and prospective applications. MicrobiologyOpen 6(1). https://doi.org/10.1002/mbo3.394
121. Bermek H, Li K, Eriksson K-E (2000) Pulp bleaching with manganese peroxidase and xylanase: a synergistic effect. Tappi J 83
122. Woolridge E (2014) Mixed enzyme systems for delignification of lignocellulosic biomass. Catalysts 4(March):1–35. https://doi.org/10.3390/catal4010001
123. Arregui L, Ayala M, Gómez-Gil X, Gutiérrez-Soto G, Hernández-Luna CE, Herrera De Los Santos M, Levin L, Rojo-Domínguez A, Romero-Martínez D, Saparrat MC, Trujillo-Roldán MA (2019) Laccases: structure, function, and potential application in water bioremediation. In: Microbial cell factories. Departamento de Ciencias Naturales, Universidad Autónoma Metropolitana, Unidad Cuajimalpa, Mexico City, Mexico. https://doi.org/10.1186/s12934-019-1248-0
124. Egan J, Wang S, Shen J, Baars O, Moxley G, Salmon S (2023) Enzymatic textile fiber separation for sustainable waste processing. Resourc Environ Sustain 13:100118. https://doi.org/10.1016/j.resenv.2023.100118
125. Singh R, Bhardwaj N, Choudhury B (2014) An experimental study of the effect of enzymeassisted refining on energy consumption and paper properties for mixed hardwood pulp. Appita J 67(September):226
126. Colonia O, Sthepani B, Woiciechowski AL, Malanski R, Letti LAJ, Soccol CR (2019) Pulp improvement of oil palm empty fruit bunches associated to solid-state biopulping and biobleaching with xylanase and lignin peroxidase cocktail produced by Aspergillus Sp. LPB-5. Biores Technol 285(August):121361. https://doi.org/10.1016/j.biortech.2019.121361
127. Kumar V, Marín-Navarro J, Shukla P (2016) Thermostable microbial xylanases for pulp and paper industries: trends, applications and further perspectives. World J Microbiol Biotechnol 32(2):1–10. https://doi.org/10.1007/s11274-015-2005-0

Nanosensors for Real-Time Monitoring in Paper Production

Chaitanya Kumar, Manvinderpreet Kaur, Neyaz Ahmed, Karan Kapoor, Vageesha Singh, Sudarshan Sahu, Gursharan Singh, Madhu Khatri, and Shailendra Kumar Arya

Abstract The paper industry is under growing pressure to produce high-quality, sustainable, and efficient products. This chapter examines the crucial role nanosensors play in overcoming these obstacles, as they enable precise control and real-time monitoring of key production parameters. At the nanoscale, nanosensors provide unmatched sensitivity and precision in identifying factors including moisture content, fiber alignment, chemical makeup, and coating thickness at different phases of the papermaking process. Their integration minimizes waste, lowers energy consumption, and enhances overall product quality by optimizing critical operations, including pulping, sheet creation, finishing, and raw material quality assessment. The operating principles of several nanosensor kinds, such as those for moisture, temperature, chemical composition, and fiber quality, as well as the materials used in their creation, such as metal oxides, graphene, and carbon nanotubes, are covered in detail in this chapter. Applications such as maintaining optimal water content, ensuring consistent coating, and detecting microbiological contamination are thoroughly examined, with a focus on their impact on accuracy and efficiency. Although nanosensors have revolutionary potential, they also have drawbacks, including the need for toughness in demanding industrial settings, high implementation costs, and difficulties in interpreting data. To address these obstacles, emerging trends are examined, including the integration of nanosensors into the Internet of Things (IoT) frameworks and advancements in nanomaterials. According to the chapter's conclusion, nanosensors have the potential to completely transform the paper sector by promoting sustainability, creativity, and compliance with upcoming regulations. They are, therefore, set to become essential components of contemporary paper production systems.

C. Kumar
Centre for Nano-Science and Nanotechnology, Panjab University, Chandigarh, India

M. Kaur · N. Ahmed · K. Kapoor · V. Singh · S. Sahu · M. Khatri · S. K. Arya (✉)
University Institute of Engineering Technology, Panjab University, Chandigarh, India
e-mail: skarya@pu.ac.in

G. Singh
Department of Medical Laboratory Sciences, Lovely Professional University, Phagwara, Punjab, India

R. K. Bachheti (ed.), *Nanotechnology in the Pulp and Paper Industry*, Smart Nanomaterials Technology, https://doi.org/10.1007/978-981-95-3473-9_11

Keywords Nanosensors · Real-time monitoring · Environmental sustainability · Nanomaterials

1 Introduction

The paper production industry has a significant role in modern society, providing essential materials for communication, packaging, and various other applications. Understanding the intricacies of paper production processes is vital for ensuring high-quality output and operational efficiency. The production process typically involves several stages, including pulping, where raw materials such as wood chips are converted into pulp; sheet formation, where the pulp is formed into thin sheets; pressing and drying, where excess water is removed; and finally, coating and finishing, which enhance the paper's properties for specific uses [1, 2]. Monitoring these processes is crucial for maintaining product quality and optimizing production efficiency. Effective monitoring enables the detection and correction of deviations in real time, thereby reducing waste, energy consumption, and production costs. Traditional monitoring methods, while useful, often lack the precision and responsiveness required for modern paper production demands [3, 4]. As the industry seeks to enhance operational efficiency and product quality, the adoption of advanced monitoring technologies is becoming increasingly important.

Nanosensors offer a promising solution for real-time monitoring in the paper production process. These sensors operate at the nanoscale, offering unprecedented sensitivity and accuracy in detecting various parameters, including temperature, pressure, humidity, and chemical composition [5]. The integration of nanosensors into paper production processes can significantly enhance monitoring capabilities, leading to more efficient and effective control of production variables.

Nanosensors can be employed at various stages of the paper production process. For instance, in the pulping stage, nanosensors can monitor the quality of raw materials, detecting impurities and ensuring consistency in the pulp mixture. During the sheet formation and pressing stages, nanosensors can measure parameters such as thickness and moisture content with high precision, enabling real-time adjustments to maintain product specifications [6]. In the coating and finishing stages, nanosensors can ensure uniform application of coatings, enhancing the final product's quality and performance.

The role of nanosensors extends beyond mere monitoring; they can also contribute to predictive maintenance by identifying early signs of equipment wear and potential failures [7]. This proactive approach can reduce downtime and extend the lifespan of production machinery, further optimizing operational efficiency.

The integration of nanosensors into paper production processes offers a transformative approach to monitoring and control. By providing real-time, high-precision data, nanosensors can help enhance product quality, reduce waste, and lower production costs. As the paper production industry continues to evolve, the adoption of nanosensor technology represents a significant step toward achieving greater efficiency and sustainability in production practices.

2 Basics of Nanosensors

2.1 Definition and Types of Nanosensors

Nanosensors are advanced instruments or systems that utilize nanomaterials to monitor, detect, and control various properties during the papermaking process. They enhance the efficiency, sustainability, and quality of paper production by providing real-time knowledge regarding variables like temperature, pressure, moisture content, chemical composition, and environmental pollutants [8].

2.2 Types of Nanosensors

By providing accurate, real-time monitoring and control over crucial parameters, nanosensors in paper production transform the manufacturing process, significantly increasing productivity, sustainability, and quality. To avoid overdrying, energy waste, or product flaws, moisture nanosensors are crucial for maintaining the ideal water content during drying and ensuring even moisture distribution throughout the process. The structural integrity of the paper is preserved, energy consumption is decreased, and operational efficiency is increased when temperature nanosensors monitor and control heat throughout pressing and drying procedures. To ensure the finished product has the appropriate brightness, smoothness, and durability while minimizing chemical waste and environmental impact, chemical composition nanosensors are crucial for assessing and regulating the amounts of bleaching agents, dyes, inks, and adhesives. To ensure consistent product quality and performance, fiber quality nanosensors assess key cellulose fiber characteristics, including length, strength, and bonding. Cleaner production methods and compliance with environmental standards are facilitated by environmental monitoring nanosensors, which detect pollutants such as particulate matter and volatile organic compounds (VOCs). Pressure nanosensors assess and control the applied force during pressing, ensuring consistent density and texture for improved mechanical properties [9]. For specific applications, such as printing, labels, and packaging, ink and coating nanosensors provide superior finishes by guaranteeing consistency, thickness, and adhesion of surface treatments. Furthermore, biological nanosensors detect microbial contamination in raw materials, preventing spoilage and ensuring the paper's durability and quality. Manufacturers can achieve unmatched precision, reduce waste, enhance product quality, comply with environmental regulations, and promote technologically sophisticated and sustainable practices in the paper sector by incorporating these advanced nanosensors into their manufacturing lines.

2.3 *Working Principles of Nanosensors*

The exact interaction between a biological identification element and a target analyte, along with a transducer that transforms this interaction into a measured signal, is the basis for the operation of biosensors used in the manufacturing of paper. To ensure product quality and process efficiency in the paper industry, biosensors are essential for monitoring a range of biological and chemical parameters, including microbial contamination, enzyme activity, and biochemical indicators.

In raw materials, intermediary products, or the final product, the biological recognition element, typically a biomolecule such as enzymes, antibodies, nucleic acids, or even entire cells, is specially selected to bind or react with the target analyte [9]. For example, it is possible to track the activity of enzymes such as cellulase during the breakdown of cellulose, a crucial step in the production of pulp. Likewise, microbial pollutants that could impair paper quality or introduce undesirable diseases can be detected using antibody-based biosensors [10]. A biochemical reaction, such as a pH shift, electron transfer, or molecule binding, is brought on by the interaction between the biological element and the analyte. The transducer, which transforms the biological signal into a measurable output such as an electrical, optical, or thermal signal, then detects this biochemical interaction. For instance, electrochemical biosensors offer real-time monitoring capabilities by measuring changes in current, potential, or impedance resulting from the interaction of the analyte [11]. When the target analyte interacts with the recognition element, optical biosensors, which are commonly employed in the manufacturing of paper, detect changes in light characteristics, including fluorescence, absorbance, or refractive index. Because they can detect even trace amounts of target molecules, they are especially useful for microbial contamination detection [12]. Another method for monitoring enzyme activity or chemical reactions during the papermaking process is the use of thermal biosensors, which detect heat changes associated with biochemical activities. To enhance sensitivity and response time, advanced biosensors are being developed using advancements in nanotechnology, incorporating materials such as graphene, carbon nanotubes, or gold nanoparticles [13]. Due to these advancements, biosensors can now detect incredibly low analyte concentrations, thereby improving the accuracy of quality control and contaminant monitoring. From pulp preparation to the finished product, biosensors are essential for optimizing different phases of paper manufacturing due to their capacity to deliver quick, real-time data. Furthermore, they save time and money while upholding strict quality standards due to their specificity and dependability, which reduces the need for costly laboratory testing. To ensure sustainable and effective manufacturing processes, biosensors are essential for modernizing the paper sector, as shown in Table 1.

Table 1 The diverse range of sensors used in the paper industry, their specific applications, working principles, and the materials

Sensor type	Application	Working principle	Material used	References
Temperature sensors	Monitoring heat during pressing and drying	Changes in resistance, voltage, or current in response to temperature variations	Graphene, carbon nanotubes, and metal oxides (ZnO, TiO_2)	[14]
Chemical composition sensors	Detecting chemical composition (inks, bleaching agents)	Changes in electrical conductivity, optical properties, or fluorescence based on chemical reactions	Gold (Au), silver (Ag), zinc oxide (ZnO), and polymers	[15]
Fiber quality sensors	Monitoring fiber length, strength, and bonding	Optical or electrochemical analysis of fiber morphology and bonding strength	Silica nanoparticles (SiO_2), carbon nanotubes (CNTs), and graphene	[16]
Environmental monitoring sensors	Detecting pollutants, VOCs, and air quality	Changes in conductivity or optical properties upon exposure to pollutants	Metal oxides (SnO_2, TiO_2), graphene, and polymers	[17]
Pressure sensors	Monitoring pressure during the pressing process	Changes in electrical resistance or capacitance due to applied pressure	Nanowires (ZnO, Ag), polymers, and CNTs	[18]
Ink and coating sensors	Monitoring ink adhesion, coating uniformity	Optical or capacitive changes based on coating thickness or uniformity	Carbon nanotubes (CNTs), graphene, polymers, and metal oxides	[19]
Biological sensors	Detecting microbial contamination in raw materials	Biorecognition elements trigger a detectable signal (colorimetric, electrical, optical)	Gold nanoparticles (Au), silver nanoparticles (Ag), and graphene	[20]
Thickness sensors	Measuring paper thickness during production	Changes in impedance or capacitance as the sensor interacts with the material thickness	Metal oxides (TiO_2), polymers, and carbon nanotubes (CNTs)	[21]
Moisture sensors	Moisture control during the paper drying process	Changes in electrical resistance, capacitance, or impedance due to moisture absorption	Carbon nanotubes (CNTs), graphene, polymers, and metal oxides	[22]

2.4 Materials Used in Nanosensors

A range of cutting-edge materials are included in nanosensors used for real-time monitoring in the paper industry to improve sensitivity, accuracy, and durability. These substances were chosen with care due to their special qualities, which include high surface area, conductivity, and the ability to interact with particular analytes. Gold (Au), silver (Ag), and titanium dioxide (TiO_2) are among the metal nanoparticles that are employed extensively because of their remarkable optical and chemical reactivity. For instance, TiO_2 nanoparticles are frequently used for environmental sensing because of their photocatalytic qualities. In contrast, gold and silver nanoparticles improve signal amplification in plasmonic sensors, allowing for the accurate detection of moisture or chemical pollutants [13]. Carbon-based nanomaterials, like graphene and carbon nanotubes (CNTs), are prized for their great surface area, mechanical strength, and superior electrical conductivity. Sensors that track variables such as temperature, pressure, and chemical composition often utilize these materials. Because CNTs and graphene may be functionalized to detect particular chemicals, they are adaptable for a range of uses in the production of paper. Gas and moisture sensors commonly utilize metal oxide nanoparticles, including copper oxide (CuO), tin oxide (SnO_2), and zinc oxide (ZnO). These materials are perfect for detecting chemical residues or pollutants in the manufacturing of paper because they change electrical conductivity when exposed to gases or volatile organic compounds (VOCs). Their performance in industrial settings is further improved by their excellent stability and sensitivity. Polymeric nanocomposites, composed of conductive polymers like polyaniline, polypyrene, and polyethylene oxide, are also employed in nanosensors for monitoring moisture, temperature, and other physical properties. These polymers are lightweight, flexible, and capable of undergoing changes in electrical properties in response to environmental stimuli. In addition, silica nanoparticles (SiO_2) are often used as carriers or support materials in sensor systems due to their stability, biocompatibility, and ease of functionalization, particularly for applications like monitoring fiber quality and surface treatments. In order to detect impurities or track the consistency of coatings and inks on paper, optical nanosensors employ quantum dots, which are semiconductor nanoparticles having size-dependent optical characteristics. High sensitivity and resolution offered by these materials allow for in-depth examination of production quality. Additionally, pressure and strain sensors employ nanowires composed of materials such as zinc oxide (ZnO) or silver (Ag) to monitor mechanical activities during the papermaking process [11]. Utilizing these cutting-edge materials, nanosensors offer precise, real-time monitoring of vital parameters, guaranteeing increased productivity, sustainability, and quality control in the papermaking sector. These tools can handle complicated problems in contemporary industrial processes thanks to a mix of varied material properties and creative sensor design.

3 Applications of Nanosensors in Paper Production

Nanosensors have become extremely effective instruments for transforming industrial processes due to their special qualities of enhanced sensitivity, selectivity, and miniaturization. At every process step, these sensors provide sophisticated capabilities in real-time monitoring, process control, and optimization [23]. Better product quality and better operational efficiency are the results of their integration. The main areas where the paper and pulp manufacturing industries have been most significantly impacted by the use of nanosensors will be examined in this section.

3.1 Monitoring Raw Material Quality

The raw materials such as wood chips, virgin and recycled fiber, filler, and other additives are in pulp and paper mill (PPM). Nanoscale molecular constituents like lignin, cellulosic, and hemicellulose [24] and their ratio present in the pulp preparation impact the composition and characteristic properties of the final product [25]. The abundance and varied source of the raw material which can be in PPM (pulp and paper mil) various paper-based product requires a quality control and a crucial check.

The following parameters are used to ensure quality checks: lignin content, moisture content, fiber composition, and contamination from pigments and undesired fibers. The production of high-quality pulp sludge, which goes through several processing stages to get the required product qualities, is achieved by monitoring these factors. Raw material moisture content must be controlled since variations can alter the composition of the pulp slurry, making drying and compositional analysis difficult. For accurate drying process management, sophisticated moisture-sensitive nanosensors are consequently necessary. Carbon-based nanosensors are very sensitive and economical, have good signal-to-noise ratios, and provide quick results—for example, carbon nanotubes (CNTs), graphite, and graphene electrodes [26].

For instance, a 1:1 graphene quantum dot (GQD)/AgNP silver nanocomposite showed shorter reaction times and a humidity response of 98.14% with exponential sensitivity throughout 25–95% relative humidity. The target analytes may be detected and measured using optical signals like absorbance and fluorescence thanks to optical nanosensors like CNTs and quantum dots, which are renowned for their exceptional fluorescence qualities. It is also critical to detect contamination, including lignin content, undesirable chemicals, and colored pigments. Nanosensors allow for instantaneous monitoring of lignin content, which has an inverse relationship with the end product's mechanical strength, therefore guaranteeing improved product quality [24].

3.2 Process Control in Pulping

The pulping process involves converting pretreated raw materials into fibrous pulp that builds the basis of papermaking as shown in Fig. 1. There are various critical steps involved as pulp undergoes a range of pretreatments of raw materials followed by cooking through chemical or mechanical processes, purification, and concentration and then approaches flow systems, pressing, bleaching, and drying [27].

Cooking of the pulp by chemical means also known as kraft pulping wood chips, which has been processed for two hours at 423–453 K with a high pH solution of NaOH and Na_2S (white liquor). CO_2-induced precipitation at low pH is used to extract lignin from black liquor [28]. In mechanical pulping, fibers are broken down by using mechanical energy. Pulp is subjected to frequent pumping and high-temperature corrosive operations, and these procedures are repeated several times. When fibers are damaged by mechanical processes, their length is reduced; occasionally, this can result in the creation of fines. However, chemical treatment causes fast changes in temperature, pH, and pressure, as well as damage to the fiber as refined chemical pulps lose density and tensile strength. These effects are reversible through refining. Recycling can generate fines and dissolve hemicellulose, reducing fiber coarseness [29]. By precisely monitoring important parameters, nanosensors play a crucial role in pulping process optimization. Fluorescent pH-sensitive nanosensors maintain precise pH control over cooking conditions [30], functionalized nanosensors with surface-enhanced Raman spectroscopy (SERS) and electrochemiluminescence (ECL) provide excellent sensitivity and selectivity for some analytes to track their chemical concentrations [31, 32], and nano-thermocouples like time–temperatures indicators (TTIs) [33] control temperature to guarantee effective lignin removal. During bleaching and refining, optical nanosensors improve recovery efficiency and decrease chemical waste by evaluating fiber properties, identifying undercooked fibers, and analyzing the composition of black liquor.

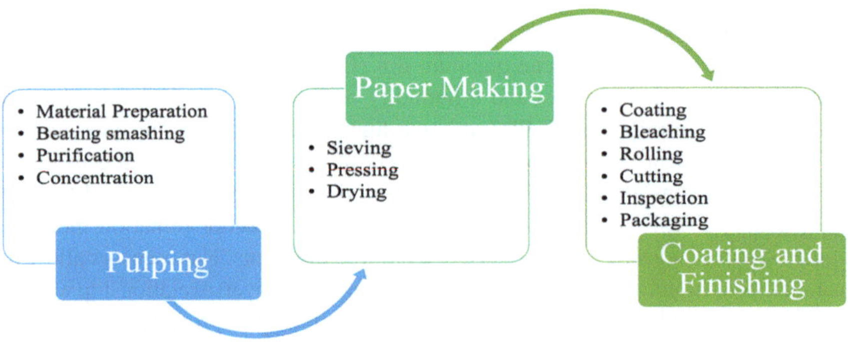

Fig. 1 Various process involved in papermaking

3.3 Real-Time Measurement in Paper Formation

The papermaking process requires fibers, water, and fillers to make the pulp which convert under immense pressure into paper. To ensure high quality of the paper necessitates maintaining constant basis weight, uniform sheet formation, and homogeneous fiber distribution; surface gloss, smoothness, and reduced ink absorber are of utmost importance [34].

Press cylinders are crucial in the papermaking process because they effectively remove excess water from the wet paper web by applying pressure; this reduces the moisture content, enhancing the structural integrity and preparing it for subsequent drying and finishing stages. Fiber orientation influences mechanical properties such as strength, stiffness, and dimensional stability significantly. The fibers must be uniformly aligned; however, uniform fiber alignment is crucial in specific applications, like printing and packaging. Variations in mechanical stress brought on by fiber alignment are detected by piezoelectric nanosensors [35] built into forming paper machines. These nanosensors enable real-time assessment of fiber orientation by transforming deformation into electrical signals. A feedback loop uses these data to modify machine settings like flow distribution and jet-to-wire ratios, which lowers anisotropy flaws and improves product quality. Pressurizing motion ensures uniform drying, press cylinders allow effective drying, and adjusted pressure settings facilitate customized product properties that enhance paper quality in general.

Paper thickness, durability, and application compatibility are basis weight or the weight of paper per unit area. Operational inefficiencies or product rejection may result from variations in basis weight. By examining changes in capacitance or dielectric characteristics, micro-electromechanical system (MEMS)-based nanosensors [36] positioned throughout the machine width quantify differences in mass distribution as illustrated in Fig. 2. With actuators that enable quick changes to headbox flow rates or slice apertures, these sensors optimize production efficiency by enabling real-time basis weight computation and continuous data. One important aspect affecting the quality of paper is the distribution of fiber and filler inside the sheet. Poor formation leads to defects including holes, streaks, and uneven density, which reduces strength and printability [24]. High-resolution pictures of the distribution of fiber and filler are obtained by optical nanosensors using sophisticated imaging methods such as infrared scanning or nanoscale fluorescence. Operators can modify forming fabric speeds or jet velocities to ensure uniform distribution by using sophisticated algorithms that evaluate the data to discover regions of uneven density.

3.4 Coating and Finishing Process Monitoring

After drying, papers undergo additional processes such as coating, whitening, rolling, cutting, and inspection to upgrade their functional and aesthetic characteristics. It

Fig. 2 Major components of MEMS

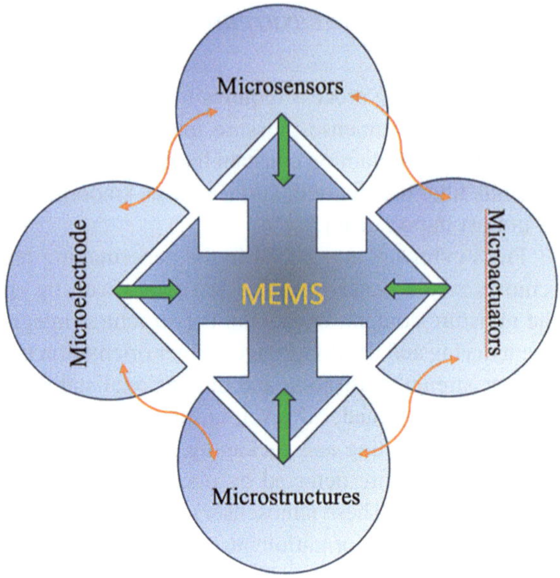

is the filling of micro-voids with pigments, such as kaolin, calcium carbonate, and binders, that improve barrier performance. It also attains properties such as gloss, opacity, and controlled drying rates. Printability, mechanical strength, and cellulose separation preventing clumping are influenced by pore size, filler addition, and dispersant, respectively. These materials are added to the pulp slurry or straight onto the sheet in tiny, concentrated doses [27, 37].

Nanosensors are essential for precise coating application and monitoring, which allows for consistent and dependable outcomes in paper treatment procedures. Nanosensors, such as surface plasmon resonance (SPR) and MEMS-based piezoelectric sensors [33, 38], have revolutionized the monitoring of coating processes. These sensors measure coating thickness with nanometer precision by detecting variations in surface topology, enhancing accuracy, speed, and resolution. Nanoscale profilometers scan paper surfaces to measure roughness, while reflective nanosensors monitor gloss by analyzing light scatter patterns. Thickness data from these sensors inform adjustments to applicator settings, such as blade pressure or nozzle output, ensuring uniform coating application. Feedback also guides modifications to polishing rollers, drying rates, or coating formulations, maintaining consistent surface properties.

Functional coatings, including moisture barriers, antimicrobial layers, and electrical conductors, demand precise application to meet performance standards as shown in Fig. 3. Electrochemical nanosensors monitor coating integrity by detecting changes in conductivity, ion permeability, or surface energy. Integrated into production lines, these sensors test coatings immediately after application, flagging deviations for corrective action to ensure optimal product quality [39].

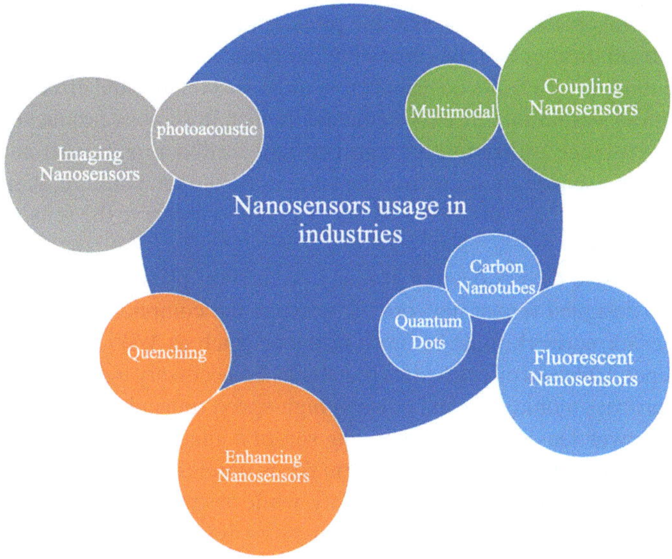

Fig. 3 Major aspect of nanosensor usage in the industries

4 Integration of Nanosensors into Paper Production Systems

Nanosensors' integration into pulp and paper processing systems represents a significant step towards sustainability, production quality, and efficiency. This involves the careful utilization of raw materials, energy efficiency, and the use of advanced techniques to meet the industries requirements. Nanosensors provide improved productivity and quality in industrial operations by enabling real-time monitoring and precise control over production processes. The key aspects of their sensor placement and network design, data acquisition and processing, and the challenges involved are discussed in this section.

4.1 Sensor Placement and Network Design

Important quality indicators, process flow, and environmental concerns should all be considered when strategically integrating nanosensors into paper production systems. Proper sensor placement is essential for both effective process control and accurate data collection. By measuring vital characteristics like fiber quality, moisture content, and chemical composition, these sensors allow for real-time monitoring and control of operations including coating thickness, digester pressure, and additive addition.

This allows for precise control over the pulping process, including digester conditions, chemical dosing, and the separation of cellulose from lignin, Kappa number, and effluent quality. Sensors contribute to a more sustainable and effective pulp and paper business by minimizing environmental impact, improving product quality, and optimizing resource utilization through constant feedback and exact adjustments. When choosing sensor sites, environmental factors like temperature, humidity, and chemical exposure must also be taken into account to preserve the sensors' lifetime and effectiveness.

Comprehensive data collection at many production phases may be made easier with a well-designed sensor network. To manage the complicated communication, a robust sensor network must be built. Protocols such as WiFi, LoRa, and Zigbee provide dependable and effective data transfer in the necessary industrial settings. For real-time process monitoring and control, high-speed transmission rates are crucial [40]. Low-power sensor nodes comprising sensors, interface circuitry, and wireless communication modules for data transfer make up wireless sensor networks (WSNs). WSNs improve quantification and selectivity, while developments in nanotechnology allow for smaller integrated circuits and transistors, which make devices portable for on-site analysis. Real-time data transfer over open frequency bands (e.g., 2.4 GHz and 5 GHz ISM) is possible with wireless capabilities and transceivers. Environmental monitoring and analysis are made more accurate and efficient by this combination. Although network structure is important, a hybrid topology that combines mesh and star configurations guarantees scalability and dependable connection—setting up very reliable data collection and processing systems for efficient use of the sensor data [41].

4.2 Data Acquisition and Processing

The advantages offered by nanosensors, their effectiveness depends on massive volumes of data may be gathered in real time thanks to reliable data processing and collecting methods. To cut down on latency and speed up decision-making, edge computing devices that are located close to sensor nodes preprocess these data [42]. To guarantee accuracy and utility, raw data from nanosensors must be preprocessed. Noise filtering efficiently eliminates interference brought on by external variables by using techniques such as Kalman filtering [43]. Cross-referencing sensor data with conventional measuring methods confirms its dependability, while routine calibration against established standards preserves sensor accuracy.

Sensor data are transformed into useful information using machine learning algorithms and advanced analytics. Based on patterns in sensor data, the machine learning model forecasts paper quality parameters, enabling proactive modifications. Signal processing systems (SPS) use noise filtering, amplification, and signal digitalization to convert electrical signals into useful information. Raw sensor data are captured by data acquisition systems (DAS) for analysis and the application of pattern recognition (PR) algorithms. In order to create prediction models from measurable data, machine learning (ML) uses computational techniques. Knowledge extraction from data, model building, anomaly detection, and future phenomena prediction are all made possible by machine learning algorithms. Methods like linear discriminant analysis (LDA), K-nearest neighbor (KNN), principal component analysis (PCA), and artificial neural networks (ANN) are crucial. Algorithms for real-time anomaly detection spot deviations, avoiding waste and problems. Raw data are transformed into insightful knowledge through the integration of machine learning with nanosensor networks, which promotes ongoing process improvement in the papermaking industry [44].

4.3 Challenges in Integration

Nanosensors have a lot of potential, but incorporating them into systems for producing paper presents a number of difficulties. For example, the durability and dependability of nanosensors are important issues. Extreme industrial settings with high humidity, temperatures, and chemical exposure are required of such devices. Sensor lifespan may be increased by creating corrosion-resistant materials and strong encapsulation procedures. Another major issue is environmental interference, which can compromise sensor accuracy due to noise and electromagnetic interference introduced by industrial environments. Electromagnetic shielding and sophisticated data processing methods can help to lessen these problems. Sensor networks' interconnectedness makes them more susceptible to cyberattacks. Sensitive manufacturing data are protected by using data encryption and secure communication methods. Furthermore, especially for small-scale paper mills, the high upfront costs associated with nanosensor deployment may be a deterrent [7].

5 Advantages and Limitations of Using Nanosensors

Utilizing nanosensors in the papermaking process offers numerous advantages, particularly in terms of real-time oversight. One of the key benefits is enhanced accuracy and precision. In the paper manufacturing industry, where maintaining consistent material quality is essential, the small size of nanosensors enables them to identify tiny alterations in chemical and physical characteristics. They possess a higher sensitivity compared to traditional sensors when it comes to tracking metrics such as fiber makeup, acidity levels, and moisture levels. This heightened accuracy not only ensures better quality management but also allows for swift adjustments in the manufacturing process, resulting in reduced waste and improved quality of the end product [45, 46]. In addition to accuracy, nanosensors contribute to making the manufacturing of paper more cost-effective and productive. Their small size and minimal energy requirements allow for high-throughput uses, enabling producers to deploy multiple sensors at different stages of production without significant costs.

The real-time information provided by these sensors can enhance decision-making by optimizing resource usage and minimizing downtime. For instance, companies can quickly identify inefficiencies or deviations from optimal conditions through constant monitoring of the production environment, allowing for prompt corrective measures and reduced operational costs. Furthermore, as nanosensor technology continues to advance and production expenses decrease, more manufacturers might adopt it, potentially revolutionizing the industry [45, 47]. In the production of paper, the application of nanosensors enhances scalability and flexibility while also improving accuracy and reducing costs. Their small size and versatility make it simple to incorporate them into existing production systems, enabling manufacturers to benefit from better oversight without needing major modifications. Nanosensors can be rapidly deployed to track new metrics or adapt to variations in raw materials as production requirements evolve. This flexibility not only enhances the responsiveness of manufacturing lines but also encourages innovation in product development. By facilitating real-time modifications grounded in precise data, nanosensors allow producers to explore new formulations and methods. This ultimately leads to the development of superior quality paper products that meet diverse market needs [48, 49].

Despite the numerous benefits, using nanosensors in paper manufacture has several limits. The initial investment needed for nanosensor system integration and installation is one of the main concerns. Specifically for smaller paper mills with tighter budgets, the price of buying nanosensors, integrating them with their current machinery, and making sure they work properly with data analysis tools can be expensive [50]. Even while there are obvious long-term advantages to lower labor costs and greater productivity, the initial financial outlay may prevent widespread acceptance, especially for businesses with narrow profit margins. Another drawback is the extreme conditions encountered in environments where paper is produced [51]. Often, paper manufacturing facilities face elevated temperatures, high humidity levels, and the utilization of chemicals, which can all affect the efficiency and durability of nanosensors. Prolonged exposure to such environments may result in the deterioration of sensors, diminished precision, and potential failures. This necessitates consistent calibration and upkeep, which complicates operations and elevates ongoing expenses [47]. Additionally, the effectiveness of sensors can be compromised by the build-up of grime, chemicals, or other pollutants, leading to the need for regular cleaning or replacements. Ultimately, even though nanosensors deliver useful real-time information, deciphering these data can be challenging [52]. Paper manufacturing plants must allocate resources for advanced data analysis tools and employ trained staff to handle and make sense of the substantial data produced by these sensors [51]. In the absence of adequate tools and knowledge, there is a chance of misreading the information, which might result in incorrect decisions or less than ideal modifications in the production workflow. Furthermore, since nanosensors frequently need recalibration to sustain their precision, the demand for specialized training and continuous maintenance can raise operational expenses and introduce additional difficulties in their application.

6 Future Trends and Developments

The prospects for nanosensors within the realm of paper manufacturing are underscored by considerable progress in their sensitivity, selectivity, and integration features. Scientists are concentrating on creating innovative nanomaterials like carbon nanotubes, graphene, and quantum dots, known for their superior electrical, mechanical, and chemical characteristics. These cutting-edge materials significantly improve the functionality of nanosensors, enhancing their durability and precision in the challenging environments of paper production facilities, which often face exposure to chemicals, high temperatures, and moisture. Moreover, efforts to miniaturize these devices will enable their smooth incorporation into manufacturing systems without the need for major infrastructural adjustments. Developments in wireless technology and low-energy solutions will also permit nanosensors to function independently, reducing the reliance on human involvement while streamlining data gathering and analysis processes. As nanosensor technology continues to progress, novel and promising applications in paper production are being introduced. In addition to assessing fundamental factors such as moisture levels and fiber uniformity, forthcoming nanosensors are anticipated to identify other variables, including temperature changes, particulate matter presence, and minute contaminants. This capability will facilitate more accurate monitoring and quality assurance, particularly for specialized products such as food wrapping or medical paper, where the risks of contamination must be kept to a minimum. Furthermore, the union of nanosensors with Internet of Things (IoT) frameworks will enable smarter manufacturing processes, providing instant insights into recycling effectiveness, chemical utilization, and water use. This synchronization will promote more sustainable methods by optimizing resource management and curtailing waste, ultimately improving the ecological footprint of paper production facilities.

The integration of nanosensors for continuous monitoring is expected to transform the standards within the paper production sector. As this technology becomes increasingly prevalent, it will establish new standards for quality assurance, operational efficiency, and environmental responsibility. Regulatory organizations may implement new standards to ensure the precision and reliability of nanosensors, especially in the production of high-quality or specialized papers. The insights gained from nanosensors will facilitate stricter quality control measures, ensuring uniformity in products and adherence to safety guidelines. Additionally, the technology's contribution to minimizing waste, energy usage, and emissions could lead to the establishment of tougher sustainability regulations within the paper industry, fostering greater innovation in environmentally friendly practices and manufacturing techniques. The future of nanosensors in paper production is promising, as ongoing progress in sensor technology is leading to more effective, precise, and environmentally friendly manufacturing methods. As these innovations advance, nanosensors are poised to become a crucial component of real-time monitoring systems, enabling manufacturers to enhance production efficiency, reduce expenses, and maintain product quality standards. The further advancement of nanosensors will not only enhance the productivity of paper mills but also shape future industry benchmarks, establishing new standards for quality, safety, and environmental impact. The scope for breakthroughs in this area is immense, and with continued research, the paper manufacturing sector is ready for considerable shifts.

7 Conclusion

In brief, the incorporation of nanosensors into paper production systems makes a giant step toward precision, efficiency, and sustainability. This is because real-time monitoring and control by nanosensors facilitate the optimization of many stages in production, including raw material assessment and the final coating and finishing of products. These sensors provide valuable insights that can enhance product quality, minimize waste, and reduce operational costs. Moreover, these sensors can contribute to compliance with environmental standards. The significant issues are high installation costs and short durability. The development in nanotechnology and the integration of sensors continue, offering a prospect of overcoming these difficulties. Future innovations, such as the acceptance of IoT and miniaturized, self-sustaining nanosensors, will certainly change the norms of industry regarding quality and sustainability. Nanosensors are a crucial factor in modernizing the paper manufacturing industry toward an intelligent, responsive, and environmentally conscious landscape.

8 Competing Interests:

The authors declare that they have no competing interest.

9 Declaration of Generative Artificial Intelligence (AI) and AI-Assisted Technologies in the Writing Process:

During the preparation of this work, the book chapter author(s) have not taken any assistance from AI and the book chapter author(s) reviewed and edited the content as needed and take(s) full responsibility for the content of the publication.

References

1. Fadel SM, Abou-Elseoud WS, Hassan EA, Ibrahim S, Hassan ML (2022) Use of sugar beet cellulose nanofibers for paper coating. Ind Crops Prod 180:114787. https://doi.org/10.1016/J.INDCROP.2022.114787

2. Haile A et al (2021) Pulp and paper mill wastes: utilizations and prospects for high value-added biomaterials. Bioresour Bioprocess 8(1):1–22. https://doi.org/10.1186/S40643-021-00385-3/FIGURES/14
3. Olejnik K (2011) Water consumption in paper industry—reduction capabilities and the consequences. NATO science for peace and security series C: environmental security. pp 113–129. https://doi.org/10.1007/978-94-007-1805-0_8
4. Abioye EA et al (2020) A review on monitoring and advanced control strategies for precision irrigation. Comput Electron Agric 173:105441. https://doi.org/10.1016/J.COMPAG.2020.105441
5. Hu H, Wang N, Liao J, Tovar-Lopez FJ (2023) Recent progress in micro- and nanotechnology-enabled sensors for biomedical and environmental challenges. Sensors 23(12):5406. https://doi.org/10.3390/S23125406
6. Huang X, Zhu Y, Kianfar E (2021) Nano biosensors: properties, applications and electrochemical techniques. J Market Res 12:1649–1672. https://doi.org/10.1016/J.JMRT.2021.03.048
7. Han D, Hosamo H, Ying C, Nie R (2023) A comprehensive review and analysis of nanosensors for structural health monitoring in bridge maintenance: innovations, challenges, and future perspectives. Appl Sci 13(20):11149. https://doi.org/10.3390/APP132011149
8. Ramesh M, Janani R, Deepa C, Rajeshkumar L (2023) Nanotechnology-enabled biosensors: a review of fundamentals, design principles, materials, and applications. MDPI. https://doi.org/10.3390/bios13010040
9. Sharma M, et al (2025) Next-generation nanomaterials-based biosensors: real-time biosensing devices for detecting emerging environmental pollutants. Elsevier Ltd. https://doi.org/10.1016/j.mtsust.2024.101068
10. Arora P, Sindhu A, Dilbaghi N, Chaudhury A (2011) Biosensors as innovative tools for the detection of food borne pathogens. https://doi.org/10.1016/j.bios.2011.06.002
11. Ronkainen NJ, Halsall HB, Heineman WR (2010) Electrochemical biosensors. Chem Soc Rev 39(5):1747–1763. https://doi.org/10.1039/b714449k
12. Nnachi RC, et al (2022) Biosensors for rapid detection of bacterial pathogens in water, food and environment. Elsevier Ltd. https://doi.org/10.1016/j.envint.2022.107357
13. Riu J, Maroto A, Rius FX (2006) Nanosensors in environmental analysis. In: Talanta. Elsevier, pp 288–301. https://doi.org/10.1016/j.talanta.2005.09.045
14. Prudente INR, dos Santos HC, Fonseca JL, de Almeida YA, de Gimenez IF, Barreto LS (2024) Graphene family (GFMs), carbon nanotubes (CNTs) and carbon black (CB) on smart materials for civil construction: self-cleaning, self-sensing and self-heating. J Build Eng 95:110175. https://doi.org/10.1016/J.JOBE.2024.110175
15. Adfar Q, Aslam M, Maktedar SS (2023) A compendium of metallic inorganic fillers' properties and applications employed in polymers. In: Nanofillers: fabrication, characterization and applications of inorganic nanofillers, pp 25–68. https://doi.org/10.1201/9781003279389-2/COMPENDIUM-METALLIC-INORGANIC-FILLERS-PROPERTIES-APPLICATIONS-EMPLOYED-POLYMERS-QAZI-ADFAR-MOHAMMAD-ASLAM-SHRIKANT-MAKTEDAR
16. Petousis M, Tzounis L, Papageorgiou D, Vidakis N (2020) Decoration of SiO_2 and Fe_3O_4 nanoparticles onto the surface of MWCNT-grafted glass fibers: a simple approach for the creation of binary nanoparticle hierarchical and multifunctional composite interphases. Nanomaterials 10(12):2500. https://doi.org/10.3390/NANO10122500
17. Belal M et al (2025) Advances in gas sensors using screen printing. J Mater Chem A Mater. https://doi.org/10.1039/D4TA06632D
18. Liu MY et al (2021) Advance on flexible pressure sensors based on metal and carbonaceous nanomaterial. Nano Energy 87:106181. https://doi.org/10.1016/J.NANOEN.2021.106181
19. da Costa TH, Choi J-W (2020) Fabrication and patterning methods of flexible sensors using carbon nanomaterials on polymers. Adv Intell Syst 2(5):1900179. https://doi.org/10.1002/AISY.201900179

20. Parkhe VS, Arpita PT, Tiwari AP (2024) Gold nanoparticles-based biosensors: pioneering solutions for bacterial and viral pathogen detection—a comprehensive review. World J Microbiol Biotechnol 40(9):1–26. https://doi.org/10.1007/S11274-024-04072-1
21. Abdulhameed A, Mahnashi Y (2023) Fabrication of carbon nanotube/ titanium dioxide nanomaterials-based hydrogen sensor using novel two-stage dielectrophoresis process. Mater Sci Semicond Process 167:107794. https://doi.org/10.1016/J.MSSP.2023.107794
22. Tulliani JM, Inserra B, Ziegler D (2019) Carbon-based materials for humidity sensing: a short review. Micromachines 10(4):232. https://doi.org/10.3390/MI10040232
23. Dhahi TS, et al (2024) Advances in nano sensors for monitoring and optimal performance enhancement in photovoltaic cells. iScience 27(4):109347. https://doi.org/10.1016/J.ISCI.2024.109347
24. Keränen JER (2025) BioResources, and undefined 2016, "Changing quality of recycled fiber material. Part 1. Factors affecting the quality and an approach for characterisation of the strength potential. bioresources.cnr.ncsu.edu, vol. 11(4). https://bioresources.cnr.ncsu.edu/wp-content/uploads/2016/11/BioRes_11_4_10404_Keranen_R_Changing_Quality_Recovered_Fiber_Mater_Pt1_Factors_8801.pdf. Accessed on 15 Jan 2025
25. Mohieldin S, Zainudin E (2011) MP-K engineering, and undefined 2011 "Nanotechnology in pulp and paper industries: a review". Trans Tech PublSD Mohieldin, ES Zainudin, MT Paridah, ZM Ainun Key engineering materials. Trans Tech Publ. https://doi.org/10.4028/www.scientific.net/KEM.471-472.251
26. Korotcenkov G (2023) Paper-based humidity sensors as promising flexible devices: state of the art: part 1. General consideration. Nanomaterials 13:1110. https://www.academia.edu/download/106026039/pdf.pdf. Accessed on 15 Jan 2025
27. Julkapli NM, Bagheri S (2016) Developments in nano-additives for paper industry. J Wood Sci 62(2):117–130 https://doi.org/10.1007/s10086-015-1532-5
28. Alén R (2015) Pulp mills and wood-based biorefineries. In: Industrial biorefineries and white biotechnology. pp 91–126. https://doi.org/10.1016/B978-0-444-63453-5.00003-3
29. Luo H, Abu-Omar MM (2017) Chemicals from lignin. In: Encyclopedia of sustainable technologies. pp 573–585. https://doi.org/10.1016/B978-0-12-409548-9.10235-0
30. Besenyei G, Tang A, Aylott J (2024) Fluorescent pH-sensitive nanosensors enable precise low-volume monitoring in high-throughput bioprocess manufacturing. biorxiv. biorxiv.org. https://doi.org/10.1101/2024.09.13.612862
31. Hu H, Wang N, Liao J, Tovar-Lopez FJ (2023) Recent progress in micro-and nanotechnology-enabled sensors for biomedical and environmental challenges. mdpi.com 23:5406. https://doi.org/10.3390/s23125406
32. Darwish MA, Abd-Elaziem W, De AE, Zayed AA (2024) Advancements in nanomaterials for nanosensors: a comprehensive review. Nanoscale Adv. https://doi.org/10.1039/d4na00214h
33. Chelliah R, et al (2021) Development of nanosensors based intelligent packaging systems: food quality and medicine. mdpi.com. https://doi.org/10.3390/nano11061515
34. Manfredi V (2024) The modern paper. In: Eucalyptus Kraft pulp refining. pp 15–36. https://doi.org/10.1007/978-3-031-47285-5_4
35. Javaid M, Haleem A, Singh R, Rab S (2025) Exploring the potential of nanosensors: a brief overview. Elsevier. https://www.sciencedirect.com/science/article/pii/S2666351121000516. Accessed on 15 Jan 2025. [Online]
36. Jena S, Gupta A (2021) Review on pressure sensors: a perspective from mechanical to micro-electro-mechanical systems. Sens Rev 41(3):320–329. https://doi.org/10.1108/SR-03-2021-0106/FULL/HTML
37. Jagaba AH, Birniwa AH, Usman AK, Mu'azu ND, Yaro NS, Soja UB, Abioye KJ, Almahbashi NM, Al-dhawi BN, Noor A, Lawal IM (2023) Trend and current practices of coagulation-based hybrid systems for pulp and paper mill effluent treatment: mechanisms, optimization techniques and performance evaluation. J Clean Prod 429:139543. https://www.sciencedirect.com/science/article/pii/S0959652623037010?casa_token=pRdonkCYFC8AAAAA:qTPsJL3XLU0Qr4Ecmqkh6JuKYLausgMMBvEGw0SUt8Ldpaz7-J4QKd7j52wpZzCIcP4pVybqYc_I. Accessed on 15 Jan 2025. [Online]

38. Das Mahapatra S, et al (2021) Piezoelectric materials for energy harvesting and sensing applications: roadmap for future smart materials. Adv Sci 8(17):2100864. https://doi.org/10.1002/ADVS.202100864
39. Hossain N et al (2024) Advances of materials science in MEMS applications: a review. Res Eng 22:102115. https://doi.org/10.1016/J.RINENG.2024.102115
40. Seferagi´c AS, et al (2020) Survey on wireless technology trade-offs for the industrial internet of things. Sensors 20(2):488. https://doi.org/10.3390/S20020488
41. Prasad P, Raut P, Goel S, Barnwal RP, Bodhe GL (2022) Electronic nose and wireless sensor network for environmental monitoring application in pulp and paper industry: a review. Environ Monit Assess 194(12). https://doi.org/10.1007/S10661-022-10479-W
42. Thomas S, Nguyen TA, Ahmadi M, Farmani A, Yasin G (2021) Nanosensors for smart manufacturing. p 634
43. Erdogan M, Berke Baytekin N, Emre Coban S, Demir A machine learning and Kalman filtering for nanomechanical mass spectrometry
44. Muñoz-García M-Á, Zhang B, Qiao Y, Leslie C, Vidal A, Govan JE (2024) Machine learning techniques for improving nanosensors in agroenvironmental applications. mdpi.com. https://doi.org/10.3390/agronomy14020341
45. Kumar A (2019) Nanosensors: Applications and challenges. Int J Sci Res (IJSR) 8(7)
46. Han D, Hosamo H, Ying C, Nie R (2023) A comprehensive review and analysis of nanosensors for structural health monitoring in bridge maintenance: innovations, challenges, and future perspectives. Multidisciplinary Digital Publishing Institute (MDPI). https://doi.org/10.3390/app132011149
47. Khanna V (2016) Nanosensors. CRC Press Boca Raton, FL
48. Javaid M, Haleem A, Singh RP, Rab S, Suman R (2021) Exploring the potential of nanosensors: a brief overview. KeAi Communications Co. https://doi.org/10.1016/j.sintl.2021.100130
49. Parameswari P et al (2024) Nanotechnology-based sensors for real-time monitoring and assessment of soil health and quality: a review. Asian J Soil Sci Plant Nutr 10(2):157–173. https://doi.org/10.9734/AJSSPN/2024/V10I2272
50. Joshi D, Shahadat M, Adnan R, Ahammad SZ, Sreekrishnan TR (2021) Chapter 22–Paper-based nanosensors for smart manufacturing. In: Thomas S, Nguyen TA, Ahmadi M, Farmani A, Yasin G (eds.) Nanosensors for smart manufacturing, in micro and nano technologies. Elsevier, pp 517–532. https://doi.org/10.1016/B978-0-12-823358-0.00024-1
51. Darwish MA, Abd-Elaziem W, Elsheikh A, Zayed AA (2024) Advancements in nanomaterials for nanosensors: a comprehensive review. R Soc Chem. https://doi.org/10.1039/d4na00214h
52. Dhahi TS, et al (2024) Advances in nano sensors for monitoring and optimal performance enhancement in photovoltaic cells. Elsevier Inc. https://doi.org/10.1016/j.isci.2024.109347

Challenges in the Use of Nanocellulose in the Pulp and Paper Industry

Akash Verma, Ishita Semwal, Muskan Bhandari, and Rakesh Kumar Bachheti

Abstract Nanocellulose is of significant interest to the pulp and paper industry due to its sustainability and high-performance properties. Due to its strength, biodegradability, and renewability, it has the potential to fundamentally improve paper quality, enhancing durability, barrier properties, and even simplifying printability. Applications could vary from light packaging materials to high-grade specialty papers. However, despite the obvious benefits, the large-scale industrial application of nanocellulose is hindered by numerous barriers. Production remains expensive and energy-intensive, with scalability remaining an issue. Uniform dispersion of nanocellulose in pulp is also a major issue, often seeming to require specialist additives or modifications to current processing methods. Questions regarding the ability of current papermaking equipment to handle it without significant modification are also an issue. Uncertainty regarding nanomaterial rules may also hinder take-up to market. This chapter examines these barriers in greater depth and proposes solutions, such as the development of advanced production technology and the encouragement of interdisciplinary collaboration. To overcome these barriers is essential to placing nanocellulose in more than a potential but as an important step toward greater sustainability and environmental responsibility in the pulp and paper industry.

Keywords Nanocellulose · Production challenges · Pulp and paper industry · Sustainability

A. Verma · I. Semwal · M. Bhandari (✉)
Cellulose and Paper Discipline, Forest Products Division, Forest Research Institute, Dehradun, India
e-mail: muskanbhandari1311@gmail.com

R. K. Bachheti
Department of Research Analytics, Saveetha Dental College & Hospitals, Saveetha Institute of Medical and Technical Sciences, Saveetha University, Chennai, Tamil Nadu, India

R. K. Bachheti (ed.), *Nanotechnology in the Pulp and Paper Industry*, Smart Nanomaterials Technology, https://doi.org/10.1007/978-981-95-3473-9_12

1 Introduction

In the term "nanocellulose (NA)," the word nano refers to the scale of order 10 to the power -9 m, and cellulose is a primary structural component of the cell wall together. Cells contain cell fibers that are decompose into many small fragments/fibers, which are not visible to naked eye [58]. Nanocellulose proves to be very important in determining the overall quality of paper. Nanocellulosic particles/fibers have high strength and smoothness and are more feasible when compared to cellulose fibers. There are three types of nanocellulose: cellulose nanocrystals (CNCs), cellulose nanofibers (CNFs), and bacterial cellulose (BC). Out of these, CNCs and CNFs play a significant role in pulp- and papermaking [15].

Out of three types of nanocellulose, CNFs are mainly used as an additive in papermaking as a coating material and to overcome dewatering issues and more. Nanocellulosic fibers are produced with the help of a mechanical process known as homogenization in which cellulose is decomposed into various small fibers. Applying high pressure, other methods can also be used in conjunction with homogenization.

Among the numerous applications of NC in papermaking, their use as dry strength additives and coating agents is significant. Some other applications that have been researched include wet strength aids, retention additives, flexographic ink removal agents, book restoration agents, lint control agents, and carriers that provide special properties such as antimicrobial properties, electrical conductivity, and fire resistance. However, the high cost of NC production has so far been a deterrent to its complete industrial-scale commercialization. Different terminologies have been applied to classify the various types of nanocellulose (NC) throughout history, including microfibrillated cellulose, nanofibrillated cellulose, nanofibers, cellulose nanowhiskers, nanorods, and microbial cellulose [15, 66]. Nevertheless, the ISO/TS 20,477:2017 standard has been instrumental in NC terminological harmonization. The most frequent NC classes—cellulose nanofibers (CNFs), cellulose nanocrystals (CNCs), and bacterial cellulose (BC)—differ in dimensions, characteristics, functions, and production methods. CNF production usually involves disruption of hydrogen bonding to allow disentanglement of fibers into nanofibrils, mostly by mechanical methods like high-pressure homogenization, microfluidization, grinding, electrospinning, and steam explosion [46, 76]. Due to the astronomical costs of these mechanical processes, chemical or enzymatic pretreatments are often applied to enhance fiber separation, increase CNF yield, and reduce production costs [13].

On the other hand, CNCs are typically obtained by acid hydrolysis [36]. Achieving high hydrolysis yields entails the use of pure cellulose or cellulose-containing raw materials. Cotton, bacterial cellulose (BC), and microcrystalline cellulose (MCC) are cellulose sources that contain nearly pure cellulose and therefore permit direct acid hydrolysis. In the instance of the use of wood or other biomass materials, extractives, hemicelluloses, lignin, and inorganic compounds must be eliminated from these through various pretreatment processes, which may involve solvent extraction (with dichloromethane, acetone, or ethanol/benzene), alkali treatments [30], kraft cooking [18], and multistage bleaching [30]. These additional procedures not only lower the total process yield but also result in higher production costs and increased environmental pressures. Interestingly, high purity does not always guarantee enhanced NC performance [10].

For BC, production is expensive due to the high quantity of nutrients required to culture acetobacteria and the low yield rates [47]. BC develops highly entangled nanofibrils when statically cultured, as bacteria move toward the surface in search of oxygen. While BC pellicles are not easily dispersible [20, 22], they can be used effectively to produce specialty paper products like fire-resistant paper and electronic and magnetic papers [12, 54, 75].

Today, industrial-scale uses of NC in papermaking remain in their infancy due to high costs and perceived risks. Additional research and development are therefore needed to bridge production cost barriers, advance characterization techniques, and clarify the governing factors of industrial applications. This review summarizes and discusses the technical issues associated with applying NC to the papermaking process. The review also outlines several new approaches from the literature that aim to facilitate industrial use, including the improvement of NC barrier properties through reduced water permeability, enhancement of NC dispersion in pulp or wet webs to improve paper properties, separation of retention and drainage effects, and reduction of NC production costs. This chapter also identifies critical issues in NC characterization and provides an overview of the market.

2 Challenges Associated with the Use of Nanocellulose in the Pulp and Paper Industry

The integration of nanocellulose into the pulp and paper industry offers significant advantages, yet it also presents several challenges. These include high production costs, scalability issues, compatibility with existing processes, and limited standardization, all of which hinder its widespread adoption (Fig. 1).

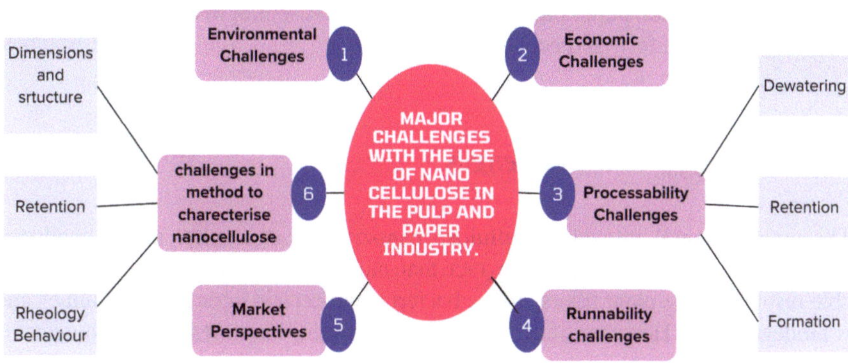

Fig. 1 Major challenges associated with the use of nanocellulose in pulp and paper industry

2.1 Market Outlook

In spite of the superior qualities, the commercial uses of nanocellulose are still low (approximately 2500(5)–10,000(6) tonn per year); the primary cause for the less commercial use of nanocellulose is the elevated manufacturing cost and the less economic viability of cellulose nanofiber as compared to the traditional technology for the large-scale use [7, 9]. The high energy usage in the production of nanocellulose is a major cause of elevated cost, thus making its large-scale application less economically feasible. Several chemical and enzymatic treatment methods can be employed to reduce energy requirements; however, their introduction may incur additional expenses [26].

Health and safety concerns are a significant factor in the less commercial use of nanocellulose, although cellulose nanomaterials are generally considered less toxic compared to other nanomaterials. However, certain health hazards are associated with cellulose nanomaterials due to their small size, high surface area, and potential for inhalation and dermal exposure. Potential health hazards may include respiratory hazards and skin and eye irritation. Respiratory hazards may be caused by the inhalation of airborne nanoparticles, especially by the workers handling cellulose nanomaterials in dry powder form. Studies suggest that CNMs may cause inflammation in the lungs and induce oxidative stress, although the severity depends on exposure duration, particle size, and surface charge. There is limited research and knowledge available on the health effects of CNMs exposure; hence, it is difficult to fully assess the possible health hazards and risks allied with CNMs [73].

Additionally, market acceptance and infrastructure also interfere with the large-scale use of nanocellulose. The paper mills are optimized for traditional raw materials, and hence, in order to integrate CNMs, significant modifications are required (Fig. 2).

These modifications involve changes in various processes, such as pulping, refining, mixing, and drying, along with the requirement of new machinery and employee training. Thus, these modifications required substantial investments, which hindered widespread acceptance [24].

Addressing these challenges is crucial for the successful integration of nanocellulose in the pulp and paper industry; therefore, it is essential to take necessary steps toward research, safety, and economic strategies.

2.2 Environmental Challenges

The environmental impact of cellulose nanocrystals is another major challenge faced by the industry, despite the fact that nanocellulose is derived from renewable resources like plant fibers, algae, bacteria, and certain agricultural residues and is biodegradable. However, it can still have negative environmental consequences.

Fig. 2 Market outlet of nanocellulose for pulp and paper industry

2.2.1 High Energy Consumption in Production:

Nanocellulose is primarily produced through acid hydrolysis, a costly and energy-intensive process. Nanocellulose production also requires a large amount of energy, particularly for mechanical fibrillation, including homogenization, ultrasonication, and ball milling methods [68], hence, large-scale production may result in high carbon emissions unless renewable energy sources are used.

2.2.2 Waste Generation:

Chemical treatments involved in the production of nanocellulose, such as acid hydrolysis and TEMPO oxidation, lead to environmental concerns [19]. Acid hydrolysis is one of the key processes used for the extraction of nanocellulose from cellulosic raw material. The major disadvantage of this process is the acid wastewater created during the washing process due to the use of sulfuric acid [68]. Tempo oxidation (2266-tetramethylpiperidin-1-yloxyl) produces halogenated oxidizing agents that have a serious impact on aquatic environments, hence limiting its large-scale implementations [82].

2.2.3 Health Hazards

With the increasing use of nanocellulose in a diverse range of industrial applications due to its numerous advantages, including a high surface area, renewable nature, and mechanical strength, there are growing concerns regarding its potential impact on human health, mainly due to increased human exposure. The nanocellulose can result in unique interactions with biological systems due to its nanoscale dimensions [77].

Several studies have shown that research on the toxicity of nanocellulose is still scarce, and knowledge gaps remain. Moreover, studies have often yielded diverging results. For example, a screening-level cycle risk assessment of cellulose nanocrystals (CNCs) and cellulose nanofibers (CNFs) has highlighted the lack of proper long-term exposure studies and data on critical health endpoints, such as carcinogenicity, neurotoxicity, and reproductive effects [73]. These gaps and lack of knowledge create obstacles in drawing firm conclusions regarding the potential impact of nanocellulose on human health.

Although in vivo studies evaluating oral and dermal exposure to cellulose nanocrystals (CNCs) usually do not report harmful effects, findings from pulmonary inhalation and cytotoxicity studies were conflicting. The study also emphasized the need for more extensive and comprehensive studies, focusing on the importance of screening samples for potential contaminants [70].

An extensive review of the impact of nanocellulose on human and environmental health has revealed that, under realistic exposure conditions, nanocellulose has a low toxicity potential; however, its toxic potential may be increased by specific physicochemical properties, although the relationship between structure and biological activity remains unclear [32].

Other studies have also highlighted some major issues, such as a lower number of peer-reviewed publications and a lack of studies on persistent or repeated exposure to nanocellulose, which appears to be a significant concern.

A recent study reviewed inhalation-related risk and estimated the quality of applicable studies. They found that most of the studies were not appropriate for reliable risk estimation. This inadequacy was due to issues such as deficient material characterization, unrealistic dosing, and the absence of proper controls. These studies concluded that short-term inhalation of cellulose nanofibers (CNFs) causes similar effects as poorly soluble, low-toxicity (PSLT) particles. However, long-term, low-dose exposure data remain lacking [31] (Fig. 3).

2.2.4 Inhalation Toxicity

In industrial environments, inhalation is the primary route of exposure to nanocellulose. Due to its small size (barometric dimensions), nanocellulose may remain airborne, especially during manufacturing operations, and thus may be inhaled, resulting in serious respiratory and other health hazards [74].

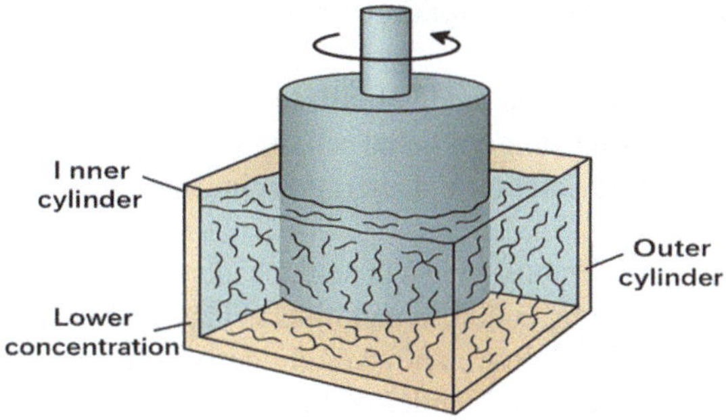

Fig. 3 Schematic representation of a Couette flow setup illustrating wall slip behavior in nanocellulose suspensions

For example, the study demonstrated the consequences of prolonged and repeated exposure to cellulose nanocrystals (CNCs) on the male reproductive system of mice. The CNC was delivered through pharyngeal aspiration, and after three months of final exposure, treated mice showed notable reduction in sperm concentration and motility, increased abnormalities in sperm structure and DNA damage, and indication of oxidative stress and inflammation in the testis and epididymis. These reproductive impairments may be due to higher levels of inflammation, cytokines, and myeloperoxidase (MPO) activities in the testis. Histological analysis shows structural damage to testicular tissue, indicating interstitial swelling and disrupted seminiferous tubules with impaired spermatocytes. Altered levels of testosterone and luteinizing hormone (LH) were also noted, indicating hormonal imbalance. Taken together, these results suggest that pulmonary exposure to cellulose nanocrystals (CNCs) leads to male reproductive toxicity (Furcs et al. 2016).

2.2.5 The Dermal Toxicity of Nanocellulose

Cellulose remains undiscovered. Some in vitro studies suggest that CNCs and CNFs may cause minor skin irritation or inflammation when exposed for a long time and at high concentrations.

For example, the study assesses the potential health and environmental effects of microfibrillated cellulose on exposure to workers. The study on toxicity, conducted by exploring mouse and human microphages, shows no indication of cytotoxicity or inflammation after sex, and a 24-h environmental safety test demonstrates low acute toxicity. Altogether, it is found to be nontoxic to immune cells and may cause low environmental risk [81].

2.2.6 Cytotoxicity and Cellular Uptake

Several studies have demonstrated the internalization of certain types of nanocellulose by human cells, including lung epithelial cells and macrophages, which may lead to cellular stress, decreased viability, and alterations in gene expression, particularly at higher concentrations.

For example, the study demonstrated a dose-dependent reduction in human alveolar epithelial cells induced by cellulose nanofibers. This study assesses the potential respiratory toxicity of cellulose nanowhiskers derived from cotton, utilizing a three-dimensional (3D) cocultured model of the human airway epithelial barrier. This model closely recreates the in vivo environment of the human lung, permitting a more realistic estimation of interactions between nanomaterials at the air–lung interface. The study suggests that surface chemistry, composition, and dimensions impact nanofiber cell interactions. Although cellulose nanocrystals exhibit low toxicity, the observed dose-dependent responses and high bioreactivity are major concerns, demonstrating the importance of toxicological evaluation before large-scale industrial applications [25].

2.2.7 Long-Term Health Risk and Data Gaps

Thesis on chronic exposure to nanocellulose is limited. It is uncertain whether prolonged and repeated exposure to nanocellulose could lead to serious health hazards like fibrosis, cancer, or systemic toxicity. The lack of proper characterization techniques, standard testing protocols, and sufficient toxicological data presents a challenge to regulatory validation and worker safety assurance (Valea et al. 2020).

2.3 Challenges in Industrial-Scale Characterization of Nanocellulose

A major challenge in the large-scale production of nanomaterials like nanocellulose is the absence of rapid, dependable, lucrative characterization tools and techniques that can make certain quality throughout the manufacturing stages [7, 9]. Although numerous methods are present in academic research that are used to analyze various properties of nanocellulose, such as size, shape, nanofiberillation degree, size distribution, mechanical strength, crystallinity, and water retention, these are not yet optimized and adopted for commercial industrial use. Several methods of characterization are carried out at a laboratory scale and are usually time-consuming, complex, and offline, which require highly significant investment. In contrast, the characterization methods for the industrial level need to be less time-consuming, simple, and steady and can be modified for online measurement [33, 42].

Furthermore, there is a lack of standardized protocols, safety standards, and regulations for nanocellulose products [7, 9]. Reliable quality control systems are pivotal to securing consistent product production and commercial viability. Moreover, they must regulate the level of characterization required for specific purposes and applications, using suitable techniques in each instance [8].

Currently, the properties and qualities of nanocellulose cannot be adequately determined due to the absence of proper technologies and rapid, accurate characterization of industrial conditions. The existing characterization method has several limitations and drawbacks, including its adaptation to offline measurements conducted in laboratory facilities, which require complex and time-consuming processes and interpretation. Furthermore, this method requires a high capital investment and highly skilled, trained professionals [8] (Table 1).

2.3.1 Dimensions and Structural Characterization

Characterizing the dimensions and structure of nanocellulose is a challenge, particularly due to the irregular rod-like shape of cellulose nanocrystals (CNCs) and network structure of cellulose nanofibers (CNFs), their high aspect ratio, wide size distributions, and their tendency to agglomerate [7, 9].

Advance microscopic techniques such as transmission electron microscopy (TEM), atomic force microscopy (AFM) and scanning electron microscopy (SEM) are commonly used and provide detailed size and morphology imagery. However, these methods are time-consuming (taking 2 h from sample preparation to image analysis), labor-intensive, prone to user bias, especially during image analysis, and yield unreliable measurements (with accuracy ranging from 50 to 60% depending on the sample due to a lack of correlation with other properties) [7, 9, 60].

Advanced scattering techniques, such as small-angle X-ray scattering (SAXS) and small-angle neutron scattering (SANS), have been explored to determine the cross-sectional dimensions of nanocellulose through statistical modeling of nanocellulose particles as high-aspect-ratio particles (such as rigid rods or parallelepipeds). Hence, it provides an alternative to direct imaging. However, their widespread adoption is limited due to the requirement for specialized equipment and complex data interpretation [8].

Although dynamic light scattering (DLS) technique is most commonly used for the analysis of nanoparticles, it faces several limitations when applied to nanocellulose. Nanocellulose has a high aspect ratio and has different parallel and perpendicular transactional diffusion constants to the particle axis. Hence, the DLS value does not correspond to the particle length or cross-direction. This technique is limited by factors such as dust sensitivity, the particle shape of nanocellulose, and the presence of agglomerates, which can distort hydrodynamic size measurements [34, 59].

Indirect techniques, such as gel-point estimation, can be used to infer the aspect ratio and dimensions of nanocellulose. This method involves tracking the sedimentation behavior of suspension which helps to determine the transaction from dilute to semi dilute phases [69].

Table 1 Main properties and techniques for characterization for nanocellulose identification

Property	Characterization	Acronym
Size and morphology: • Length (L) • Width (w) • Diameter • Aspect ratio (L/w) • Average particle size/size distribution	Atomic force microscopy Scanning electron microscopy Transmission electron microscopy Optical microscope Fiber analyzer Fractionators Gel point (for aspect ratio) Viscosity (for aspect ratio) Light scattering techniques: – Dynamic light scattering – Depolarized DLS	AFM SEM TEM OM DLS DDLS
Physical properties: • Crystallinity (mainly for NCC) • Degree of polymerization (DP) • Specific surface area (SSA) • Density	Crystallinity and dimensions of cellulose crystals: – X-ray diffraction – Raman spectroscopy – Infrared spectroscopy – 13C nuclear X-ray diffraction magnetic resonance – Solid-state nuclear magnetic resonance Viscosity method X Determination of SSA: – N2 adsorption, Brunauer–Emmett–Teller isotherms – Congo red method – Small-angle X-ray scattering Network density (on nanopapers): -Air permeability (Vase de Mariotte, Bendsen) -Porosity (calculation from basis weight and thick)	– XRD – 13C NMR -ssNMR BET SAXS
Elemental analysis	Carbon hydrogen nitrogen elemental analysis Secondary ion mass spectrometry X-ray photoelectron Auger electron spectroscopy Energy dispersive X-ray spectroscopy Inductively coupled plasma: – Mass spectrometry – Atomic emission spectroscopy	CHN CHNS CHNSO CHNSOX SIMS XPS AES EDS (OR EDX) ICP-MS ICP-AES
Water retention	Water retention value	WRV

(continued)

Table 1 (continued)

Property	Characterization	Acronym
Fibrillation degree	Atomic force microscopy Scanning electron microscopy Transmission electron microscopy Optical microscope (for low fibrillation degree) Mechanical fractionation by combination of sieves and membranes Field/tube flow fractionation Centrifugation Transmittance by UV–vis spectroscopy Turbidity Transmittance by UV–vis spectroscopy X X Turbidity	AFM SEM TEM OM
Dissolved and colloidal substances (amount and quality)	Atomic force microscopy Scanning electron microscopy Transmission electron microscopy Gel permeation chromatography Size-exclusion chromatography High-performance liquid chromatography	AFM SEM TEM GPC SEC HPLC
Surface charge and chemistry: • Surface modifications • Charge determination	Surface chemically modified by adsorption: – Fourier-transform infrared spectroscopy – Elemental analysis Surface chemically modified by covalent bonding: – Fourier-transform infrared spectroscopy – Solid-state nuclear magnetic resonance – X-ray photoelectron spectroscopy – Elemental analysis Inverse gas chromatography Conductimetric titration Charge determination: – Cationic demand – Zeta potential	
Rheology: • Viscosity • Yield stress • Gel point •Aspect ratio	Low shear viscosity by viscometers Gel strength, viscoelastic properties by rheometers	

Such techniques offer a simplified approach (used for microfibrilled cellulose); hence, their applicability for highly fibrillated nanocellulose is limited and requires further clarification.

In spite of all the research endeavor, there is still no effective and standard method available for measuring size and fibrillation degree of nanocellulose. Further, there is no method available that can estimate the dispersion–aggregation state of nanocellulose at the industrial scale [59].

2.3.2 Rheology and Thermal Stability

Nanocellulose hydrogen causes various challenges in industrial applications due to their high viscosity and complex flow behavior [40]. The rheological properties of nanocellulose suspensions are primarily governed by the interactions between the nanofibers themselves and with the surrounding water matrix. Dispersion quality, morphology, concentration, and surface chemistry of the nanocellulose are the major characteristics on which this interaction are dependent [35].

Nanocellulose suspensions may exhibit one of two rheological behaviors that is either plastic or Newtonian behavior depending on concentration and fibrillar network structure. Nanocellulose suspensions above the critical concentration threshold and with intact network structure behave as hydrogels with plastic characteristics. On the other hand, suspensions below the threshold concentration or with a disrupted network structure (due to shear forces) may behave as Newtonian fluids [45, 50]. Many researchers have presented an indirect method for characterizing nanocellulose using rheological properties, due to the correlation between morphology, degree of nanofibrillation, and viscosity of the suspensions.

Several issues complicate the rheological characterization of nanocellulose suspensions; some of which are listed below.

Wall Slip

Strong interactions between fibrils and smooth measurement surfaces create a lubricating liquid layer with dramatically lower viscosity than bulk viscosity at low shear rates, leading to wall slip. This allows for accurate readings of viscosity, as only the water layer close to the wall flows, while the bulk of the suspension remains stationary [14, 40] (Fig. 4).

Shear Bending

Laminar shear stress at low viscosity conditions results in the formation of Taylor vortices, which may cause uneven distribution of fibrils, causing zones of differing

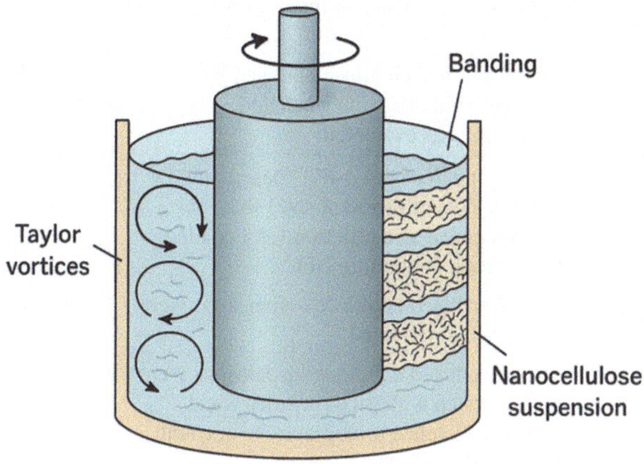

Fig. 4 An image depiction of nanocellulose suspended material at the walls (wall slip) when suspensions are exposed to laminar shear flow especially when the cylinder surfaces are smooth and there are no vanes experienced into the annual region

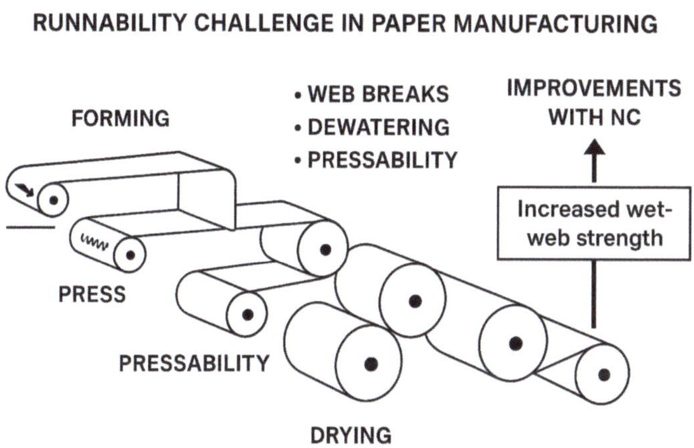

Fig. 5 A schematic diagram of coquette transitional flow in rotating viscometer, showing formation of Taylor vortices and tendency for formation of bands enriched with suspended particles

solid concentrations under shear; the network disintegrates, resulting in fiber flocculation and release of free water between flocks and hence altering rheological properties and disrupting the measurement accuracy [7, 9, 40] (Fig. 5).

Table 2 Publications demonstrating the use of rheological measurements as a means of characterizing nanocellulose

Bucket-vane viscosity	Nanofibrillated cellulose obtained by grinding, with TEMPO-mediated oxidation, was evaluated	0.25 to 1%	[63]
Apparent viscosity	A low "apparent viscosity" (based on flow through slots of different size) was recommended as a criterion for a coating formulation containing NFC	–	Kumar et al. 2016
Pipe rheometer	Microfibrillated cellulose suspensions were monitored at lab scale	0.2 to 1.5%	Haavisto et al. 2015
Pipe rheometer and optical coherence	Microfibrillated cellulose suspensions were further tested with pipe rheometry and optical coherence tomography to account for wall slip effects	0.2 to 1.5%	Haavisto et al. 2015
Rolling ball	Dilute aqueous CNC could be evaluated with just 1 ml of suspension with an automated test	0.06 to 2%	Gonzalez-Labrada and Gray 2012
Rotational rheometer (Paar)	Nanofibrillated cellulose quality was monitored using viscoelastic and steady-state measurements	1 to 15%	Gruneberger et al. 2014 [28]
Other than viscosity	Though these authors considered viscosity test methods, they found reliable results only for turbidity, centrifugation, and light transmittance	0.01 to 0.2%	Moser et al. 2015

Fiber Alignment

Fiber may align in the direction of flow under prolonged shear or relatively high solid concentrations, leading to rheopectic behavior, that is, increased viscosity of fluid with time under constant shear. This results in the occurrence of various phenomena like shear thinning, preferential orientation, formation of flocks, and entangled groups of fibers.

Thixotropic and Rheopexy

Nanocellulose suspensions show time-dependent viscosity changes in fluid under shear stress. Nanocellulose suspension exhibits thixotropic behavior when viscosity decreases over time under shear, while rheopexy occurs when viscosity increases overtime under shear stress, thus complicating reproducibility [11, 41] (Table 2).

2.4 Runnability Problems

In the context of paper production, runnability describes the ability of machinery to operate at full capacity while consistently producing high-quality paper. Ensuring optimal machine performance is, therefore, essential for efficient papermaking operations [16]. Several studies have investigated the impact of nanocellulose (NC) on the runnability of paper machines [21, 44, 56].

For example, [44] assessed the feasibility of nanofibrillated cellulose (NFC) as a strength additive through trials on a high-speed pilot paper machine. The runnability was observed to be good during trials with no web breaks in terms of paper strength or water removal issues. The levels of retention were consistent, and sheet formation was satisfactory, requiring no additional optimization. Additionally, water removal during the forming and pressing stages was within acceptable levels [44]. Lu et al. [56] centered their research on base paper pressability and wet-web strength and their effect on sheet break rates and machine runnability in general. According to their findings, the addition of 5% NFC enhanced tensile energy absorption from 6.32 to 10.93 J/m^2. They concluded that the addition of cellulose nanofibers (CNFs) to papermaking improves wet-web strength and, consequently, machine runnability. Likewise, Charani and Moradian [21] determined that incorporating a lower concentration of cationic starch (0.5%) blended with 3% NFC significantly enhanced the breaking length of the paper and minimized process issues associated with starch utilization [21]. In another study, incorporation of 3%, 6%, and 9% NFC into hand sheets led to a linear increase in the tensile index, up to approximately 55 N·m/g at 35°SR, which was determined to be appropriate for machine runnability [80]. Ankerfors et al. [4] conducted runnability and strength tests of paper as influenced by microfibrillated cellulose (MFC) on the FEX pilot paper machine. Their tests did not reveal any severe runnability problems. They noted, however, that MFC influenced dewatering, especially in the wire section. The slowed drainage rate required lower machine speeds and the use of countermeasures to mitigate this effect. Methods such as the use of retention and drainage aids were recommended to mitigate the slowed drainage and enhance the retention of NC particles and fines, thereby improving overall machine runnability [43]. Several research studies have also investigated the application of NC in paper coating [6, 23, 37, 51]. The water release resistance of the NC suspension into the substrate is one of the key parameters that determine the runnability of the coating. Success in wet film formation and the coating process depends heavily on the retention or release of water. The sudden release of water from MFC suspensions causes swelling and debonding of the fibers, leading to process issues such as web breaks [51]. Thus, a good understanding and control of NC suspension water retention behavior are necessary for successful coating applications [49]. Furthermore, the high water content of NFC coatings may lead to expensive drying, which only pays off if the resulting improvement in product properties covers the expense [17]. Drying also increases the concentration of NC, with implications for drying dynamics and the assembly process [83]. The other major runnability problem is the problem of depositing NC suspensions onto substrates with standard coating

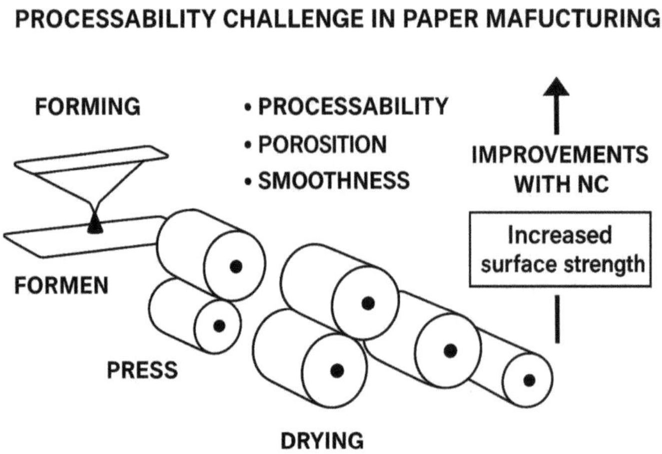

Fig. 6 Illustration of runnability challenges in papermaking

and metering equipment. This is due to clogging caused by fiber agglomeration and the naturally high viscosity of NC suspensions [49]. NFC is also associated with high viscosity at low solid content, which reduces the coating operating parameters [17]. The management of NFC-based coating formulation rheological properties is, therefore, a significant challenge. Reducing the water content at the same uniform coating demands stringent control of the properties [17]. Alterations in nip pressure, blade angle, or coating pressure are, therefore, usually necessary when NFC is incorporated into the formulation [72] (Fig. 6).

2.5 *Processability Issues*

See Fig. 7.

2.5.1 Dewatering

Nanocellulose (NC) still needs to gain extensive use in papermaking, primarily due to challenges in its practical application [7, 9]. Overall, nanofibrillated cellulose (NFC) retards drainage, whereas papermaking requires rapid dewatering to achieve low energy usage and production costs [39]. Some of the reasons that have been found for retarding dewatering include formation of dense layers, blocking of drainage channels by cellulose particles in soaked web, and poor vacuum response due to fiber flocculation. A self-healing property repairs thin regions in the paper sheet. A rewet surface sheet membrane was created when it comes into contact with the wet-press felt [41]. In water dispersion, NFC creates a strongly entangled, gel-like

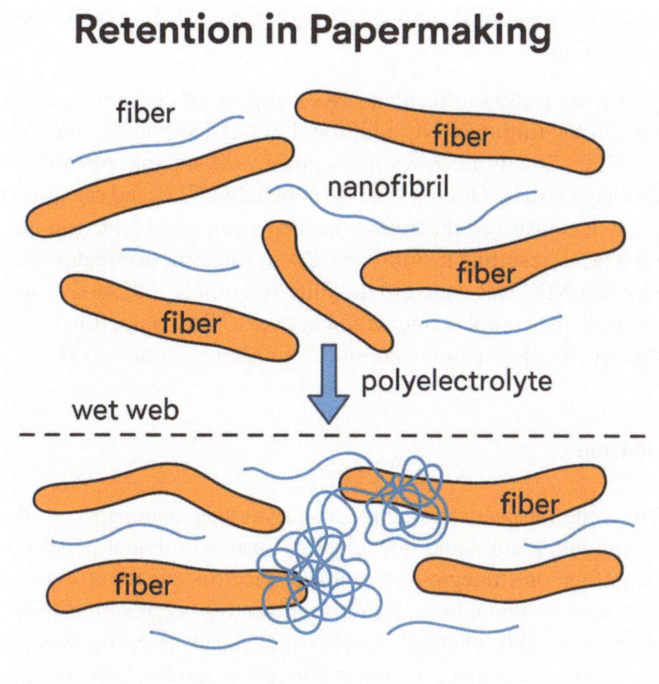

Fig. 7 Illustration of processability challenges in papermaking

network at as low as 1% of solid concentrations. The surface charge of NFC has a significant influence on both the rheological properties and dewatering behavior, as it governs fiber swelling and water retention. Fiber wall swelling reduces the frequency of fiber–fiber contacts, which in turn reduces floc size and frictional forces, ultimately altering rheology and flow-driven dewatering behavior [28]. Dewatering performance is also consistent with the shear-thinning behavior of the furnish. When shear destroys the structure of the suspension, water passes more easily through the sheet. Several cycles of shear, therefore, can enhance dewatering efficiency [28].

Some research suggests that cationic NFC (CNFC) can enhance wet-web strength and dewatering [57]. Diab et al. [27] did, however, observe that cationic microfibrillated cellulose (CMFC) failed to enhance drainage and suggested further tests on more highly cationized CMFC to determine its potential as a drainage aid. Lu et al. [57] suggested that CNFC enhances fine retention and sheet density for a multitude of reasons:

- Its nanoscale nature provides a large surface area that effectively traps fines in the pulp.
- Its high aspect ratio makes it capable of entangling fines and creating CNFC-fine networks.

- Its high surface charge, due to quaternary ammonium groups, facilitates the adsorption of fines.

Other significant factors influencing NC's impact on dewatering are the type of retention aid and the timing of NC addition. Li et al. [53] opined that adding small quantities of NFC directly to the pulp together with suitable retention agents can minimize drainage issues. Ottesen et al. [67] blended NFC and retention chemicals with fillers and demonstrated that retention agents enhanced NFC and filler bonding to available surfaces, resulting in improved dewatering. These effects were even more pronounced when NFC was added to the filler fraction before mixing with the bulk pulp [67]. Overall, the complex interaction between NFC, pulp, fillers, and additives must be properly investigated to overcome drainage constraints [53].

2.5.2 Retention

Although NC's high aspect ratio helps create stronger nanopaper, it also presents challenges in agglomeration and flocculation. From a colloidal perspective, one of the common mitigation strategies is to enhance electrostatic repulsion by increasing the surface charge of nanofibrils, thereby preventing aggregation [52]. In papermaking, however, weakly charged small particles can be easily washed away by shear and turbulence, lowering retention efficiency during sheet formation. Better retention is primarily achieved through charge neutralization and retention agent-induced flocculation [1], indicating a necessary balance between flocculation and retention. To enhance retention and reduce drainage time, polyelectrolytes are typically used. These bind fines to fibers via flocculation, trapping them within the wet web. In this way, polyelectrolytes not only contribute to the retention of fines and fillers but also influence the reinforcing character of NFC [39]. The action of retention agents determines the manner of association and adsorption of fibrils during papermaking and the formed sheet structure. Handsheets made by using polyvinyl amine as a retention additive were less dense, for example, than handsheets produced in the absence of an additive. This may be because fibrils become attached to the fiber surface before dewatering instead of filling voids during drainage [38]. Thus, the interactions between NFC and retention additives and optimal dosing play an important role [61].

Taipale et al. [79] reported that good operating conditions and materials could enhance paper strength without affecting drainage. In particular, they had reported that a thin NC layer could be created on adsorbed cationic starch, covering the fibers instead of filling voids. Merayo et al. [61] again tested different retention additives—polyvinylamine, chitosan (CH), cationic starch (CS), and two cationic polyacrylamides (C-PAM and C-PAM-B)—to enhance the strength of recycled paper and minimize drainage problems. According to their findings, C-PAM and C-PAM-B were the strongest at low dosages, followed by CH. While polyvinylamine was also very effective, it needed a much higher dose. CS was the weakest, particularly at lower dosages.

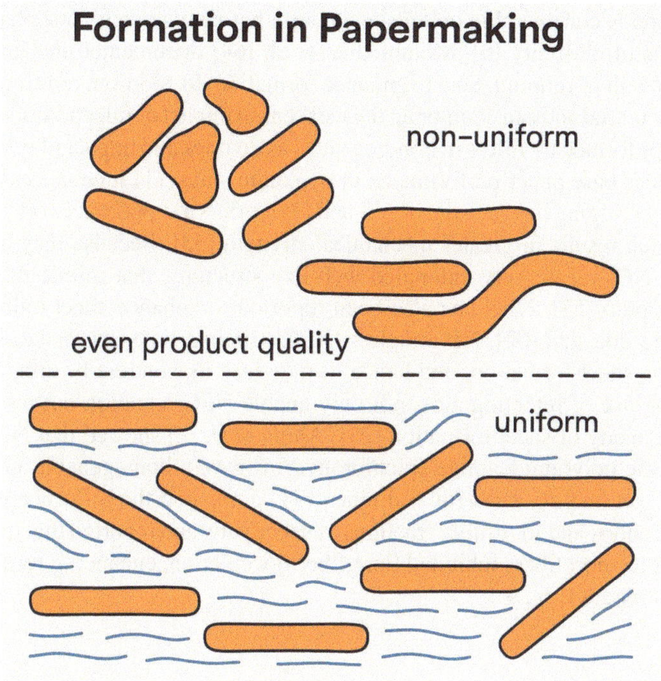

Fig. 8 Illustration of retention in papermaking

These results indicate that retention aids influence the interaction between fibrillated materials and pulp fibers, which, in turn, affects paper strength and drainage [84]. It is also proposed that a single retention agent may not be sufficient, therefore, designing dual-component retention systems may be more effective [55]. As retention aids, NFC's high surface area allows it to flocculate cationic particles by bridging. Once shear disintegrates these flocs, secondary retention agents can cause reflocculation by patching or network reforming. Still, chemical modification of NFC is usually required to enhance its electrostatic compatibility with other system components [48] (Fig. 8).

2.5.3 Formation

The formation, or formation uniformity, refers to uniformity in the distribution of fibers on the paper surface [65]. It is quantified by measuring the distribution of mass in the sheet. The objective is to achieve a balanced spread of pulp on the wire, ensuring even product quality. As explained by Dodson and Serafino [29], fiber flocculation and dispersion are conflicting forces in papermaking. A uniform structure is critical to visual quality and mechanical performance, as strength derives from fiber bonding during formation, consolidation, and drying [78]. The impact of formation is considerable on tensile strength but not as much on such properties as modulus of elasticity [5]. Motamedian et al. [64] documented that finer fibers, being the result of refining, tend to enhance formation. In addition, refining increases bonding potential and can counteract the weakening impact of filler addition [4]. NFC can fill gaps formed by fillers in bonding sites, as do fines and impact fiber matrix [3]. NC enhances base paper performance in two main ways: (1) It enhances inter-fiber bonding by bridging between the fibers, and (2) it offers a support network among big fibers, which results in greater mechanical strength [53]. Because they are fibrous in nature, NCs create very entangled web-like structures that mimic the behavior of regular pulp [53]. NFC has also been reported to enhance sheet uniformity by minimizing floc size [62]. Nevertheless, NFC's fine particle size makes it difficult to retain on sheet formation, and hence, it may lose its bonding benefit. Solutions, such as the use of retention aids, not only enable NFC retention but also improve the homogeneity of sheet formation [27]. Ahola et al. [2] showed that NFC mixing with cationic poly(amideamine epichlorohydrin) gave a homogeneous coating that improved wet and dry strength. Salminen [71] compared the influence of C-PAM and CS on short- and long-fiber fractions, respectively, and reported that the addition of C-PAM to short fibers inhibited long-fiber flocculation, enhancing formation and wet-web strength (Fig. 9).

3 Conclusion

Nanocellulose (NC), especially cellulose nanofibers (CNFs), has attracted significant attention as a functional additive to improve the quality and performance of paper products. CNFs offer unique properties, such as high surface area, tunable surface chemistry, and excellent mechanical reinforcement, which enhance paper strength and density. They are especially beneficial in compensating for the loss of mechanical properties caused by mineral fillers in paper formulations. CNFs have shown reinforcing effects on various pulp types, with chemical pulps demonstrating the most notable improvements. Beyond reinforcement, CNFs are being increasingly explored as coating agents to enhance barrier properties, including water and gas permeability, thereby offering opportunities for sustainable packaging. However, despite their potential, widespread commercial use of CNFs in papermaking is limited by key challenges. These include the high cost of CNF production, the negative impact on drainage rates, and increased energy demands due to their high water retention.

To address these barriers, researchers are investigating green surface modification techniques, such as covalent bonding, noncovalent hydrophobization, plasma etching, and cationic functionalization. These strategies can enhance CNF compatibility, water resistance, and flocculation, improving paper strength while maintaining drainage performance. Nonetheless, most of these modifications are still confined to laboratory-scale demonstrations. Industrial-scale application remains difficult due to high production costs, complex rheological behavior, and integration challenges with existing papermaking processes. Further work is needed to develop cost-effective

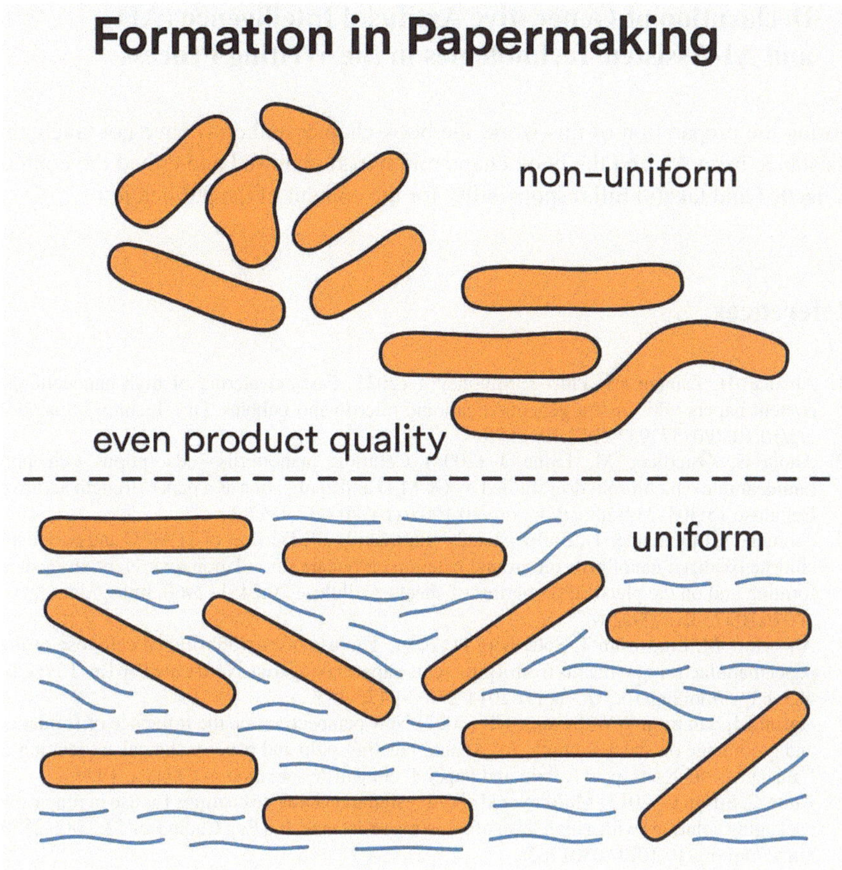

Fig. 9 Illustration of formation in papermaking

CNF production methods, optimize dosages, and balance enhancements in paper properties without compromising opacity or porosity.

In summary, CNFs hold considerable promise for improving the sustainability and functionality of paper products. However, addressing the economic, technical, and scalability challenges is crucial for their successful industrial adoption.

4 Competing Interests

The authors declare that they have no competing interest.

5 Declaration of Generative Artificial Intelligence (AI) and AI-Assisted Technologies in the Writing Process

During the preparation of this work, the book chapter author(s) have not taken any assistance from AI, and the book chapter author(s) reviewed and edited the content as needed and take(s) full responsibility for the content of the publication.

References

1. Ahadian H, Zamani ES, Phiri J, Maloney T (2021) Fast dewatering of high nanocellulose content papers with in-situ generated cationic micro-nano bubbles. Dry Technol https://doi.org/10.1080/07373937.2021.1942898
2. Ahola S, Österberg M, Laine J (2008) Cellulose nanofibrils—adsorption with poly (amideamine) epichlorohydrin studied by QCM-D and application as a paper strength additive. Cellulose 15:303–314. https://doi.org/10.1007/s10570-007-9167-3
3. Ämmälä A, Liimatainen H, Burmeister C, Niinimäki J (2013) Effect of TEMPO and periodate-chlorite oxidized nanofibrils on ground calcium carbonate flocculation and retention in sheet forming and on the physical properties of sheets. Cellulose 20:2451–2460. https://doi.org/10.1007/s10570-013-0012-6
4. Ankerfors M, Lindström T, Söderberg D (2014) The use of microfibrillated cellulose in fine paper manufacturing—results from a pilot scale papermaking trial. Nord Pulp Pap Res J 29:476–483. https://doi.org/10.3183/npprj-2014-29-03-p476-483
5. Anson SJ, Sampson WW, Sevajee CR (2007) New perspectives on the influence of formation and grammage on sheet strength. In: Annual meeting-pulp and paper technical association of Canada, vol 93 No B, p 141. Pulp and Paper Technical Association of Canada; 1999
6. Aulin C, Ström G (2013) Multilayered alkyd resin/nanocellulose coatings for use in renewable packaging solutions with a high level of moisture resistance. Ind Eng Chem Res 52:2582–2589. https://doi.org/10.1021/ie301785a
7. Balea A, Fuente E, Monte MC, Merayo N, Campano C, Negro C, Blanco A (2020a) Industrial application of nanocelluloses in papermaking: a review of challenges, technical solutions, and market perspectives. Molecules 25(3):526
8. Balea A, Blanco A, Delgado-Aguilar M, Monte MC, Tarrés Q, Fuente E, Mutje P, Negro C (2021) Nanocellulose characterization challenges. BioResources 16(2):4382
9. Balea A, Monte MC, Merayo N, Campano C, Negro C, Blanco A (2020b) Industrial application of nanocelluloses in papermaking: a review of challenges, technical solutions, and market perspectives. Molecules 25(3):526. https://doi.org/10.3390/molecules25030526
10. Balea A, Merayo N, De La Fuente E, Negro C, Blanco Á (2017) Assessing the influence of refining, bleaching and TEMPO-mediated oxidation on the production of more sustainable cellulose nanofibers and their application as paper additives. Ind Crops Prod 97:374–387
11. Barnes HA (1997) Thixotropy—a review. J Nonnewton Fluid Mech 70(1–2):1–33
12. Basta AH, El-Saied H (2009) Performance of improved bacterial cellulose application in the production of functional paper. J Appl Microbiol 107:2098–2107
13. Bharimalla AK, Deshmukh SP, Patil PG, Vigneshwaran N (2015) Energy efficient manufacturing of nanocellulose by chemo-and bio-mechanical processes: a review. World J Nano Sci Eng 5:204
14. Blanco A, Negro C, Fuente E, Tijero J (2007) Rotor selection for a searle-type device to study the rheology of paper pulp suspensions. Chem Eng Process 46(1):37–44
15. Blanco A, Monte MC, Campano C, Balea A, Merayo N, Negro C (2018) Nanocellulose for industrial use: cellulose nanofibers (CNF), cellulose nanocrystals (CNC), and bacterial cellulose (BC). In: Handbook of nanomaterials for industrial applications. Elsevier, pp 74–126

16. Bondoc, E. (2018). Runnability-the condition monitoring perspective. App Technol Innov Manuf Environ 71(2):114–119
17. Brodin FW, Gregersen ØW, Syverud K (2014) Cellulose nanofibrils: challenges and possibilities as a paper additive or coating material–a review. Nord Pulp Pap Res J 29:156–166. https://doi.org/10.3183/npprj-2014-29-01-p156-166
18. Brännvall E (2017) The limits of delignification in kraft cooking. BioResources 12:2081–2107
19. Cai C, Wang G, Bai X, Xu D, Yan C, Yang Y (2024) Optimizing carboxylated nanocellulose preparation: a kinetic and mechanistic study on the enhancement of TEMPO-mediated oxidation via swelling treatment. Int J Biol Macromol 274:133342
20. Campano C, Merayo N, Negro C, Blanco A (2018) Low-fibrillated bacterial cellulose nanofibers as a sustainable additive to enhance recycled paper quality. Int J Biol Macromol 114:1077–1083
21. Charani PR, Moradian MH (2019) Utilization of cellulose nanofibers and cationic polymers to improve breaking length of paper. Cellul Chem Technol 53:767–774. https://doi.org/10.35812/CelluloseChemTechnol.2019.53.75
22. Chen GQ, Wu GC, Alriksson B, Wang W, Hong FF, Jonsson LJ (2017) Bioconversion of waste fiber sludge to bacterial nanocellulose and use for reinforcement of CTMP paper sheets. Polymers 9:14
23. Chowdhury RA, Nuruddin M, Clarkson C, Montes F, Howarter J, Youngblood JP (2019) Cellulose nanocrystal (CNC) coatings with controlled anisotropy as high-performance gas barrier films. ACS Appl Mater Interfaces 11:1376–1383. https://doi.org/10.1021/acsami.8b16897
24. Çiçekler M, Akın M (2025) Advances and challenges in the integration of cellulose nanocrystals in paper manufacturing
25. Clift MJ, Foster EJ, Vanhecke D, Studer D, Wick P, Gehr P, Weder C (2011) Investigating the interaction of cellulose nanofibers derived from cotton with a sophisticated 3D human lung cell coculture. Biomacromol 12(10):3666–3673
26. Das AK, Islam MN, Ashaduzzaman M, Nazhad MM (2020) Nanocellulose: its applications, consequences and challenges in papermaking. J Pack Technol Res 4:253–260
27. Diab M, Curtil D, El-shinnawy N, Hassan ML, Zeid IF, Mauret E (2015) Biobased polymers and cationic microfibrillated cellulose as retention and drainage aids in papermaking: COMPARISON between softwood and bagasse pulps. Ind Crops Prod 72:34–45. https://doi.org/10.1016/j.indcrop.2015.01.072
28. Dimic-Misic K, Puisto A, Paltakari J, Alava M, Maloney T (2013) The influence of shear on the dewatering of high consistency nanofibrillated cellulose furnishes. Cellulose 20:1853–1864. https://doi.org/10.1007/s10570-013-9964-9
29. Dodson CTJ, Serafino L (1993) Flocculation, dispersion and dynamic scenarios for formation. Nord Pulp Pap Res J 8(2):264–272
30. Dufresne A (2017) Nanocellulose: from nature to high. performance tailored materials. Walter de Gruyter, GmgH & CoKG, Berlin, Germany
31. Ede JD, Ong KJ, Goergen M, Rudie A, Pomeroy-Carter CA, Shatkin JA (2019) Risk analysis of cellulose nanomaterials by inhalation: current state of science. Nanomaterials 9(3):337
32. Endes C, Müller S, Schmid O, Vanhecke D, Foster EJ, Petri-Fink A, Clift MJ (2013) Risk assessment of released cellulose nanocrystals–mimicking inhalatory exposure. In: Journal of physics: conference series, vol 429, No 1, p 012008. IOP Publishing
33. Foster EJ, Moon RJ, Agarwal UP, Bortner MJ, Bras J, Camarero-Espinosa S, Chan KJ, Clift MJ, Cranston ED, Eichhorn SJ (2018) Current characterization methods for cellulose nanomaterials. Chem Soc Rev 47:2609–2679
34. Fraschini C, Chauve G, Le Berre JF, Ellis S, Méthot M, O'Connor B, Bouchard J (2014) Critical discussion of light scattering and microscopy techniques for CNC particle sizing. Nord Pulp Pap Res J 29(1):31–40
35. Gong G (2014) Rheological properties of nanocellulose materials. In: Handbook of green materials: 1 Bionanomaterials: separation processes, characterization and properties. pp 139–157

36. Habibi Y, Lucia LA, Rojas OJ (2010) Cellulose nanocrystals: chemistry, self-assembly, and applications. Chem Rev 110:3479–3500
37. He Y-Q, Li H, Fei X, Peng L-C (2021) Carboxymethyl cellulose/cellulose nanocrystals immobilized silver nanoparticles as an effective coating to improve barrier and antibacterial properties of paper for food packaging applications. Carbohydr Polym 252:117156. https://doi.org/10.1016/j.carbpol.2020.117156
38. Hollertz R, Durán VL, Larsson PA, Wågberg L (2017) Chemically modified cellulose micro- and nanofibrils as paper-strength additives. Cellulose 24:3883–3899. https://doi.org/10.1016/j.ijbiomac.2018.11.214
39. Hu FG, Zeng J-S, Cheng Z, Wang X-J, Wang B, Zeng Z-T, Chen K-F (2021) Cellulose nanofibrils (NFC) produced by different mechanical methods to improve mechanical properties of recycled paper. Carbohydr Polym 254:117474. https://doi.org/10.1016/j.carbpol.2020.117474
40. Hubbe MA, Tayeb P, Joyce M, Tyagi P, Kehoe M, Dimic-Misic K, Pal L (2017) Rheology of nanocellulose-rich aqueous suspensions: a review. BioResources 12(4):9556–9661
41. Hubbe MA, Sjöstrand B, Nilsson L, Koponen A, McDonald JD (2020) Rate-limiting mechanisms of water removal during the formation, vacuum dewatering, and wet-pressing of paper webs: a review. BioResources 15:9672–9755. https://doi.org/10.15376/biores.15.4.Hubbe
42. Ioelovich M (2017) Characterization of various kinds of nanocellulose. In: Handbook of nanocellulose and cellulose nanocomposites, vol 1. pp 51–100
43. Jin L-Q, Wei Y-W, Xu Q-H, Yao W-R, Cheng Z-L (2014) Cellulose nanofibers prepared from TEMPO-oxidation of kraft pulp and its flocculation effect on kaolin clay. J Appl Polym Sci 131. https://doi.org/10.1002/app.40450
44. Kajanto I, Kosonen M (2012) The potential use of micro-and nanofibrillated cellulose as a reinforcing element in paper. J Sci Technol For Prod Proc 2(6):42–48
45. Karppinen A, Saarinen T, Salmela J, Laukkanen A, Nuopponen M, Seppälä J (2012) Flocculation of microfibrillated cellulose in shear flow. Cellulose 19:1807–1819
46. Khalil H, Davoudpour Y, Islam MN, Mustapha A, Sudesh K, Dungani R, Jawaid M (2014) Production and modification of nanofibrillated cellulose using various mechanical processes: a review. Carbohydr Polym 99:649–665
47. Klemm D, Kramer F, Moritz S, Lindström T, Ankerfors M, Gray D, Dorris A (2017) Nanocelluloses: A New family of nature-based materials. Angew. Chem. Int. Ed. 2011, 50, 5438–5466. [CrossRef] [PubMed]
48. Korhonen M, Laine J (2014) Flocculation and retention of fillers with nanocelluloses. Nord Pulp Pap Res J 29:119–128. https://doi.org/10.3183/npprj2014-29-01-p119-128
49. Kumar V (2018) Roll-to-roll processing of nanocellulose into coatings [Ph.D. Dissertation]. Åbo Academy University
50. Lasseuguette E, Roux D, Nishiyama Y (2008) Rheological properties of microfibrillar suspension of TEMPO-oxidized pulp. Cellulose 15:425–433
51. Lengowski EC, Bonfatti Júnior EA, Kumode MMN, Carneiro ME, Satyanarayana KG (2019) Nanocellulose in the paper making. In: Sustainable polymer composites and nanocomposites. Springer. https://doi.org/10.1007/978-3-030-05399-4_36
52. Li M-C, Wu Q-L, Song K-L, Lee S-Y, Qing Y, Wu Y-Q (2015) Cellulose nanoparticles: structure–morphology–rheology relationships. ACS Sustain Chem Eng 3:821–832. https://doi.org/10.1021/acssuschemeng.5b00144
53. Li A, Xu D-Z, Luo L, Zhou Y, Yan W, Leng X, Dai D-S, Zhou Y-H, Ahmad H, Jiuping Rao J-P (2021) Overview of nanocellulose as additives in paper processing and paper products. Nanotechnol Rev 10:264–281. https://doi.org/10.1515/ntrev-2021-0023
54. Lim G-H, Lee J, Kwon N, Bok S, Sim H, Moon K-S, Lee S-E, Lim B (2016) Fabrication of flexible magnetic papers based on bacterial cellulose and barium hexaferrite with improved mechanical properties. Electron Mater Lett 12:574–579
55. Lourenço AF, Godinho D, Gamelas JAF, Sarmento P, Ferreira PJT (2019) Carboxymethylated cellulose nanofibrils in papermaking: influence on filler retention and paper properties. Cellulose 26:3489–3502. https://doi.org/10.1007/s10570-019-02303-5

56. Lu Z-H, An X-Y, Zhang H, Guan M, Liu J, Sun Y-W, Nie S-X, Cao H-B, Lu B, Liu H-B (2019) Study on the wet-web strength and pressability of paper sheet during the press process with the addition of nano-fibrillated cellulose (NFC). Carbohyd Polym 210:332–338. https://doi.org/10.1016/j.carbpol.2019.01.083
57. Lu Z-H, An X-Y, Zhang H, Liu L-Q, Dai H-Q, Cao H-B, Lu B, Liu H-B (2020) Cationic cellulose nano-fibers (CCNF) as versatile flocculants of wood pulp for high wet web performance. Carbohyd Polym 229:115434. https://doi.org/10.1016/j.carbpol.2019.115434
58. Mariano M, El Kissi N, Dufresne A (2014) Cellulose nanocrystals and related nanocomposites: review of some properties and challenges. J Polym Sci Part B Polym Phys 52(12):791–806
59. Martín-Sampedro R, Eugenio ME, Ibarra D, Ruiz-Hitzky E, Aranda P, Darder M (2022) Tailoring the properties of nanocellulose-sepiolite hybrid nanopapers by varying the nanocellulose type and clay content. Cellulose 29(9):5265–5287
60. Marway H (2017) Investigation of nanocellulose mechanical properties and interactions in salt and surfactant solutions measured by atomic force microscopy (Doctoral dissertation)
61. Merayo N, Balea A, de la Fuente E, Blanco A, Carlos Negro C (2017b) Synergies between cellulose nanofibers and retention additives to improve recycled paper properties and the drainage process. Cellulose 24:2987–3000. https://doi.org/10.1007/s10570-017-1302-1
62. Merayo N, Balea A, de la Fuente E, Blanco A, Carlos Negro C (2017a) Interactions between cellulose nanofibers and retention systems in flocculation of recycled fibers. Cellulose 24:677–692. https://doi.org/10.1007/s10570-016-1138-0
63. Mohtaschemi M, Dimic-Misic K, Puisto A, Korhonen M, Maloney T, Paltakari J, Alava MJ (2014) Rheological characterization of fibrillated cellulose suspensions via bucket vane viscometer. Cellulose 21:1305–1312
64. Motamedian HR, Armin E. Halilovic AE, Kulachenko A (2019) Mechanisms of strength and stiffness improvement of paper after PFI refining with a focus on the effect of fines. Cellulose 26(6):4099–4124
65. Na YJ, Muhlstein CL (2019) Relating nonuniform deformations to fracture in uniaxially loaded non-woven fiber networks. Exp Mech 59:1127–1144. https://doi.org/10.1007/s11340-019-00527-x
66. Nishiyama Y (2014) Structure and physical properties of cellulose: micro-to nanoscale. In: Handbook of green materials: 1 Bionanomaterials: separation processes, characterization and properties. pp 5–17
67. Ottesen V, Syverud K, Gregersen ØW (2016) Mixing of cellulose nanofibrils and individual furnish components: effects on paper properties and structure. Nord Pulp Pap Res J 31(3):441–447
68. Phanthong P, Reubroycharoen P, Hao X, Xu G, Abudula A, Guan G (2018) Nanocellulose: extraction and application. Carbon Resourc Convers 1(1):32–43
69. Raj P, Batchelor W, Blanco A, de la Fuente E, Negro C, Garnier G (2016) Effect of polyelectrolyte morphology and adsorption on the mechanism of nanocellulose flocculation. J Colloid Interface Sci 481:158–167
70. Roman M (2015) Toxicity of cellulose nanocrystals: a review. Ind Biotechnol 11(1):25–33
71. Salminen K (2010) The effects of some furnish and paper structure related factors on wet web tensile and relaxation characteristics. Lappeenranta University of Technology
72. Sharma PR, Sharma SK, Lindström T, Hsiao BS (2020) Nanocellulose-enabled membranes for water purification: Perspectives. Adv Sustain Sys 4(5):1900114
73. Shatkin JA, Kim B (2015) Cellulose nanomaterials: life cycle risk assessment, and environmental health and safety roadmap. Environ Sci Nano 2(5):477–499
74. Shatkin JA, Kim B (2017) Environmental health and safety of cellulose nanomaterials and composites. Handb Nanocellul Cellul Nanocompos 2:683–729
75. Skoˇcaj M (2019) Bacterial nanocellulose in papermaking. Cellulose 26:6477–6488
76. Spence KL, Venditti RA, Rojas OJ, Habibi Y, Pawlak JJ (2011) A comparative study of energy consumption and physical properties of microfibrillated cellulose produced by different processing methods. Cellulose 18:1097–1111

77. Stoudmann N, Schmutz M, Hirsch C, Nowack B, Som C (2020) Human hazard potential of nanocellulose: quantitative insights from the literature. Nanotoxicology 14(9):1241–1257
78. Su J-L, Mosse WKJ, Sharman S, Batchelor W, Garnier G (2012) Paper strength development and recyclability with polyamideamine-epichlorohydrin (PAE). BioResources 7(1):913–924
79. Taipale T, Österberg M, Nykänen A, Ruokolainen J, Laine J (2010) Effect of microfibrillated cellulose and fines on the drainage of Kraft pulp suspension and paper strength. Cellulose 17:1005–1020. https://doi.org/10.1007/s10570-010-9431-9
80. Vallejos ME, Felissia FE, Area MC, Ehman NV, Tarrés Q, Mutjé P (2016) Nanofibrillated cellulose (CNF) from eucalyptus sawdust as a dry strength agent of unrefined eucalyptus handsheets. Carbohyd Polym 139:99–105. https://doi.org/10.1016/j.carbpol.2015.12.004
81. Vartiainen J, Pöhler T, Sirola K, Pylkkänen L, Alenius H, Hokkanen J, Tapper U, Lahtinen P, Kapanen A, Putkisto K, Hiekkataipale P (2011) Health and environmental safety aspects of nanofibrillated cellulose. Nanocon. 5p
82. Vijay P, Batchelor W, Saito K (2022) One-pot treatment of cellulose using iron oxide catalysts to produce nanocellulose and water-soluble oxidised cellulose. Carbohyd Polym 282:119060
83. Wang Q-Q, Yao Q, Liu J, Sun JZ, Zhu QQ, Chen H-L (2019) Processing nanocellulose to bulk materials: a review. Cellulose 26:7585–7617. https://doi.org/10.1007/s10570-019-02642-3
84. Zambrano F, Starkey H, Wang Y-H, Abbati de Assis C, Venditti R, Pal L, Jameel H, Hubbe MA, Rojas OJ, Gonzalez R (2020) Using micro-and nanofibrillated cellulose as a means to reduce weight of paper products: a review. BioResources 15:4553–4590. https://doi.org/10.15376/biores.15.2.Zambrano

Life Cycle Assessment of Nanomaterials in Paper Production

Sina Ardabili, Aziz Babapoor, and Seyyed Mojtaba Mousavi

Abstract The paper industry is a notable and widely used industry in the world. On the other hand, this industry is facing major challenges. Solutions, such as using nanoparticles (NPs), have been proposed to address some of these challenges. Therefore, due to various inputs and NPs in this industry, papermaking is dealing with environmental challenges. Based on this, investigating, evaluating, and identifying hot spots from an environmental investigation perspective and obtaining effective input can clarify the perspective for future policies in this industry. This chapter takes a comprehensive approach to providing a taxonomy in the field of NP application in papermaking, ensuring that the reader is well-informed. The following explains the importance of life cycle assessment (LCA) in this field. The following examines various inputs in the environmental subset and reviews the process from the perspective of life cycle assessment (LCA). This chapter highlights the challenges of using nanoparticles (NPs) in papermaking using a life cycle assessment (LCA) approach.

Keywords Paper industry · Life cycle assessment · Life cycle inventory · Nanomaterial · Nanotechnology

S. Ardabili
Department of Engineering Sciences, Faculty of Advanced Technologies, University of Mohaghegh Ardabili, Ardabil, Iran
e-mail: sina.faiz@uma.ac.ir

A. Babapoor (✉)
Department of Chemical Engineering, Faculty of Engineering, University of Mohaghegh Ardabili, Ardabil, Iran
e-mail: babapoor@uma.ac.ir

S. M. Mousavi
Department of Chemical Engineering, National Taiwan University of Science and Technology, Taipei, Taiwan

R. K. Bachheti (ed.), *Nanotechnology in the Pulp and Paper Industry*, Smart Nanomaterials Technology, https://doi.org/10.1007/978-981-95-3473-9_13

1 Introduction

In recent years, microparticle (MP) and nanoparticle (NP) technologies have witnessed significant development and expansion and have gradually penetrated all scientific, industrial, and technological fields [4, 27]. The public has noticed it as today's most modern technology [7]. Figure 1 shows an approximate statistic of various fields in which nanotechnology (NT) has penetrated and caused transformation and achievement. This rapid growth and transformation in NT offer a promising future for the paper industry.

The paper industry is no exception to this rule and has witnessed the penetration and application of MP and NP technologies. The application of NT in the paper industry has led to extensive developments with positive effects.

The use of nanotechnology in paper production, although leading to positive developments, can also pose serious environmental challenges in this industry [5, 32, 36, 38]. On the other hand, if there is no proper policy in the field of applying nanotechnology in paper production, the future of this scientific field may be limited and even have a weak impact. Based on this, it is necessary to conduct a comprehensive study to reveal the challenges and environmental limitations of using nanotechnology in papermaking and to direct the specialized minds of this field in this direction.

Life cycle assessment (LCA) is a standard method for evaluating the environmental impacts of a production process [19, 25, 26]. LCA estimates and presents mid-term and long-term environmental effects within the process's boundary. Many researchers have discussed LCA in different applications, such as wastewater treatment [40], biorefinery [51], gasification [29], construction [14], agriculture [31], and the paper industry [9, 28].

According to the introduction, this chapter examines the role of life cycle assessment (LCA) in the application of nanomaterials in papermaking. It discusses the existing challenges and limitations from the perspective of experts in this field.

Fig. 1 Various fields of NT applications

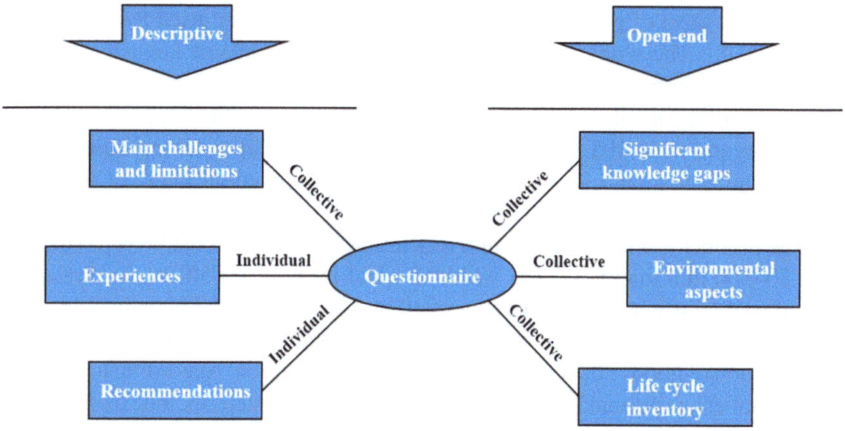

Fig. 2 The main scope of the questionnaire

Considering that this study's primary focus is on the experiences and opinions of experts in papermaking and LCA of the production process, it was decided to form a focus group. This group consisted of 15 experts (40% women and 60% were men) with an average age of 43 $\pm$ 2. A focus group is an interview-based technique developed to facilitate interaction with group members and encourage deeper discussion [15, 34]. One of the characteristics of focus group interviews is the interaction between the group interviewees, which arouses the desire to think and exchange attitudes and ideas.

Focus groups are superior to other research methods. Their primary goal is to uncover people's attitudes, feelings, beliefs, experiences, and reactions, which cannot be fully understood through observation, individual interviews, and survey questionnaires [23]. Focus groups discover and infer the opinions and emotional processes within the group context. Researchers in focus groups can obtain more information in a shorter time than those in individual interviews. This study used Morgan's method to form a focus group [33]. Figure 2 shows the direction of the questionnaire from the focus group.

2 Fundamentals of Paper Production

Before the appearance of language, humanity first showed its tendency to write and create pictures with the pictures on the walls of the caves. With the emergence of language, the so-called clay molds were the first paper that humanity tried to write on, and even the orders of the kings. The Old Testament was written on clay molds. After clay molds, leather from animal skins became a writing medium until paper was born.

It is a thin sheet of paper, usually made of vegetable fibers, used for writing and printing [1, 8]. The word "paper" originated from the Sanskrit language and was later translated into the Persian language. The European word for paper, in turn, derives from the Greek word "papyrus." For the first time, the Chinese succeeded in inventing paper. In 105 AD, Cai Lun invented paper [50], reaching Japan around 610 AD and Samarkand in 751 AD. Paper is a sheet material made of a network of natural cellulose fibers formed by sedimentation from an aqueous suspension [5, 39]. The product is a network of intertwined fibers with an almost layered structure and a 30–300 mm thickness. The width of a leaf is about 10–50 mm as a result. A sheet of writing paper with a thickness of 10 mm is about 5–10 fibers thick in cross-section. The two main applications of paper are the transmission of information and the protection of goods during transportation and distribution. According to the focus group members, papermaking has six general steps. These steps include tree selection and peeling, chipping, pulping, bleaching, chemical processes in paper production, paper shaping, quality control, and packaging. Each of these is considered one of the leading paper production processes.

2.1 Processes

According to the focus group, paper production begins with the selection and cutting of wood. Selecting and harvesting wood for paper production is one of the most important steps in production.

In order to manufacture paper, it is important that the tree variety selected consists of fibers that are of good tissue quality for pulp manufacture. The process of wood extraction is also conducted by the principles of sustainable forest management to minimize the negative impacts on the forests. Wood can be extracted in a manner that promotes maximization of production gearing—while ensuring that the forest ecosystems are not negatively affected. Such care and attention to the wood required, as well as how it is cut, ensure that the finished product is high-quality paper and also contribute to environmental conservation.

Prior to making the pulp, the upper bark of the trees is scraped and cut into pieces using a rotary knife that rotates around the tree trunk. Subsequently, in the manufacture of paper, all these wood chips have to be turned into pulp. This intermediate stage of the wood–paper cycle is known as the papermaking process. It mainly employs two systems—mechanical and chemical—for the manufacture of paper pulp. In the mechanical process, the wood is treated with heat and force in order to disintegrate the wood into its fibrous components. This technology is quick and inexpensive; however, the paper produced by such technology typically has lower quality properties and degrades more quickly. Unlike this process, in the chemical passing process, high concentrations of caustic sodas and sulfites are used to dissolve away noncellulosic parts of the wood, resulting in a slurry mix in which the adhesive fibers are of a longer and cleaner nature. That pulp obtained using the chemical method can be utilized for the manufacture of all highest quality paper grades, including printing and writing papers. Nowadays, both techniques are implemented in production plants to minimize adverse environmental effects and enhance the utilization of materials. The method used is determined by the paper grade and the properties required from the product.

At every stage of paper manufacture, many chemical processes are employed; these processes are more pronounced during bleaching and pulp refining. At this stage, chemicals such as sodium hydroxide and hydrogen peroxide are added to the pulp fibers to absorb impurities and whiten the fibers. The use of baking soda aids in the removal and separation of lignin, which is an adhesive used to hold wood pulp in suspension. Hydrogen peroxide, by itself, is a bleaching compound that helps oxidize plant pigments, ensuring that the end product is a brighter, whiter paper without any dark pigments. These substances should be avoided or used in an efficient manner and within controlled environmental factors to enhance the quality of the paper produced, its longevity, and resistance to color fading. These chemical processes not only enhance the paper's physical and decorative attributes but also mitigate the environmental impact of papermaking, as less waste is generated and the resources employed can be recycled and reused.

The subsequent process involves forming paper from the pulp, which entails converting the pulp into thin sheets of paper. This operation is predominantly performed in papermaking machines where the pulp is processed over a number of screens and rollers to extract the excess water and shape the pulp into a thin even band. This band, in turn, is directed to the sections, where pressure and temperature are applied for the purpose of drying and final strengthening of the band. The height of the paper, degree of softness, etc., can all be customized at this stage, and also, dyes or other special chemicals can be introduced in order to enhance the overall quality of the paper. The final shaping of paper is a crucial factor associated with the purpose and quality of the product; hence, this step calls for stringent controls.

Quality control and adherence to paper standards are among the critical steps in paper production that ensure the production of a quality product that meets the market's needs. The produced paper is subjected to various tests at this stage, including checking the tensile strength, tear resistance, whiteness, and ink absorption. These tests enable quality teams to identify and rectify any deviations from the desired standards. In addition, the paper is evaluated to ensure compliance with environmental standards, including reducing energy and chemical consumption, as well as increasing recyclability. Continuous efforts to improve quality control processes are significant to ensure that the paper produced is of high quality and sustainability.

2.2 *Challenges*

Table 1 presents the main challenges discussed by the focus group members. According to Table 1, focus group members, on average, believe that environmental impacts have the most important role in the paper production challenges. The second stage is related to energy consumption, followed by technological challenges.

Table 1 Results of the statistical analysis

Statistic parameters	Parameter	Challenge	Section
0.87	Correlation coefficient	Chemical pollution	Environmental impact
0.001	Significance level		
15	Number		
0.74	Correlation coefficient	Waste production	
0.001	Significance level		
15	Number		
0.73	Correlation coefficient	Input materials	Energy consumption
0.001	Significance level		
15	Number		
0.71	Correlation coefficient	Electricity	
0.001	Significance level		
15	Number		
0.44	Correlation coefficient	Material cost	Cost and economy
0.003	Significance level		
15	Number		
0.43	Correlation coefficient	Energy efficiency	
0.003	Significance level		
15	Number		
0.48	Correlation coefficient	Recycling efficiency	Waste management
0.003	Significance level		
15	Number		
0.62	Correlation coefficient	Digitalization	Technological challenges
0.002	Significance level		
15	Number		
0.59	Correlation coefficient	Automation	
0.002	Significance level		
15	Number		
0.61	Correlation coefficient	Lack of policy	
0.002	Significance level		
15	Number		

3 Nanomaterials in Paper Production

NT has revolutionized materials science, and nanostructures have led to the development of a range of new and improved materials [2]. These developments can be vital for the paper industry [2].

By reducing the dimensions of NMs below the critical length, completely different behaviors can be observed [2]. Additionally, nanotechnology (NT) has led to the development of a range of new and improved materials utilizing nanostructures [2].

When papermakers use NPs and NMs, they refer to additive programs that can improve water drainage during paper formation and help retain soft particles.

Before 1980, the subject of NT was not well known. Therefore, the papermakers were unaware of the potential of using additives such as colloidal silica and bentonite, and these particles were only occasionally used in water treatment or to control the friction characteristics of paper. However, today, at least one of these two types of minerals is used with cationic starch or nanopolymers to improve water absorption and shelf life. In addition, many papers have used or are currently using retention and dehydration programs with anionic polymers or lignin by-products that play the role of nanoparticles.

The most important and common observation related to the provision of NP applications in the papermaking process is a significant increase in the dehydration rate [2]. Although there are more common chemical approaches to improve water absorption in the papermaking process, the effects caused by the presence of NPs are more intense and precise. Table 2 presents the reasons for encouraging the use of NPs in the paper production process from the perspective of the focus group. According to Table 2, in addition to improving durability, it reduces pollution in factory effluent, which can deal with the main challenge related to paper production.

Table 3 presents the types of NPs according to their application in the papermaking process.

3.1 Challenges

This section presents the main challenges of NPs in the papermaking industry, as identified by the focus group. Figure 3 presents a diagram of the main challenges, including their respective sub-challenges. Table 4 presents the statistical results of each sub-challenge evaluated by the focus group.

According to the results presented in Table 4, the focal group members emphasize challenges based on environmental impacts and waste management scenarios so that the economic and cost issues are on the second step, which, according to the members of the focus group, has a significant distance from the environmental effects in terms of the explanatory factor in prioritizing the challenges.

The results (Tables 1 and 4) clearly show that environmental effects are integral to using nanoparticles. This effect is particularly noticeable in the paper industry (Table 4). Based on this, addressing the environmental impacts of using nanoparticles in the paper production industry, investigating the factors that affect it, and weighing their aspects can be significant.

Table 2 Results of the statistical analysis of NPs' effects on papermaking

Statistic parameters	Parameter	Effect on:
0.91	Correlation coefficient	Durability improvement
0.001	Significance level	
15	Number	
0.81	Correlation coefficient	Increasing the rate of dehydration
0.001	Significance level	
15	Number	
0.79	Correlation coefficient	Formation improvement
0.001	Significance level	
15	Number	
0.76	Correlation coefficient	Increased dry resistance
0.001	Significance level	
15	Number	
0.69	Correlation coefficient	Reduced drying energy
0.002	Significance level	
15	Number	
0.65	Correlation coefficient	Reduced production line stoppages
0.002	Significance level	
15	Number	
0.94	Correlation coefficient	Reduced pollution in factory effluent
0.001	Significance level	
15	Number	
0.61	Correlation coefficient	Reduced changes in paper machine operations
0.002	Significance level	
15	Number	

4 Environmental Aspect of Nanomaterials in Paper Production

NT can transform the entire landscape of the paper production sector by enhancing the operational characteristics and efficiency of its products. Nonetheless, the possible environmental hazards must be assessed thoroughly. The manipulation of materials at nanoscale levels comes with one major benefit: the conservation of raw materials, if applicable. Incorporating NPs in paper products results in stronger and more durable products, thus making it possible for manufacturers to create thinner paper. This lowers the use of wooden pulp and other natural materials, thus protecting the forests and reducing carbon shedding. On the other hand, this advantage is, to a certain degree, counterbalanced by the energy-intensive processes involved in the production

Table 3 Types of NPs in papermaking

Type of NPs	Application	Benefits	References
Silica	Coating paper surface	Improving ink absorbance and printability	Samyn et al. [42], Alikhah et al. [2]
Kaolinite	Filler for paper smoothness	Improving barrier properties	Alikhah et al. [2], Tajik et al. [45]
Calcium carbonate	Filler and Paper brightness	Improving properties	Tao et al. [46]
Alkali montmorillonite	Separation of colloidal substances	Improving chemical reaction	Tajik et al. [45]
Lignin	Adding to paper pulp	Improving forming and dehydration	Focus group
TiO_2	Paper coating	Improving paper opacity and whiteness, boosting of antibacterial performance	Maślana et al. [30], Tao et al. [46]
Cellulose nanocrystals	Reinforcement agent	Improving mechanical properties	Salam et al. [41], Sheikhi and Van De Ven [43]
Magnesium hydroxide nanoparticles	Flame retardancy	Improving flame retardancy	Wang et al. [47], Kang et al. [22]
Graphene	Thermal and electronic applications	Enhancing mechanical strength	Wu and Drzal [48]
Carbone nanotube	Specialty papers	Enhancing electrical conductivity	Anderson et al. [3], Pang et al. [37]
Zink oxides	Specialty papers and antimicrobial	Enhancing UV resistance	Xie et al. [49]

of nanoparticles, which can also compromise environmental benefits if fossil fuels are used [20].

Energy efficiency is another helpful aspect that NT can provide in papermaking. NPs have the potential to enhance drying, dewatering, and coating stages, which can greatly reduce energy usage. Such efficiency can ease the emission of greenhouse gases in the sector. The production of NPs is also energy-demanding, thus making the situation more complex. Though this may not be the case if green energy is not harnessed for the production of NPs, the anticipated savings in power may not be as much as commonly thought [24].

From another perspective, there are also environmental concerns regarding the disposal of NPs incorporated into paper production. The distribution of such particles in treated effluent and the environment cannot be fully accounted for, and thus, they may endanger nonterrestrial organisms. Dangerous NPs like titanium dioxide (TiO_2) and silver can slip through many normal filtration systems because of their size and

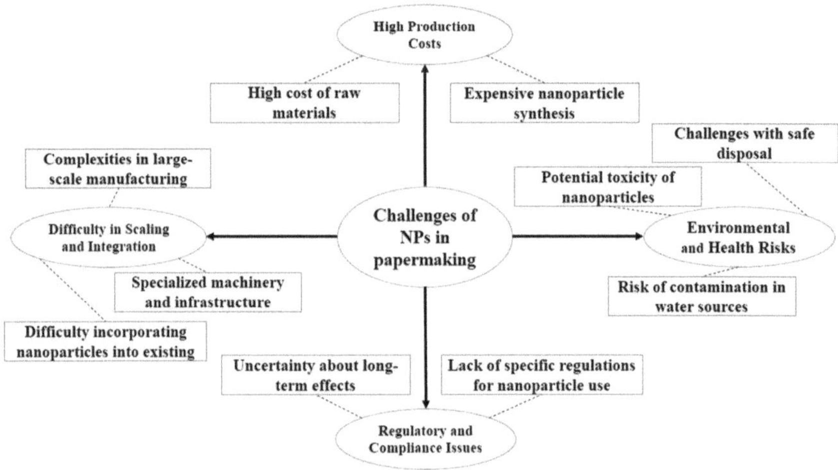

Fig. 3 Challenges of NPs in papermaking

size. They are released into the environment while researchers investigate the impacts such NPs cause in ecosystems [21].

Water contamination is an additional concern during manufacturing processes or when waste is disposed of. Metallic NPs utilized in the papermaking process can leach into water sources. Because they tend to remain suspended in water without being biodegradable, such particles can endanger aquatic organisms and their environments. This suggests a high necessity for better wastewater treatment and filtration systems to mitigate the pollution from such NPs [6].

Recycling is another field in which the use of NPs complicates the existing methods. Paper products with NPs are more likely to become waste than ordinary paper, causing problems in the recovery processes. Additionally, NPs in composite products made from recycled paper can degrade their quality or create pollution during the recycling process. This makes it difficult for the industry to achieve the relevant sustainability targets since recycling constitutes an important aspect of the paper life cycle.

Another overwhelming issue is the health hazards of workers. Because of their size, NPs can be easily breathed or absorbed through the skin, causing respiratory or other health-related issues, particularly for factory workers. Safety measures include wearing protective clothing and installing exhaust fans to reduce such dangers [44]. There are also issues likely to be encountered at the end of the life cycle of the product in use, paper laced with NPs. Most likely, banned used paper will not be disposed of correctly and necessarily. Policymakers acknowledge the potential risk posed by anthropogenic NPs or NMs—materials engineered at the nanometer scale and integrated into products—in altering the environment for the worse [35].

Considering the importance of environmental impact assessment, this study discusses the life cycle assessment method as a standard and reliable method.

Table 4 Results of the statistical analysis of NPs' challenges in papermaking

Statistic parameters	Parameter	Challenge	Section
0.65	Correlation coefficient	Expensive nanoparticle synthesis	High production costs
0.002	Significance level		
15	Number		
0.62	Correlation coefficient	High cost of raw materials	
0.002	Significance level		
15	Number		
0.93	Correlation coefficient	Potential toxicity of nanoparticles	Environmental and health risks
0.001	Significance level		
15	Number		
0.95	Correlation coefficient	Challenges with safe disposal	
0.001	Significance level		
15	Number		
0.89	Correlation coefficient	Risk of contamination in water sources	
0.001	Significance level		
15	Number		
0.45	Correlation coefficient	Complexities in large-scale manufacturing	Difficulty in scaling and integration
0.003	Significance level		
15	Number		
0.42	Correlation coefficient	Need for specialized machinery and infrastructure	
0.003	Significance level		
15	Number		
0.41	Correlation coefficient	Difficulty incorporating nanoparticles into existing papermaking processes	
0.003	Significance level		
15	Number		
0.58	Correlation coefficient	Lack of specific regulations for nanoparticle use	Regulatory and compliance issues
0.002	Significance level		
15	Number		
0.57	Correlation coefficient	Uncertainty about long-term effects	
0.002	Significance level		
15	Number		
0.55	Correlation coefficient	Need for certifications and compliance with health standards	
0.002	Significance level		
15	Number		

5 Life Cycle Assessment

An LCA study comprises four key stages: defining goals and scope, conducting a life cycle inventory (LCI) analysis, performing a life cycle impact assessment (LCIA), and the interpretation phase [17, 18]. In the initial stage, the study's objectives and boundaries are determined [10]. The second stage focuses on gathering data related to the inputs, outputs, and processes throughout the product's life cycle [10]. The third stage evaluates the potential environmental impacts based on the LCI [47], and in the final step, the findings are analyzed.

Defining the goal is essential as it identifies the purpose of the study, its intended application, and the target audience. The scope definition outlines how the LCA will be conducted, including the establishment of system boundaries and the definition of a functional unit. This functional unit measures the specific functions the product or service must provide in an LCA, incorporating the quantity and quality requirements for the alternatives [11–13].

There are several approaches to LCA quantification, often referred to as LCA variants (see Fig. 4). Popular variants include cradle-to-gate, cradle-to-grave, and cradle-to-cradle approaches, which guide inventory preparation and scenario design. The cradle-to-grave approach evaluates the entire life cycle, while cradle-to-gate approach focuses only on the stages from raw material acquisition to when the product leaves the factory. Cradle-to-cradle approach, in contrast, involves recycling at the end of a product's life, ensuring that the product is reused rather than discarded [10, 17].

Conducting and reporting LCA results can be an extensive process, particularly in areas such as NP production, where unique factors must be considered. Specific criteria, such as input ranges (inventory analysis), LCA variants, and LCIA methodologies, should be addressed in order to review these studies [47] comprehensively. After inventory analysis and identifying the LCA variants, the next step is performing LCIA, which assesses the significance of potential environmental impacts based on the LCI data (see Table 5). This process begins with selecting impact categories,

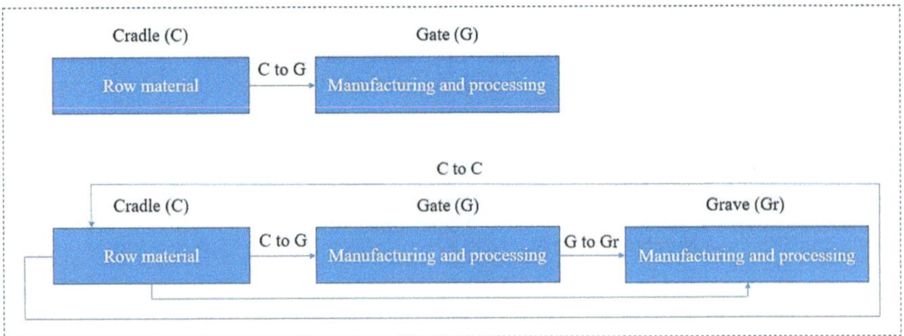

Fig. 4 Approaches of LCA

Table 5 LCA procedure

Environmental interventions				
Production inputs	Extraction of inputs	Creation of inputs	Outputs	Emissions
	Collection of inputs	Energy		Noise
Impact category				
Midpoint impacts	Climate change	Water use	Ozone creation	Acidification
	Resource depletion	Human toxic effects	Ecotoxic effects	Biodiversity
	Land use	Ozone depletion	Eutrification	
Damage category				
Endpoint impacts	Human health	Resource depletion	Ecosystem quality	Climate change

making a decision during the goal and scope phase, and revisiting it after data collection [17].

6 Applications of Life Cycle Assessment

According to the focus group's experiences, LCA is a valuable and proposed tool for assessing the environmental impacts of NPs in the papermaking process.

By evaluating the results of the interview, it was concluded that LCA includes a comprehensive assessment of the environmental impact of applying NPs in the papermaking industry, examining the entire life cycle of the product, from the production of raw materials (e.g., producing NPs) to the disposal of solid waste. It considers the factors such as energy use, water consumption, emissions, and environmental effects. This assists the producers in evaluating the environmental viability of using NMs in paper products. LCA can assess the environmental impact of paper infused with NPs relative to conventional or alternative materials. Various processes and types of NPs (such as titanium dioxide, silica, and cellulose nanofibers) can be compared, allowing practitioners to select more eco-friendly processes for use.

Utilizing LCA enables one to identify the appropriate types and optimal amounts of NPs to minimize the environmental impacts and energy use during papermaking processes. It also strategically balances enhanced performance (high strength, brightness, and water resistance) with ecological costs to achieve economic resource efficiency. LCA looks for environmental "hot spots" in NP production and papermaking. These hot spots include energy-intensive processes in synthesizing NPs or using certain evaluation materials that emit poisonous gases, and their ideal reduces the net environmental impact.

LCA can assess whether NPs are released during the recycling of paper and how such release affects the environment and public health. In addition, LCA provides information on the environmental efficiency of products containing nanotech additives, which helps paper manufacturers comply with environmental safety and certification requirements (e.g., EU Ecolabel, ISO 14040/44). Considering the environmental concerns at each process step, LCA can foster creativity in producing NPs or their applications—for instance, finding more environmentally friendly materials, optimizing energy-intensive processes, or synthesizing NMs from bioresources. NPs may also affect human health and the environment at toxicological levels. This is particularly useful for papermaking, as it helps make the practice safer when undertaking operations that could otherwise have harmful effects.

Table 6 presents the opinions of the focal group members regarding the potential and applications of LCA in NPs usage in papermaking, as evaluated using statistical parameters.

As shown in Table 6, evaluating environmental impacts followed by optimizing the inputs and waste management scenarios has the highest correlation coefficient among all the LCA applications from the focus group point of view. Accordingly, using sensitivity analysis, Fig. 5 sorts the inventory based on the importance score of each input on LCA outputs.

Sensitivity analysis is all about analyzing the sources of uncertainty in the inputs to a mathematical model or system and their effects on the model's or system's output [16].

According to Fig. 5a, NM usage, recycling, and end-of-life scenarios are the most critical inputs in this process, which affect the environmental impacts of NM usage in papermaking applications from the viewpoint of LCA. By considering Fig. 5b, NP discharge followed by greenhouse gas (GHG) emission and sludge management can be the most important outputs of NM usage in papermaking applications from the viewpoint of LCA. These results are taken based on the experiences of the focus group members.

7 Strategies for Preventing Challenges and Improving Limitations

This section presents a general attitude toward the primary strategy for avoiding challenges, as recommended by the focus group members. Figure 6 presents the main strategies for preventing challenges and improving limitations using visualization software (X-mind software).

According to the recommendations of focus group members, addressing these six strategies can successfully prevent the main challenges of using LCA in NP applications during the papermaking process.

Table 6 Results of the statistical analysis of LCA and NPs' applications in papermaking

Statistic parameters	Parameter	Section
0.96	Correlation coefficient	Evaluating environmental impacts
0.001	Significance level	
15	Number	
0.79	Correlation coefficient	Comparing alternative scenarios
0.002	Significance level	
15	Number	
0.93	Correlation coefficient	Optimizing the inputs
0.001	Significance level	
15	Number	
0.88	Correlation coefficient	Identifying the hot spots
0.001	Significance level	
15	Number	
0.92	Correlation coefficient	Waste management scenarios
0.001	Significance level	
15	Number	
0.77	Correlation coefficient	Eco-labeling
0.002	Significance level	
15	Number	
0.58	Correlation coefficient	Innovating for sustainability
0.003	Significance level	
15	Number	
0.59	Correlation coefficient	Assessing human and ecological health
0.003	Significance level	
15	Number	

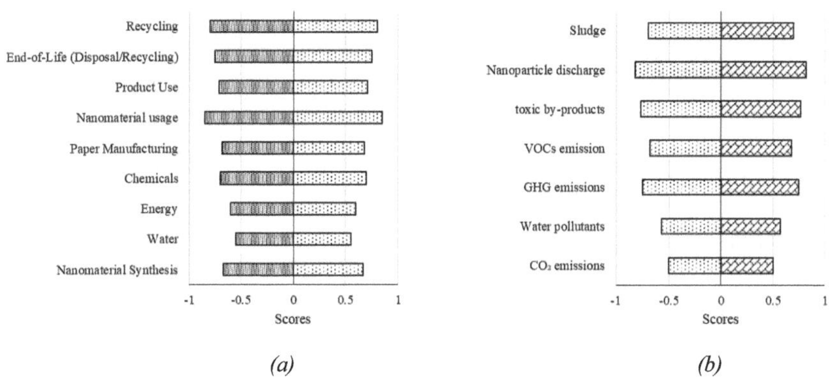

Fig. 5 Sensitivity analysis of **a** inputs, and **b** outputs

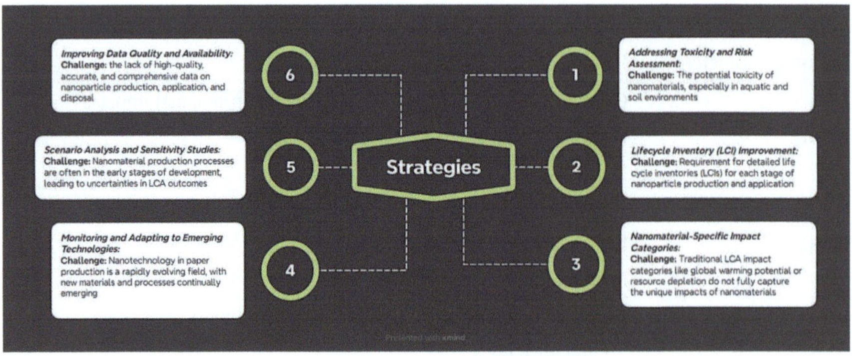

Fig. 6 The main strategies for preventing challenges and improving limitations

8 Emerging Trends and Future Directions

It is understandable when it becomes evident how such new practices, such as using NPs in polymeric materials, can be transformative, especially in paper production and other processes undergoing sustainable changes. A focus group that integrated members from the industry, academia, and environmental studies identified several trends and prospects for the evolution of the paper industry. The introduction of NT to paper products enhanced their functionality, but concerns about safety and regulation have been raised. Figure 7 presents the main trends and future directions from the discussions of this chapter.

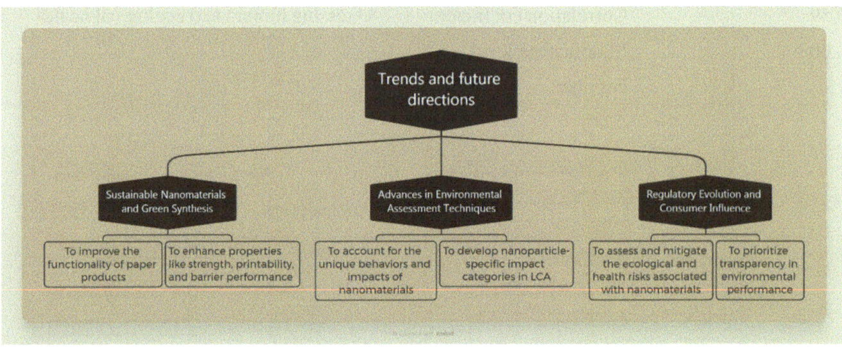

Fig. 7 The main trends and future directions

9 Conclusion

This chapter provides a general evaluation of LCA employment in NMs for the paper production industry. This study formed a focus group of expert members in this field and did an interview. In the following, opinions, suggestions, and findings, according to the expertise of this focus group, were reviewed and presented about the statement of the problem of this study. Accordingly, in general, the following observations were obtained:

- The environmental impact of chemical pollution and waste production is the primary and significant challenge for the paper production industry.
- NMs are employed in paper production industries to enhance durability and improve the dehydration rate.
- In addition to improving durability, the addition of NMs reduces pollution in factory effluent.
- Employing NMs presents challenges related to environmental impacts and waste management scenarios.
- LCA can impact NM employment in papermaking industries by evaluating environmental impacts and optimizing the inputs by proposing the proper waste management scenarios.
- NM usage, recycling, and end-of-life scenarios are the most important inputs in this process which affect the environmental impacts of NM usage in papermaking applications from the view point of LCA.
- Improving data quality and availability, life cycle inventory (LCI) improvement, nanomaterial-specific impact categories, addressing toxicity and risk assessment, scenario analysis and sensitivity studies and monitoring and adapting to emerging technologies are the proposed strategies for preventing challenges and improving limitations.

According to the recommendations of the focus group, future trends should be directed toward enhancing performance through NT, advances in environmental assessment techniques, regulatory evolution, and consumer influence.

Competing Interests The authors declare that they have no competing interest.

DataAvailability Declaration of Generative Artificial Intelligence (AI) and AI-Assisted Technologies in the Writing Process During the preparation of this work, the book chapter author(s) have not taken any assistance from AI, and the book chapter author(s) reviewed and edited the content as needed and take(s) full responsibility for the content of the publication.

References

1. Abd El-Sayed ES, El-Sakhawy M, El-Sakhawy MA-M (2020) Non-wood fibers as raw material for pulp and paper industry. Nord Pulp Pap Res J 35:215–230
2. Alikhah T, Afra E, Sarayian A, Ghasemian A (2015) Types of nano particles used in cellulose industries and proper ways of keeping them in paper (in Persian). Nano world 11:52–54

3. Anderson RE, Guan J, Ricard M, Dubey G, Su J, Lopinski G, Dorris G, Bourne O, Simard B (2010) Multifunctional single-walled carbon nanotube–cellulose composite paper. J Mater Chem 20:2400–2407
4. Astruc D (2020) Introduction: nanoparticles in catalysis. ACS Publications
5. Bajpai P (2011) Environmentally friendly production of pulp and paper. John Wiley & Sons
6. Benn TM, Westerhoff P (2008) Nanoparticle silver released into water from commercially available sock fabrics. Environ Sci Technol 42:4133–4139
7. Bhushan B (2017) Springer handbook of nanotechnology. Springer
8. Deshwal GK, Panjagari NR, Alam T (2019) An overview of paper and paper based food packaging materials: health safety and environmental concerns. J Food Sci Technol 56:4391–4403
9. Dias A, Carvalho R, Marques C, Bértolo RA, Carta AM, Machado LM (2024) Life cycle assessment of tissue paper: effects of bioenergy integration. SSRN 4713899
10. Faizollahzadeh Ardabili S (2021) Improving the combustion process of biodiesel using additives. Ph.D. thesis fundamental, University of Mohaghegh Ardabili
11. Finnveden G, Hauschild MZ, Ekvall T, Guinée J, Heijungs R, Hellweg S, Koehler A, Pennington D, Suh S (2009) Recent developments in life cycle assessment. J Environ Manag 91:1–21
12. Finnveden G, Johansson J, Lind P, Moberg Å (2005) Life cycle assessment of energy from solid waste—part 1: general methodology and results. J Clean Prod 13:213–229
13. Finnveden G, Potting J (2014) Life cycle assessment. Encycl Toxicol 74–77
14. Gay IC, Hvam L, Haug A, Huang GQ, Larsson R (2024) A digital tool for life cycle assessment in construction projects. Dev Built Environ 20:100535
15. Greenbaum TL (1998) The handbook for focus group research. Sage
16. Hashemi-Nejhad A, Najafi B, Ardabili S, Jafari G, Mosavi A (2023) The effect of biodiesel, ethanol, and water on the performance and emissions of a dual-fuel diesel engine with natural gas: sustainable energy production through a life cycle assessment approach
17. Hashemi F (2021) Modeling and investigation of the life cycle of the hybrid power generation process from the diesel engine. Master thesis, University of Mohaghegh Ardabili
18. Hashemi F, Pourdarbani R, Ardabili S, Hernandez-Hernandez JL (2023) Life cycle assessment of a hybrid self-power diesel engine. Acta Technol Agric 26:17–28
19. Hauschild MZ, Rosenbaum RK, Olsen SI (2018) Life cycle assessment. Springer
20. Hubbe MA, Rojas OJ, Lucia LA, Sain M (2008) Cellulosic nanocomposites: a review. BioResources 3:929–980
21. Kah M, Hofmann T (2014) Nanopesticide research: current trends and future priorities. Environ Int 63:224–235
22. Kang M, Wang G, Liu W, Yu D, Li G, Song Z, Liu X, Wang H (2023) Fabrication of highly flame-retardant paper by in situ loading of magnesium hydroxide/basic magnesium chloride onto cellulose fibers. Cellulose 30:7295–7312
23. Kitzinger J, Barbour R (1999) Developing focus group research: politics, theory and practice. Sage
24. Klemm D, Kramer F, Moritz S, Lindström T, Ankerfors M, Gray D, Dorris A (2011) Nanocelluloses: a new family of nature-based materials. Angew Chem Int Ed 50:5438–5466
25. Klöpffer W (2014) Background and future prospects in life cycle assessment. Springer Science & Business Media
26. Klöpffer W, Curran MA, Frankl P, Heijungs, R, Köhler A, Olsen SI (2007) Nanotechnology and life cycle assessment: a systems approach to nanotechnology and the environment
27. Logothetidis S (2012) Nanomedicine and nanobiotechnology. Springer Science & Business Media
28. Mainardis M, Ferrara C, Cantoni B, Di Marcantonio C, De Feo G, Goi D (2024) How to choose the best tertiary treatment for pulp and paper wastewater? Life cycle assessment and economic analysis as guidance tools. Sci Total Environ 906:167598
29. Manesh MHK, Davadgaran S, Rabeti SAM (2024) Gasification potential of municipal solid waste in Iran: application of life cycle assessment, risk analysis, and machine learning. J Clean Prod 434:140177

30. Maślana K, Żywicka A, Wenelska K, Mijowska E (2021) Boosting of antibacterial performance of cellulose based paper sheet via TiO_2 nanoparticles. Int J Mol Sci 22:1451
31. Medel-Jiménez F, Krexner T, Gronauer A, Kral I (2024) Life cycle assessment of four different precision agriculture technologies and comparison with a conventional scheme. J Clean Prod 434:140198
32. Mohieldin S, Zainudin E, Paridah M, Ainun Z (2011) Nanotechnology in pulp and paper industries: a review. Key Eng Mater 471:251–256
33. Morgan DL (1996) Focus groups. Ann Rev Sociol 22:129–152
34. Morgan DL, Krueger RA, King JA (1998) The focus group guidebook. Sage
35. Nowack B, Bucheli TD (2007) Occurrence, behavior and effects of nanoparticles in the environment. Environ Pollut 150:5–22
36. Ogunwusi A, Ibrahim H (2014) Advances in pulp and paper technology and the implication for the paper industry in Nigeria. Ind Eng Lett 4:3–11
37. Pang Z, Sun X, Wu X, Nie Y, Liu Z, Yue L (2015) Fabrication and application of carbon nanotubes/cellulose composite paper. Vacuum 122:135–142
38. Puurunen K, Vasara P (2007) Opportunities for utilising nanotechnology in reaching near-zero emissions in the paper industry. J Clean Prod 15:1287–1294
39. Rantanen J (2016) The manufacturing potential of micro and nanofibrillated cellulose composite papers. Aalto University
40. Ribeiro JP, Sarinho L, Nunes MI (2024) Application of life cycle assessment to Fenton processes in wastewater treatment–a review. J Water Process Eng 57:104692
41. Salam A, Lucia LA, Jameel H (2013) A novel cellulose nanocrystals-based approach to improve the mechanical properties of recycled paper. ACS Sustain Chem & Eng 1:1584–1592
42. Samyn P, Barhoum A, Öhlund T, Dufresne A (2018) Nanoparticles and nanostructured materials in papermaking. J Mater Sci 53:146–184
43. Sheikhi A, Van De Ven TG (2018) Colloidal starch and cellulose nanocrystals unite to improve the mechanical properties of paper: from enhanced coatings to reinforced nanocomposites. ACS Appl Nano Mater 1:1841–1852
44. Shvedova AA, Kagan VE, Fadeel B (2010) Close encounters of the small kind: adverse effects of man-made materials interfacing with the nano-cosmos of biological systems. Annu Rev Pharmacol Toxicol 50:63–88
45. Tajik M, Resalati H, Jalali H (2014) Applied nanoparticles in pulp and paper industries and their effects (in Persian). In: The second national conference of new technologies in wood and paper industries
46. Tao H, He Y, Zhao X (2015) Preparation and characterization of calcium carbonate–titanium dioxide core–shell ($CaCO_3@TiO_2$) nanoparticles and application in the papermaking industry. Powder Technol 283:308–314
47. Wang S, Yang X, Li Y, Gao B, Jin S, Yu R, Zhang Y, Tang Y (2022) Colloidal magnesium hydroxide nanoflake: one-step surfactant-assisted preparation and paper-based relics protection with long-term anti-acidification and flame-retardancy. J Colloid Interface Sci 607:992–1004
48. Wu H, Drzal LT (2012) Graphene nanoplatelet paper as a light-weight composite with excellent electrical and thermal conductivity and good gas barrier properties. Carbon 50:1135–1145
49. Xie H, Zhang H, Liu X, Tian S, Liu Y, Fu S (2021) Design and preparation of multiple function-integrated lignin/tannin/ZnONP composite coatings for paper-based green packaging. Biomacromol 22:3251–3263
50. Yi X (2023) Development and evolution of papermaking in ancient China based on the uses and forms of paper. Pap Biomater 8:38–46
51. Yusoff MA, Mohammadi P, Ahmad F, Sanusi NA, Hosseinzadeh-Bandbafha H, Vatanparast H, Aghbashlo M, Tabatabaei M (2024) Valorization of seafood waste: a review of life cycle assessment studies in biorefinery applications. Sci Total Environ 952:175810

Nanotechnology in Specialty Paper Production and Applications

Naser Hamid, Aziz Babapoor, and Seyyed Mojtaba Mousavi

Abstract The paper industry is one of the most critical industrial sectors internationally due to its economic benefits. The present-day status of nanotechnology development, particularly in the context of additives used in the paper industry, can be the main topic of this overview. Nanotechnology is increasingly being used in the paper industry to improve production processes with a strong emphasis on sustainability. This study primarily examines the role of nanofillers, nanofibers, chemical compounds and nanocomposites in paper applications. The incorporation of nanocomposites has led to significant improvements in various properties of the produced sheets, including printability, mechanical strength, gloss and gas barrier properties. This development has attracted considerable attention from scientists and researchers and highlights the potential of nanotechnology to revolutionize the paper manufacturing sector. Strength, low weight, high stiffness, large surface area, and enhanced stability are just a few of the claimed advantages of nano additives. However, several challenges, such as high costs, material incompatibilities, and knowledge gaps, have hindered the large-scale adoption and commercialization of nanotechnology in the paper industry

Keywords Paper application · Paper industry · Nanotechnology · Paper · Economic benefits · Nanocellulose

N. Hamid · A. Babapoor (✉)
Department of Chemical Engineering, Faculty of Engineering, University of Mohaghegh Ardabil, Ardabil, Iran
e-mail: babapoor@uma.ac.ir

S. M. Mousavi
Department of Chemical Engineering, National Taiwan University of Science and Technology, Taipei, Taiwan

R. K. Bachheti (ed.), *Nanotechnology in the Pulp and Paper Industry*, Smart Nanomaterials Technology, https://doi.org/10.1007/978-981-95-3473-9_14

1 Introduction

Wood is a key component in the forest product industry, and it comprises hemicellulose, cellulose, ash, extractives, and lignin [30]. The chemical composition varies between species, with hardwoods having lower lignin and higher cellulose and extractive contents than softwoods [3, 21]. However, global fiber supplies are growing, with Asia being the largest fiber-deficit region [96, 100] This can lead to reducing the dependence of the paper and pulp industry on virgin fibers with the growth of nonwood plant fibers in Asia and the expansion of the use of recycled paper [12, 49, 89]. The paper industry is increasingly focusing on clean or green technology, introducing nonwoody raw materials to replace conventional resources like woody raw materials [49, 68]. This cleaner technology reduces environmental impact and steadiness hazards, decreases dumping expenses and resource costs, and increases revenues in the industry [40, 51, 78] Nonwood fibers are not unusual in nations with wood shortages and are considered the first-class method for replacing wood in the industry. Several research has investigated the ability of nonwood substances, including wheat straw, canola straw, tobacco stalks, Tunisian alfa, giant reed, and vine stems, as alternatives to wood in the pulp and paper industry [38]. Paper is a crucial material used globally, with global production of paper and cardboard reaching 407 million metric tons in 2015 [31]. Over half of this was packaging paper, while almost one-third was graphic paper [29]. By 2020, 500 million tons of paper were estimated to be consumed worldwide, representing an annual rise of 1.6%. Paper use has grown by nearly half since 1980 [38]. With half of the world's production, the three major paper-producing nations are China, Japan, and the United States. The top two nations in terms of both importing and exporting paper goods are Germany and the United States. The packaging business, in particular, has seen a surge in demand for paper-based products due to rapid economic expansion [5, 82]. Nano-based additives are used to enhance paper performance and to make it functional with unique properties, such as gloss, mechanical strength, recyclability and writing quality [35, 54, 97, 98, 101]. Nanocellulose (NC)-based additives are paper enhancers, retention and filtration reagents, and coating aids. NC is also applied to specialty papers, such as waterproofing, freshness, insulation, blotting, and wrapping [35]. Especially in the industrial world, nanotechnology has rapidly taken over various sectors, and in the last 10 years, nanoscale markets have emerged rapidly [15, 54, 63, 67, 94]. Small-scale molecular nanosystems, active and passive nanoassemblies, and general nanosystems represent four generations of nanomaterials today [34, 41, 62, 84]. Due to the rapid development of nanoscience, almost every field of science and technology will eventually involve nanoscale manufacturing [9, 36, 62].

2 General Review of Paper Industry and Its Production

The pulp and paper industry is a global production sector with operations across numerous industries. Paper is a biodegradable material that can replace plastic bags and applications, making it crucial for the industry to develop rapid production procedures that maintain environmental factors, cost-effectiveness, and high product value [80, 33, 67, 79]. The paper industry faces challenges, including the increasing need for high-quality materials, maintaining exceptional recycled fibers, and implementing price-competitive manufacturing methods [10, 22, 46, 75]. Record digitalization, sustainable development, and the rising need for packaging across industries are contemporary issues. Food and beverage packaging is in high demand, while cosmetics, health care, and industrial packaging are in lower demand [56, 60, 85]. The paper industry must address these challenges to ensure the sustainability and quality of its products. In recent years, the paper manufacturing industry has focused on developing new techniques to enhance the quality and overall performance of its products [22, 46, 59, 77]. The production of recycled paper is of paramount significance due to accelerated ecological consciousness and a higher demand for packaging papers [13, 10, 12, 103]. There are challenging situations related to the deterioration of fabric properties, notably caused by hornification phenomena which affect the material integrity, as well as the accumulation of dissolved and colloidal substances that further complicate the situation [18, 20, 42]. Several methods, including dry strengthening agents, soft mechanical refining of recycled fibers, and biorefining techniques utilizing enzymes, have been employed to address these challenges. Using lignocellulosic nanofibers (LCNF), microfibrillated cellulose (MFC), or cellulose nanofibers (CNF) in the production of paper is one such approach [9, 79, 106]. Cellulosic pulp with different amounts of lignin and hemicelluloses is the source of these nanofibers. Unbleached pulp with a high lignin concentration yields LCNF, whereas bleached pulp yields CNF [80, 33]. Because these nanofibers increase paper's quality and performance and strengthen its barrier qualities, they are advantageous in food packaging. Filtration reagents improve the speed of wet paper drainage and dehydration by neutralizing the charge on fiber and filler surfaces [22, 80]. This reduces polarity, making it difficult for water molecules to wet or align. Enhancing slurry drainage performance is essential for papermaking [50, 57]. The pulp and paper industries make use of a ramification of filtering reagents, the principal types of which are mentioned in Table 1. Enhancing the effectiveness of slurry drainage is vital to the papermaking technique. Reagents for filtration function as polymer surfactants, decreasing the surface tension of fillers and fibers. They alleviate floor anxiety and call angles, which make it more difficult for water molecules to penetrate the fiber floor. Potential drainage issues may arise from the grid structure that adequate fibers provide. The concentration of cellulose nanofibrils (CNF) increases the drainage time due to the potential clogging of pores in paper sheets. Researchers have explored the complex interactions between cellulose pulp, various components, mineral fillers, and CNF to tackle these dewatering challenges. Utilizing retention aids and administering CNF in small quantities can help prevent these issues, while

Table 1 The use of filter reagents in pulp and paper production

Type	Representative compounds	References
Nonionic	Polyacrylamide (PAM), poly mannitol galactose, polyethylene oxide	[88]
Cationic	Polyvinylamine, cationic starch, polyethyleneimine (PEI), polyamide polyamine epichlorohydrin (PAE)	[46, 85]
Anionic	Hydrolyzed polyacrylamide, carboxymethyl starch (CMS)	[56, 59]
Zwitterionic	Amphoteric starch, amphoteric polyacrylamide	[13, 77]

the use of retention and drainage aids (RSS and NC) can improve drainage efficiency [11].

To address the challenges of drainage and retention in paper production, Retention and Drainage Aids (RSS) play a crucial role. These agents help optimize the formation of paper by enhancing fiber retention during the papermaking process and improving water removal from the pulp. By optimizing retention, the amount of fiber lost during production is minimized, which ultimately leads to more efficient paper production. Additionally, RSS helps improve the overall strength and quality of the paper, ensuring that the final product meets the desired standards. The use of retention aids, when combined with cellulose nanofibrils (CNF) in small quantities, ensures a balance between maintaining the integrity of the fiber network and enhancing the drainage properties, thus improving the overall efficiency of the dewatering process. Moreover, the application of RSS and nanocellulose (NC) in the production process also helps to reduce the energy required for drying, contributing to more sustainable and cost-effective paper manufacturing

In conclusion, the strategic use of retention and drainage aids, such as RSS and NC, can significantly enhance the efficiency of paper production, reduce operational costs, and improve the quality of the final product [65].

Paper strength relies on three primary parameters: fiber strength, bonding strength between fibers, and fiber entanglement within the matrix. Reinforcing agents increase paper strength by increasing fiber bonding strength, particularly through hydrogen bonds. The thin diameter of CNF and its large length-to-diameter ratio lead to a large specific surface area, resulting in high hydrogen bonding potential [13, 10, 18, 103].

3 Nanotechnology in Paper Production

Nanotechnology is a rapidly growing field expected to revolutionize material use in the twenty-first century [2, 20]. It provides practical and eco-friendly substitutes for existing technologies, enabling the exploration of new substances, procedures, and phenomena at the nanoscale [8, 62]. Novel approaches to study, both theoretical and practical, provide new avenues for the creation of novel nanosystems and nanostructured materials [37, 50, 55, 93]. Because of its benefits, which include reduced size, speed, energy and material consumption, unique phenomena and qualities, high production efficiency, and ability to combine living and nonliving things, nanotechnology is significant [57]. Malaysia has been growing in nanotechnology since the early millennium, with the authorities initiating microtechnology and nanotechnology development in 2001. Malaysia's nationwide nanotechnology initiatives, launched in 2006, are a crucial aspect of the country's growing nanotechnology sector, which has been actively supported by numerous organizations [86]. Nanotechnology is a rapidly growing discipline with applications in medicine, biofuel manufacturing, wastewater engineering, and drug delivery. The nañoparticle, a fundamental unit of nanotechnology, is a crucial element, ranging from 1 to 100 nm in size [4, 90]. The U.S. Environmental Protection Agency (EPA) and the U.S. Food and Drug Administration (FDA) have one-of-a-kind opinions on nanomaterials, with the EPA pointing out that nanomaterials have unique properties compared to non-nano forms [7, 24]. Nanomaterials are used in diverse fields including electronics, optical communication, cancer treatment, and biofuel manufacturing, where their performance increases with physical properties and higher surface-to-volume ratio.[61, 66, 81]. Nanotechnology is likewise utilized in bioenergy, enzymology, and prescribed drugs, enhancing productivity, long-term stability, and cost-effectiveness. Recent research indicates that when nanomaterials are combined with algal culture medium, they enhance biodiesel productivity, lipid accumulation, and cellular growth, while also reducing production costs across various fields (Fig. 1) [92].

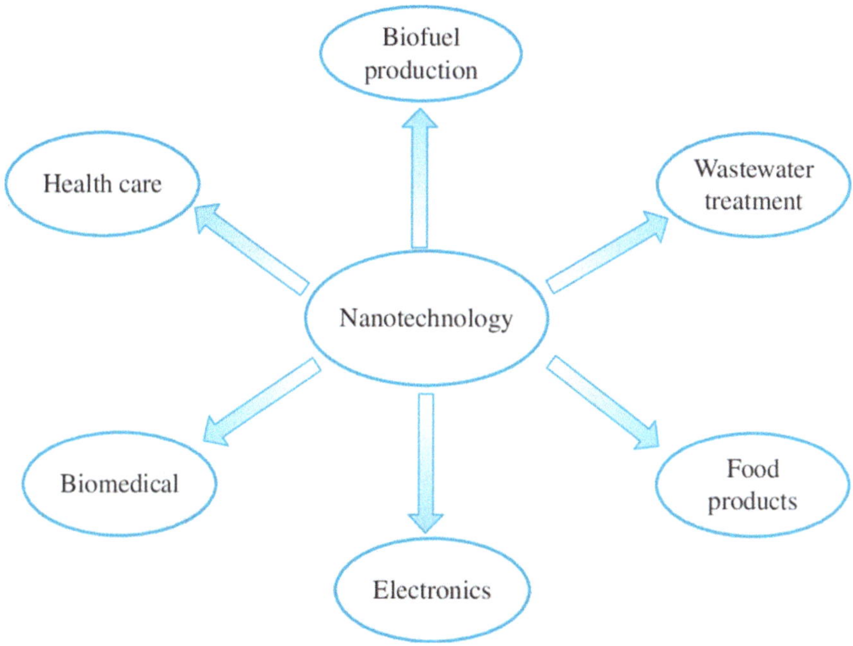

Fig. 1 Diverse applications of nanotechnology across various fields [53]

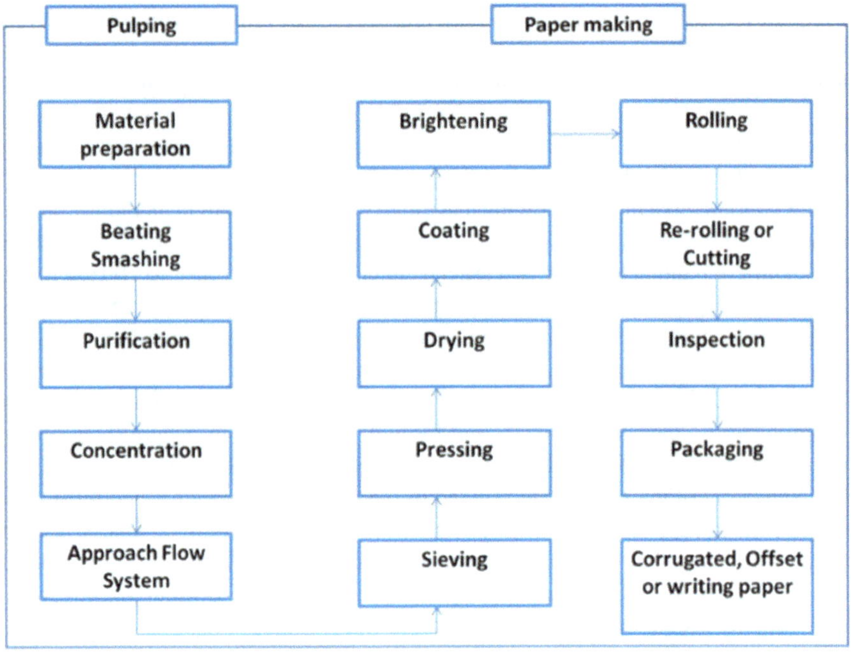

Fig. 2 General flow system of papermaking [110]

Today, the world produces approximately over 400 million tons of paper and paperboard annually, with growth in Asia more than offsetting the decline in the United States and Europe [25, 45]. Over 40% of the output originates from Asia, while 30% comes from Europe and 25% from North America [43, 44, 108]. It has been predicted that this total volume would best reveal a significant real (rather than inflated) increase in 2014, as the Asian market starts to offset the decline in the American and EU markets [26]. The processing of raw materials, pulp manufacturing, pulp bleaching, paper production, and recycling of fibers are all included in the first step of the paper manufacturing process [14]. The process entails preparing delicate wood pulp as a thin slurry with various additives, mainly fillers, to achieve the preferred paper quality, if necessary (Fig. 2).

4 Strategies to Improve the Efficiency of Paper Production

Nano-components that are smaller in particle size and have a uniform distribution can enhance the paper industry by providing unique attributes and functionalities [62, 71, 72]. Traditional components, such as adsorption, have a high floor vicinity, while those components, due to their shorter particle length, offer higher performance. Incorporating nano-components into paper slurry may also offer high smoothness and improved appearance. These advantages are predicted to draw worldwide papermakers to discover this promising research area and work toward tremendous breakthroughs [95]. The renewable and green polymer enhancer known as herbal carbon (NC) exhibits confined adsorption functionality and single-use impact. To boost paper strength, researchers have combined NCs with cationic polymers [6, 58]. CNF with cationic polyelectrolyte polyamide polyamine epichlorohydrin (PAE) is one such system. The CNF/PAE device significantly improves the paper's dry and wet tensile strength, however, when introduced in nanoaggregates, the improvement is mild [58]. This describes using functionalized nanocellulose (NC) in a two-layer papermaking process to improve paper quality. By adding functionalized NC, which acts as a binder, you can increase bonding between fibers and fillers, leading to enhanced paper strength and potentially reducing the need for other additives. This technique can improve paper's aesthetics, mechanical strength, and other properties, making it a sustainable approach [6, 19, 58]. Research observed that adding CNF–carboxymethylation and CNF–TEMPO-mediated oxidation additives to bleached eucalyptus kraft pulp is more desirable for the upkeep of mineral filler and the paper's properties.

Carboxymethylated or TEMPO-oxidized CNF enhances the paper's optical and mechanical properties. To enhance the mechanical, physical, and barrier properties of paper products, nanostructured cellulose-based composite materials can network and bridge long and short fibers [15, 16]. The shortage of wood resources and the increasing utilization of secondary and nonwood fibers, along with waste paper pulp, make this especially important. Components based entirely on NC can satisfy these specifications, resulting in increased recycled paper use [63]. Nanocellulose (NC) to recycled eucalyptus pulp enhances recycled paper by creating a stronger, more uniform fiber network, improving paper properties like strength and durability. This highlights the crucial role of papermaking additives, such as nanocellulose, in modifying and enhancing recycled fibers throughout pulping, processing, and papermaking [17, 63]. The shortage of woodland sources necessitates the improvement of non-timber fibers and secondary fibers, together with waste paper pulp, as raw substances in papermaking. Powerful components are needed for their retention and enhancement. NC-based total components can enhance the use of recycled paper, particularly mechanized NC, thereby increasing the consistency of recycled eucalyptus pulp within the paper matrix. These additives, also called papermaking chemicals, play a role in papermaking, pulping, and processing techniques. Common papermaking auxiliaries include maintaining aids, sizing agents, enhancers, filter aids, dispersants, and defoamers, which notably enhance paper performance. Table 2 shows various types of auxiliary materials used in papermaking.

NC, a cellulose derivative, is an essential papermaking additive known for its high degradability, biocompatibility, and swelling ability [23, 39]. NC, with its high surface area, tensile strength, and high aspect ratio, has a Young's modulus of 138 GPa. It contains several hydroxyl groups that form hydrogen bonds with fibers, contributing to several effects, including improved drainage, retention, and enhancement. The study and production of NC have been essential for the rapid advancement of the papermaking and packaging industries [15, 16, 69]. Due to its unique nanoproperties, NC can form stronger hydrogen bonds with the hydroxyl groups

Table 2 Different types of auxiliary materials for papermaking [28]

Category	Main purpose	Common additives	Classification
Papermaking additives	Improve paper performance	Sizing agents, dry and wet strengthening agents, etc	Functional additives
	Improve the performance of paper stock in the wet end	Retention aid, filter aid, defoamer, deinking agent, etc	Process additives

in fibers, which makes it an excellent paper enhancer for reinforcing mechanical strength. The pulp fibers and numerous hydroxyl groups on their surface can form hydrogen bonds, filling gaps and strengthening the paper. NC is a bio-based substance that shows great promise and has excellent nanostructures [99].

Techniques for Enhancing the Efficiency of Paper Production

Paper manufacturing is a critical aspect of daily life, with the United States being a significant patron and producer. As the sector becomes increasingly environmentally aware, there is a growing demand for strategies to enhance paper production performance while minimizing environmental impacts [48]. Essential techniques include adopting renewable energy resources and optimizing manufacturing processes. These steps can lessen the industry's carbon footprint, increase cost savings, and enhance competitiveness. By implementing these strategies, paper producers can reduce their environmental impact and improve their operations, ultimately contributing to a more sustainable future [52].

5 Strategies to Improve the Efficiency of Specialty Paper Products

Nano-components can enhance paper properties and advantages through various factors, which include gloss and opacity, light scattering attributes, paper penetration, calendaring steps, and drying rate [70]. The scale of nano-components is vast due to various factors, including the propensity for aggregation. However, larger diameter nano-additives are challenging to acquire due to insufficient unique surface [91, 102]. For example, surfaces with tiny particles nano-components are extra proof against internal sizing agents due to more significant surface regions [1]. The pulp and paper industry uses nano-bleaching agents to decolorize or whiten papers, which damages the absorbing properties of chromophores, which absorb visible light and transmit a portion of it [27, 111]. Nano-additives, which are smaller in particle size and distribution, can enhance paper industries by offering higher overall performance than traditional additives like adsorbents [47, 73]. They may offer specific paper attributes and functionalities with high smoothness and enhanced appearance [64]. The incorporation of nano-additives into paper slurry can also provide new features.

The advantages of these additives are anticipated to draw global papermakers to explore this promising research area and work toward high-quality breakthroughs. This may lead to new functionalities in paper manufacturing [87, 109]. Herbal Carbon is used as a reinforcing agent for base papers. It increases the strength and performance of the paper by improving adhesion and forming structures that increase the load-bearing capacity of coarse fibers [107]. Because of its robust polarity and the complementary interaction between the pulp fiber and surface, NC is utilized as a reinforcing agent in paper. It has a layered structure that relies on hydrogen bonds and intermolecular bonding [76]. When used as additives in unbleached kraft paper, MFC and NFC extracted from cotton lint offer superior tensile strength compared to kraft sheets without these components. Bleached eucalyptus pulp can be converted into suitable printing and writing paper by adding CNF. However, depending on the stability of the interaction between NCs and fibers, an excessive amount of NC may have negative effects, reducing fiber-to-fiber bonding strength [32, 63]. Of their evaluation of the effects of four specific NCs at the houses of paper, Hollertz et al. found that although NC can substantially growth stress fracture, Young's modulus, and tensile electricity, the addition of CNF oxidized with the aid of periodic acid, in aggregate with polyvinyl amine (PVAm) or poly-dimethyl diallyl ammonium chloride (pDADMAC), ought to beautify the reinforcing effect notably. By enhancing the mechanical properties and capabilities of paper, NC can improve its performance.

For the production and consumption of products internationally, the pulp and paper sector is essential; it confronts difficulties due to the increasing demand for paper and environmental issues related to conventional wood-based feedstock [32, 105]. The emerging interest in alternative feedstocks, particularly lignocellulosic fibers from nonwood sources such as agricultural residues and bamboo, addresses concerns about pulp yields and paper mechanical properties due to their precise chemical and physical characteristics [63, 74]. Integrating nonwood fibers offers the potential for environmental sustainability, reduced deforestation, and implementation of circular economy concepts. Improving industry resilience through feedstock diversification and tailor-made paper properties can be achieved through collaboration, research, and technological advancements, regardless of challenges [104]. Specialty paper is a paper with specific properties for specialized packages. Its capabilities encompass software-based features, specific performance, small-scale manufacturing, quantitative demand, restricted application, low-cost entry devices, and added value, as well as a high technical threshold. The unexpectedly rapid growth of the home paper market is having a significant effect on home printing, design, and the paper industry. Specialty papers can be categorized based on their unique characteristics and their applications across various fields [83]. Takeda et al. [105] reviewed the current status of nanobiotechnology, an emerging research area in nanotechnology. It focuses on the structure, research areas, and performance of countries and organizations. The main research areas include drug delivery, nanostructures, bioimaging, biomedical applications, biosensors, and carbon nanotubes. The United States is a leader in nanobiotechnology, and China is also a significant contributor. Most institutions with a significant impact on nanobiotechnology are located in the United States.

A study by Bardet et al. (Zhao et al. [112]) on nanocellulose found that cellulose nanocrystals (CNCs) and cellulose nanofibers (CNFs) have distinct properties that can be used for different applications. CNFs can act as a network for scattering particles, while the self-assembly of CNCs can be used to create iridescent films. The research was scaled up to a papermaking process, recommending CNF coatings for lightweight paper to hold pigments and create iridescent anti-counterfeiting properties.

6 Conclusion

The paper industry has greatly benefited from integrating nanotechnology into specialty paper production. Nano-additives, such as nanofibers, nanofillers, nanocomposites, and nanoscale chemicals, have enhanced the mechanical, gloss, printability, and gas barrier properties of paper sheets. Key advantages include exceptional strength, high surface area, high stiffness, low weight, and sustainable properties. However, challenges remain, including compatibility issues, higher costs, and knowledge gaps in optimizing nano-additive performance and scalability. Further research and development are necessary to overcome these hurdles and facilitate the widespread commercial adoption of nanotechnology in the paper industry. As global demand for specialty papers continues to grow, nano-enabled solutions can offer enhanced product performance and sustainability.

Generally, there are strategies for maintaining pulp stability and quality:

- The pulp and paper industry focuses on strategies to improve product quality and performance.
- Lignocellulosic nanofibers (LCNF), microfibrillated cellulose (MFC), or cellulose nanofibers (CNF) are used in paper production to increase paper quality and performance.
- Filtration material improves the discharge rate of wet and dehydrated paper by neutralizing the load on the fiber and filler surfaces.
- Researchers are investigating the interactions between cellulose pulp, additives, CNF, and mineral fillers to solve dewatering problems.
- Nanotechnology provides practical and environmentally friendly alternatives to existing technologies and enables the discovery of new materials, processes, and phenomena at the nanoscale.
- Malaysia has been developing nanotechnology since the beginning of the millennium and the government started its microtechnology and nanotechnology development in 2001.
- Nano additives can enhance papermaking industries by providing unique properties and capabilities.
- Natural carbon (NC), a renewable and environmentally friendly polymer reinforcement, is combined with cationic polymers to increase paper strength.

- NC-based additives can increase the utilization of recycled paper and strengthen recycled eucalyptus pulp.
- Excessive NC can have negative consequences and reduce fiber–fiber bond strength.
- Interest in alternative raw materials, such as lignocellulosic fibers from nonwood sources, addresses concerns about pulp yield and mechanical properties of paper.

Competing Interests The authors declare that they have no competing interest.

DataAvailability Declaration of Generative AI and AI-Assisted Technologies in the Writing Process During the preparation of this work, the book chapter author(s) did not take any assistance from AI, and the book chapter author(s) reviewed and edited the content as needed and take(s) full responsibility for the content of the publication.

References

1. Abbasi R, Shineh G, Mobaraki M, Doughty S, Tayebi L (2023) Structural parameters of nanoparticles affecting their toxicity for biomedical applications: a review. J Nanopart Res 25(3):43. https://doi.org/10.1007/s11051-023-05690-w
2. Abdallah AF, Jawaid M, Mohamed AZ, Tahir PM, Abdullah UH (2023) Performance of nanofibrillated cellulose with chitosan as a wet-end additive for paper applications. Ind Crops Prod 203:117219. https://doi.org/10.1016/j.indcrop.2023.117219
3. Abolore RS, Jaiswal S, Jaiswal AK (2023) Green and sustainable pretreatment methods for cellulose extraction from lignocellulosic biomass and its applications: a review. Carbohydr Polym Technol Appl 100396
4. Ahire SA, Bachhav AA, Pawar TB, Jagdale BS, Patil AV, Koli PB (2022) The augmentation of nanotechnology era: a concise review on fundamental concepts of nanotechnology and applications in material science and technology. Results Chem 4:100633. https://doi.org/10.1016/j.rechem.2022.100633
5. Al-Azkawi A, Elliston A, Al-Bahry S, Sivakumar N (2019) Waste paper to bioethanol: current and future prospective. Biofuels Bioprod Biorefin 13(4):1106–1118
6. Aljabali AA, Obeid MA, Al Zoubi MS, Charbe NB, Chellappan DK, Mishra V, Dureja H et al (2022) Nanocelluloses in sensing technology: design, preparation, and applications. In: Handbook of nanocelluloses: classification, properties, fabrication, and emerging applications, pp 745–74. Springer. https://doi.org/10.1007/978-3-030-89621-8_44
7. Allan J, Belz S, Hoeveler A, Hugas M, Okuda H, Patri A, Rauscher H et al (2021) Regulatory landscape of nanotechnology and nanoplastics from a global perspective. Regul Toxicol Pharmacol 122:104885. https://doi.org/10.1016/j.yrtph.2021.104885
8. Arantes V, Las-Casas B, Dias IKR, Yupanqui-Mendoza SL, Nogueira CFO, Marcondes WF (2024) Enzymatic approaches for diversifying bioproducts from cellulosic biomass. Chem Commun 60(72):9704–9732. https://doi.org/10.1039/d4cc02114b
9. Atakan B (2014) Molecular communications and nanonetworks. Mol Commun Nanonetworks. https://doi.org/10.1007/978-1-4939-0739-7
10. Bajpai P (2010) Environmentally friendly production of pulp and paper. Wiley & Sons, Environmentally friendly production of pulp and paper. https://doi.org/10.1002/9780470649657
11. Bajpai P (2011) Environmentally friendly production of pulp and paper. Wiley & Sons
12. Bajpai P (2024) Recycling and Deinking of Recovered Paper. Elsevier, Recycling and Deinking of Recovered Paper. https://doi.org/10.1016/C2023-0-00941-4

13. Bajpai P (2005) ECF and TCF bleaching. In: Environmentally benign approaches for pulp bleaching, pp 177–92. https://doi.org/10.1016/b978-044451724-1/50005-8
14. Bajpai P (2015) Basic overview of pulp and paper manufacturing process. In: Green chemistry and sustainability in pulp and paper industry, pp 11–39. https://doi.org/10.1007/978-3-319-18744-0_2
15. Balea A, Elena Fuente M, Monte C, Merayo N, Campano C, Negro C, Blanco A (2020) Industrial application of nanocelluloses in papermaking: a review of challenges, technical solutions, and market perspectives. Molecules 25(3):526. https://doi.org/10.3390/molecules 25030526
16. Balea A, Fuente E, Monte MC, Merayo N, Campano C, Negro C, Blanco A (2020) Industrial application of nanocelluloses in papermaking: a review of challenges, technical solutions, and market perspectives. Molecules 25(3). https://doi.org/10.3390/molecules25030526.
17. Balea A, Merayo N, Fuente E, Delgado-Aguilar M, Mutje P, Blanco A, Negro C (2016) Valorization of corn stalk by the production of cellulose nanofibers to improve recycled paper properties. BioResources 11(2):3416–31. https://doi.org/10.15376/biores.11.2.3416-3431
18. Barhoum A, Rastogi VK, Mahur BK, Rastogi A, Abdel-Haleem FM, Samyn P (2022) Nanocelluloses as new generation materials: natural resources, structure-related properties, engineering nanostructures, and technical challenges. Mater Today Chem 26:101247. https://doi.org/10.1016/j.mtchem.2022.101247
19. Basak S, Dangate MS, Samy S (2024) Oil-and water-resistant paper coatings: a review. Prog Org Coat 186:107938
20. Benselfelt T, Kummer N, Nordenström M, Fall AB, Nyström G, Wågberg L (2023) The colloidal properties of nanocellulose. Chemsuschem 16(8):e202201955. https://doi.org/10.1002/cssc.202201955
21. Bhaskar T, Sera A, Muto A, Sakata Y (2008) Hydrothermal upgrading of wood biomass: influence of the addition of K_2CO_3 and cellulose/lignin ratio. Fuel 87(10–11):2236–2242. https://doi.org/10.1016/j.fuel.2007.10.018
22. Blanco A, Negro C, Monte C, Fuente E, Tijero J (2004) The challenges of sustainable papermaking. ACS Publications, Environmental science and technology
23. Boufi S, González I, Delgado-Aguilar M, Tarrès Q, Mutjé P (2017) Nanofibrillated cellulose as an additive in papermaking process. Cellul-Reinf Nanofibre Compos Prod Prop Appl 154:153–173. https://doi.org/10.1016/B978-0-08-100957-4.00007-3
24. Boverhof DR, Bramante CM, Butala JH, Clancy SF, Lafranconi WM, West J, Gordon SC (2015) Comparative assessment of nanomaterial definitions and safety evaluation considerations. Regul Toxicol Pharmacol 73(1):137–150. https://doi.org/10.1016/j.yrtph.2015.06.001
25. Bumgardner M, Johnson S, Luppold W, Maplesden F, Pepke E (2013) Markets and market forces for lumber. Glob For Sect Change Pract, Prospect 41–76. https://doi.org/10.1201/b16186.
26. Cashin P, Mohaddes K, Raissi M (2017) China's slowdown and global financial market volatility: is world growth losing out? Emerg Mark Rev 31:164–175. https://doi.org/10.1016/j.ememar.2017.05.001
27. Castellan A, Da Silva Perez D, Nourmamode A, Grelier S, Terrones MGH, Machado AEH, Ruggiero R (1999) The improvement of the bleaching of peroxyformic sugar cane bagasse pulp by photocatalysis and photosensitization. J Braz Chem Soc 10(3):197–202. https://doi.org/10.1590/S0103-50531999000300007
28. Chauhan VS, Bhardwaj NK (2015) Amphoteric starch preflocculated talc filler for improving composite tensile index and filler bondability of paper. Nord Pulp Pap Res J 30(3):436–442. https://doi.org/10.3183/npprj-2015-30-03-p436-442
29. Da Silva PC, de Oliveira Neto GC, Correia JM, Tucci HNP (2021) Evaluation of economic, environmental and operational performance of the adoption of cleaner production: survey in large textile industries. J Clean Prod 278:123855. https://doi.org/10.1016/j.jclepro.2020.123855

30. Das KC, Singh K, Bibens B, Hilten R, Baker SA, Greene WD, Peterson JD (2011) Pyrolysis characteristics of forest residues obtained from different harvesting methods. Appl Eng Agric 27(1):107–113
31. Daskalopoulos E, Badr O, Probert SD (1997) Economic and environmental evaluations of waste treatment and disposal technologies for municipal solid waste. Appl Energy 58(4):209–255. https://doi.org/10.1016/S0306-2619(97)00053-6
32. Deivayanai VC, Yaashikaa PR, Senthil Kumar P, Rangasamy G (2022) A comprehensive review on the biological conversion of lignocellulosic biomass into hydrogen: pretreatment strategy, technology advances and perspectives. Biores Technol 365:128166. https://doi.org/10.1016/j.biortech.2022.128166
33. Deshpande Sarma S, Anand M (2012) Status of nano science and technology in India. Proc Natl Acad Sci India Sect B Biol Sci 82(S1):99–126. https://doi.org/10.1007/s40011-012-0077-2
34. Devasahayam S (2018) Nanotechnology and nanomedicine in market: a global perspective on regulatory issues. A global perspective on regulatory issues. In: Characterization and biology of nanomaterials for drug delivery: nanoscience and nanotechnology in drug delivery, pp 477–522. Elsevier. https://doi.org/10.1016/B978-0-12-814031-4.00017-9
35. Dhiman J, Anupam K, Kumar V, Saruchi (2024) Varieties, characteristics, and applications of nanofillers in pulp and paper industry. In: Handbook of nanofillers, pp 1–32. Springer. https://doi.org/10.1007/978-981-99-3516-1_43-1
36. Domingues C, Santos A, Alvarez-Lorenzo C, Concheiro A, Jarak I, Veiga F, Barbosa I, Dourado M, Figueiras A (2022) Where is nano today and where is it headed? A review of nanomedicine and the dilemma of nanotoxicology. ACS Nano 16(7):9994–10041. https://doi.org/10.1021/acsnano.2c00128
37. Ehman NV, Lourenço AF, McDonagh BH, Vallejos ME, Felissia FE, Ferreira PJT, Chinga-Carrasco G, Area MC (2020) Influence of initial chemical composition and characteristics of pulps on the production and properties of lignocellulosic nanofibers. Int J Biol Macromol 143:453–461. https://doi.org/10.1016/j.ijbiomac.2019.10.165
38. El-Sayed A, Essam S, El-Sakhawy M, El-Sakhawy MAM (2020) Nonwood fibers as raw material for pulp and paper industry. Nord Pulp Pap Res J 35(2):215–230. https://doi.org/10.1515/npprj-2019-0064
39. Etale A, Onyianta AJ, Turner SR, Eichhorn SJ (2023) Cellulose: a review of water interactions, applications in composites, and water treatment. Chem Rev 123(5):2016–2048. https://doi.org/10.1021/acs.chemrev.2c00477
40. Fore S, Mbohwa CT (2010) Cleaner production for environmental conscious manufacturing in the foundry industry. J Eng Des Technol 8(3):314–333. https://doi.org/10.1108/17260531011086180
41. Forster SP, Olveira S, Seeger S (2011) Nanotechnology in the market: promises and realities. Int J Nanotechnol 8(6–7):592–613. https://doi.org/10.1504/IJNT.2011.040193
42. Geffert A, Vacek O, Jankech A, Geffertová J, Milichovský M (2017) Swelling of cellulosic porous materials-mathematical description and verification. BioResources 12(3):5017–30. https://doi.org/10.15376/biores.12.3.5017-5030
43. Gordon RJ (2003) Europe chasing the american frontier. University of Oviedo, Department of Economics, Oviedo Efficiency Group (OEG)
44. Gordon RJ (2004) Two centuries of economic growth: Europe chasing the American frontier. NBER working paper. National Bureau of Economic Research Cambridge, Mass., USA
45. Gronfors J (2010) Use of fillers in paper and paperboard grades, pp 1–36. https://theseus.fi/bitstream/handle/10024/16226/Gronfors_Jarkko.pdf?sequence=1
46. Han N, Zhang J, Hoang M, Gray S, Xie Z (2021) A review of process and wastewater reuse in the recycled paper industry. Environ Technol Innov 24:101860. https://doi.org/10.1016/j.eti.2021.101860
47. Han L, Zhang X, Ji J, Ma K (2022) Research progress on the influence of nano-additives on phase change materials. J Energy Storage 55:105807. https://doi.org/10.1016/j.est.2022.105807

48. Hariram NP, Mekha KB, Suganthan V, Sudhakar K (2023) Sustainalism: an integrated socio-economic-environmental model to address sustainable development and sustainability. Sustain (Switz) 15(13):10682. https://doi.org/10.3390/su151310682
49. Harish MS, Bhuker A, Chauhan BS (2024) Millet production, challenges, and opportunities in the Asia-Pacific region: a comprehensive review. Front Sustain Food Syst 8:1386469. https://doi.org/10.3389/fsufs.2024.1386469
50. Hashemzehi M, Mesic B, Sjöstrand B, Naqvi M (2022) A comprehensive review of nanocellulose modification and applications in papermaking and packaging: challenges, technical solutions, and perspectives. BioResources 17(2): 3718–80. https://doi.org/10.15376/biores.17.2.Hashemzehi
51. Hawanis HSN, Ilyas RA, Jalil R, Ibrahim R, Majid RA, Hamid NHA (2024) Insights into lignocellulosic fiber feedstock and its impact on pulp and paper manufacturing: a comprehensive review. Sustain Mater Technol 40:e00922. https://doi.org/10.1016/j.susmat.2024.e00922
52. He L, Wang H, Liu F (2024) Emission abatement in low-carbon supply chains with government subsidy and information asymmetry. Int J Prod Res 1–29
53. Hellström P, Heijnesson-Hultén A, Paulsson M, Håkansson H, Germgård U (2014) Fenton pre-treated microfibrillated cellulose evaluated as a strength enhancer in the middle ply of paperboard. Nord Pulp Pap Res J 29(4):732–740. https://doi.org/10.3183/npprj-2014-29-04-p732-740
54. Hornyak, GL, Moore JJ, Tibbals HF, Dutta J (2008) Fundamentals of nanotechnology. CRC press. https://doi.org/10.2174/97898152388151240100002
55. Horstkotte B (2023) The automation technique lab-in-syringe: developments and applications. M.Sc. thesis, Habilitation
56. Ibrahim NI, Shahar FS, Sultan MTH, Ain Umaira M, Shah SN, Safri A, Yazik MHM (2021) Overview of bioplastic introduction and its applications in product packaging. Coatings 11(11):1423. https://doi.org/10.3390/COATINGS11111423
57. Jing S, Lianping W, Siciliano AP, Chen C, Li T, Liangbing H (2023) The critical roles of water in the processing, structure, and properties of nanocellulose. ACS Nano 17(22):22196–22226. https://doi.org/10.1021/acsnano.3c06773
58. Karthika R, Jayanthi B, Aruna A, Selvankumar T (2022) Nanocellulosic materials for papermaking and paper coating industry. In: Handbook of nanocelluloses: classification, properties, fabrication, and emerging applications, pp 1001–33. Springer. https://doi.org/10.1007/978-3-030-89621-8_1
59. Kaur D, Bhardwaj NK, Lohchab RK (2017) Prospects of rice straw as a raw material for paper making. Waste Manage 60:127–139. https://doi.org/10.1016/j.wasman.2016.08.001
60. Kerry JP (2013) New packaging technologies, materials and formats for fast-moving consumer products, 2nd edn. In: Innovations in food packaging, pp 549–84. Elsevier. https://doi.org/10.1016/B978-0-12-394601-0.00023-0
61. Kumari S, Raturi S, Kulshrestha S, Chauhan K, Dhingra S, András K, Thu K, Khargotra R, Singh T (2023) A comprehensive review on various techniques used for synthesizing nanoparticles. J Market Res 27:1739–1763. https://doi.org/10.1016/j.jmrt.2023.09.291
62. Kundra SN (2008) Toward the emergence of nanoneurosurgery: part III-nanomedicine: targeted nanotherapy, nanosurgery and progress toward the realization of nanoneurosurgery. Neurosurgery 62(6):1009–1026
63. Li A, Dezhong X, Luo L, Zhou Y, Yan W, Leng X, Dai D et al (2021) Overview of nanocellulose as additives in paper processing and paper products. Nanotechnol Rev 10(1):264–281. https://doi.org/10.1515/ntrev-2021-0023
64. Little AC, Jones BC, Debruine LM (2011) Facial attractiveness: evolutionary based research. Philos Trans R Soc B Biol Sci 366(1571):1638–1659. https://doi.org/10.1098/rstb.2010.0404
65. Małachowska E (2025) Impact of retention agents on functional properties of recycled paper in sustainable manufacturing. Appl Sci 15(2):875
66. Malik S, Muhammad K, Waheed Y (2023) Nanotechnology: a revolution in modern industry. Molecules 28(2):661. https://doi.org/10.3390/molecules28020661

67. Malik S, Singh J, Goyat R, Saharan Y, Chaudhry V, Umar A, Ibrahim AA, Akbar S, Ameen S, Baskoutas S (2023) Nanomaterials-based biosensor and their applications: a review. Heliyon 9(9). https://doi.org/10.1016/j.heliyon.2023.e19929
68. Mboowa D (2024) A review of the traditional pulping methods and the recent improvements in the pulping processes. Biomass Convers Biorefinery 14(1):1–12. https://doi.org/10.1007/s13399-020-01243-6
69. Miyashiro D, Hamano R, Umemura K (2020) A review of applications using mixed materials of cellulose, nanocellulose and carbon nanotubes. Nanomaterials 10(2):186. https://doi.org/10.3390/nano10020186
70. Morsy FA, El-Sherbiny S, Samir M, Fouad OA (2016) Application of nanostructured titanium dioxide pigments in paper coating: a comparison between prepared and commercially available ones. J Coat Technol Res 13(2):307–316. https://doi.org/10.1007/s11998-015-9735-7
71. Muhd Julkapli N, Bagheri S (2016) Developments in nano-additives for paper industry. J Wood Sci 62(2):117–130. https://doi.org/10.1007/s10086-015-1532-5
72. Muhd Julkapli N, Bagheri S, Mansouri N (2017) Nanomaterials: an advanced and versatile nanoadditive for kraft and paper industries. Handb Compos Renew Mater 1–8:305–326. https://doi.org/10.1002/9781119441632.ch159
73. Muhd Julkapli N, Nurhidayatullaili MJ, Bagheri S (2016) Developments in nano-additives for paper industry
74. Mujtaba M, Fraceto LF, Fazeli M, Mukherjee S, Savassa SM, de Medeiros GA, Santo Pereira AD, Mancini SD, Lipponen J, Vilaplana F (2023) Lignocellulosic biomass from agricultural waste to the circular economy: a review with focus on biofuels, biocomposites and bioplastics. J Clean Prod 402:136815. https://doi.org/10.1016/j.jclepro.2023.136815
75. Perera D (2023) Paper recycling, vol 4, pp 164–186. https://www.researchgate.net/publication/372365680_Paper_Recycling_for_a_Sustainable_Future_Global_Trends
76. Przybysz P, Dubowik M, Kucner MA, Przybysz K, Buzała KP (2016) Contribution of hydrogen bonds to paper strength properties. PLoS ONE 11(5):e0155809. https://doi.org/10.1371/journal.pone.0155809
77. Psarommatis F, Sousa J, Mendonça JP, Kiritsis D (2022) Zero-defect manufacturing the approach for higher manufacturing sustainability in the era of Industry 4.0: a position paper. Int J Prod Res 60(1):73–91. https://doi.org/10.1080/00207543.2021.1987551
78. Quintana E, Valls C, Blanca Roncero M (2024) Dissolving-grade pulp: a sustainable source for fiber production. Wood Sci Technol 58(1):23–85. https://doi.org/10.1007/s00226-023-015 19-w
79. Rangasamy M (2011) Nano technology: a review. J Appl Pharm Sci (no. Issue):8–16
80. Del Rio F, Dylan D, Sovacool BK, Griffiths S, Bazilian M, Kim J, Foley AM, Rooney D (2022) Decarbonizing the pulp and paper industry: a critical and systematic review of sociotechnical developments and policy options. Renew Sustain Energy Rev 167:112706. https://doi.org/10.1016/j.rser.2022.112706
81. Saleh HM, Hassan AI (2023) Synthesis and characterization of nanomaterials for application in cost-effective electrochemical devices. Sustain (Switz) 15(14):10891. https://doi.org/10.3390/su151410891
82. Samanta KK, Basak S, Chattopadhyay SK (2015) Recycled fibrous and nonfibrous biomass for value-added textile and nontextile applications. Environ Footpr Eco-Des Prod Process 167–212. https://doi.org/10.1007/978-981-287-643-0_8
83. Sampson WW (2009) Materials properties of paperas influenced by its fibrous architecture. Int Mater Rev 54(3):134–156. https://doi.org/10.1179/174328009X411154
84. Saxl O (2013) Nanotechnology: applications and markets, present and future. In: Ellipsometry at the nanoscale, pp 705–30. Springer. https://doi.org/10.1007/978-3-642-33956-1
85. Schaefer D, Cheung WM (2018) Smart packaging: opportunities and challenges. Procedia Cirp 72:1022–1027
86. Sciences, National Academies of, Division on Engineering, Physical Sciences, National Materials, Manufacturing Board, Committee on National Nanotechnology Initiative, and A

Quadrennial Review (2020) A quadrennial review of the national nanotechnology initiative: nanoscience, applications, and commercialization, pp 1–126
87. Selvamaharajan G, Chauhan Vipul S, Bhardwaj Nishi K, Varadhan R (2013) Impact of different fillers during papermaking using multi-layer fabrics. IPPTA Q J Indian Pulp Pap Tech Assoc 25(4):135–39
88. Senarathna C, Maneth D, Weerasinghe J, Gamage M, Thennakoon T, Hewage HTM, Panagoda S, Sandunika I, Perera D (2023) Paper recycling for a sustainable future: global trends, pp 169–86
89. Sengupta S (2024) Plant fibres & its potential for sustainable development. J Andaman Sci Assoc 29(1):1–15
90. Shahcheraghi N, Golchin H, Sadri Z, Tabari Y, Borhanifar F, Makani S (2022) Nano-biotechnology, an applicable approach for sustainable future. 3 Biotech 12(3):65. https://doi.org/10.1007/s13205-021-03108-9
91. Shan L, Tan CY, Shen X, Ramesh S, Zarei MS, Kolahchi R, Hajmohammad MH (2023) The effects of nano-additives on the mechanical, impact, vibration, and buckling/post-buckling properties of composites: a review. J Market Res 24:7570–7598
92. Singh V, Yadav P, Mishra V (2020) Recent advances on classification, properties, synthesis, and characterization of nanomaterials. Green Synth Nanomater Bioenergy Appl 83–97. https://doi.org/10.1002/9781119576785.ch3
93. Solala I, Iglesias MC, Peresin MS (2020) On the potential of lignin-containing cellulose nanofibrils (LCNFs): a review on properties and applications. Cellulose 27(4):1853–1877. https://doi.org/10.1007/s10570-019-02899-8
94. Song Y, Yuan Y, Zhu J (2024) A review on applications of fine particles integrated with fluidization technologies. Can J Chem Eng. https://doi.org/10.1002/cjce.25260
95. Su J, Zhang L, Batchelor W, Garnier G (2014) Paper engineered with cellulosic additives: effect of length scale. Cellulose 21(4):2901–2911. https://doi.org/10.1007/s10570-014-0298-z
96. Suota MJ, Alessandre T, da Silva S, Zawadzki F, Sassaki GL, Hansel FA, Paleologou M, Ramos LP (2021) Chemical and structural characterization of hardwood and softwood LignoForce^{TM} lignins. Ind Crops Prod 173:114138. https://doi.org/10.1016/j.indcrop.2021.114138
97. Szabó L, Soria A, Forsström J, Keränen JT, Hytönen E (2009) A world model of the pulp and paper industry: demand, energy consumption and emission scenarios to 2030. Environ Sci Policy 12(3):257–269. https://doi.org/10.1016/j.envsci.2009.01.011
98. Taib M, Mohamad NA (2023) Nanofillers in pulp and paper. In: Nanofillers for sustainable applications, pp 198–213. CRC Press. https://doi.org/10.1201/9781003400998-10
99. Tajik M, Torshizi HJ, Resalati H, Hamzeh Y (2021) Effects of cellulose nanofibrils and starch compared with polyacrylamide on fundamental properties of pulp and paper. Int J Biol Macromol 192:618–626. https://doi.org/10.1016/j.ijbiomac.2021.09.199
100. Telmo C, Lousada J (2011) The explained variation by lignin and extractive contents on higher heating value of wood. Biomass Bioenerg 35(5):1663–1667. https://doi.org/10.1016/j.biombioe.2010.12.038
101. Teseme BW (2020) Review on the manufacturing of biodegradable plastic packaging film from root and tuber starches. Am J Nano Res Appl 8(1):1. https://doi.org/10.11648/j.nano.20200801.11
102. Thu HE, Haider MA, Khan S, Sohail M, Hussain Z (2023) Nanotoxicity induced by nanomaterials: a review of factors affecting nanotoxicity and possible adaptations. OpenNano 14:100190. https://doi.org/10.1016/j.onano.2023.100190
103. Villanueva A, Wenzel H (2007) Paper waste-recycling, incineration or landfilling? A review of existing life cycle assessments. Waste Manage 27(8):S29-46. https://doi.org/10.1016/j.wasman.2007.02.019
104. Vogiantzi C, Tserpes K (2023) On the definition, assessment, and enhancement of circular economy across various industrial sectors: a literature review and recent findings. Sustain (Switz) 15(23):16532. https://doi.org/10.3390/su152316532

105. Wang Q, Oshita K, Takaoka M (2021) Flocculation properties of eight microalgae induced by aluminum chloride, chitosan, amphoteric polyacrylamide, and alkaline: life-cycle assessment for screening species and harvesting methods. Algal Res 54:102226. https://doi.org/10.1016/j.algal.2021.102226
106. Worku LA, Tadesse MG, Bachheti RK*, Bachheti A, Husen A (2024) Synthesis of carboxylated cellulose nanocrystal/ZnO nanohybrids using Oxytenanthera abyssinica cellulose and zinc nitrate hexahydrate for radical scavenging, photocatalytic, and antibacterial activities. Int J Biol Macromol 267:131228. https://doi.org/10.1016/j.ijbiomac.2024.131228
107. Yang X, Chen Y, Zhang C, Duan G, Jiang S (2023) Electrospun carbon nanofibers and their reinforced composites: preparation, modification, applications, and perspectives. Compos B Eng 249:110386. https://doi.org/10.1016/j.compositesb.2022.110386
108. Yeats AJ (2023) Just how big is global production sharing? Fragmentation 108–43. https://doi.org/10.1093/oso/9780199243310.003.0007
109. Yu S, Sun J, Shi Y, Wang Q, Jian W, Liu J (2021) Nanocellulose from various biomass wastes: its preparation and potential usages towards the high value-added products. Environ Sci Ecotechnology 5:100077. https://doi.org/10.1016/j.ese.2020.100077
110. Yu H, Yuqin X, Ni Y, Qiong W, Shiwei Liu L, Li SY, Ji Z (2018) Enhanced enzymatic hydrolysis of cellulose from waste paper fibers by cationic polymers addition. Carbohyd Polym 200:248–254. https://doi.org/10.1016/j.carbpol.2018.07.079
111. Zhang H, Shiyu F, Chen Y (2020) Basic understanding of the color distinction of lignin and the proper selection of lignin in color-depended utilizations. Int J Biol Macromol 147:607–615. https://doi.org/10.1016/j.ijbiomac.2020.01.105
112. Zhao X, Liu L, Wang Y, Dai H, Wang D, Cai H (2008) Influences of partially hydrolyzed polyacrylamide (HPAM) residue on the flocculation behavior of oily wastewater produced from polymer flooding. Sep Purif Technol 62(1):199–204. https://doi.org/10.1016/j.seppur.2008.01.019

Recent Advances in Wood Pulping, Lignification, Extraction, Isolation Processes, and Their Applications

Mohamed Azlan Suhot, Mohamad Zaki Hassan, and Nicholas Kuan Hoo Tien

Abstract This chapter discusses the potential for recovering lignin from the pulp and paper industry as an industrial resource for various applications by transforming it from the status of waste. An overview of traditional and innovative ways of lignin recovery, including the Kraft process and organosolv recovery, together with the most recent trends, such as ultrasonic-assisted recovery and ionic liquid methods, is given at the beginning. This chapter covers various applications for which the recovered lignin can be used, ranging from environmental remediation as an excellent adsorbent to developing energy storage solutions, like batteries and supercapacitors, and construction materials for improving structural properties. It enlists the challenges associated with scale-up of lignin recovery and application based on the variability in structure and composition of lignin, economic viability, and establishment of standardized protocols. The chapter concludes with the note of future research directions, focusing on advanced recovery methods and the optimization of lignin properties for specific applications, coupled with sustainable technologies in enhancing the role of lignin toward a circular economy. Specific research is thus deemed necessary for the development of advanced recovery methods, the adjustment of lignin properties to suit the requirements of its various applications, and the sustainable technologies that could enhance the role of lignin within a circular economy.

Keywords Lignin recovery · Industrial applications · Sustainable technologies · Circular economy

M. A. Suhot (✉) · M. Z. Hassan
Faculty of Artificial Intelligence, Universiti Teknologi Malaysia, Kuala Lumpur, Malaysia
e-mail: azlans.kl@utm.my

N. K. H. Tien
Faculty of Engineering, Universiti Malaysia Sarawak, Kota Samarahan, Malaysia

R. K. Bachheti (ed.), *Nanotechnology in the Pulp and Paper Industry*, Smart Nanomaterials Technology, https://doi.org/10.1007/978-981-95-3473-9_15

1 Introduction

The emergence of advanced technologies in the pulp and paper industry has opened great opportunities to recover lignin, the industry's major by-product, which is generally treated as waste with low value. Recovery and subsequent use of lignin align with environmental needs and hold great potential for industrial innovation and related economic benefits. The pulp and paper industry is one of the largest industrial polluters. The industry produces vast quantities of lignocellulosic waste, most of which is lignin. Some conventional processes include Kraft pulping, sulfite pulping, and some relatively new methods, such as organosolv, which have been used for separating cellulose fiber from lignin [1, 2]. While it has been able to remove lignin, most of the process uses hazardous chemicals that threaten the environment and require high-energy resources [3–5]. For instance, the Kraft pulping process generates black liquor, which is toxic and has high chemical oxygen demand (COD) and biological oxygen demand (BOD) and, therefore, leads to water pollution if not treated before being released into the environment [6]. In addition, the complexity and hard nature of lignin make its recovery difficult [5]. Lignin has a heterogeneous and highly branched polymer structure; therefore, it is highly stable and resists degradation under harsh conditions for recovery and conversion processes. This results in increased energy usage and additional production costs, highly complicating lignin recovery in the pulp and paper industry [6].

Recent advances in technology for recovering lignin have been aimed at being more efficient and safer for the environment. An example is the organosolv pulping process, which recovers relatively pure lignin by extracting it with organic solvents, whereas supercritical fluid recovery extracts lignin using less toxic solvents [7, 8]. New methods in this direction include microwave-assisted recovery and ionic liquids (ILs) to increase the efficiency and selectivity in the lignin recovery process [7].

Some applications of the recovered lignin include high-value products such as biofuels, biopolymers, phenols, and advanced materials like carbon fibers and adsorbents [9, 10]. In addition, due to its aromatic structure and high carbon content, lignin possesses suitable characteristics that allow for its use in dispersants, adhesives, and flocculants [11]. Due to different pulping processes, lignin's heterogeneous nature presents some challenges in the recovery process. This factor calls for further research, especially in setting up standardized and controlled modifications to increase the use of lignin across industries [12]. Besides that, the economic viability and technological scalability of lignin recovery should also be addressed for its industrial use [13].

2 Structure of Lignin

Lignin is a complex aromatic biopolymer. It contains mainly phenylpropanoid units containing three main monomers: p-coumaryl alcohol, coniferyl alcohol, and sinapyl alcohol. The monomers are linked in random, highly branched chains, forming complex, heterogeneous, and three-dimensional structures. The aromatic property of lignin arises from the monomers containing aromatic rings, to which it is owed that this macromolecule is so rigid and resistant to degradation [14]. The majority of the lignocellulosic biomass in the pulp and paper industry is contributed by lignin, which has shown great potential for being a sustainable source to provide renewable materials, chemicals, and energy (Fig. 1). Recovery processes from biomass are important operations for valorizing plant residues and enhancing the economic viability of biorefineries [15, 16].

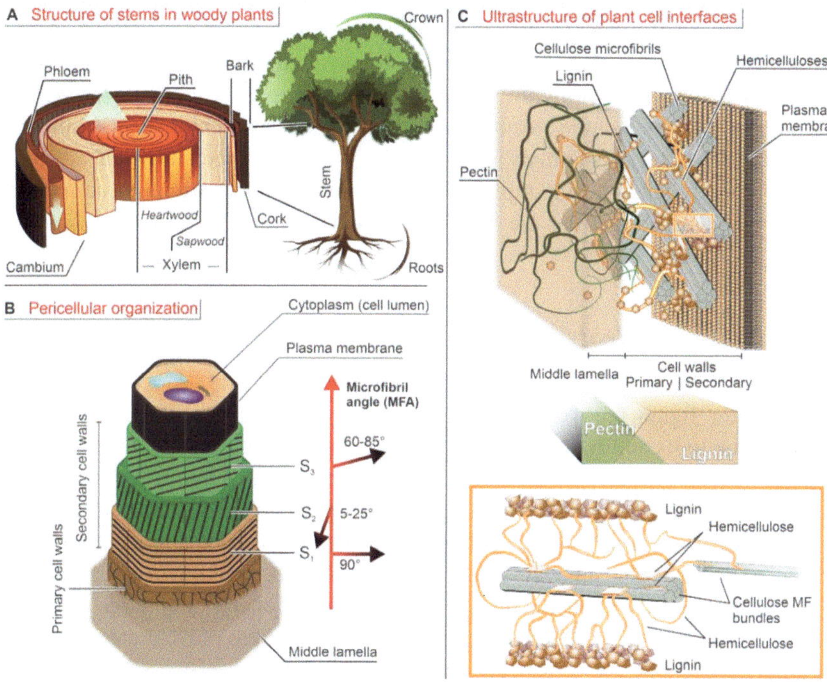

Fig. 1 A typical cross-section of **A** stems in woody plants, **B** a multilayer structure encircling the lumen, and **C** hemicelluloses and pectin embedding cellulose microfibrils in cell walls [17]

3 Lignin Recovery: Methods and Advances

Lignin recovery is a critical step in the lignocellulosic valorization process in the pulp and paper industry, providing a way for producing high-added-value materials, chemicals, or biofuels. Numerous methods have been developed to isolate lignin from biomass efficiently, each displaying a different mechanism and having various advantages and limitations. Some of the processes currently applied in the industry include the Kraft, sulfite, and soda processes, as these have proved their high efficiency in the decomposition of complex biomasses; nevertheless, they involve harsh chemicals and have severe environmental impacts [3, 18]. Compared to these, newer techniques like the organosolv process, ionic liquid recovery, and deep eutectic solvent (DES) recovery are better sustainable alternatives because they provide lignin with high purity and very minimal structural degradation, making an important difference in the environmental footprint [7, 19]. Advanced methods further improve lignin recovery by optimizing conditions and enhancing selectivity, including enzymatic hydrolysis, microwave-assisted recovery, supercritical fluid recovery, and nonthermal plasma treatment [7, 18]. Other than these, pretreatment methodologies were developed to maximize the yield of lignin while still preserving its functional properties. These pretreatments include steam explosion, acid pretreatment, and hydrothermal processes [20]. Selected methods for lignin recovery, recent technological advances in these methods, and relevant implications for industrial applications, sustainability, and future research directions are discussed in this section.

3.1 Alkaline Hydrolysis

The basis for lignin recovery is alkaline hydrolysis. The use of NaOH is most preferred due to its ability to break down a large part of the lignocellulose structure during alkaline treatment without altering the aromatic core of lignin to a large extent [15]. In this method, lignin can be isolated with a low degree of degradation, ensuring that it remains functional. This is key to developing high-value applications—such as in phenolic resins and adhesives [3]. However, this process is environmentally less attractive, with the use of chemicals like NaOH and the generation of large quantities of alkaline wastewater, which require proper management to prevent environmental contamination [7]; hence, effective improvements in recovery processes are required [1].

3.2 Ultrasonic-Assisted Alkaline Treatment

In the current application of ultrasonic-assisted recovery for lignin from biomass, there is a possibility of improving lignin yield while reducing the environmental footprint during the recovery process. High-frequency ultrasound can facilitate bond breakdown within the biomass, which makes this technique attractive for lignin recovery [15]. This technique led to the recovery of 19.02% lignin using ultrasound irradiation at 500 W power in an alkaline solution (2 M of sodium hydroxide) at 90 °C. The process was carried out in an ethanol solution, and the pH of the mixture was adjusted to 5.5 by using HCl before precipitating it. The yield of lignin increased with increased sonication times to a maximum of 84.9%, which implies that most of the lignin was extracted within the first 30 min of sonication. Alkaline recovery with ultrasound resulted in an increase in yield by 12.3% in comparison with conventional methods. Such improvements are crucial for technological development in industrial applications, especially given the challenges of equipment wear and energy use [4].

3.3 *Kraft Process*

The Kraft process is a well-established technique for isolating lignin from wood in the pulp and paper industry. While the technology shows high efficiency in isolating lignin from cellulose fibers, there have been strong concerns about the environmental impact of black liquor rich in sulfur and the energy-intensive recovery process [2]. Black liquor mainly consists of Kraft lignin; it represents approximately 85% of the technical lignins available worldwide and is commonly burned to regain energy and chemicals (Fig. 2). The energy content of black liquor thus becomes quite critical in pulp mill energy demands [2].

Innovations in new technologies, like LignoBoost and LignoForce, enable better lignin recovery from black liquor. LignoBoost technology increases the yield of extracted lignin by acidification of black liquor, which leads to a pH decrease, precipitating lignin with further washing and filtration steps to obtain improved purity of the recovered lignin. This greatly reduces the content of inorganic impurities and sulfur

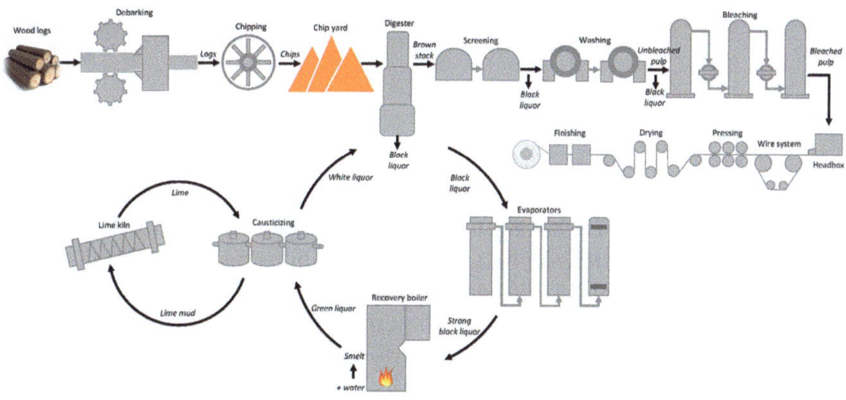

Fig. 2 A summary of the Kraft pulping plant [22]

to a low level, thus making the lignin more suitable for high-value applications such as composite materials and renewable raw materials for chemicals [4].

Similarly, the LignoForce technology uses oxidation followed by acidification to control the formation of odorous sulfur compounds and even improve the safety and the environmental profile of the recovery process. The LignoForce technology is also equipped with a stage of filtration that helps achieve high purity and minimal environmental effects from the recovery of lignin [4]. These technologies should be considered for implementation in traditional pulp mills to improve overall sustainability [21].

3.4 Organosolv Recovery

Organosolv recovery is the process of delignification using organic solvents, usually in combination with an acid or alkali catalyst and at a lower temperature than the Kraft process. The end product is high-purity lignin that finds its value in a number of applications, from high-performance composites to pharmaceuticals [16]. The lignin that the organosolv process can obtain is of great interest and value for such applications due to its relatively narrow molecular weight range and low condensation, making it more reactive and more usable in further processes [16].

The recoverability and reusability of the solvents used in the process are features of high sustainability with respect to environmental concerns, and these features can make the process economically feasible compared to conventional methods [3]. For instance, solvents such as ethanol in organosolv processes can be easily recovered through distillation, thus avoiding both economic and environmental costs [16]. Besides, the organosolv process often requires less severe reaction conditions than the Kraft method, resulting in energy saving and reduced generation of detrimental by-products (Fig. 3). This makes the organosolv method greener and more attractive as an alternative to Kraft recovery techniques for sustainable biorefinery operation [16].

3.5 Emerging Techniques

New innovations in lignin recovery include ionic liquids (ILs) and deep eutectic solvents (DESs). The ILs belong to a group of organic salts with low melting points. They are characterized by tailor-made solubility for lignin by dissolving the biomass either completely or selectively according to the type of IL used. This tailored solubility for lignin allows for high-purity lignin recovery, enabling advanced applications in pharmaceuticals and high-performance composites [7]. ILs have low volatility, low flammability, and reduced toxicological profiles compared to conventional solvents due to their high thermal stability; thus, their recovery process is more environmentally friendly [7].

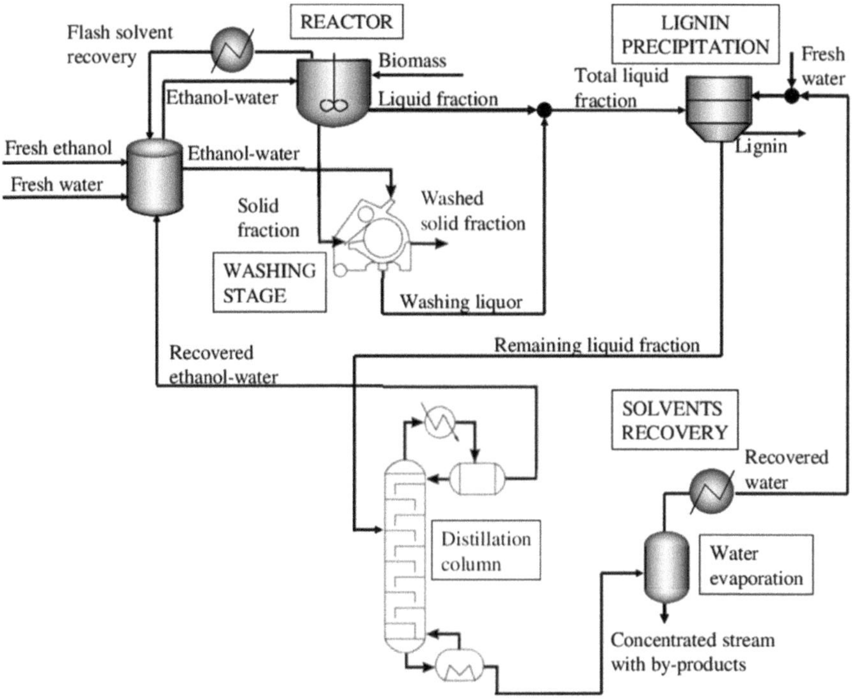

Fig. 3 A diagram of the organosolv-ethanol process [23]

Similarly, being a combination of hydrogen bond donors (HBD) and acceptors (HBA), DES emerges as a versatile and low-cost recovery method for lignin. Because DES can be fine-tuned in its viscosity, polarity, and solvency, it has the potential to provide adaptable criteria for the use of lignin [7]. In addition, designing DESs with desired properties can optimize recovery processes for a given lignin structure, opening new ways of valorizing lignin for application in bioplastics, resins, carbon fiber, etc. [7]. Figure 4 shows the schematic diagram of (A) sulfite [24], (B) soda [25], and (C) enzymatic hydrolysis [26] biorefinery processes.

Comprehensive knowledge of the strengths and weaknesses of lignin-recovery methods will allow for the best selection of technology according to operational needs and sustainability goals. Table 1 gives an in-depth comparison of some major methods of lignin recovery, including a preview of their suitability in different applications, potentially showing a way to promote the circular economy in the pulp and paper industry.

Significant progress has been made in the field of lignin recovery thus far, yet several issues remain to be addressed. Economic, effective, and eco-friendly recovery means are still needed and provide a driving force for research even today. Furthermore, how various recovery methodologies impact the molecular structure of the lignin needs to be well understood because this will certainly dictate its future applications [18].

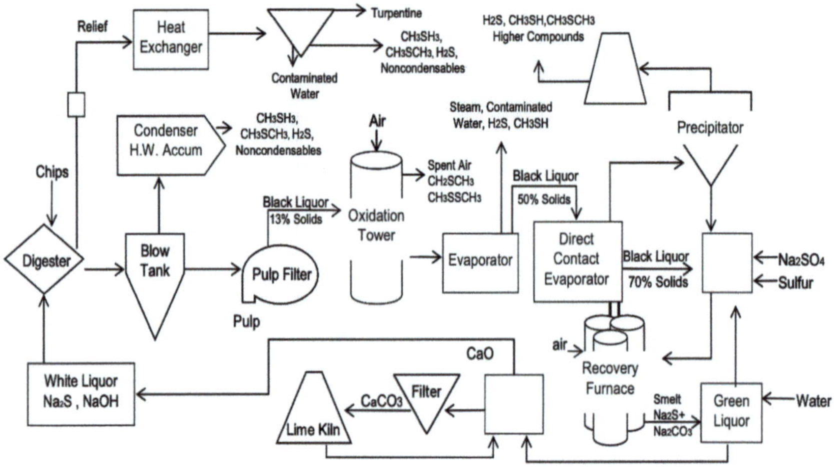

(A)

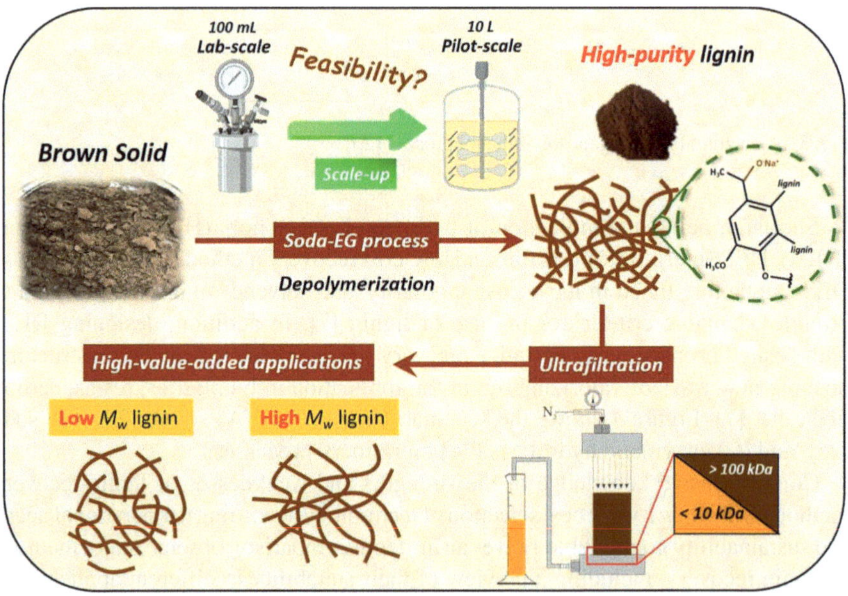

(B)

Fig. 4 Typical process flow diagram of **A** sulfite [24], **B** soda [25], and **C** enzymatic hydrolysis [26] pulping processes

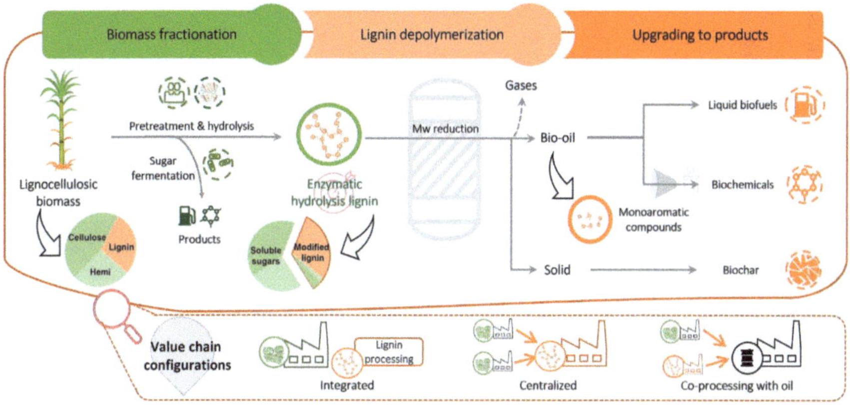

(C)

Fig. 4 (continued)

In summary, exploring methods for lignin recovery has illustrated numerous innovative processes and applications for industry. Methods like ultrasonic-assisted recovery and organosolv enhance the environmental efficiency, increase compatibility, and provide high-purity lignin for uses in the future. The following section on applications of recovered lignin looks into the ability of technologies to retain and enhance the properties of lignin for tailor-made purposes, transforming it from a by-product into a valuable resource in feasible solutions for fields such as construction, automotive, and energy. This shows that the future lies in aligning recovery technologies with end-use applications to fully exploit the potential of lignin for a greener future.

4 Applications of Recovered Lignin

The recovered lignin, long treated as a low-value by-product of the pulp and paper industry, has attracted much interest in recent years due to its increasing potential in many high-value applications arising from its unique aromatic structure, high carbon content, and diverse functional groups [9]. Much progress has recently been achieved in lignin recovery and modification techniques to convert lignin into a range of functional materials that could find applications across a wide variety of sectors. Specifically, lignin shows excellent potential in developing nanocarriers for agrochemicals and pharmaceutical products, which could ensure sustainable approaches in developing an effective delivery system [15]. Inherent properties also make it suitable for applications related to energy storage devices, such as batteries and supercapacitors, with a cost-effective and eco-friendly alternative to traditional materials [27–29]. Further highlights of its versatility include its potential as an adsorbent

Table 1 Comparative analysis of lignin recovery methods

Recovery method	Advantages	Disadvantages	Yield and purity	Energy consumption	Environmental impact	Scalability	References
Kraft process	Provides high yield and efficiency in lignin extraction; used by most pulp and paper industries and enables cost-effective recovery and reuse of chemicals such as sodium hydroxide and sodium sulfide	Generates sulfur-rich black liquor, necessitating substantial energy use and creating environmental concerns due to toxic by-products	Moderate yield with low-purity lignin containing high sulfur content limits its applications	High energy requirement due to the need for high temperatures and pressures	Significant environmental impact due to sulfur emissions; requires treatment for black liquor	High scalability as it is widely used in large-scale industrial operations	[3, 4, 18]
Sulfite process	Efficient in producing water-soluble lignosulfonates for use in applications such as dispersants and adhesives; breaks down lignin into smaller, more reactive molecular components	Low efficiency in recovering lignin, limited primarily to softwood biomass; involves the use of hazardous chemicals	Low to moderate yield, with variable purity depending on process conditions	High energy consumption due to specialized equipment and management of hazardous waste by-products	High environmental impact due to sulfur emissions and the generation of acidic waste	Moderate scalability, primarily limited to specific applications	[3, 18]

(continued)

Table 1 (continued)

Recovery method	Advantages	Disadvantages	Yield and purity	Energy consumption	Environmental impact	Scalability	References
Soda process	Produces sulfur-free lignin with fewer contaminants, suitable for high-purity applications; lower environmental impact than sulfur-based processes; effective for nonwoody biomass	Lower extraction efficiency than other processes; requires high pH conditions, which can lead to potential equipment corrosion and increased maintenance costs	Low to moderate yield with varying lignin purity depending on the biomass type	Moderate energy consumption, lower than Kraft and sulfite processes	Lower environmental impact compared to sulfur-based methods; requires management of alkaline waste	Moderate scalability, depending on the type of biomass used	[3]
Organosolv process	Produces high-purity lignin suitable for advanced applications; organic solvents used can be recycled, reducing the environmental footprint; also less energy-intensive than traditional methods	High costs associated with organic solvents and their recovery; less effective for highly recalcitrant biomass; requires careful handling and recycling of solvents	High yield with high-purity lignin	Moderate energy consumption due to the need for solvent recovery, which partially reduces energy use	Lower environmental impact due to solvent recycling, which reduces waste generation	Moderate to high scalability, with complexity and costs linked to solvent recovery systems	[7, 16, 18]

(continued)

Table 1 (continued)

Recovery method	Advantages	Disadvantages	Yield and purity	Energy consumption	Environmental impact	Scalability	References
Enzymatic hydrolysis	Provides high-purity, sulfur-free lignin under mild process conditions, making it environmentally friendly; generates no toxic intermediates, and the process is highly selective	Slow process with high costs for enzymes; limited scalability due to current technological constraints	High yield with high-purity lignin	Low energy consumption under mild processing conditions	Low environmental impact as biodegradable enzymes is used	Low to moderate scalability, still in developmental stages	[2, 18]
Ionic liquid (IL) recovery	Allows for tailored solubility for specific lignin types, achieving high purity; solvents are recyclable, reducing waste	High cost and potential toxicity of ionic liquids; complex recovery process for solvents, increasing operational challenges	High yield with high-purity lignin	Moderate to high energy consumption depending on the efficiency of ionic liquid recovery	Moderate environmental impact due to the toxicity and cost of ionic liquids	Low to moderate scalability due to the costs and challenges associated with solvent management	[3, 7, 19]
Deep eutectic solvent (DES) recovery	Utilizes low-cost, biodegradable solvents with adjustable properties tailored to specific lignin types; more sustainable than traditional solvents	Lower yield compared to other methods; some DESs may be corrosive, causing equipment degradation	Moderate yield with moderate to high-purity lignin	Moderate energy consumption, depending on the efficiency of DES recovery	Lower environmental impact due to the use of biodegradable solvents	Moderate scalability, with potential for scaling up as costs decrease	[7]

(continued)

Table 1 (continued)

Recovery method	Advantages	Disadvantages	Yield and purity	Energy consumption	Environmental impact	Scalability	References
Microwave-assisted extraction (MAE)	Reduces extraction time and enhances yield; requires less solvent, making it more energy-efficient than conventional methods	Requires specialized equipment; potential for uneven heating may affect the consistency of lignin quality; limited industrial adoption	High yield with moderate to high-purity lignin	Low to moderate energy consumption due to shorter processing times	Lower environmental impact than traditional methods, with reduced solvent and energy use	Moderate scalability, limited by the high cost of specialized equipment	[7]
Supercritical fluid extraction	Solvent-free process offering high selectivity while preserving the structural integrity of lignin; environmentally friendly due to minimal use of harmful chemicals	High operational costs due to the need for high-pressure and high-temperature conditions; requires specialized equipment, limiting accessibility	High yield and high-purity lignin	High energy consumption, due to the necessity of maintaining high pressure and temperature	Low environmental impact, contingent on the energy source and equipment used; avoids harmful solvents	Low to moderate scalability, constrained by high costs and technical complexity	[7]
Nonthermal plasma (NTP) treatment	Simplifies the extraction process and enhances lignin solubility and reactivity with minimal solvent use	High energy consumption, currently limited to laboratory-scale applications; requires specialized equipment, increasing cost	Moderate to high yield, with variability in purity depending on the conditions used	High energy consumption due to the demands of plasma generation	Low to moderate environmental impact due to minimal solvent use	Low scalability, as it remains largely experimental	[7]

(continued)

Table 1 (continued)

Recovery method	Advantages	Disadvantages	Yield and purity	Energy consumption	Environmental impact	Scalability	References
Ultrasonic-assisted extraction	Increases lignin yield and improves reaction efficiency while reducing processing time; uses less chemical input than traditional methods	Potential for equipment degradation over time; high energy use for prolonged operations	Moderate to high yield with moderate-purity lignin	Moderate to high energy consumption, influenced by the power and duration of ultrasound	Lower environmental impact than conventional methods due to reduced chemical usage	Moderate scalability, dependent on equipment durability and costs	[15]
Steam explosion	Enhances the accessibility of cellulose and hemicellulose for further processing; reduces the need for additional chemicals	Produces heterogeneous lignin with variable quality and requires specialized equipment to manage high-pressure and high-temperature conditions; typically used as a pretreatment method	Low yield; minimal use of chemicals	Low energy consumption compared to other methods	Low environmental impact due to minimal chemical use; challenges in handling explosive conditions	Moderate scalability, mostly used in combination with other methods	[3]

(continued)

Table 1 (continued)

Recovery method	Advantages	Disadvantages	Yield and purity	Energy consumption	Environmental impact	Scalability	References
Aldehyde-assisted fractionation (AAF)	Generates lignin with a less condensed structure, aiding in depolymerization; offers high selectivity and efficiency, preventing undesirable condensation reactions	Use of aldehydes, presenting safety and handling concerns; requires specific conditions and reagents, adding to the process complexity	Moderate yield with high-purity lignin	Moderate energy consumption based on reagent recovery and reuse	Moderate environmental impact; requires proper handling of aldehydes to mitigate safety risks	Low scalability due to limited data on large-scale applications	[3]
Alkaline hydrolysis	Effectively removes hemicellulose, leaving behind relatively pure lignin; uses common reagents like sodium hydroxide, making it economically feasible	Time-consuming process; generates large volumes of alkaline wastewater, posing environmental challenges	High yield of lignin with minimal damage to its aromatic structure	Moderate energy consumption, as it operates at low temperatures and pressures	High environmental impact due to the generation of large volumes of alkaline wastewater	High scalability due to the use of widely available reagents	[15]

(continued)

Table 1 (continued)

Recovery method	Advantages	Disadvantages	Yield and purity	Energy consumption	Environmental impact	Scalability	References
Binary solvent system	Offers flexibility in tuning solvent properties for optimized lignin extraction; adaptable to different types of biomasses	Complex solvent management and recycling, requiring careful optimization of solvent ratios and conditions, increasing operational complexity	High-purity lignin, achievable with appropriate solvent combinations	Moderate energy consumption, depending on solvent recovery processes	Moderate environmental impact, depending on the specific solvents used and their recyclability	Moderate scalability, affected by solvent management and recycling challenges	[15]
Glycerolysis	Uses renewable, biodegradable solvents, making the process environmentally friendly	Limited efficiency of lignin recovery due to reaction equilibrium; scalability, challenging due to the high temperatures and extended reaction times requirement	Moderate yield with environmentally friendly solvent use	High energy consumption due to high temperatures and prolonged reaction times	Lower environmental impact due to the use of biodegradable solvents	Low to moderate scalability, hindered by energy costs and reaction conditions	[15]

of pollutants, a solar protector, and a precursor of advanced carbon materials [15, 30, 31]. Other applications of lignin include biofuels, chemicals, hydrogels, and adhesives, contributing to the production of sustainable material development relevant to construction, soil stabilization, and water treatment [9, 11, 32, 33]. This will be expanded further in this section, discussing selected applications of recovered lignin, current developments, potential benefits, and related challenges that need to be addressed for successful commercialization.

4.1 Lignin in Environmental Remediation

Among the major applications that show the most promise for the future is the potential use of lignin as a material in the field of environmental remediation, especially in the adsorption of pollutants from wastewater. It has been evidenced through research studies that functionalized lignin reveals high adsorption capacities for heavy metals and organic pollutants owing to its aromatic structure and enriched functional groups [30, 33]. Functionalization of lignin with amines or carboxylates enhances hydrophilicity and interaction with aqueous pollutants, hence offering a cost-effective alternative to conventional adsorbents.

It is proven to be very effective in wastewater treatment and shows remarkable removal of dyes from industrial effluents. Song et al. [31] showed that these adsorbents could remove dyes at efficiencies as high as 99.1%, especially in extracting malachite green from wastewater. High efficiencies in dye removal suggest that lignin-derived materials may serve as an alternative to conventional adsorbents used in environmental remediation. Furthermore, the applications of lignin have expanded from water treatment to land management. As Khajeh et al. [33] pointed out, lignin has also been considered for roles in soil stabilization and mechanical improvement. The soil with increased incorporation of lignin has proved to increase bearing capacity and, therefore, provide greater stability against different environmental stresses. Such vast applications of lignin establish its versatility and potential to lend itself to many more environmentally sustainable practices.

4.2 Lignin in Energy Storage

The high carbon content and complex phenolic structure of lignin make it a very prospective material for developing carbon-based substances for energy storage devices, like batteries and supercapacitors [28]. Lignin-derived porous carbons have been employed as electrodes because they present high surface area and high electrical conductivity, improving the performances of the energy storage systems.

Several recent works have shown the huge potential of lignin-derived materials as cathode and anode materials in some types of batteries. These lignin-based materials show improvements in cycle stability and energy capacity that can be promising in the future based on lignin and its exploitation in energy storage [34]. Moreover, the development of nitrogen-doped carbon microspheres with excellent electrochemical properties has been proven to be particularly suitable for high-performance supercapacitors. Overall, this broadens the scope of lignin applications and emphasizes its role in developing green technology by offering an alternative, sustainable way of storing energy with high efficiency [27, 29].

4.3 Lignin in Building and Construction

Lignin also finds application as a natural binder, additive, and reinforcing agent in the building and construction industry. For concrete, lignin-based additives provide better workability, reduced water use, and improved concrete mixture strength [13]. It is also used to produce ecologically benign insulating materials, assure heat resistance, and render building materials sustainable.

Recent studies such as those by Patel et al. [9] have corroborated using modified lignin, particularly lignin sulfonates, to improve concrete quality. These modified lignins significantly improve the compressive strength and durability of concrete structures, hence becoming an important building ingredient in the construction industry. This use enhances structural integrity and sustainability in buildings through involvement in reusing by-products. Similarly, Sun et al. [35] studied the innovative use of lignin to enhance kapok fibers in vacuum insulation panels (VIPs). They found that the fibers exhibited significant enhancement in performance for VIPs from both thermal and mechanical aspects, which makes it viable as a possible component in advanced insulation technologies. These developments give an example of the transformation that lignin allows, not only in the materials used for construction and insulation, but also in the improvement of efficiency and the reduction of environmental impact.

5 Challenges and Future Perspectives in Lignin Recovery and Utilization

While recovering and applying lignin from the pulp and paper industry provides numerous opportunities for sustainable development, it also faces significant technical, economic, and environmental challenges. This section attempts to present these issues, future directions of research, and opportunities for developing new applications of lignin, which could finally result in its wider use as a valuable resource.

5.1 Technical Challenges in Scaling Lignin Recovery and Application

Several technical difficulties are encountered in lignin recovery and its application at the industrial scale. The main obstacles, for example, lie in the complexity due to high variability in the structure of lignin, which may depend significantly on the source of biomass and the method of recovery [18]. Lignins recovered from various pulping methods, including Kraft and sulfite, as well as organosolv and enzymatic processes, are chemically distinct. For instance, the lignin obtained from Kraft and lignosulfonate processes does not have the same sulfur content as that obtained from organosolv and hydrolytic processes, and this significantly affects their applications in areas such as hydrogel production, adhesives, and dispersants [18].

Another important challenge is that the cost and technical challenges are relatively higher in the recovery of lignin. Traditional processes such as Kraft and sulfite pulping use harsh chemicals that are energy-intensive and hazardous to the environment. Alternative methods, such as the organosolv method, supercritical liquid recovery, and ionic liquids, offer superior selectivity and less environmentally impacting features. However, these processes still need to be developed further to lower costs and make them more optimized for application on a commercial scale [7]. In addition, the nonuniform recovery processes in industry lead to differences in lignin quality; hence, it is an enormous challenge to maintain product quality consistently in the final product [18].

However, the contents and composition of lignin vary greatly between different types of biomasses, making the uniformity of the recovery process quite low. Modifying the recovery methods to suit a given feedstock is one way to ensure the sustainability and economic viability of the whole process [7]. Other environmental challenges include handling wastewater and by-products from the recovery of lignin. It is equally important to develop green methods for controlling and recycling these materials if sustainability is to be incorporated and implemented [7].

Apart from that, the scalability of lignin-derived materials is currently hampered by their relatively low level of technological readiness and their limited performance compared to more advanced carbon materials [28]. To overcome these obstacles, future research should optimize synthesis methods, fine-tune material properties, and prioritize sustainable production practices. Additionally, it is essential to establish standardized protocols for evaluating lignin-based materials across different environmental conditions and pollutant types [36].

6 Future Research Directions and Technology Needs

To address these challenges, future research should target several critical areas. First, developing advanced structural designs and achieving better control over the properties of lignin molecules are essential to enhance their efficiency of use. Researchers should also focus on stabilizing and standardizing lignin quality from various sources to maintain consistent properties in lignin-derived materials [37]. Moreover, there is a need for comprehensive characterization frameworks that can match the structural properties of lignin with its functional applications, such as in photothermal materials [37].

Second, greater control over the molecular design of lignin is necessary to enhance its properties for specific applications. For instance, optimizing its photothermal characteristics for solar steam or thermoelectric generators demands precise adjustments of molecular distances and grafted groups [37]. Additionally, it is important to develop environmentally friendly and energy-efficient recovery methods, like green chemical modifications, to lower environmental impact and promote sustainability [37].

Third, using advanced technologies like artificial intelligence (AI) can significantly optimize recovery processes and improve the functional performance of lignin-based materials [36]. AI offers a more efficient approach by predicting the best recovery conditions, which can enhance applications such as contaminant removal. For lignin-based adsorbents, future research should focus on creating methods that allow their regeneration without compromising adsorption efficiency—an essential factor for maintaining sustainability [36].

Fourth, to broaden the range of applications for lignin-based materials, exploring synergies with other materials that can help unlock lignin's full potential is important. For instance, lignin-based materials show significant promise in energy storage applications, such as batteries and supercapacitors, due to their high surface area, electrical conductivity, and porosity [28]. Future research should tailor the properties of lignin-derived materials for specific uses while promoting sustainable production practices [28].

7 Conclusions

This chapter has explored the potential for recovery and utilization of lignin within the pulp and paper industry, highlighting various recovery methods and the associated challenges and opportunities. Recent technological advancements have paved the way to recover lignin, which was once an underutilized by-product, contributing to economic growth and environmental sustainability. The chapter examined traditional methods like the Kraft process alongside newer techniques such as organosolv and ultrasonic-assisted recovery, outlining the advantages and drawbacks of each in terms of purity, efficiency, and environmental impact.

This chapter also explored the diverse applications of recovered lignin, ranging from high-value products like biofuels and biopolymers to more innovative uses in energy storage and environmental remediation. Due to its aromatic structure and high carbon content, lignin is well suited for developing advanced materials, including carbon fibers, adsorbents, and biocomposites. However, the heterogeneous nature of lignin, which varies with the recovery process type and source, presents significant challenges in ensuring consistent quality and performance across different applications.

To fully exploit lignin's potential, this chapter emphasizes the need for more comprehensive research to tackle the technical challenges of scaling lignin recovery and its applications. The discussion highlights the importance of aligning recovery technologies with specific end uses to enhance cost-effectiveness, scalability, and environmental sustainability. The chapter also identifies new opportunities for lignin to serve as a sustainable alternative across various industrial sectors, from construction to energy storage.

Overall, advancing lignin applications and maximizing their value will depend on integrating innovative recovery methods and establishing standardized frameworks. By tackling current challenges and embracing new technologies, lignin recovery can become crucial in driving a circular economy and fostering a more sustainable future. As the potential uses for recovered lignin continue to be explored in greater depth, ongoing progress in this field will help transform lignin from a waste by-product into a versatile and valuable resource, enhancing both the sustainability and competitiveness of the pulp and paper industry.

Acknowledgements We gratefully acknowledge the funding support from the Malaysian Ministry of Higher Education and Universiti Teknologi Malaysia under grant number FRGS/1/2020/TK0/UTM/02/104.

Competing Interests The authors declare that they have no competing interest.

Declaration of Generative AI and AI-Assisted Technologies in the Writing Process During the preparation of this work, the book chapter author(s) did not take any assistance from AI, and the book chapter author(s) reviewed and edited the content as needed and take(s) full responsibility for the content of the publication.

References

1. Rumpf J et al (2021) Types of lignin, properties, and structural characterization techniques. Lignin-based materials for biomedical applications. Elsevier, pp 105–158
2. Andeme Ela RC, Heiden PA (2024) Lignin smart nanoparticles for effective and eco-friendly resource recovery from wastewater: status, challenges, and opportunities. ACS Sustainable Resource Management
3. Libretti C, Correa LS, Meier MA (2024) From waste to resource: advancements in sustainable lignin modification. Green Chem
4. Kohl O (2024) Synthetic biology towards lignin utilization
5. Chettri D et al (2023) Lignin depolymerization for its valorization. BioEnergy Res 16(3):1264–1279
6. Kumar V et al (2024) Valorization of pulp and paper industry waste streams into bioenergy and value-added products: An integrated biorefinery approach. Renew Energy 120566

7. Alam MM et al (2024) Efficient and environmentally friendly techniques for extracting lignin from lignocellulose biomass and subsequent uses: a review. Clean Mater 100253
8. Brienza F et al (2024) A guide to lignin valorization in biorefineries: traditional, recent, and forthcoming approaches to convert raw lignocellulose into valuable materials and chemicals. RSC Sustain 2(1):37–90
9. Patel R et al (2023) Transforming lignin into renewable fuels, chemicals, and materials: a review. Bioresour Technol Rep 22:101463
10. Bilal M et al (2024) Bioprospecting lignin biomass into environmentally friendly polymers—applied perspective to reconcile sustainable circular bioeconomy. Biomass Convers Biorefinery 14(4):4457–4483
11. Wang T et al (2023) Lignin to dispersants, adsorbents, flocculants and adhesives: a critical review on industrial applications of lignin. Ind Crops Prod 199:116715
12. Shorey R et al (2024) Valorization of lignin for advanced material applications: a review. RSC Sustain
13. Mittal T et al (2024) Recent advancement and emerging applications of lignin. Indones J Chem
14. Kumar D, Gupta RK (2024) Natural polymers-humic acid and lignin based hydrogels: in agriculture, environment and energy storage. Ind Crops Prod 219:119029
15. Mondal S et al (2023) Lignin modification and valorization in medicine, cosmetics, environmental remediation and agriculture: a review. Environ Chem Lett 21(4):2171–2197
16. Margarida Martins M, Carvalheiro F, Gírio F (2024) An overview of lignin pathways of valorization: from isolation to refining and conversion into value-added products. Biomass Convers Biorefinery 14(3):3183–3207
17. Balk M et al (2023) Lignin, the lignification process, and advanced, lignin-based materials. Int J Mol Sci 24(14):11668
18. Pandit S et al (2024) A comprehensive review on technical lignin, lignin hydrogels, properties, preparation, applications & challenges in lab to market transition. Ind Crops Prod 211:118262
19. Bagh FSG, Ray S (2022) Lignin extraction from black liquor using acetate based ionic liquids: extraction kinetics and product characterization. J Ion Liq 2(1):100028
20. Tong Y et al (2024) A review on the overall process of lignin to phenolic compounds for chemicals and fuels: from separation and extraction of lignin to transformation. J Anal Appl Pyrolysis 106663
21. Sheridan E et al (2024) A systematic study on the processes of lignin extraction and nanodispersion to control properties and functionality. Green Chem 26(6):2967–2984
22. Rahman M, Avelin A, Kyprianidis K (2020) A review on the modeling, control and diagnostics of continuous pulp digesters. Processes 8(10):1231
23. Alriols MG et al (2010) Lignocellulosic biorefinery processes energy integration. Chem Eng 21
24. Gopalakrishnan K, Kim S, Ceylan H (2010) Lignin recovery and utilization. Bioenergy Biofuel Biowastes Biomass 12:247–274
25. Chen Y-L et al (2023) Lignin extraction and fractionation from rice straw biorefinery residues. Sep Purif Technol 326:124778
26. Menezes FF et al (2023) Depolymerization of enzymatic hydrolysis lignin: review of technologies and opportunities for research. Fuel 342:127796
27. Zhang J et al (2023) Efficient depolymerization of alkali lignin to monophenols using one-step synthesized Cu–Ni bimetallic catalysts inlaid in homologous biochar. Biomass Bioenerg 175:106873
28. Munir MT et al (2024) From waste to watts: emerging role of waste lignin-derived materials for energy storage. J Energy Storage 82:110477
29. Liu C et al (2023) Dissolution of alkali lignin in ionic liquid to fabricate nitrogen-doped carbon microspheres for electrochemical supercapacitors. J Electroanal Chem 951:117921
30. Heo JW et al (2024) Green synthesis of silver nanoparticles-capped aminated lignin as a robust active catalyst for dye discoloration. Int J Biol Macromol 274:133211
31. Song C et al (2023) Influence of structure and functional group of modified kraft lignin on adsorption behavior of dye. Int J Biol Macromol 240:124368

32. Chandna S et al (2024) Lignin upconversion by functionalization and network formation. Angew Chem Int Ed 63(8):e202313945
33. Khajeh A et al (2024) A state-of-the-art review on the application of lignosulfonate as a green alternative in soil stabilization. Sci Total Environ 173500
34. Huang Y et al (2024) Advances in application of sustainable lignocellulosic materials for high-performance aqueous zinc-ion batteries. Nano Energy 109416
35. Sun Q et al (2024) Functionalising kapok fibre with lignin to enhance the structural and thermal performance of vacuum insulation panels. Ind Crops Prod 220:119277
36. Zhang H et al (2024) Advances in lignin-based biosorbents for sustainable wastewater treatment. Bioresour Technol 130347
37. Guan Y et al (2024) Unlocking the photothermal conversion capacity of lignin and lignin-derived materials. Chem Eng J 154985

Nanotechnology for Sustainable Paper and Forest Management

Diksha Singh, Ankit Kumar Singh, and Ziaul Hasan

Abstract Nanotechnology, with its innovative approaches to enhancing production, mitigating environmental impacts, and improving material functionality, has emerged as a key enabler for sustainable research. Forests and paper, valued in the billions, are considered a vital sector among major industries. This chapter highlights the numerous methods involving nanotechnologies to manage forests and manufacture wood-based products, especially paper, for the benefit of future generations. The study draws attention to different forestry applications like nanosensors for real-time monitoring and nanofertilizers, which increase nutrient-use efficiency by 30%. In the subsequent sections, it is discussed that nanopesticides and antimicrobial agents are used in pest and disease control. At the same time, flame-retardant nanocoatings and nanosensors are used to manage wildfires. Different applications of sustainable nanotechnology in the paper industry include smart paper with conductive nanomaterials, nanocellulose as a sustainable substitute, nanofiltration systems for lower water consumption and pollutant emissions, and nano-engineered coatings for increased product durability. Although nanotechnological applications offer numerous advantages, including enhanced efficiency, reduced environmental impacts, improved

D. Singh
Department of Forensic Science, Institute of Excellence in Higher Education, Barkatullah University, Bhopal, Madhya Pradesh, India
e-mail: diksha.phd2025@gmail.com

A. K. Singh
Department of Agriculture Science, Mandsaur University, Mandsaur 458001, Madhya Pradesh, India

Department of Entomology, Birsa Agricultural University, Ranchi, Jharkhand, India

A. K. Singh
e-mail: ankit7232999999@gmail.com

Z. Hasan (✉)
Department of Biosciences, Faculty of Life Sciences, Jamia Millia Islamia, New Delhi 110025, India
e-mail: zhasan.biochem@gmail.com

Department of Life Science, Faculty of Life Sciences, Mandsaur University, Mandsaur 458001, Madhya Pradesh, India

R. K. Bachheti (ed.), *Nanotechnology in the Pulp and Paper Industry*, Smart Nanomaterials Technology, https://doi.org/10.1007/978-981-95-3473-9_16

material performance, and support for circular economy principles, challenges such as nanoparticle toxicity concerns, regulatory gaps, and public skepticism persist and require attention. To conclude, this chapter highlights the various applications of nanotechnology and nanomaterials currently being utilized by the forest and paper industries worldwide and advocates for rigorous risk assessments and transparent policies to ensure their safe and eco-friendly adoption in these sectors.

Keywords Nanotechnology · Nanofertilizers · Nanosensors · Nanopesticides · Nanocellulose · Smart paper · Environmental sustainability

1 Introduction

Nanotechnology, which enables the control of matter at the atomic and molecular level (1–100 nm), has become a powerful tool for various industries, including forestry and paper manufacturing. It could make materials stronger, more efficient, and less harmful to the environment, which could lead to improved ways to manage forests and produce paper in a more environmentally friendly manner. The paper discusses its applications, the growing demand for sustainability, and how it can help address some of the most pressing environmental issues. Nanotechnology is transforming the way we manage forests, from controlling pests to monitoring the health of the trees. Various sustainable nanotechnology applications in forestry and the paper industry are listed in Table 1, including key findings and advancements in these areas.

Copper-carbon core–shell nanoparticles (CCCSNs) are a type of technology that effectively kill termites and fungal infections. They are also an eco-friendlier alternative to traditional wood preservatives [58]. Another product in this series is "Nano-Herbicides." These are much more precise and targeted than traditional herbicides, which means they have the potential to be much less harmful to other plant and animal life (and to us). So far, there have been no harmful effects on nontarget organisms from nanoherbicides.

In the paper industry, nanotechnology increases durability and functionality. Engineered coatings, such as titanium dioxide and nanosilica, make paper more resistant to moisture, UV damage, and microbial decay [65]. Conductive nanomaterial-embedded smart paper promises novel packaging and electronic uses [3]. Nanocellulose, obtained from renewable wood pulp, is revolutionizing packaging by developing an ultra-strong yet lightweight material [48]. Apart from enhancing performance, nanotechnology also contributes to sustainability by minimizing waste, maximizing resource utilization, and reducing the use of harmful chemicals. For example, pulp mill nanofiltration systems significantly reduce water usage and pollutant emissions [57]. These innovations demonstrate the potential of nanotechnology to drive environmentally friendly innovation in the paper and forestry industries. Forests are critical for carbon sequestration, water cycling, and providing raw materials; yet, deforestation, chemical abuse, and unsustainable logging endanger these forests [33]. The

Table 1 Sustainable nanotechnology applications in forestry and the paper industry

Category	Application	Key findings/data	Citations
Forest growth and protection	Nanofertilizers for nutrient efficiency	Improve nutrient use efficiency (NUE) by 30%, reducing runoff and pollution	Singh et al. [64]
	Nanosensors for soil and tree health monitoring	Carbon nanotubes (CNTs) and hyperspectral sensors detect stress, pathogens, and nutrients in real time	Kaushik et al. [40], Khan and Rizvi [41]
Pest and disease control	Nanopesticides (e.g., copper-carbon core–shell nanoparticles)	Target pathogens while minimizing ecological harm; effective against termites and fungi	Qi et al. [58]
	Antimicrobial nanoparticles (e.g., silver nanoparticles)	Disrupt fungal cell membranes, protecting wood from decay (e.g., *Trametes versicolor*)	Kumar et al. [44]
Forest fire prevention	Flame-retardant nanocoatings (e.g., Fire Mesh)	Create protective barriers to suppress flames and delay ignition	Singh et al. [64]; McCrank [48]
	Nanosensors for wildfire detection	Detect temperature, humidity, and gas changes for early warning	Singh and Singh [63]
Paper production	Nanocellulose (CNF/CNC) for high-strength paper	Tensile strength of 145 GPa; reduces fiber usage by 20–30%	Beecher [8], Wagay et al. [65]
	Nanocoatings (e.g., TiO_2, nanosilica)	Enhance water resistance, UV protection, and antimicrobial properties	Yang et al. [70], Wagay et al. [65]
Waste reduction	Nanofiltration systems in pulp mills	Reduce water consumption by 50% and pollutant emissions by 60%	Puurunen and Vasara [57], McCrank, [48]
	Nanoenzymes for de-inking	Improve recycling efficiency by breaking down ink particles	Atalla et al. [5]

paper industry, which relies heavily on forests, faces the dilemma of meeting production demands while conserving natural resources. Traditional paper manufacturing consumes high levels of energy, water, and harmful chemicals, such as chlorine-based bleaches [48]. To mitigate these impacts, the industry must adopt greener alternatives, such as sustainable coatings and recycled fibers. Because nanocellulose-based materials are biodegradable substitutes for synthetic plastics and nanocoatings reduce the need for excessive chemical treatments, nanotechnology is significant [75]. Nanosensors enable precision forestry by allowing forest conditions to be monitored in real-time, resources to be utilized more efficiently, and waste to be minimized [64]. Nanotechnology also supports reforestation. Nanofertilizers and seed treatments promote germination and harden young saplings, allowing for recovery in the ecosystem [37].

Through the union of nanotechnology and environmentally friendly measures, the forest and paper industries can be economically feasible and ecologically sound. From stronger, smarter paper products to healthier forests under nanosensor protection, these technologies usher in a world where nature and industry coexist. The key lies in constant research, moral implementation, and commitment to keeping the world's vital resources intact.

Nanotechnology provides innovative solutions to the pressing environmental issues in papermaking and forestry. One of the primary issues is pollution due to chemical treatments and industrial waste. Contaminants are converted into less harmful forms by nanoparticles (NPs), such as zero-valent iron, which effectively remediates polluted soils and water [14]. Nanofiltration technologies in paper production facilitate compliance with environmental laws through the reduction of toxicity in effluent [57]. Climate change exacerbates existing issues, including forest fires, pest infestations, and habitat destruction. Fire-resistant nanomaterials such as Fire Mesh, which extinguish fires and promote ecosystem recovery, are one example of how nanotechnology contributes to developing climate resilience. In addition, proactive management practices are facilitated by nanosensors monitoring climate-caused stress within forests [64]. Another key area where nanotechnology excels is in resource efficiency. For example, nanofertilizers enhance nutrient uptake in plants, reducing runoff and excessive fertilizer use [20]. Likewise, low-value biomass is utilized to produce nanocellulose, reducing resource waste and extracting maximum value [48]. Despite its benefits, nanotechnology poses environmental hazards that must be effectively managed, including the toxicity of nanoparticles and potential ecosystem disruption. For ensuring safe and durable applications, research efforts and regulatory systems are essential. Nanotechnology has considerable potential, but it also presents some challenges, including a lack of regulations and public distrust. Only 20% of nanotechnology articles talk about the risks to health and the environment [60]. Full toxicity investigations and clear policies must be implemented to ensure that the process is safe. Nanotechnology connects new ideas with long-term viability in the fields of forestry and paper production. Combining nanomaterials and nanosensors will lead to a more sustainable future by enabling the industry to produce wood products that last longer, smarter paper solutions, and longer-lasting pest control. Deforestation remains a significant issue, with an estimated 10 million hectares of forest lost annually from 2015 to 2020 [24]. Unsustainable forest logging driven by the demand for timber and wood products leads to habitat loss, biodiversity loss, and increased carbon emissions. The issue is exacerbated by illegal logging, which accounts for 15–30% of the world's timber trade [35]. Furthermore, clear-cutting reduces forest regeneration and soil quality [55]. Because they are more expensive and have fewer immediate returns, sustainable alternatives, such as agroforestry and reduced-impact logging, are often underutilized. The paper industry is water-intensive, using around 54,000 L of water per ton of paper [7]. The pulp and paper industry is one of the largest consumers of industrial energy, making energy use a significant concern as well [34]. The Kraft process and other traditional pulping processes rely on the use of fossil fuels, thereby increasing greenhouse gas emissions. Wastewater from paper mills often contains toxic materials that are harmful to aquatic environments, including dioxins and chlorinated compounds [30]. Poisonous chemicals, such as chromated copper arsenate (CCA) and creosote, are typically used in wood preservation; these chemicals can pollute soil and water over a prolonged period [39]. These chemicals persist in the environment, accumulate in wildlife, and harm human health [29]. Due to economic constraints, many developing countries continue to use hazardous wood preservatives, despite being banned by the European Union under the Registration, Evaluation, Authorisation and Restriction of Chemicals (REACH) regulation [21]. According to Lovett et al. [46], invasive pests and diseases such as *Phytophthora ramorum* and the emerald ash borer, which result in billions of dollars' worth of damage annually, are increasingly becoming a threat to forests. In spite of their effectiveness, traditional pesticides harm nontarget organisms and promote pesticide resistance [18].

Economic loss is caused by timber quality deterioration due to fungal infection by white rot (Trametes versicolor) [61]. Sludge, lignin residues, and refuse fibers are among the major waste products generated by the paper industry, which often end up in landfills [50]. Although paper is increasingly recycled (up to approximately 65% in Europe [17]), there is a limitation related to fiber degradation after multiple recycling cycles of paper [31]. Laminated and plastic-coated packaging are among the nonrecyclable paper-based products that contribute to the increase in waste volume [23]. This chapter aims to highlight the important aspects of nanotechnological applications in the paper and forest industries, utilizing sustainable means, as shown in Fig. 1. Throughout the following sections of this research, we will gain a comprehensive understanding of the various fields of management in the forest and paper industries that have a direct or indirect impact from nanotechnology.

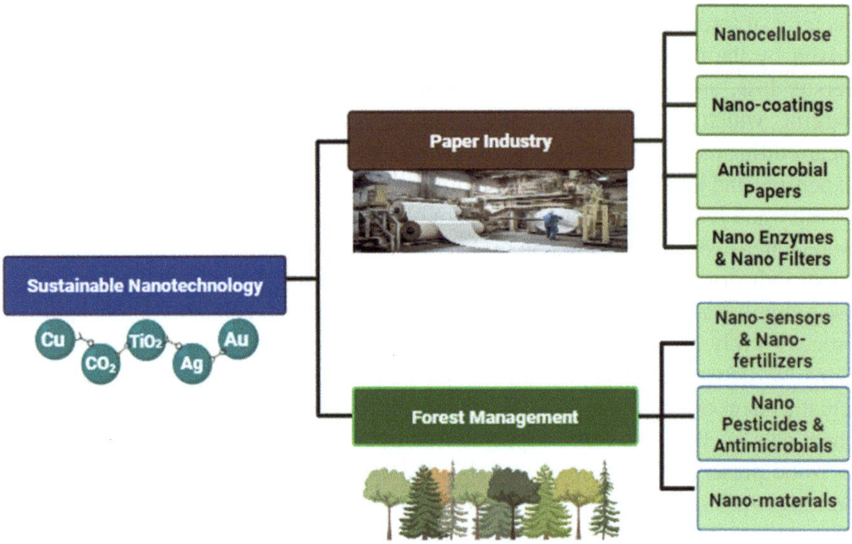

Fig. 1 Sustainable applications of nanotechnology in forest management and the paper industry

2 Nanotechnology in Forest Management

2.1 *Nano-Enhanced Forest Growth and Protection*

Nanotechnology has also introduced unparalleled possibilities to modern-day forestry, unveiling sophisticated solutions to enhance tree health, boost growth, and maintain healthier ecosystems. By utilizing attributes that are distinct to nanoparticles, forestry processes can increasingly become more efficient, whether it's to deliver nutrients, apply pesticides, or manage environmental conditions.

One of the areas with significant potential, where nanotechnology is making a breakthrough, is the use of nanofertilizers. Most fertilizers are unsafe for the environment because they are hard for plants to absorb, and they will run off and poison water and soil. Nanofertilizers are unique in their structure. Nanofertilizers utilize tiny particles that slowly release required constituents such as nitrogen, phosphorus, and potassium, which are slowly released to trees for long periods of time. Through their tiny size, they are also more easily taken up by trees, being more efficient in general. With reduced runoff of nutrients, there are fewer pollutants distributed to other soil and water [59].

This targeted nutrient delivery system could enhance tree growth, biomass production, and overall forest productivity, making it extremely useful for reforestation and afforestation activities.

Forests are always at risk of pathogens, insects, and fungal diseases, resulting in high ecological and economic losses. Higher volumes of conventional pesticides are sometimes required, which are also poisonous to nontarget organisms.

A more environmentally friendly option is nanotechnology, which includes nanoinsecticides and nanofungicides. These products are more targeted and adhere better to pests, requiring fewer quantities while improving their effectiveness. Encapsulation technology—active moieties are encapsulated in nanoparticles, which are released after contact with pests, reducing wasteful chemical deposition. Lower toxicity—biodegradable or biocompatible nanomaterials are used to minimize the environmental footprint of pest management [62]. Silver nanoparticles (AgNPs), for example, have demonstrated robust antifungal and antibacterial activity, which has protected trees from diseases such as *Phytophthora* and Dutch elm disease, without harming beneficial microorganisms in the soil.

One of the strengths of utilizing nanotechnology in forestry is that it minimizes chemical pollution while achieving high productivity. Classical agrochemicals are likely to pollute watercourses and soil, thereby destroying biodiversity. Nanoparticle technologies counteract this by reducing the application rate due to their long-term staying power. Reducing off-target movement because of precise delivery systems. Increasing the biodegradability of agrochemicals prevents long-term build-up in ecosystems. This chimes well with principles of sustainable forest management, ensuring that forestry operations are not being done at the expense of soil quality, water quality, or wildlife in their immediate vicinity.

While there are tremendous prospects for nanotechnology, mass application in forestry requires More research on long-term impacts on the environment. Cost-effective production for mass application of nanoproducts. Regulatory processes for safer and responsible application.

2.1.1 Nanoparticles for Controlled-Release Fertilizers

Traditional fertilizers often suffer from low nutrient use efficiency (NUE), leading to environmental pollution through runoff and leaching. Nanofertilizers solve this issue by packaging nutrients into microscopic carriers, which release them steadily as plants need them. Research indicates that nitrogen nanofertilizers enhance nutrient uptake by nearly a third, resulting in healthier crops with reduced runoff [64].

2.1.2 Nanosensors for Monitoring Tree Diseases and Soil Health

Real-time monitoring is necessary for the long-term health of forest ecosystems. Carbon nanotubes (CNTs) and surface acoustic wave (SAW) devices are examples of nanosensors that can identify early signs of stress, including pathogens, nutrient deficiencies, and soil moisture levels [40]. For example, hyperspectral sensors, such as the Hyperion-EO1, measure the levels of nitrogen and chlorophyll, providing information about the health of trees [41]. Electronic noses, or e-noses, can also detect volatile organic compounds released by diseased trees, enabling early intervention [6]. These technologies optimize resource utilization and minimize waste, making precision forestry more efficient.

2.2 *Nanotechnology in Pest and Disease Control*

Traditional pesticides often harm nontarget organisms and persist in the environment. Targeted delivery from nanopesticides, such as nanogels and capsules, reduces the dosage required and lessens environmental damage. Methyl eugenol nanogels, for instance, effectively control pests while lowering chemical exposure [10]. Similarly, copper-carbon core–shell nanoparticles (CCCSNs), which possess strong antifungal properties, protect trees from diseases such as *Ophiostoma minus* [58]. These developments promote sustainable forestry by increasing efficacy and reducing chemical loads.

2.3 *Nanomaterials for Forest Fire Prevention*

2.3.1 Protecting Trees with Fire-Resistant Nanocoatings

As wildfires intensify due to climate change, scientists are using advanced materials to help save forests. One cutting-edge technology that exhibits promise is flame-retardant nanocoatings, like Fire Mesh, which form a protective barrier on tree surfaces.

2.3.2 Nanosensors for Early Wildfire Detection

Effective fire management requires early wildfire detection. Satellite networks can deliver real-time updates from sophisticated nanosensors [63]. For instance, sensors based on quantum dots can identify even the tiniest concentrations of gases, like carbon monoxide, enabling prompt action [71]. Drones and these sensors have the potential to revolutionize forest monitoring by providing vital information to help extinguish large wildfires.

3 Nanotechnology in Sustainable Paper Production

3.1 *Nanocellulose*

Plant cellulose is the source of nanocellulose, which is transforming the paper industry due to its remarkable strength, biodegradability, and renewable nature. The two primary forms of cellulose are cellulose nanofibrils (CNF) and cellulose nanocrystals (CNC). While CNC produces stiff, rod-like structures with exceptional stiffness, CNF is composed of long, flexible fibers with a high tensile strength [68].

This multifunctional material offers a high-performing and ecologically friendly alternative to papermaking.

3.1.1 Properties and Extraction of Nanocellulose

Nanocellulose is an extraordinary biomaterial that may be manufactured in a mechanical, chemical, or enzymatic way from wood pulp or agricultural waste as a renewable material. For example, chemical fibrillation disintegrates cellulose fibers into nanometer-sized fibrils, and acid hydrolysis removes the amorphous portions of cellulose, resulting in highly crystalline CNC [11].

3.2 Nanocoatings for Improved Paper Quality

By providing water repellency, antimicrobial, and better printability properties, nanocoating on paper is an excellent technique to add functionalities to paper. The films are deposited at the nanoscale to achieve homogeneous coverage and minimum material consumption.

3.2.1 Water-Resistant and Grease-Proof Nanopaper for Packaging

Hydrophobic coatings are increasingly being utilized throughout various industries because of their ability to create superhydrophobic surfaces that repel water and oils. Generally, such coatings are constructed from either silica or polymer nanomaterials, as well as provide a waterproof protective surface coating. For instance, "Nanoseal® Wood" uses nanoparticles to form an invisible, long-lasting lattice-like structure that blocks water while preserving the wood's original look [48]. These coatings are crucial in food packaging, where they are used to create grease-proof fast-food bags that prevent oil from penetrating and keep food fresh. Furthermore, they offer external signage with enhanced weather resistance, making them durable and visible even in the most adverse conditions. Yang et al. (2006) highlighted, at the time of their study, the development of these coatings. Everyday uses of materials can be improved through the application of nanotechnology, providing both economic benefits and enhanced functionality.

3.2.2 Antimicrobial Paper for Medical and Food Industries

Silver nanoparticles (AgNPs) have emerged as a highly effective antimicrobial agent in paper coatings due to their significant reduction in bacterial growth. This makes them particularly suitable for use in medical bandages and food packaging. Their strong biocidal properties help prolong the shelf life of packaged foods and prevent wound dressings from becoming infected by reducing microbial contamination [44]. Additionally, zinc oxide (ZnO) and titanium dioxide (TiO_2) nanoparticle-containing composite coatings offer dual functionality by being UV and antimicrobial. These coatings make paper-based products more sterile and durable, making them ideal for long-term packaging and medical applications, while also shielding them from UV rays that can degrade them [65]. The combination of these advanced nanomaterials in paper coatings represents a significant innovation in sustainable packaging and healthcare materials, ensuring both product safety and longevity.

4 Environmental and Economic Benefits

4.1 Lower Chemical and Water Usage in Paper Processing

Conventional paper manufacturing consumes vast amounts of water and toxic chemicals (e.g., chlorine for bleaching). Nanotechnology introduces eco-friendly alternatives (Fig. 2).

Nanotechnology has revolutionized the pulp and paper industry by introducing sustainable and efficient solutions that minimize environmental impact while enhancing productivity. One key advancement is the use of nanofiltration systems, which use nanoporous membranes to purify water in pulp mills, significantly reducing freshwater consumption by up to 50% and enabling the recycling of effluents [57].

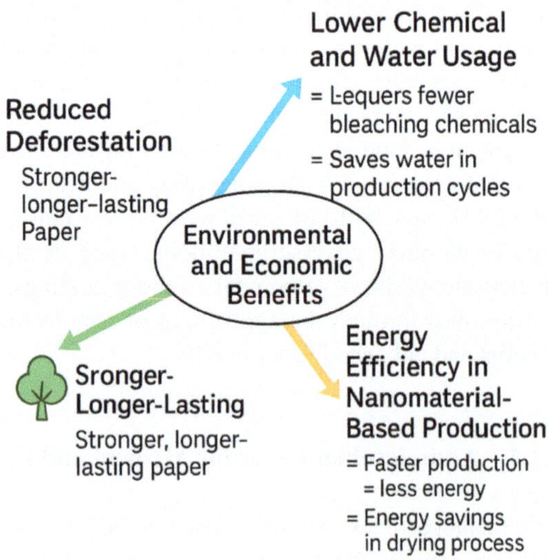

Fig. 2 Integrated environmental and economic benefits of nanotechnology-enhanced paper production

For instance, nanocellulose-based filtration systems have demonstrated remarkable efficiency, removing 99% of contaminants from wastewater and significantly reducing the discharge of harmful chemicals [65]. Another breakthrough is the application of nanocatalysts, such as nano-zeolites and clays, which replace traditional harsh bleaching agents, resulting in a 40% reduction in chemical usage [65]. A notable example is the Oji Paper Company in Japan, which utilizes titanium dioxide (TiO_2) nanoparticles for chlorine-free pulp bleaching, substantially lowering toxicity levels in the process [22]. Additionally, smart nanocoatings have transformed paper finishing techniques; layer-by-layer (LbL) nanocoatings incorporating halloysite clay have reduced dye and ink requirements by 30% while improving printability and surface quality [47].

4.2 Energy Efficiency in Nanomaterial-Based Production

4.2.1 Nanomaterials Streamline Production Processes, Lowering Energy Demands

The unique characteristics of nanocellulose significantly improve the sustainability and utility of paper products. Nakagaito and Yano [51] found that adding nanocrystalline cellulose (CNC) to paper pulp can improve the paper's tensile strength by as much as 40% and strengthen the paper's structure. Paper can now be made thinner and lighter without compromising strength thanks to this advancement, which makes it ideal for high-performance applications. According to Wang et al. [66], environmentally friendly packaging is also changing as a result of nanocellulose's exceptional resistance to oxygen and moisture.

4.3 Implementation of Nanotechnological Applications

The cautious application of nanotechnology in the forestry industry, along with its allied businesses, is evolving at varying rates due to the impact of technological advancements, regulatory readiness, and social and environmental concerns. Table 2 shows that products derived from wood, such as nanocellulose, can be realistically used in the near future, and we expect their use to increase in the coming decades. Nevertheless, their advancement is hampered by unsolved health, environmental, and regulatory issues.

Table 2 Implementation timeline, current readiness, key barriers, and environmental considerations of various nanotechnologies

Nanotechnology category	Current readiness	Implementation timeline	Key barriers	Environmental and social considerations	References
Wood-based nanomaterials (nanocellulose)	Near-implementation	Next decade (by 2050)	Health/environmental risks, regulatory gaps, social acceptance	Must assess lifecycle impacts, composability, and public awareness of sourcing from wood/agriculture	Bengston et al. [9]
Nanostructured biocide delivery systems	Conceptual/experimental	Medium to long-term	Environmental/health impact assessment, technical challenges	Monitoring of nanopesticide release and uptake, soil and water safety needed	Borges et al. [13]
Metal/metal oxide nanoparticles for wood protection	Experimental	Medium to long-term	Risk assessment, regulatory approval, and cost	Potential for leaching and ecotoxicity; requires robust fate and biodegradation data	Borges et al. [13]
Nanosensors and nanobionics	Conceptual/experimental	Medium to long-term	High cost, technical complexity, and biosafety	Privacy issues, forest wildlife disturbance, data ownership, and equitable access concerns	Abbas et al. [1]
Nanofertilizers and soil/water management	Conceptual/experimental	Not specified	Lack of risk assessment, regulatory frameworks	Potential risks of nanoparticle accumulation in ecosystems and food chains requiring transparent risk frameworks	Kalaiyarasi et al. [38]
Value-added nanoforest products (smart paper, coatings)	Conceptual/early experimental	Medium to long-term	Technical challenges, market readiness	Issues around end-of-life disposal, recyclability, and consumer acceptance of active packaging	Porwal et al. [56]

5 Challenges and Risks

5.1 Potential Toxicity of Nanoparticles in Ecosystems

Despite their broad benefits, the cytotoxicity and genotoxicity of NPs in ecosystems are a crucial issue. Meanwhile, the minute size of NPs and their strong reactivity enable them to permeate biological membranes easily, accumulate in organisms, and disrupt the balance of the ecosystem [64]. For instance, silver nanoparticles (AgNPs) intended for antimicrobial coating can be released into the soil and/or water, which is detrimental to beneficial microbes and aquatic organisms [44]. The nanoparticles that have been implemented in agri-products, namely nanopesticides, nanoherbicides, and nanosensors, present an ambiguous relationship with the environment and impose risks. Improper disposal of nanomaterials after their use is also a problem, affecting not only humans but also the environment, forests, and ecosystems [28]. Additionally, the further application of copper-based nanoparticles, such as CCCSNs, a potential termites and fungi killer, should be considered with great caution due to the potential hazards to nontarget organisms in the absence of effective control strategies [58]. A lot has been reported on the two-sided effects of NPs, i.e., they can improve plant growth (e.g., nanofertilizer), but they can also lead to the generation of reactive oxygen species (ROS) in plants at high doses [2]. The diverse destiny of nanomaterials in the environment hinders risk assessments due to the fact that their behavior is influenced by parameters such as pH, organic matter (OM), and duration of exposure [26]. These uncertainties must be incorporated into the regulatory landscape to ensure the safe deployment of these technologies.

Nanotechnology has transformed sustainable forestry and paper manufacturing; however, the environmental risks associated with the use of nanoparticles (NPs) must be carefully assessed. NPs can penetrate organisms and interfere with the structure and function of their ecosystems, and can accumulate in organisms due to their small size (1–100 nm) and strong surface reactivity [64]. The environmental risk assessment of NPs is complex because the fate of NPs depends on their composition, concentration, as well as on abiotic factors such as pH and organic matter [26]. This section discusses the ecological implications of NPs, particularly for soils, aquatic systems, and plant conditions, emphasizing relevant regulatory issues.

5.1.1 Mechanisms of Nanoparticle Toxicity

The properties of nanoparticles (NPs) are significantly different from the corresponding bulk materials, and the differences are attributed to the size effect, the large surface area to volume ratio, as well as the increased reactivity. A major property is bioaccumulation in living animals. Due to their small size, they can readily cross cell membranes and are retained in tissue. For instance, AgNPs taken up by aquatic invertebrates can be biotransferred along the food chain, ultimately impacting predators of high trophic levels, as evidenced by Kumar et al. [44]. Another important issue is their ability to elicit oxidative stress by producing reactive oxygen species (ROS), which can damage cellular components, including lipids, proteins, and DNA. The efficacy of one of these nanocoating agents, copper-carbon core–shell nanoparticles (CCCSNs), in inhibiting wood-decay fungi has been demonstrated; however, it also induces oxidative stress in nontarget organisms, such as earthworms, at concentrations higher than 50 mg/kg [58]. Furthermore, NPs may undergo trophic transfer in ecosystems, especially in aquatic ecosystems. When sorbed to microplastics or organic matter, these particles may be taken up by filter-feeding organisms, such as mussels, and can subsequently enter higher trophic levels, as demonstrated by Nel et al. [52]. Such activity raises concerns about its long-term ecological impact and underscores the need for further research into its environmental effects, toxicity, and potential hazards to biodiversity.

5.2 *Case Studies of Nanoparticle Impacts*

The use of engineered NPs in industrial and agricultural sectors has raised a great deal of ecological concerns attributed to their negative impacts on both terrestrial and aquatic biota. The exposure to silver nanoparticles (AgNPs) used as paper coating in terrestrial environments has led to a reduction of *Nitrosomonas spp.* that are involved in nitrogen cycling due to their role as beneficial soil microorganisms. It has been proposed that AgNPs exposure may decrease the efficacy of the nitrogen cycle by 40%, thereby disrupting soil fertility and reducing nutrient uptake into plants [15]. Furthermore, nanofertilizers such as zinc oxide nanoparticles (ZnO NPs) have been reported to have the capacity to increase crop yields, although they have adverse effects on some plants at very high concentrations (>500 ppm). For example, retardation of root development in wheat (Triticum aestivum) is caused by the production of ROS, leading to oxidative damage [2], which is exacerbated by high levels of ZnO NPs.

Since the 2010s, NP pollution, particularly regarding the environment of aquatic systems, has been the most alarming among marine and freshwater organisms. Titania nanoparticles (TiO_2 NPs) are a component of what is discarded in effluents from paper mills, and were reported to cause gill inflammation and dyspnea in rainbow trout (Oncorhynchus mykiss) at 100 μg/L, and elevated fish mortality [25]. Additionally, copper oxide nanoparticles (CuO NPs) have been reported to reduce phytoplankton growth at concentrations greater than 10 mg/L, as they act as a fungicide, disturbing primary production and leading to changes in the aquatic food web. This inhibition may lead to a "chain effect," in which food availability for higher trophic levels and the possibility of an algal bloom will decrease, which will make aquatic ecosystems unstable [4]. These results underscore the importance of rigorous control and management of nanoparticles to mitigate their ecological hazards and harness their technological benefits.

5.3 *Factors Influencing Nanoparticle Behavior*

The cytotoxicity of nanoparticles (NPs) depends on various parameters, such as pH, type of organic matter, and the duration of exposure. Under acidic conditions (pH < 5), the solubility of NPs increases, and the enhanced release of metal ions—e.g., Ag^+ from Ag nanoparticles (AgNPs)—may lead to increased bioavailability and potential toxicity [45]. OM, and notably humic acids, has an influential role by producing coatings on NPs that may decrease their immediate toxicity and, at the same time, facilitate settlement on sediments, and in doing so, slowing down the degradation of these materials and favoring their long-term retention in the environment [43]. Moreover, the interaction of NPs with the ecosystem is strongly dependent on the length of exposure. For instance, chronic low-dose exposure to cerium oxide nanoparticles (CeO_2 NPs) over just 6 months has now been shown to alter soil enzyme activities due to its disruption of microbial functions involved in nutrient cycling [69]. Together, these factors change the effects, actions, and dangers of NPs. Thorough assessments that account for varying environmental conditions with exposure scenarios are therefore needed. For anticipating NP effects and creating methods to minimize possible harm to ecosystems, understanding these dynamics is vital.

5.4 *Regulatory Gaps and Mitigation Strategies*

Current risk characterization of industrial nanoparticles (NPs) highlights major deficits in ecotoxicity screening due to <5% of NPs being subjected to standardized testing [30]. This limited testing series leaves open questions about their ecotoxicity and environmental persistence, particularly in soil, where they may accumulate and impact earthworms and microbial communities. However, studies have reported that mitigation measures, such as surface coating of NPs, can significantly alleviate toxicity. For example, coating carbon core–shell nanoparticles (CCCSNs) with a biodegradable polymer, chitosan, has reduced the toxicity of prokaryotic and eukaryotic organisms by 60%, as presented in Fig. 3 [58]. Likewise, lignin-coated iron oxide (Fe_3O_4) NPs exhibit improved biodegradability, so that they reduce their persistence within soil and minimize ecological risks [26]. These findings emphasize the need for more ecotoxicity testing, while also highlighting the potential of eco-friendly coatings to make nanotechnology more sustainable, thereby ensuring safer NP applications in industry and agriculture. Researchers could conduct further research into alternative coatings and degradation pathways. This could enhance risk management and aid regulators in developing frameworks for nanomaterials.

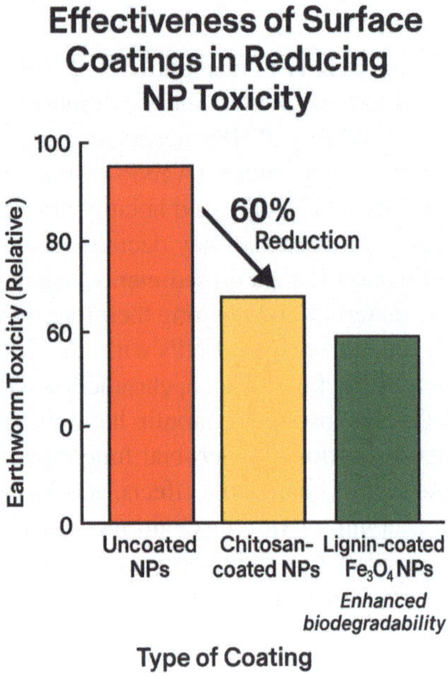

Fig. 3 Effectiveness of surface coatings in reducing nanoparticle (NP) toxicity to earthworms

5.5 High Production Costs and Scalability Issues

Producing new materials with those technologies is extremely expensive, as it requires specialized machines and a significant amount of energy. For example, making tiny fibers from plants through a process called electrospinning is more expensive than using conventional techniques [68]. However, while scientists have found ways to deploy these small materials to benefit plant growth, it has been difficult to produce large volumes of them for use in a wide variety of applications due to high costs and challenges in scaling production [49]. Some treatments, using tiny particles, can help paper last a long time, but it is challenging to find a way to use them that is both inexpensive and effective. To overcome these challenges, businesses, scientists, and the government need to collaborate and share ideas [57].

5.6 Concerns About Safety and Regulation

Nanotechnology is advancing so rapidly that it has outpaced the regulations intended to protect people from harm. Nanoparticles' particularities concerning their toxicity. Since the toxicity of nanoparticles differs from that of the corresponding bulk materials, nanoparticles require specific regulations [16]. For example, micronutrients are regulated for bulk copper, and we need to regulate nanocopper more strictly as it is more reactive [58]. Another concern is exposure at work. Researchers who handle nanomaterials can be exposed to them through inhalation or skin contact, and safety practices often lag [54]. The public's doubt, created by the opacity of the technology deployment process itself (for instance, 80% of articles on nanotechnology do not address risks), is an obstacle to its take-off [60]. Establishing unified global standards and engaging in proactive risk communication are crucial for building trust and promoting responsible use.

6 Future Trends

As we move forward with sustainable paper and forest management, nanotechnology, which bridges the gap between conventional materials science and cutting-edge environmental solutions, will play an increasingly larger role. Paper can thus act like an electronic interface that "knows" when it is being touched or struck, as well as a dynamic paper display platform [3]. It is the effect induced by the conductive nanocoatings of carbon nanotubes and silver nanowires. This development becomes the spanking new strategy for smart packaging to bloom in one direction (pH-sensitive nanosensors monitor food freshness [67]), to boom in another way (UV-light-induced fluorescence nanoscale Quick Response (QR) lines as anti-counterfeiting [73]), and to bloom in one new way (humidity-regulating nanoporous membranes as conservation tools for archival documents [70]). In addition to these, self-healing coatings can automatically repair minor mechanical damage to prolong the service life of paper products [75], while antimicrobial films containing silver nanoparticles can prevent microbial growth under high humidity storage conditions [19]. Such tech gains were simultaneous. People are deliberately transitioning toward using bio-based nanomaterials to fabricate biodegradable, along with high-performance paper. Nanocellulose exhibits mechanical strength comparable to that of steel and carbon nanotubes [8], and is extracted from wood pulp and agricultural residues. Due to its resistance to both oxygen and moisture, its barrier properties make it particularly suitable for sustainable food packaging [11], while its transparency allows for applications in flexible electronic displays [53]. Lignin, a major byproduct of the pulping process, is also contributing to a closed-loop material economy. It is being valorized into lignin-based nanoparticles that serve as UV-protective coatings [5] and biodegradable adhesives for composite paperboard products [46].

Nanotechnology is enabling the digitization and precision monitoring of forest systems, both ecologically and broadly. Hyperspectral nanosensors facilitate early pest infestation detection by identifying changes in chlorophyll fluorescence [41], and carbon nanotube sensors can also measure soil pH and nutrient concentrations in real-time [40]. The data harvested from these nanosensors, when combined with AI-driven predictive analytics, can inform models that assess wildfire risk based on climatic parameters such as temperature with humidity [63], so they optimize water with fertilizer usage, which yields a 20–30% reduction in input waste [69].

In nanoscale sensing, artificial intelligence, and real-time data transmission, these aspects converge to improve resource efficiency. This merging also changes the way people manage, observe, and protect forest ecosystems.

7 Conclusion

To summarize, nanotechnology is paving the way for the development of advanced wood-based materials that are efficient, eco-friendly, and biodegradable. This innovation is transforming sustainability in both the paper and forestry sectors. These materials have various applications in fields such as energy storage and water purification, effectively replacing plastics, glass, and metals. By utilizing both top-down wood processing techniques and bottom-up assembly of nanocellulose, we are advancing towards actual circular economy practices with reduced environmental impact. At the same time, nanotechnology is introducing green methods to mitigate the harmful pollutants produced by the pulp and paper industry, thereby enhancing the forests' ability to withstand climate challenges such as drought and disease. Solutions like nanoparticles, nanocomposites, and nanoadsorbents are leading the way. However, for widespread implementation to occur, we must navigate technical obstacles and regulatory issues, boost research funding, and foster strong collaborations in research to unlock the full potential of nanotechnology for sustainable resource management.

Competing Interests The authors declare that they have no competing interests.

DataAvailability **Declaration of Generative AI and AI-Assisted Technologies in the Writing Process** During the preparation of this work, the book chapter author(s) have not taken any assistance from AI, and the book chapter author(s) reviewed and edited the content as needed and take(s) full responsibility for the content of the publication.

References

1. Abbas M, Yan K, Li J, Zafar S, Hasnain Z, Aslam N, Iqbal N, Hussain SS, Usman M, Abbas M, Tahir MM, Abbas S, Abbas S, Qiulan H, Zhao X, El-Sappah AH (2022) Agri-nanotechnology and tree nanobionics: augmentation in crop yield, biosafety, and biomass accumulation. Front Bioeng Biotechnol 10:Article 853045. https://doi.org/10.3389/fbioe.2022.853045
2. Adisa IO, Pullagurala VLR, Peralta-Videa JR, Dimkpa CO, Elmer WH, Gardea-Torresdey JL (2019) Recent advances in nano-enabled fertilizers and pesticides: a critical review of mechanisms of action. Environ Sci Nano 6(7):2002–2030. https://doi.org/10.1039/C9EN00265K
3. Agarwal M, Lvov Y, Varahramyan K (2006) Conductive wood microfibres for smart paper. Nanotechnology 17(21):5319–5325. https://doi.org/10.1088/0957-4484/17/21/006
4. Aruoja V, Dubourguier HC, Kasemets K, Kahru A (2009) Toxicity of nanoparticles of CuO, ZnO and TiO_2 to microalgae *Pseudokirchneriella subcapitata*. Sci Total Environ 407(4):1461–1468. https://doi.org/10.1016/j.scitotenv.2008.10.053
5. Atalla RH, Crowley MF, Himmel ME (2006) Lignin-based nanomaterials for sustainable coatings. Green Chem 8(4):328–336. https://doi.org/10.1039/B517131G

6. Baietto M, Wilson AD, Bassi D, Ferrini F (2010) Evaluation of three electronic noses for detecting incipient wood decay. Sensors 10(2):1062–1092. https://doi.org/10.3390/s100201062
7. Bajpai P (2018) Biermann's handbook of pulp and paper: volume 2: paper and board making, 3rd edn. Elsevier
8. Beecher JF (2007) Nanocellulose: a renewable material for advanced applications. Nat Mater 6(12):957–960. https://doi.org/10.1038/nmat2057
9. Bengston DN, Dator J, Dockry MJ, Yee AM (2016) Alternative futures for forest-based nanomaterials: an application of the Manoa School's alternative futures method. World Futur Rev 8(4):197–221. https://doi.org/10.1177/1946756716659650
10. Bhagat D, Samanta S, Bhattacharya S (2013) Efficient management of fruit pests by pheromone nanogels. Sci Rep 3:1294. https://doi.org/10.1038/srep01294
11. Bhatnagar A, Sain M (2005) Processing of cellulose nanofiber-reinforced composites. J Reinf Plast Compos 24(12):1259–1268. https://doi.org/10.1177/0731684405049864
12. Bhatnagar P, Sharma S, Sri P, Alias NH, Ibrahim NA (2025) Nanocrystalline cellulose in tissue regeneration: innovations, challenges, and future directions. Regen Eng Transl Med 1–20. https://doi.org/10.1007/s40883-025-00428-7
13. Borges CC, Tonoli GHD, Moreira Cruz T, Duarte PJ, Antunes Junqueira T (2018) Nanoparticles-based wood preservatives: the next generation of wood protection? Cerne 24(4):490–501. https://doi.org/10.1590/01047760201824042244
14. Chinnamuthu CR, Boopathi PM (2009) Nanotechnology and agroecosystem. Madras Agric J 96(1–6):17–31
15. Colman BP, Arnaout CL, Anciaux S, Gunsch CK, Hochella MF Jr, Kim B, Lowry GV, McGill BM, Reinsch BC, Richardson CJ, Unrine JM, Bernhardt ES (2013) Low concentrations of silver nanoparticles in biosolids cause adverse ecosystem responses under realistic field scenario. PLoS ONE 8(2):e57189. https://doi.org/10.1371/journal.pone.0057189
16. Colvin VL (2003) The potential environmental impact of engineered nanomaterials. Nat Biotechnol 21(10):1166–1170. https://doi.org/10.1038/nbt875
17. Confederation of European Paper Industries (2023) Press release: the paper value chain reached a 70.5% recycling rate in 2022. CEPI – Confederation of European Paper Industries. Available at: https://www.cepi.org/press-release-the-paper-value-chain-reached-a-705-recycling-rate-in-2022/. Accessed 16 Nov 2025
18. Damalas CA, Eleftherohorinos IG (2011) Pesticide exposure, safety issues, and risk assessment indicators. Int J Environ Res Public Health 8(5):1402–1419. https://doi.org/10.3390/ijerph8051402
19. Dankovich TA, Gray DG (2011) Bactericidal paper impregnated with silver nanoparticles for point-of-use water treatment. Environ Sci Technol 45(5):1992–1998. https://doi.org/10.1021/es103302t
20. Dimkpa CO, Bindraban PS (2018) Nanofertilizers: new products for the industry? J Agric Food Chem 66(26):6462–6473. https://doi.org/10.1021/acs.jafc.7b02150
21. ECHA (2019) Restriction of hazardous substances in wood preservatives. European Chemicals Agency
22. El-Sherbiny S, Morsy F, Samir M, Fouad OA (2014) Synthesis, characterization and application of TiO_2 nanopowders as special paper coating pigment. Appl Nanosci 4(3):305–313. https://doi.org/10.1007/s13204-013-0196-y
23. Van Ewijk S, Stegemann JA, Ekins P (2021) Global paper consumption and sustainability trends. Nat Rev Earth & Environ 2(6):448–461. https://doi.org/10.1038/s43017-021-00169-5
24. FAO (2020) Global forest resources assessment 2020. Food and Agriculture Organization
25. Federici G, Shaw BJ, Handy RD (2007) Toxicity of titanium dioxide nanoparticles to rainbow trout (Oncorhynchus mykiss): Gill injury, oxidative stress, and other physiological effects. Aquat Toxicol 84(4):415–430. https://doi.org/10.1016/j.aquatox.2007.07.009
26. Ghorbanpour M, Hadian J (2017) Multi-walled carbon nanotubes stimulate callus induction, secondary metabolites biosynthesis and antioxidant capacity in medicinal plant Satureja khuzestanica grown in vitro. Carbon 118:434–444. https://doi.org/10.1016/j.carbon.2017.03.014

27. Han L, Li Y, Hou Y, Liang H, Feng P, Hu K (2023) Improvement of water and nitrogen use efficiencies by alternative cropping systems based on a model approach. Plants 12(3):597. https://doi.org/10.3390/plants12030597
28. Hasan Z, Lateef M, Agrawal N, Khanam A, Prasai JR, Husain A (2024) Uncertainties, phototoxicity, health impacts, and agricultural and environmental concerns of nanomaterials in the food industry. In: Nanotechnology and nanomaterials in the agri-food industries, pp 361–390. Elsevier. https://doi.org/10.1016/B978-0-323-99682-2.00013-X
29. Hingston JA, Collins CD, Murphy RJ, Lester JN (2001) Leaching of chromated copper arsenate wood preservatives: a review. Environ Pollut 111(1):53–66. https://doi.org/10.1016/S0269-7491(00)00030-0
30. Holden PA, Nisbet RM, Lenihan HS, Miller RJ, Cherr GN, Schimel JP, Gardea-Torresdey JL (2016) Ecological nanotoxicology: integrating nanomaterial hazard considerations across the subcellular, population, community, and ecosystems levels. Acc Chem Res 49(5):663–672. https://doi.org/10.1021/acs.accounts.5b00481
31. Hubbe MA, Zhang M (2019) Recycling and deinking of recovered paper. In: Paper and wood-based materials, pp 1–28. Springer
32. Hubbe MA, Rojas OJ, Lucia LA, Sain M (2008) Cellulosic nanocomposites: a review. BioResources 3(3):929–980. https://doi.org/10.15376/biores.3.3.929-980
33. Hurteau MD (2021) The role of forests in the carbon cycle and in climate change. In: Climate change, pp 561–579. Elsevier. https://doi.org/10.1016/B978-0-12-821575-3.00027-X
34. IEA (2021) Energy Efficiency 2021. OECD Publishing, Paris. https://doi.org/10.1787/cfb80e1e-en
35. INTERPOL (2020) Illegal logging and forest crime. International Criminal Police Organization
36. Islam S, Bhat G (2021) Progress and challenges in self-healing composite materials. Mater Adv 2(6):1896–1926. https://doi.org/10.1039/D0MA00873G
37. Jakhar AM, Aziz I, Kaleri AR, Hasnain M, Haider G, Ma J, Abideen Z (2022) Nano-fertilizers: a sustainable technology for improving crop nutrition and food security. NanoImpact 27:100411. https://doi.org/10.1016/j.impact.2022.100411
38. Kalaiyarasi M, Karthi S, Muthusamy S, Muniraj R, Saravanan S, Karunanithi K (2025) Nanotech for sustainable afforestation: enhancing soil, water, and plant systems. E3S Web Conf 619:04005. https://doi.org/10.1051/e3sconf/202561904005
39. Kartal SN, Green F, Clausen CA (2015) Do the unique properties of nanomaterials affect leaching and toxicity? Wood Sci Technol 49(6):1125–1137. https://doi.org/10.1007/s00226-015-0755-3
40. Kaushik A, Kumar R, Jayant RD, Nair M (2015) Nanostructured gas sensors for health care: an overview. J Pers Nanomedicine 1(1):10–23. https://doi.org/10.1016/j.nano.2014.12.003
41. Khan MR, Rizvi TF (2014) Nanotechnology: scope and application in plant disease management. Plant Pathol J 13(3):214–231. https://doi.org/10.3923/ppj.2014.214.231
42. Kim J, Grate JW, Wang P (2006) Nanostructures for enzyme stabilization. Chem Eng Sci 61(3):1017–1026. https://doi.org/10.1016/j.ces.2005.05.067
43. Klaine SJ, Alvarez PJJ, Batley, GE, Fernandes TF, Handy RD, Lyon DY, Mahendra S, McLaughlin MJ, Lead JR (2008) Nanomaterials in the environment: behavior, fate, bioavailability, and effects. Environ Toxicol Chem 27(9):1825–1851. https://doi.org/10.1897/08-090.1
44. Kumar A, Vemula PK, Ajayan PM, John G (2008) Silver-nanoparticle-embedded antimicrobial paints based on vegetable oil. Nat Mater 7(3):236–241. https://doi.org/10.1038/nmat2099
45. Levard C, Hotze EM, Lowry GV, Brown GE (2012) Environmental transformations of silver nanoparticles: impact on stability and toxicity. Environ Sci Technol 46(13):6900–6914. https://doi.org/10.1021/es2037405
46. Lovett GM, Weiss M, Liebhold AM (2016) Nonnative forest insects and pathogens in the United States. Ecol Appl 26(5):1437–1455. https://doi.org/10.1890/15-1176
47. Lu Z, Eadula S, Zheng Z, Xu K, Grozdits G, Lvov Y (2007) Layer-by-layer nanoparticle coatings on lignocellulose wood microfibers. Colloids Surf A 292(1):56–62. https://doi.org/10.1016/j.colsurfa.2006.06.008

48. McCrank J (2009) Nanotechnology applications in the forest sector. Natural Resources Canada. https://cfs.nrcan.gc.ca/publications?id=29162
49. Mohieldin SD, Zainudin ES, Paridah MT, Ainun ZM (2011) Nanotechnology in pulp and paper industries: a review. Key Eng Mater 471:251–256. https://doi.org/10.4028/www.scientific.net/KEM.471-472.251
50. Moon RJ, Martini A, Nairn J, Simonsen J, Youngblood J (2011) Cellulose nanomaterials review: structure, properties and nanocomposites. Chem Soc Rev 40(7):3941–3994. https://doi.org/10.1039/C0CS00108B
51. Nakagaito AN, Yano H (2004) The effect of morphological changes from pulp fiber towards nano-scale fibrillated cellulose on the mechanical properties of high-strength plant fiber-based composites. Appl Phys A 78(4):547–552. https://doi.org/10.1007/s00339-003-2453-5
52. Nel A, Xia T, Mädler L, Li N (2006) Toxic potential of materials at the nanolevel. Science 311(5761):622–627. https://doi.org/10.1126/science.1114397
53. Nogi M, Iwamoto S, Nakagaito AN, Yano H (2009) Optically transparent nanofiber paper. Adv Mater 21(16):1595–1598. https://doi.org/10.1002/adma.200803174
54. Oberdörster G, Oberdörster E, Oberdörster J (2005) Nanotoxicology: an emerging discipline evolving from studies of ultrafine particles. Environ Health Perspect 113(7):823–839. https://doi.org/10.1289/ehp.7339
55. Payn T, Carnus JM, Freer-Smith P, Kimberley M, Kollert W, Liu S, Orazio C, Rodriguez L, Silva LN, Wingfield MJ (2015) Changes in planted forests and future global implications. For Ecol Manage 352:57–67. https://doi.org/10.1016/j.foreco.2015.06.021
56. Porwal P, Porwal S, Singh SP, Husen A, Singh AK (2020) Improving futuristic nanomaterial researches in forestry sector: an overview. In: Husen A, Siddiqi KS (eds) Nanomaterials for agriculture and forestry applications, pp 505–518. Elsevier. https://doi.org/10.1016/B978-0-12-817852-2.00021-4
57. Puurunen K, Vasara P (2007) Opportunities for utilising nanotechnology in reaching near-zero emissions in the paper industry. J Clean Prod 15(13–14):1287–1294. https://doi.org/10.1016/j.jclepro.2006.07.018
58. Qi Y, Lian K, Wu Q, Li Y, Danzy M, Menard R, Klepzig K, Oliveira F (2011) Potentials of nanotechnology application in forest protection. In: TAPPI international conference on nanotechnology for renewable materials
59. Sarkar N, Chaudhary S, Kaushik M (2021) Nano-fertilizers and nano-pesticides as promoters of plant growth in agriculture. In: Nanotechnology: principles and applications, pp 229–244. Springer. https://doi.org/10.1007/978-3-030-66956-0_10
60. Scheufele DA, Corley EA, Dunwoody S, Shih T-J, Hillback E, Guston DH (2007) Scientists worry about some risks more than the public. Nat Nanotechnol 2(12):732–734. https://doi.org/10.1038/nnano.2007.392
61. Schwarze FWMR, Engels J, Mattheck C (2000) Fungal strategies of wood decay in trees. Springer. https://doi.org/10.1007/978-3-642-57302-6
62. Sethuraman V, Kumar PS, Kirubakaran D, Shivakumar MS (2024) Nanomaterials for improving health of agricultural crops and forest trees. In: Nanoscience in sustainable agriculture, pp 39–56. CRC Press. https://doi.org/10.1201/9781003375104-3
63. Singh R, Singh P (2019) AI and IoT in precision forestry. J Environ Manage 245(1):501–510. https://doi.org/10.1016/j.jenvman.2019.05.065
64. Singh D, Arya S, Gupta B, Kaushik D, Arya VS, Priyanka, Kumars U, Singh K (2021) Applications of nanotechnology in forest management. J Nanosci Nanotechnol 21(6):3466–3480. https://doi.org/10.1166/jnn.2021.19071
65. Wagay OA, Khan S, Rafeeq J, Pala NA, Bhat GM, Dutt V, Peerzada IA, Malik AR, Sofi PA, Mushtaq T, Mugloo JA (2023) Nanotechnology and its potential application in forest and forest-based industries: a review. SKUAST J Res 25(4):527–537. https://doi.org/10.5958/2349-297X.2023.00065.X
66. Wang L, Li X, Zhang G, Dong J, Eastoe J (2007) Oil-in-water nanoemulsions for pesticide formulations. J Colloid Interface Sci 314(1):230–235. https://doi.org/10.1016/j.jcis.2007.04.079

67. Wang Z, Skidmore AK, Darvishzadeh R, Wang T (2023) Mapping forest canopy nitrogen content by inversion of coupled leaf-canopy radiative transfer models from airborne hyperspectral imagery. Agric For Meteorol 253–254:247–260. https://doi.org/10.1016/j.agrformet.2018.02.018
68. Wegner TH, Jones PE (2006) Advancing cellulose-based nanotechnology. Cellulose 13(2):115–118. https://doi.org/10.1007/s10570-006-9056-1
69. Xie C, Guo Z, Zhang P, Yang J, Zhang J, Ma Y, He X, Lynch I, Zhang Z (2022) Effect of CeO_2 nanoparticles on plant growth and soil microcosm in a soil-plant interactive system. Environ Pollut 300:118938. https://doi.org/10.1016/j.envpol.2022.118938
70. Yang B, Aksak B, Lin Q, Sitti M (2006) Compliant and low-cost humidity nanosensors using nanoporous polymer membranes. Sens Actuators B Chem 114(1):254–262. https://doi.org/10.1016/j.snb.2005.04.044
71. Yao J, Yang M, Duan Y (2009) Chemistry, biology, and medicine of fluorescent nanomaterials and related systems: new insights into biosensing, bioimaging, genomics, diagnostics, and therapy. Chem Rev 114(12):6130–6178. https://doi.org/10.1021/cr200359p
72. Yuan Q, Bian J, Ma M-G (2020) Advances in biomedical application of nanocellulose-based materials: a review. Curr Med Chem 28(40):8275–8295. https://doi.org/10.2174/0929867328666201130124501
73. Zhang L, Li Y, Sun J, Shen J (2017) Layer-by-layer deposition of polyelectrolyte complexes for the fabrication of foam coatings with high loading capacity. ACS Appl Mater Interfaces 9(8):7424–7432. https://doi.org/10.1021/acsami.6b14408
74. Zhu Q, Chua MH, Ong PJ, Lee JJC, Chin KLO, Wang S, Kai D, Ji R, Kong J, Dong Z, Xu J, Loh XJ (2022) Recent advances in nanotechnology-based functional coatings for the built environment. Mater Today Adv 15:100270. https://doi.org/10.1016/j.mtadv.2022.100270
75. de Lima GG, Zakaluk ICB, Artner MA, Pedro AC, Gonzalez de Cademartori PH, Muniz GIBD, Magalhães WLE (2025) Enhancing barrier and antioxidant properties of nanocellulose films for coatings and active packaging: a review. ACS Appl Nano Mater. https://pubs.acs.org, https://doi.org/10.1021/acsanm.4c04805

Role of Nanotechnology in Treating Wastewater from Pulp and Paper Industries

Arbab Husain, Afreen Khanam, Ghayyum Husain, Hamda Khan, Adil Husain, Muneera Lateef, and Ziaul Hasan

Abstract The pulp and paper industry is a significant contributor to global wastewater pollution, discharging large volumes of effluents laden with organic matter, suspended solids, chlorinated compounds, and recalcitrant lignin derivatives. Traditional treatment methods often fall short in completely removing these persistent pollutants, leading to environmental degradation and ecological risks. In recent years,

A. Husain (✉) · A. Khanam
Department of Biotechnology and Life Sciences, Faculty of Sciences, Mangalayatan University, Aligarh, India
e-mail: arbabhusain1@gmail.com

A. Khanam
e-mail: afreen.iul@gmail.com

G. Husain
Department of Post Harvest Engineering and Technology, Faculty of Agricultural Sciences, Aligarh Muslim University, Aligarh, India
e-mail: ghayyumh@gmail.com

H. Khan
Department of Biochemistry, Faculty of Medicine, Jawahar Lal Nehru Medical College, Aligarh Muslim University, Aligarh, India
e-mail: hamdakhan2@gmail.com

A. Husain
Department of Biochemistry, Babu Banarasi Das College of Dental Sciences, Babu Banarasi Das University, Lucknow, India
e-mail: adilhusain.ad70@gmail.com

M. Lateef
Department of Agricultural Genetic Engineering, Faculty of Agricultural Sciences and Technology, Niğde Ömer Halisdemir University, Niğde, Turkey
e-mail: muneeralateef@gmail.com

Z. Hasan (✉)
Department of Life Science, Faculty of Life Sciences, Mandsaur University, Mandsaur 458001, Madhya Pradesh, India
e-mail: zhasan.biochem@gmail.com

Department of Biosciences, Faculty of Life Sciences, Jamia Millia Islamia, New Delhi 110025, India

R. K. Bachheti (ed.), *Nanotechnology in the Pulp and Paper Industry*, Smart Nanomaterials Technology, https://doi.org/10.1007/978-981-95-3473-9_17

nanotechnology has emerged as a promising solution to address these challenges due to its superior surface reactivity, high adsorption capacity, and tunable physicochemical properties. This chapter explores the advancements in nanomaterials, such as metal and metal oxide nanoparticles, carbon-based nanostructures, and nano-enabled membranes for the treatment of pulp and paper mill effluents. The mechanisms of contaminant removal, including adsorption, photocatalysis, and membrane filtration, are critically analyzed. Furthermore, the chapter discusses the environmental implications, scalability, and prospects of integrating nanotechnology into industrial wastewater management systems. Emphasis is placed on the development of sustainable, cost-effective, and efficient nanotechnological approaches to support the transition toward cleaner production and the circular economy in the pulp and paper sector.

Keywords Nanotechnology · Wastewater treatment · Pulp and paper industry · Nanomaterials · Photocatalysis · Membrane filtration

1 Introduction

The pulp and paper industry is fundamental to the global manufacturing sector, generating vital products such as paper, cardboard, and packaging materials. This business is acknowledged as a major user of freshwater and a substantial contributor to industrial wastewater [5]. The production process encompasses a sequence of operations, pulping, bleaching, and papermaking, that produce substantial quantities of effluents laden with suspended particles, organic matter, chlorinated compounds, and heavy metals [44].

The ecological consequences of wastewater from pulp and paper mills are significant. The release of untreated or insufficiently treated effluents into water bodies can result in oxygen depletion, toxicity to aquatic animals, bioaccumulation of persistent chemicals, and disruption of ecosystems [18]. Numerous pollutants, including lignin derivatives, resin acids, and organochlorines, exhibit resistance to biodegradation and endure in the environment for prolonged durations. Conventional wastewater treatment techniques used in the pulp and paper sector, such as primary sedimentation, activated sludge processes, and chemical precipitation, have demonstrated inadequate efficacy in eliminating specific recalcitrant compounds and meeting increasingly stringent environmental regulations [52]. Additionally, these methods can be quite energy-intensive and generate large volumes of sludge that require further handling and disposal.

The development of efficient, cost-effective, and environmentally compatible treatment technologies is crucial in this regard. Nanotechnology, defined as the manipulation of matter at atomic, molecular, and supramolecular levels (1–100 nm), offers novel methods and tools with which these complications can be addressed. Nanomaterials such as titanium dioxide (TiO_2) nanoparticles, carbon nanotubes (CNTs), zero-valent iron nanoparticles (nZVI), and nanocomposites have shown remarkable effectiveness in the removal of organic as well as inorganic contaminants through adsorption, photocatalysis, and redox reactions [27, 46].

Current findings have established that nanotechnology can significantly enhance the removal efficiencies of conventional systems, presenting various benefits, including higher surface area-to-volume ratios, tunable surface chemistry, and multi-functionality [32]. Due to nano-enabled membranes, only specific pollutants are separated while fouling is prevented, both of which enhance the longevity of the membranes and ease of cleaning.

However, the utilization of nanotechnology in industrial wastewater treatment remains in its infancy, and certain hazards exist, such as the potential toxicity of nanomaterials to aquatic organisms and uncertainties regarding their environmental fate and behavior [41]. Therefore, it is essential to thoroughly evaluate the benefits and risks.

This chapter discusses the application of nanotechnology for the purification of pulp and paper wastewater, providing an in-depth analysis of various nanomaterials, including their mechanisms, applications, and future perspectives. The discussion is supported by new data, images, and models, as well as case studies from the industry and techno-economic analysis, which provide a comprehensive picture of the potential for introducing new materials and novel technologies into wastewater management strategies in this key sector (Fig. 1).

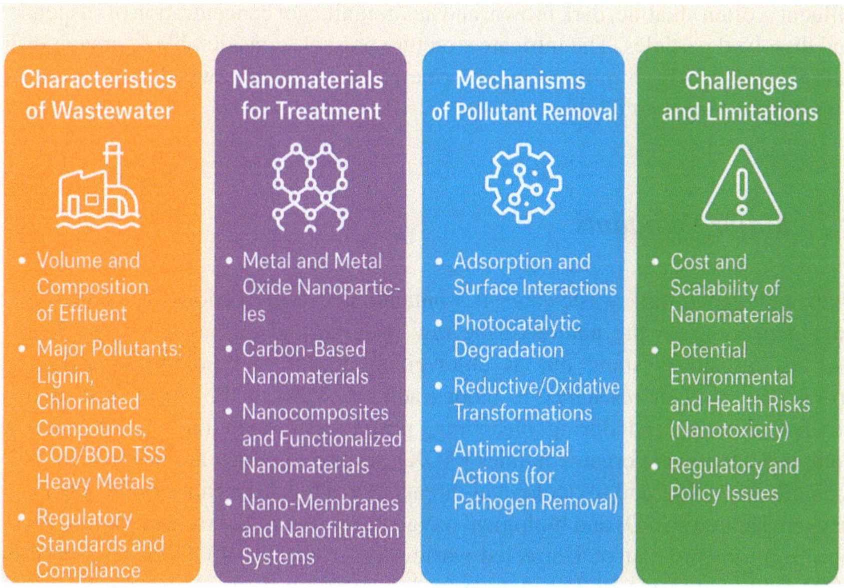

Fig. 1 Role of nanotechnology in pulp and paper wastewater treatment. Flow chart summarizing wastewater features, key nanomaterials, pollutant removal mechanisms, and major challenges, each represented with icons

2 Characteristics of Wastewater from Pulp and Paper Industries

It takes 250–300 cubic meters of water to make just one ton of paper in the pulp and paper sector [44]. For this reason, the industry generates large amounts of effluent, which exhibits considerable diversity in its physical, chemical, and biological characteristics, primarily dependent on the pulping method, the types of raw materials used, and the operations performed by the mill.

2.1 Volume and Composition of Effluent

The pulp and paper industry generates substantial quantities of wastewater, primarily due to its extensive water usage during pulping, bleaching, and paper manufacturing processes. Typically, 250–300 cubic meters of water are used to manufacture each ton of paper. In contrast, wastewater production ranges from 50 to 200 cubic meters per ton, depending on the type of process and water recycling methods used [44]. This effluent exhibits significant variability in its composition, encompassing both organic and inorganic components. Typical constituents comprise fibrous substances, lignin, carbohydrates, resin acids, inorganic salts, and residual processing chemicals. The effluent is often alkaline, dark brown, and has a significant concentration of suspended and dissolved particles. The intricate structure presents considerable treatment difficulties, particularly in mills using chlorine-based chemicals, which exacerbate the toxicity and longevity of contaminants.

2.2 Major Pollutants

Pulp and paper mill wastewater is contaminated with various substances that vary depending on the nature of the raw material and the process used. Lignin and its derivatives stand out as color-giving toxins and pollutants to aquatic species due to their resistance to biodegradation. Most of the chlorinated organic compounds produced during bleaching, primarily through the use of chlorine, comprise adsorbable organic halides (AOX), several of which are deemed environmentally persistent and toxicological threats [36]. High concentrations of chemical oxygen demand (COD) and biological oxygen demand (BOD) indicate severe pollution from organic matter. Untreated wastewaters can reach COD levels exceeding 5000 mg/L, alongside BOD values exceeding 1000 mg/L. Such untreated wastewaters disrupt the balance of dissolved oxygen (DO) in water bodies [65]. Total suspended solids (TSS), such as wood fibers, pulp remnants, and other fragmented debris, increase cloudiness and deposit layers of silt in water bodies. Additionally, small quantities of heavy metals, such as cadmium, lead, and mercury, can enter the system due to contaminants in raw materials or certain additives. These metals are highly toxic and can bioaccumulate, leading to long-term environmental pollution issues.

2.3 Regulatory Standards and Compliance Challenges

Discharge of effluents from pulp and paper mills is regulated by stringent environmental protection policies at both domestic and international levels. The Central Pollution Control Board (CPCB) of India enforces regulatory limits for BOD $\leq$ 30 mg/L, COD $\leq$ 250 mg/L, TSS $\leq$ 100 mg/L as well as color and toxic limit rules [23, 77] (Table 1). The disallowance of persistent pollutant discharges, such as AOX and suspended matter, is equally enforced by the U.S. Environmental Protection Agency (EPA) and the European Union. Under these regulatory policies, many mills still face challenges in achieving full compliance. Outdated treatment facilities, along with varying wastewater characteristics, pose significant hindrances to the full adoption of advanced technologies, especially in developing countries where costs are more prohibitive. Conventional biological processes tend to focus on the degradation of nonbiodegradable chemicals, which require the development of more robust elimination frameworks (Fig. 2). In this regard, advanced treatment methodologies, such as nanotechnology-based systems, are gaining attention due to their potential to reduce the regulatory gap while minimizing negative impacts [1, 31].

3 Fundamentals of Nanotechnology in Wastewater Treatment

3.1 Definition and Principles of Nanotechnology

Nanotechnology is the science and engineering of materials and technologies at the nanoscale, 1–100 nm in size. At this scale, materials exhibit different physical, chemical, and biological properties compared to their bulk equivalents. The high surface area-to-volume ratio, quantum effects, and tailorable surface properties of nanomaterials enable greater reactivity, adsorptive capacity, and catalytic activity, making them extremely attractive for environmental applications. Nanotechnology offers advanced tools for detecting, removing, and treating pollutants in wastewater treatment. It operates on the principles of surface interaction, molecular-scale reactivity, and the design of materials for targeted pollution removal. These features make nanotechnology a means to overcome the limitations of traditional methods, particularly for complex industrial effluents such as those from pulp and paper mills [54].

Table 1 Key aspects of effluent discharge from the pulp and paper industry

Section	Key aspect	Detail	References
Volume and composition of effluent	Water usage	Each ton of paper utilizes 250–300 m^3 of water	Ali and Sreekrishnan [5] Covinich et al. [13] Liang et al. [35]
	Effluent volume	50–200 m^3 per ton, depending on pulping process and water recycling	Pokhrel and Viraraghavan [44]
	Composition	Fibrous solids, organic matter (lignin, carbohydrates, resin acids), inorganic salts (sulfates, chlorides), process chemicals (chlorine dioxide, H_2O_2)	Singh and Tripathi [58]
Major pollutants	Lignin	Nonbiodegradable; forms toxic chlorinated organics	
	Chlorinated compounds	AOX from chlorine bleaching is mutagenic, persistent	Verma and Chandra [70]
	COD/BOD	COD: up to 6000 mg/L; BOD: >1000 mg/L causes oxygen stress	
	Total suspended solids (TSS)	Wood particles, fibers increase turbidity and sedimentation	Vítězová et al. [71]
	Heavy metals	Trace amounts of Pb, Cd, Hg, and ecological risk from raw material impurities	
Regulatory standards and compliance challenges	Global standards	US: EPA guidelines, EU: Industrial Emissions Directive, India: CPCB (BOD $\leq$ 30 mg/L, COD $\leq$ 250 mg/L, TSS $\leq$ 100 mg/L)	Thompson et al. [65]
	Compliance issues	Effluent variability, Inadequate infrastructure, economic barriers, weak enforcement in developing nations	
	Industry response	Push toward advanced, eco-friendly technologies (e.g., nanotech) for sustainable compliance	

3.2 *Properties of Nanomaterials Relevant to Wastewater Treatment*

Wastewater treatment is beneficial when pollutants are transformed into less harmful substances that are physico-chemically adsorbed. This process primarily occurs due to the unique properties of nanoparticles (NPs). It is known that a large number of functional groups on the surface creates enough binding sites for pollutants.

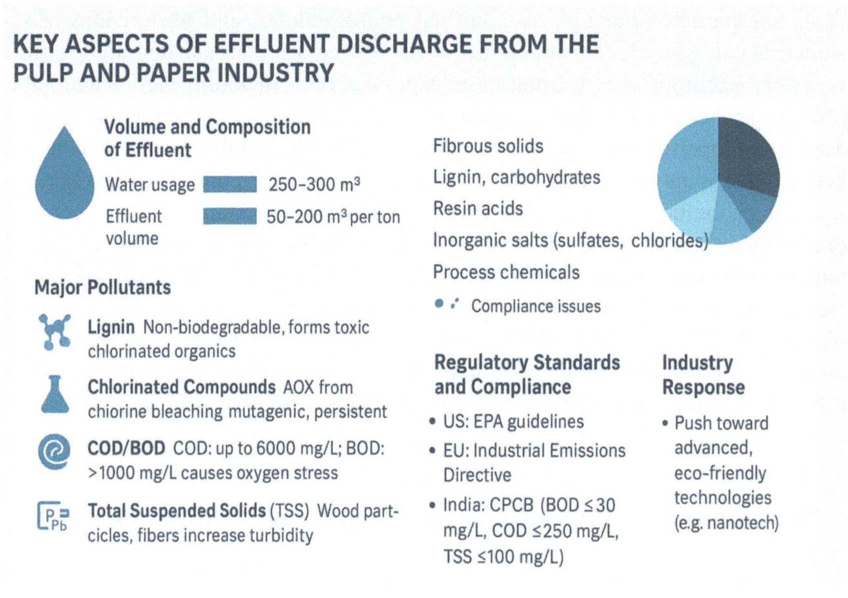

Fig. 2 An overview of effluent characteristics, major pollutants, and regulatory standards and compliance challenges in the pulp and paper industry

Nanomaterials also enhance the overall efficiency, hardness, and reduced porosity, with controllable active particle surfaces, leading to the retention of reagent species in catalysts. As one such application, water purification has emerged as a growing area that encompasses catalysts, such as ion exchange polymer membranes, which selectively enhance adsorption.

Specific nanoparticles exhibit inherent antibacterial characteristics that aid in regulating microbial proliferation in treatment systems. Moreover, the capacity to alter or functionalize nanoparticles with chemical moieties or polymers facilitates customization for the targeted elimination of pollutants, including the binding of heavy metals, degradation of persistent organic compounds, or inactivation of hazardous chlorinated species. The features jointly enhance the efficiency, speed, and specificity of nanomaterial-based water treatment methods, which are crucial for addressing the complex matrices of pulp and paper mill effluents [64].

3.3 *Classification of Nanomaterials Used*

In wastewater treatment, nanomaterials can be classified according to their structure and composition. NPs are usually metallic nanoparticles, which can appear in the form of silver (Ag), iron (Fe), titanium dioxide (TiO_2), or zinc oxide (ZnO).

These nanoparticles function as catalysts, photocatalysts, and bactericides. Zero-valent iron nanoparticles are effective in reducing heavy metals and degrading chlorinated organic compounds. Carbon-based nanomaterials, including carbon nanotubes (CNTs), graphene, and graphene oxide (GO), are characterized by their exceptional adsorption properties and electrical conductivity, enabling both physical and electrochemical treatment techniques [46]. Nanocomposites, which integrate diverse nanomaterials or include nanoparticles with polymers or membranes, offer synergistic benefits such as enhanced stability, multifunctionality, and recyclability. Supplementary categories include dendrimers, quantum dots, and nano-membranes, each offering unique advantages contingent upon the specific pollutant and treatment objective. The diverse array of nanomaterials available provides a versatile toolkit for developing tailored treatment systems that can address the extensive spectrum of pollutants found in effluents from the pulp and paper industry [73].

4 Nanomaterials for Wastewater Treatment Applications

4.1 Metal and Metal Oxide Nanoparticles

TiO_2, ZnO, and Fe_3O_4 nanoparticles are being widely studied for their ability to eliminate contaminants from wastewater due to their unique properties. As TiO_2 nanoparticles are exposed to UV light, they generate reactive oxygen species (ROS) radicals such as hydroxyl radicals that can decompose dyes, pesticides, and medicines. Due to their stability, nontoxicity, and high photocatalytic activity, they can be used for various purposes. When zinc oxide (ZnO) nanoparticles are used as photocatalysts, they are effective in regulating and degrading microbial wastes. Such particles decompose organic pollutants and play a role in disinfection. Fe_3O_4, a type of iron oxide also known as magnetite, is very efficient in eliminating heavy metals such as lead, arsenic, and cadmium through reactions at its surface. Fe_3O_4 nanoparticles can be easily removed from water using magnets, which makes them useful in processes that are repeated or carried out continuously. Wastewater treatment systems that use metal and metal oxide nanoparticles are considered sophisticated [16, 68].

4.2 Carbon-Based Nanomaterials

Carbon nanomaterials, such as carbon nanotubes (CNTs), graphene oxide (GO), and activated carbon, are highly sought after due to their powerful adsorption properties and their ability to be reused multiple times. CNTs' ability to absorb pollutants is based on their large surface area, unique shape, and the ability to modify the outer layer so that it can bind to specific contaminants. Carbon nanotubes can effectively treat industrial wastewater due to their ability to bind with various substances, including heavy metals, oil, and dyes. Thanks to the oxygen groups on GO, it exhibits a high affinity for pollutants and is highly effective in removing heavy metal ions, as well as certain organic contaminants, from water. With hydrophilic properties, GO supports the removal of many pollutants, while its surface can also be modified with polymers or metal ions to enhance its absorption capacity further. Although activated carbon is not a nanomaterial by strict definition, it is often used in conjunction with nanoparticles to enhance the adsorption process [55]. Due to its large surface area and pores, the material excels at removing dissolved chemicals, volatile organic compounds (VOCs), and other pollutants from wastewater. Additionally, they can be regrown when heated or chemicals are applied and can be reused several times, making them more useful and environmentally friendly in water treatment [38].

4.3 Nanocomposites and Functionalized Nanomaterials

Nanocomposites are hybrid materials that combine nanoparticles with other materials, such as polymers, metals, or ceramics, to produce better composites with better properties. In wastewater treatment, nanocomposites can be engineered to extract specific pollutants, providing a more tailored and effective solution. TiO_2 nanoparticles in composite materials can remove organic pollutants and pathogens more efficiently as the composite structure enhances TiO_2's photocatalytic activity and its stability in water [43]. Functionalized nanomaterials, where nanoparticles are modified with specific chemical groups or organic molecules, can be more selective to certain pollutants. Functionalization can enhance the material's affinity to specific pollutants, increase adsorption efficiency, and improve interaction with targeted contaminants, such as heavy metals or organic pollutants. Nanocomposites and functionalized nanoparticles enhance treatment efficiency, selectivity, and sustainability. Additionally, they are often more stable and reusable, allowing them to be used for a long time in water filtration.

4.4 Nano-Membranes and Nanofiltration Systems

Nano-membranes and nanofiltration systems are revolutionizing water treatment through separation techniques that operate efficiently on a molecular or nanometer scale. Such systems are used in procedures including ultrafiltration, reverse osmosis (RO), and nanofiltration. They serve as discriminating obstacles isolating contaminants in water by size, charge, or molecular interactions. Nano-membranes feature large surface area, improved permeability, and greater selectivity compared with conventional membranes. Considering these attributes, nanomembranes remain suitable for processing thorough wastewater, including discharges from sectors such as pulp and paper mills [30]. Nano-membranes withstand fouling substantially, which benefits them beyond ordinary filtration systems, where impurities amass, occlude the membrane surface, and lessen effectiveness. Nanomaterials, including carbon nanotubes (CNTs) and graphene-based membranes, withstand fouling substantially due to their smooth surface architecture, and they filter more effectively by eliminating contaminants such as bacteria, heavy metals, and organic compounds. Nanomembranes exhibit improved energy efficiency in processes such as reverse osmosis, requiring reduced pressure to achieve equivalent filtration compared to typical membranes. Nano-membranes, along with nanofiltration systems, progress steadily and refine constantly, and may achieve wastewater treatment that seems more tenable, practical, and thrifty [12, 67].

5 Mechanisms of Pollutant Removal

The efficacy of nanoparticles in wastewater treatment largely stems from their unique ability to eradicate pollutants. These mechanisms include adsorption, photocatalytic degradation, surface contacts, as well as oxidative and reductive conversions. Antimicrobial action is incorporated, too. Nanomaterials effectively handle a diverse range of pollutants, including organic chemicals, heavy metals, and pathogens, in industrial wastewater due to their adjustable nature [3, 39, 55].

5.1 Adsorption and Surface Interactions

Adsorption is a widely studied and used technique for removing contaminants using nanomaterials. This process happens when impurities cling to the surfaces of nanoparticles. The high surface area-to-volume ratio of these tiny particles, particularly carbon materials like carbon nanotubes and graphene oxide, along with metal oxides such as TiO_2 and Fe_3O_4, provides numerous active spots for pollutants to stick. As a result, researchers such as Ratnam et al. [50] and Thakur et al. [61] have emphasized its effectiveness. Interestingly, the unique properties of nanoparticles make them suitable for this application. Furthermore, their properties can be tailored to specific contaminants. As a result, their adaptability significantly improves their performance. Furthermore, multiple studies have shown that different nanoparticles are more effective against different pollutants, which complicates the situation. Adsorption selectivity and efficacy are determined by surface contact properties, including electrostatic attraction, van der Waals forces, hydrogen bonding, and coordination bonds. For example, nanomaterials with a high affinity for metal ions can effectively remove heavy metals such as lead, cadmium, and arsenic. Similarly, organic pollutants, such as dyes and pharmaceuticals, can be adsorbed onto the surfaces of nanomaterials through hydrophobic interactions. Consequently, this leads to better effectiveness in wastewater treatment processes [34]. Thus, when you modify these materials, the results can be impressive; you see more contaminants removed efficiently. So, it's crucial to consider these changes for optimal results.

5.2 Photocatalytic Degradation

Photocatalytic degradation is a dominant mechanism used by the nanomaterials in the degradation of organic compounds from wastewater, especially the metal oxide nanoparticles, including ZnO and TiO_2. Under UV irradiation, these photocatalysts can generate reactive oxygen species (ROS), like superoxide anions (O2•−) and hydroxyl radicals (•OH). These ROS possess the ability to break down a wide range of organic pollutants, such as dyes, phenolic compounds, insecticides, and pharmaceutical residues, into simpler, harmless products like carbon dioxide and water. The photocatalytic reaction process involves the excitation of conduction band electrons in the material, which then react with molecular oxygen to form reactive oxygen species (ROS) [28, 69]. Photocatalytic degradation offers numerous benefits, such as high efficiency, sustainability (especially with sunlight activation), and the capability to break down recalcitrant pollutants that cannot be destroyed during conventional treatment processes. The photocatalytic performance of nanomaterials can be enhanced through metal doping or by manipulating particle size and morphology, thereby enhancing their performance and functional stability as wastewater treatment agents [47].

5.3 Reductive/Oxidative Transformations

Reductive and oxidative conversions play major roles in the breakdown of inorganic and organic contaminants in wastewater. Metal nanoparticles, such as zero-valent iron (Fe^0) and copper nanoparticles (Cu), can facilitate conversions through electron-transfer processes. During reductive transformation, NPs like Fe^0 transfer electrons and convert contaminants, including chlorinated organics, pesticides, and heavy metals, into less dangerous or easily degraded forms [72]. Zero-valent iron nanoparticles are commonly used in the remediation processes for heavy metal pollutants, the best example being the reduction of hexavalent chromium (Cr^{6+}) into its less toxic trivalent form (Cr^{3+}). Conversely, oxidative conversions are obtained when particles such as TiO_2 or MnO_2 act as electron acceptors, catalyzing the oxidation of pollutants and the degradation of complex organic molecules. Such conversions are extremely efficient for the destruction of nonbiodegradable or persistent organic pollutants (POPs) that are otherwise refractory to conventional processes. By virtue of the reductive and oxidative procedures, the nanomaterials act as a double-edged sword for efficient pollutant removal, which means that they could potentially be used to address a variety of contaminants in wastewater treatment [75, 74].

5.4 Antimicrobial Mechanisms (for Pathogen Removal)

Nanomaterials exhibit higher antibacterial activity, which is very efficient in eliminating pathogens. Such activity is particularly beneficial in wastewater treatment facilities because pathogens, including bacteria, viruses, and protozoa, can pose significant risks to human health if not eliminated. Nanomaterials, such as silver (Ag) nanoparticles and copper (Cu) nanoparticles, exhibit enhanced bactericidal activity by disrupting the bacteria's cell membranes and interacting with their biological structures. Most people use silver nanoparticles. These are small pieces of silver that are smaller than a grain of rice and are in the nanoscale. They are important because they can kill a wide range of bacteria by releasing silver ions (Ag^+), which inhibit microbial enzyme activity, disrupt deoxyribonucleic acid (DNA) stability, and damage cell membrane structures [25, 42]. More importantly, researchers have found that carbon nanomaterials, such as graphene oxide and carbon nanotubes, can be used to kill bacteria because they can enter the cells and disrupt their normal metabolic processes. Nanoparticles are also effective at cleaning water contaminated with harmful microorganisms, as they kill bacteria. This means that treated wastewater is safe to drink. Nanomaterials also have the advantage of being reusable and not becoming resistant to microbes, unlike antibiotics and other traditional antibacterial agents, which can become resistant over time [53].

6 Case Studies and Recent Developments

Nanotechnology in wastewater treatment has been proven effective by recent research, including the treatment of raw material industry waste, which has had great results. Laboratory and pilot-scale research has provided a wealth of information about the effectiveness and viability of nanomaterials. Similarly, case studies demonstrate how these advanced technologies can be integrated into practical applications, with significant differences in the amounts or ratios relevant for on-site operation, compared to laboratory experiments. While nanomaterial-based approaches are being used with increasing frequency and in new areas, including environmental engineering, they have received both criticism and praise [10, 33].

6.1 Laboratory and Pilot Studies

A laboratory and pilot-scale study is needed to develop information regarding the real-world applications of nanomaterials in wastewater treatment. This literature typically involves the use of a variety of nanomaterials, including metal and metal oxide nanoparticles, carbon nanomaterials, and nanocomposites, to compare the effectiveness of contaminant removal in controlled experiments. A lab-scale test by Zhang et al. [77] demonstrated that TiO_2 and ZnO nanoparticles can serve as photocatalysts for the degradation of organic pigments in model wastewater, which may have led to them being the most widely used metal oxide nanomaterials in wastewater treatment [63].

The results were promising, with degradation efficiencies high and a very good effluent quality obtained. Pilot-scale studies serve as the second stage in scaling up laboratory results, where the effectiveness of nanomaterials is evaluated under larger and more complex conditions that simulate real-life industrial wastewater environments. This stage supports demonstration and evaluation [6]. Pilot-scale demonstration in the treatment plant of a European paper mill of the reuse of carbon nanotubes (CNTs) for heavy metals removal.

This research has reported that CNTs effectively adsorb lead, cadmium, and other heavy metals from wastewater under a wide range of operating conditions, with excellent removal efficiency [8, 45]. These works investigate the efficiency of contaminant removal and discuss the problems related to the regeneration and reuse of nanomaterials, as well as reaction kinetics and operational stability. Practical upscaling of research into large-scale industrial applications requires that the synthesis and processing of nanomaterials be scalable from lab-scale to pilot-scale.

6.2 Applications of Industrial Cases

A challenge for validating the use of nanotechnology in wastewater treatment is scaling the technology from laboratory and pilot scales to demonstrated industrial use. Nanomaterials have already begun to be used in various sectors for water treatment, and the pulp and paper industry is a particularly interesting option. In one recent industrial application, wastewater from a large US-based pulp and paper mill was treated using TiO_2-based photocatalytic systems. The application of a photocatalytic reactor with TiO_2 nanoparticles led to a significant decrease in organic pollutants, including lignin and chlorinated compounds, thereby improving the quality of the treated effluent and enabling the plant to meet environmental regulations [66].

Another case study examined the application of graphene oxide membranes for water treatment at a textile dyeing plant in India. The graphene oxide (GO) membranes demonstrated superior removal of dyes and other contaminants from wastewater, resulting in a significant improvement in the performance of the water treatment system. Industrial application of the nano-particles has led to energy saving and improved permeability, and antifouling performance of the nano-membranes, less energy consumption of the conventional type of filtration system, for example, RO. The above examples demonstrate how nanotechnology can enhance the efficiency, economics, and environmental sustainability of wastewater treatment across various industries [74].

6.3 Evaluation of Performance Relative to Traditional Methods

Comparison reveals several notable advantages of nanotechnology-based treatment systems over conventional methods, including activated sludge, coagulation-flocculation, and adsorption. As a general rule, nanomaterials are more efficient in pollutant removal, have shorter treatment times, and can be designed to target specific pollutants. Studies have shown that nanoparticles made of iron oxide (Fe_3O_4) exhibit superior adsorption capacities and can effectively remove heavy metals, such as lead and cadmium, from wastewater compared to conventional activated carbon. For organic pollutants, photocatalytic systems involving TiO_2 nanoparticles increase the rate of degradation and yield complete mineralization of the pollutants, whereas conventional techniques yield incomplete degradation or require longer treatment times [21].

One of the most significant advantages of nanotechnology-based processes is that they can simultaneously treat a broad spectrum of contaminants, including organic compounds, heavy metals, and microorganisms, whereas conventional processes often require multiple chemicals or steps to remove diverse contaminants. Moreover, nanoparticles are reusable or can be regenerated, thus enhancing their economic feasibility and sustainability for repeated use. Nonetheless, it should be noted that therapeutic processes utilizing nanotechnology also have some drawbacks, including high initial costs for material synthesis, potential environmental issues associated with the disposal of nanoparticles, and the requirement to optimize operating conditions suitably [29, 37].

Recent advances have focused on mitigating these disadvantages, with studies aimed at enhancing the stability, reusability, and financial feasibility of nanomaterials, as well as identifying potential environmental hazards associated with their use in wastewater treatment systems. Hybrid treatment systems that integrate conventional techniques with nanotechnology are also on the horizon as a potential solution, aiming to combine the advantages of both techniques [2].

7 Obstacles and Constraints

Despite the considerable potential of nanoparticles for wastewater treatment, various challenges and limitations must be addressed to utilize nanoparticles at a commercial scale. These limitations and challenges include the high cost and scalability issues of nanoparticles, environmental and health risks, regulatory problems, and waste disposal [14, 22]. It is essential to understand these limitations in detail to develop further nanotechnology for wastewater treatment, including its safety, efficiency, and sustainability [14, 22].

7.1 *Financial Implications and Scalability of Nanomaterials*

The primary drawback of using nanoparticles for wastewater treatment is their high production cost. Their manufacture requires intricate, energy-intensive processes, making their material more expensive [60]. In large-scale applications, their cost can discourage the use of more affordable technologies, such as activated carbon, membrane filtration, and chemical coagulation. Scalability of nanomaterials. Although the constructs described in this study exhibit many advantageous characteristics, the problem of the scalability of nanomaterials remains. In laboratory tests, they have proven effective for treating wastewater; however, when scaled up to an industrial process, issues such as particle distribution, synthesis on an industrial scale, and stability in field conditions and reuse arise. We believe that, in today's climate of energy, environmental, and economic crises, the ability to develop cost-effective methods for production (e.g., "green" synthesis from waste or biogenic sources) would also serve to lower costs and enhance the economic feasibility of using nanotechnology for wastewater treatment. For the commercial use of nanomaterials in treating industrial effluent, these scalability and cost concerns must be addressed [14].

7.2 *Possible Environmental and Health Hazards (Nanotoxicity)*

Although their inventive and futuristic uses are becoming more widely acknowledged, nanomaterials do pose some difficulties. Nanotoxicity is still a major problem. Because they are so tiny, these particles have the ability to enter biological systems in ways that are still unclear. Nanomaterials may remain in tissues after exposure through ingestion, inhalation, or skin contact, which could have an impact on biological functions [26].

Nanomaterials may evade filtration in wastewater treatment and end up in rivers or soil, endangering aquatic life and possibly influencing human health via the food chain [51]. The risk associated with nanomaterials depends on several factors, including particle composition, size, surface properties, and other characteristics.

Metallic nanoparticles, such as silver and copper, can release toxic ions, which are harmful to fish and other aquatic life [17, 51].

7.3 *Regulatory and Policy Concerns*

The regulatory framework for nanotechnology is still being shaped, as several countries lack regulations and guidelines regarding the application of nanoparticles in wastewater treatment. Due to the unique properties, concerns exist regarding the environmental, health, and safety regulations (EHS) implications of nanoparticles, for which regulations are in place yet are largely limited; the focus is mostly on bulk materials. The behavior of nanoparticles in wastewater treatment systems, their bioaccumulation potential, and the effects on downstream water quality warrant the establishment of new standards and testing approaches. Regulatory agencies, such as the US Environmental Protection Agency and the European Chemicals Agency, are developing frameworks for evaluating and regulating nanomaterials; however, a need for coherent and uniform international standards remains.

If nanomaterials are not properly regulated, there is a significant risk of environmental and human health impacts. Potential consequences include contamination of rivers, harm to aquatic animals, and human exposure to hazardous nanoparticles. Regulators need to collaborate with scientists, industry professionals, and even environmental experts to develop evidence-based policies and guidelines. Without this oversight, the uncontrolled utilization of nanomaterials may cause global environmental and public health concerns [48].

7.4 Waste Management and Life Cycle Assessment

Nanoparticle post-use wastewater treatment disposal is of prime concern. Here, despite the efficiency of nanomaterials in pollutant removal, disposal often leads to environmental pollution through the accumulation of sludge or waste. The release of such substances into the environment via effluent or sludge may result in their deposition in water, soil, and sediments, leading to long-term ecological effects [56, 76].

Life cycle analysis (LCA) is crucial in determining the environmental footprint of nanomaterials throughout their entire life cycle, from manufacturing to waste disposal. LCA analyzes energy consumption, resource usage, and environmental impact of nanomaterial-based wastewater systems in their entire life cycle. LCA assists scientists in determining sustainable materials and processes, enhancing the regeneration and recycling of nanomaterials to mitigate environmental damage. Additionally, the development of safe disposal or recovery techniques for nanoparticles, such as sedimentation or filtration, is essential to ensure the long-term sustainability of nanotechnology in wastewater treatment [40].

8 Prospective Outlook

Nanotech in wastewater treatment has serious potential—especially when addressing the effluents coming out of pulp and paper plants. Researchers are deeply involved, and it is not just about creating nanoparticles. There is a growing push for more eco-friendly ways to produce materials, and researchers are exploring the combination of nanotechnology with other advanced treatment techniques. There is also interest in "smart" nanomaterials—tiny particles that can adapt to conditions. Moreover, there is attention on the economic aspect: how to take this technology from the lab and make it work in the real world cost-effectively. Still, there are some significant gaps in the research—big enough to drive a truck through—so if this technology is to be used at scale, those gaps need to be addressed [62].

8.1 Sustainable Fabrication of Nanomaterials

The creation of new environmentally friendly processes for creating nanomaterials is the true game-changer in nanotechnology for wastewater treatment. Even though they work well, the conventional approaches have drawbacks in terms of pollution and energy consumption. They use a lot of dangerous chemicals, consume a lot of energy, and leave a big carbon footprint.

Researchers are using materials like plant extracts, algae, or microbes to create nanoparticles through green synthesis, which circumvents conventional techniques. While still creating useful nanoparticles, nature helps with the synthesis process, lessening its negative effects on the environment [49].

Carbon-based materials like graphene oxide or metal oxide nanoparticles that are green-synthesised (like ZnO or TiO_2) are not only very efficient but also environmentally friendly. Their stability, high reactivity, and large surface areas make them ideal for eliminating contaminants from water. Furthermore, customized nanoparticles that are optimized for particular contaminants can be produced using green methods. Researchers are now trying to increase particle production, keep consistency, and scale up the process without harming the environment. This technology seems to have a far less harmful future for the environment [59].

8.2 Integration with Additional Advanced Therapeutic Approaches

Nanoparticles can extract pollutants, but advanced treatment technologies do not necessarily transform or upgrade wastewater systems. The integration of more elaborate advanced oxidation processes (AOPs) and biotreatments with nanotechnology is an effective strategy for treating the complex pollutants in pulp and paper effluents. AOPs, such as ozonation, Fenton's reaction, and photocatalysis, can facilitate the oxidation process using nanomaterials [24]. UV light-irradiated TiO_2 nanoparticles produce hydroxyl radicals that degrade organic pollutants in wastewater. Nanomaterials in combination with AOPs can enhance efficiency, saving energy and chemicals for treatment [24].

Biological treatments, such as activated sludge and biofilm reactors, can be enhanced by nanomaterials that stimulate microbial activity and improve pollutant degradation. Nanoparticles act as transporters for bacteria, facilitating both surface adhesion and contaminant degradation. The incorporation of nanotechnology into biological processes can be cost-effective and more sustainable in wastewater treatment, especially for industries with challenging effluents, such as pulp and paper [11].

8.3 Intelligent and Reactive Nanomaterials

Innovation in wastewater treatment with nanoparticles relies on the development of intelligent materials that can respond to variable conditions, such as pH, temperature, and contaminant concentration, to enable efficient treatment. Intelligent nanoparticles can be engineered to selectively recognize and degrade specific pollutants in wastewater, including heavy metals and organic pollutants [9, 15]. Their structure can be altered when they are in contact with pollutants, and this can enhance both adsorption and degradation. The sensors are combined with nanomaterials to achieve real-time monitoring of the treatment progress and provide useful information on treatment efficiency and the concentration of pollutants. Such responsiveness can be leveraged to optimize the operation of wastewater treatment plants (WWTPs), thereby minimizing energy demand and chemical use [7, 19, 20].

8.4 Prospects for Commercialization and Research Deficiencies

Although there exists a vast potential for the application of nanotechnology in treating wastewater, commercialization of nanomaterial-based treatments remains a challenge. The high cost of nanomaterials, scalability concerns, and environmental hazard concerns are significant barriers to their widespread application. Nevertheless, with advancements in research and manufacturing processes, the cost of nanomaterials is expected to decrease, making them more affordable for mass application. Additionally, as regulatory systems become more adept at incorporating nanotechnology, the path to commercialization is likely to become clearer [68]. Numerous gaps in research areas need to be addressed to understand the potential of nanotechnology in wastewater treatment fully. More research is needed to investigate the long-term environmental and health effects of nanoparticles, particularly in real-world applications. Research needs to focus on improving the regeneration and reutilization of nanomaterials, as well as economic synthesis processes that are amenable to industrial upscaling. Much more needs to be done to integrate nanomaterials into existing treatment technologies, to advance further hybrid systems that capitalize on both the strengths of conventional and innovative treatment strategies [57].

Collaboration among research centers, industrial institutions, and regulatory agencies will be crucial in overcoming these challenges and accelerating the commercialization of nanotechnology-based wastewater treatment technologies. As research continues and innovations are discovered, the application of nanotechnology in wastewater treatment will also become a key element of sustainable water management strategies in the future [4].

9 Conclusion

Nanotechnology is a powerful and rapidly emerging field that holds promise as a solution to treat wastewater from the pulp and paper industry, which generates substantial quantities of polluted water. This chapter aimed to illuminate how nanomaterials, which are minute-scale materials such as metal oxide nanoparticles, carbon-based materials, and nanocomposites, can remove harmful substances more efficiently than conventional methods of wastewater treatment. These materials are characterized by several features, including a large surface area and high reaction propensity, which make them highly efficient in removing both organic and inorganic pollutants, such as lignin, chlorine-based chemicals, heavy metals, and suspended solids.

Nanomaterials act in various ways. The adsorption process occurs when a contaminant is trapped on the surface of nanomaterials. In contrast, photocatalytic degradation involves the breakdown of contaminants by light, and chemical reactions heat the water for cleansing. Other nanomaterials destroy bacteria and pathogens, thus making the water safe.

When combined with other advanced methods, such as advanced oxidation or biological processes, nanomaterials can be used to enhance the completion and improve the efficiency of wastewater treatment. However, there remain challenges. Hurdles that need to be overcome include high costs, limited large-scale use, environmental concerns, and inadequate regulations.

Nanotechnology also opens up new avenues for treating recalcitrant pollutants, such as pharmaceuticals and personal care products, which are typically difficult to remove using conventional methods. It also boosts the recycling of water, which is critical given the global concern over water scarcity. To conclude, despite some challenges that remain, nanotechnology has the potential to significantly improve industrial wastewater treatment, making it more efficient, sustainable, and environmentally friendly.

Competing Interests The authors declare that they have no competing interest.

DataAvailability Declaration of Generative AI and AI-Assisted Technologies in the Writing Process During the preparation of this work the book chapter author(s) have not taken any assistance from AI, and the book chapter author(s) reviewed and edited the content as needed and take(s) full responsibility for the content of the publication.

References

1. Abouzeid RE, Khiari R, El-Wakil N, Dufresne A (2019) Current state and new trends in the use of cellulose nanomaterials for wastewater treatment. Biomacromol 20(2):573–597. https://doi.org/10.1021/ACS.BIOMAC.8B00839/ASSET/IMAGES/LARGE/BM-2018-00839M_0006.JPEG
2. Adel M, Ahmed MA, Elabiad MA, Mohamed AA (2022) Removal of heavy metals and dyes from wastewater using graphene oxide-based nanomaterials: a critical review. Environ Nanotechnol Monit Manag 18. https://doi.org/10.1016/J.ENMM.2022.100719
3. Ahmad IZ, Ahmad A, Tabassum H, Kuddus M (2017) Applications of nanoparticles in the treatment of wastewater. Handb Ecomater 1–25. https://doi.org/10.1007/978-3-319-48281-1_37-1
4. Ajith MP, Rajamani P (2021) Nanotechnology for water purification–current trends and challenges. J Nanotechnol Nanomater 2(Issue 2):88–91. https://doi.org/10.33696/NANOTECHNOL.2.025
5. Ali M, Sreekrishnan TR (2001) Aquatic toxicity from pulp and paper mill effluents: a review. Adv Environ Res 5(2):175–196. https://doi.org/10.1016/S1093-0191(00)00055-1
6. Amini M, Ashrafi M (2016) Photocatalytic degradation of some organic dyes under solar light irradiation using TiO_2 and ZnO nanoparticles. Nanochemistry Res 1(1):79–86. https://doi.org/10.7508/NCR.2016.01.010
7. Annu Mittal M, Tripathi S, Shin DK (2024) Biopolymeric nanocomposites for wastewater remediation: an overview on recent progress and challenges. Polymers 16(2):294. https://doi.org/10.3390/POLYM16020294
8. Arora B, Attri P (2020) Carbon nanotubes (CNTs): a potential nanomaterial for water purification. J Compos Sci 4(3):135. https://doi.org/10.3390/JCS4030135
9. Asghar N, Hussain A, Nguyen DA, Ali S, Hussain I, Junejo A, Ali A (2024) Advancement in nanomaterials for environmental pollutants remediation: a systematic review on bibliometrics analysis, material types, synthesis pathways, and related mechanisms. J Nanobiotechnology 22(1):1–28. https://doi.org/10.1186/S12951-023-02151-3
10. Bulskaya I, Fytianos K (2022) Oxidative treatment of paper mill effluent by TiO_2 photocatalysis and photo-fenton reagent. Water Air Soil Pollut 233(12):1–11. https://doi.org/10.1007/S11270-022-06003-Y/FIGURES/7
11. Chandran P, Suresh S, Balasubramain B, Gangwar J, Raj AS, Aarathy UL, Meyyazhagan A, Pappuswamy M, Sebastian JK (2023) Biological treatment solutions using bioreactors for environmental contaminants from industrial waste water. J Umm Al-Qura Univ Appl Sci 11(2):185–207. https://doi.org/10.1007/S43994-023-00071-4
12. Covaliu-Mierlă CI, Păunescu O, Iovu H (2023) Recent advances in membranes used for nanofiltration to remove heavy metals from wastewater: a review. Membranes 13(7):643. https://doi.org/10.3390/MEMBRANES13070643
13. Covinich LG, Bengoechea DI, Fenoglio RJ, Area MC (2014) Advanced oxidation processes for wastewater treatment in the pulp and paper industry: a review. Am J Environ Eng 2014(3):56–70. https://doi.org/10.5923/j.ajee.20140403.03
14. Digiulio DC, Jackson RB (2016) Impact to underground sources of drinking water and domestic wells from production well stimulation and completion practices in the Pavillion, Wyoming, field. Environ Sci Technol 50(8):4524–4536. https://doi.org/10.1021/ACS.EST.5B04970/ASSET/IMAGES/LARGE/ES-2015-049706_0006.JPEG
15. Epelle EI, Okoye PU, Roddy S, Gunes B, Okolie JA (2022) Advances in the applications of nanomaterials for wastewater treatment. Environments 9(11):141. https://doi.org/10.3390/ENVIRONMENTS9110141
16. Fadlalla MI, Senthil Kumar P, Selvam V, Ganesh Babu S (2019) Recent advances in nanomaterials for wastewater treatment, pp 21–58. https://doi.org/10.1007/978-3-030-04477-0_2

17. Francis DV, Abdalla AK, Mahakham W, Sarmah AK, Ahmed ZFR (2024) Interaction of plants and metal nanoparticles: exploring its molecular mechanisms for sustainable agriculture and crop improvement. Environ Int 190:108859. https://doi.org/10.1016/J.ENVINT.2024.108859
18. Gavrilescu D, Puitel AC, Dutuc G, Craciun G (2012) Environmental impact of pulp and paper mills. Environ Eng Manag J 11(1):81–86. https://doi.org/10.30638/EEMJ.2012.012
19. Gehrke I, Geiser A, Somborn-Schulz A (2015) Innovations in nanotechnology for water treatment. Nanotechnol Sci Appl 8:1. https://doi.org/10.2147/NSA.S43773
20. Ghosh N, Das S, Biswas G, Haldar PK (2022) Review on some metal oxide nanoparticles as effective adsorbent in wastewater treatment. Water Sci Technol 85(12):3370–3395. https://doi.org/10.2166/WST.2022.153
21. Glažar D, Simončič B (2023) TiO_2 and ZnO as advanced photocatalysts for effective dye degradation in textile wastewater. Tekstilec 66(3):178–198. https://doi.org/10.14502/TEKSTILEC.66.2023045
22. Godwin HA, Chopra K, Bradley KA, Cohen Y, Harthorn BH, Hoek EMV, Holden P, Keller AA, Lenihan HS, Nisbet RM, Nel AE (2009) The University of California Center for the environmental implications of nanotechnology. Environ Sci Technol 43(17):6453–6457. https://doi.org/10.1021/es8034544
23. Gupta GK, Sailwal M, Shukla P (2024) Sustainable nanotechnology based techniques for mitigating the pollutants from pulp and paper industry. ACS Omega. https://doi.org/10.1021/ACSOMEGA.4C06022/ASSET/IMAGES/LARGE/AO4C06022_0002.JPEG
24. Hairom NHH, Soon CF, Mohamed RMSR, Morsin M, Zainal N, Nayan N, Zulkifli CZ, Harun NH (2021) A review of nanotechnological applications to detect and control surface water pollution. Environ Technol Innov 24:102032. https://doi.org/10.1016/J.ETI.2021.102032
25. Hamid LL, Ali AY, Ohmayed MM, Ramizy A, Mutter TY (2024) Antimicrobial activity of silver nanoparticles and cold plasma in the treatment of hospital wastewater. Kuwait J Sci 51(2):100212. https://doi.org/10.1016/J.KJS.2024.100212
26. Hasan Z, Lateef M, Agrawal N, Khanam A, Prasai JR, Husain A (2024) Uncertainties, phototoxicity, health impacts, and agricultural and environmental concerns of nanomaterials in the food industry. In: Nanotechnology and nanomaterials in the agri-food industries. Elsevier, pp 361–390
27. Iqbal HMN, Kyazze G, Keshavarz T (2013) Advances in the valorization of lignocellulosic materials by biotechnology: an overview. BioResources 8(2):3157–3176. https://doi.org/10.15376/BIORES.8.2.3157-3176
28. Jiang Q, Han Z, Yuan Y, Cai C, Li J, Cheng Z (2022) Controlled preparation and photocatalytic performance of TiO_2/ZnO phase-mixed nanotubes-based nano-spheres. Mater Chem Phys 279:125737. https://doi.org/10.1016/J.MATCHEMPHYS.2022.125737
29. Kadhim RJ, Al-Ani FH, Al-Shaeli M, Alsalhy QF, Figoli A (2020) Removal of dyes using graphene oxide (GO) mixed matrix membranes. Membranes 10(12):366. https://doi.org/10.3390/MEMBRANES10120366
30. Kamalesh R, Karishma S, Saravanan A, Yaashikaa PR (2024) Emerging breakthroughs in membrane filtration techniques and their application in agricultural wastewater treatment: reusability aspects. Sustain Chem Environ 8:100183. https://doi.org/10.1016/J.SCENV.2024.100183
31. Kamali M, Persson KM, Costa ME, Capela I (2019) Sustainability criteria for assessing nanotechnology applicability in industrial wastewater treatment: current status and future outlook. Environ Int 125:261–276. https://doi.org/10.1016/J.ENVINT.2019.01.055
32. Khin MM, Nair AS, Babu VJ, Murugan R, Ramakrishna S (2012) A review on nanomaterials for environmental remediation. Energy Environ Sci 5(8):8075–8109. https://doi.org/10.1039/C2EE21818F
33. Kumar P, Kumar S, Bhardwaj NK (2011) Titanium dioxide photocatalysis for the pulp and paper industry wastewater treatment. Indian J Sci Technol 4(3):25–29. http://www.indjst.org
34. Lee KM, Wong CPP, Tan TL, Lai CW (2018) Functionalized carbon nanotubes for adsorptive removal of water pollutants. Mater Sci Eng B 236–237:61–69. https://doi.org/10.1016/J.MSEB.2018.12.004

35. Liang X, Xu Y, Yin L, Wang R, Li P, Wang J, Liu K (2023) Sustainable utilization of pulp and paper wastewater. Water (Switz) 15(23). https://doi.org/10.3390/W15234135
36. McMaster ME, Hewitt LM, Parrott JL (2006) A decade of research on the environmental impacts of pulp and paper mill effluents in Canada: field studies and mechanistic research. J Toxicol Environ Health Part B CritAl Rev 9(4):319–339. https://doi.org/10.1080/152873905 00195869
37. Mubarak NM, Sahu JN, Abdullah EC, Jayakumar NS (2014) Removal of heavy metals from wastewater using carbon nanotubes. Sep Purif Rev 43(4):311–338. https://doi.org/10.1080/154 22119.2013.821996
38. Nasrollahzadeh M, Sajjadi M, Iravani S, Varma RS (2021) Carbon-based sustainable nanomaterials for water treatment: state-of-art and future perspectives. Chemosphere 263. https://doi. org/10.1016/J.CHEMOSPHERE.2020.128005
39. Nishu, Kumar S (2023) Smart and innovative nanotechnology applications for water purification. Hybrid Adv 3:100044. https://doi.org/10.1016/J.HYBADV.2023.100044
40. Nizam NUM, Hanafiah MM, Woon KS (2021) A content review of life cycle assessment of nanomaterials: current practices, challenges, and future prospects. Nanomaterials 11(12):3324. https://doi.org/10.3390/NANO11123324
41. Nowack B, Bucheli TD (2007) Occurrence, behavior and effects of nanoparticles in the environment. Environ Pollut 150(1):5–22. https://doi.org/10.1016/J.ENVPOL.2007.06.006. (Barking, Essex : 1987)
42. Olawade DB, Wada OZ, Fapohunda O, Egbewole BI, Ajisafe O, Ige AO (2024) Nanoparticles for microbial control in water: mechanisms, applications, and ecological implications. Front Nanotechnol 6:1427843. https://doi.org/10.3389/FNANO.2024.1427843/XML/NLM
43. Panigrahy SK, Nandha A, Chaturvedi M, Mishra PK (2024) Novel nanocomposites with advanced materials and their role in waste water treatment. Next Sustain 4:100042. https:// doi.org/10.1016/J.NXSUST.2024.100042
44. Pokhrel D, Viraraghavan T (2004) Treatment of pulp and paper mill wastewater—a review. Sci Total Environ 333(1–3):37–58. https://doi.org/10.1016/J.SCITOTENV.2004.05.017
45. Qasem NAA, Mohammed RH, Lawal DU (2021) Removal of heavy metal ions from wastewater: a comprehensive and critical review. Npj Clean Water 4(1):1–15. https://doi.org/10.1038/ s41545-021-00127-0
46. Qu X, Alvarez PJJ, Li Q (2013) Applications of nanotechnology in water and wastewater treatment. Water Res 47(12):3931–3946. https://doi.org/10.1016/J.WATRES.2012.09.058
47. Raes A, Minja AC, AG KR, Verbruggen SW (2024) Recent advances in metal-doped defective TiO_2 for photocatalytic CO_2 conversion. Curr Opin Chem Eng 44:101013. https://doi.org/10. 1016/J.COCHE.2024.101013
48. Rana S, Kalaichelvan PT (2013) Ecotoxicity of nanoparticles. ISRN Toxicol 2013:1–11. https:// doi.org/10.1155/2013/574648
49. Rathod S, Preetam S, Pandey C, Bera SP (2024) Exploring synthesis and applications of green nanoparticles and the role of nanotechnology in wastewater treatment. Biotechnol Rep 41:e00830. https://doi.org/10.1016/J.BTRE.2024.E00830
50. Ratnam MV, Akilamudhan P, Kumar KS, Reddy SN, Rao KN, Shaik F, Prasad DMR (2023) Carbon-based nanoadsorbents for the removal of emerging pollutants. Adsorpt Sci & Technol. https://doi.org/10.1155/2023/3579165
51. Ray PC, Yu H, Fu PP (2009) Toxicity and environmental risks of nanomaterials: challenges and future needs. J Environ Sci Health Part C Environ Carcinog & Ecotoxicol Rev 27(1):1. https://doi.org/10.1080/10590500802708267
52. Saeed MU, Hussain N, Sumrin A, Shahbaz A, Noor S, Bilal M, Aleya L, Iqbal HMN (2022) Microbial bioremediation strategies with wastewater treatment potentialities–a review. Sci Total Environ 818:151754. https://doi.org/10.1016/J.SCITOTENV.2021.151754
53. Sajid M, Asif M, Baig N, Kabeer M, Ihsanullah I, Mohammad AW (2022) Carbon nanotubes-based adsorbents: properties, functionalization, interaction mechanisms, and applications in water purification. J Water Process Eng 47:102815. https://doi.org/10.1016/J.JWPE.2022. 102815

54. Savage N, Diallo MS (2005) Nanomaterials and water purification: opportunities and challenges. https://doi.org/10.1007/s11051-005-7523-5
55. Sheoran K, Kaur H, Siwal SS, Saini AK, Vo DVN, Thakur VK (2022) Recent advances of carbon-based nanomaterials (CBNMs) for wastewater treatment: synthesis and application. Chemosphere 299:134364. https://doi.org/10.1016/J.CHEMOSPHERE.2022.134364
56. Singh BJ, Chakraborty A, Sehgal R (2023) A systematic review of industrial wastewater management: evaluating challenges and enablers. J Environ Manage 348:119230. https://doi.org/10.1016/J.JENVMAN.2023.119230
57. Singh G, Yadav N, Yadav P, Singh SK, Kumar A (2024) Incorporation of nanotechnology in wastewater remediation: advancement and challenges. Adv Chem Pollut Environ Manag Prot 10:363–382. https://doi.org/10.1016/BS.APMP.2023.08.006
58. Singh US, Tripathi YC (2020) Characteristics and treatment of pulp and paper mill effluents-a review. Int J Eng Tech Res 11:454–4698. www.erpublication.org
59. Sofia G, Koventhan C, Kanmani S, Lo AY (2025) Green synthesized rGO/TiO_2/g-C_3N_4 nanocomposites via Plectranthus amboinicus extract for efficient photocatalytic degradation of methylene blue: RSM optimization, antimicrobial and phytotoxicity assessment. J Environ Chem Eng 13(1):115267. https://doi.org/10.1016/J.JECE.2024.115267
60. Strifling D, Dalton J, Folvarska V (2023) Nanotechnology in drinking water treatment systems: risk and regulatory compliance
61. Thakur A, Kumar A, Singh A (2024) Adsorptive removal of heavy metals, dyes, and pharmaceuticals: carbon-based nanomaterials in focus. Carbon 217:118621. https://doi.org/10.1016/J.CARBON.2023.118621
62. Thamarai P, Kamalesh R, Saravanan A, Swaminaathan P, Deivayanai VC (2024) Emerging trends and promising prospects in nanotechnology for improved remediation of wastewater contaminants: present and future outlooks. Environ Nanotechnol Monit Manag 21. https://doi.org/10.1016/J.ENMM.2024.100913
63. Thambiliyagodage C (2022) Efficient photocatalysis of carbon coupled TiO_2 to degrade pollutants in wastewater–a review. Environ Nanotechnol Monit & Manag 18:100737. https://doi.org/10.1016/J.ENMM.2022.100737
64. Theron J, Walker JA, Cloete TE (2008) Nanotechnology and water treatment: applications and emerging opportunities. Crit Rev Microbiol 34(1):43–69. https://doi.org/10.1080/10408410701710442
65. Thompson G, Swain J, Kay M, Forster CF (2001) The treatment of pulp and paper mill effluent: a review. Biores Technol 77(3):275–286. https://doi.org/10.1016/S0960-8524(00)00060-2
66. Tripathi M, Singh S, Pathak S, Kasaudhan J, Mishra A, Bala S, Garg D, Singh R, Singh P, Singh PK, Shukla AK, Pathak N (2023) Recent strategies for the remediation of textile dyes from wastewater: a systematic review. Toxics 11(11):940. https://doi.org/10.3390/TOXICS11110940
67. Tripathy DB, Gupta A (2023) Nanomembranes-affiliated water remediation: chronology, properties, classification. Chall Futur Prospects Membr 13(8):713. https://doi.org/10.3390/MEMBRANES13080713
68. Tripathy J, Mishra A, Pandey M, Thakur RR, Chand S, Rout PR, Shahid MK (2024) Advances in nanoparticles and nanocomposites for water and wastewater treatment: a review. Water 16(11):1481. https://doi.org/10.3390/W16111481
69. Turkten N, Bekbolet M (2020) Photocatalytic performance of titanium dioxide and zinc oxide binary system on degradation of humic matter. J Photochem Photobiol A 401:112748. https://doi.org/10.1016/J.JPHOTOCHEM.2020.112748
70. Verma P, Chandra R (2024) Removal of residual organic pollutants from pulp paper industry wastewater using bacterial strains Acinetobacter baumannii (OK582199) PC4 and Bacillus cereus (OK582201) PC10 via biostimulation and bioaugmentation process. Total Environ Adv 10:200101. https://doi.org/10.1016/J.TEADVA.2024.200101
71. Vítězová M, Kohoutová A, Vítěz T, Hanišáková N, Kushkevych I (2020) Methanogenic microorganisms in industrial wastewater anaerobic treatment. Processes 8(12):1546. https://doi.org/10.3390/PR8121546

72. Xia L, Liao Q, Zhao C, Wang J, Ren Y, Xie H, Chen J, Shang J (2024) Sulfidated nano zero-valent iron activated peracetic acid for ciprofloxacin degradation: performance and mechanism. Chem Eng J 491:151861. https://doi.org/10.1016/J.CEJ.2024.151861
73. Yang J, Hou B, Wang J, Tian B, Bi J, Wang N, Li X, Huang X (2019) Nanomaterials for the removal of heavy metals from wastewater. Nanomaterials 9(3):424. https://doi.org/10.3390/NANO9030424
74. Yang Y, Ji W, Li X, Lin H, Chen H, Bi F, Zheng Z, Xu J, Zhang X (2022) Insights into the mechanism of enhanced peroxymonosulfate degraded tetracycline using metal organic framework derived carbonyl modified carbon-coated Fe0. J Hazard Mater 424:127640. https://doi.org/10.1016/J.JHAZMAT.2021.127640
75. Yang J, Shojaei S, Shojaei S (2022) Removal of drug and dye from aqueous solutions by graphene oxide: Adsorption studies and chemometrics methods. Npj Clean Water 5(1):1–10. https://doi.org/10.1038/s41545-022-00148-3
76. Yaqoob AA, Parveen T, Umar K, Ibrahim MNM (2020) Role of nanomaterials in the treatment of wastewater: a review. Water 12(2):495. https://doi.org/10.3390/W12020495
77. Zhang Q, Volker RE, Lockington DA (2002) Experimental investigation of contaminant transport in coastal groundwater. Adv Environ Res 6(3):229–237. https://doi.org/10.1016/S1093-0191(01)00054-5

Future Prospects and Emerging Trends in Nanotechnology for Pulp and Paper

Akash Verma, Akanksha Chauhan, Yana Gupta, and Rakesh Kumar Bachheti

Abstract We are aware that nanotechnology possesses maximum potential in the pulp and paper sector by improving the strength, functionality, and sustainability of paper products. This chapter provides an overview of the future opportunities and trends in nanotechnology within the industry, with a special focus on key areas such as cellulose nanofibrils (CNFs), cellulose nanocrystals (CNCs), nanoclays, and metallic nanoparticles. These nanomaterials are designed to produce numerous lightweight and high-strength papers, moisture barrier films and gas barrier coatings, antimicrobial packaging, and even printable electronics. Present industrial applications involve nano-reinforced packaging, smart labels, and 3D-printed paper structures. But whereas such promising use is on the go, mass-level adoption is still hindered by such limitations as high production costs, technical upscaling, regulatory loopholes, and environmental risk issues related to engineered nanoparticles. Such barriers are explicated by the chapter and reinforced with opportunities for future development. Innovative solutions like artificial intelligence-based process optimization, the introduction of bioactive paper through integrating biotechnology to make paper bioactive, and circular economy models to facilitate sustainable procurement of nanomaterials are also considered. Synergistic partnerships among academia, industry, and policy-makers as enablers of commercial applicability and safety implementation are viewed. Overall, nanotechnology is a revolutionary approach to propelling the pulp and paper industry into the future, bringing material innovation into an active partnership with worldwide goals for sustainability, efficiency, and performance.

Keywords Nanotechnology · Nanocellulose · Cellulose nanofibrils (CNFs) · Smart packaging · Sustainable paper products

---

A. Verma · A. Chauhan · Y. Gupta (✉)
Cellulose and Paper Discipline, Forest Products Division, Forest Research Institute, Dehradun, India
e-mail: Yanagupta2004@hotmail.com

R. K. Bachheti
Department of Research Analytics, Saveetha Dental College & Hospitals, Saveetha Institute of Medical and Technical Sciences, Saveetha University, Chennai, Tamil Nadu, India

R. K. Bachheti (ed.), *Nanotechnology in the Pulp and Paper Industry*, Smart Nanomaterials Technology, https://doi.org/10.1007/978-981-95-3473-9_18

1 Introduction

Nanotechnology is now a transformative technology for the pulp and paper industry with new solutions beyond traditional processing and material performance. This multidisciplinary technology leverages the use of materials on the nanoscale—typically smaller than 100 nm—to enhance physical, chemical, and functional performances of paper-based products. Nanomaterials such as cellulose nanofibrils (CNFs), cellulose nanocrystals (CNCs), nanoclays, metal nanoparticles (NPs), and carbon-based nanomaterials are increasingly being studied to improve paper strength, barrier properties, antimicrobial properties, and even incorporate smart functions such as sensing and data transfer [1, 2].

In a world where environmental sustainability, circular economies, and connectivity are becoming increasingly important industry drivers, nanotechnology is the intersection point where performance enhancement and environmental stewardship meet. The introduction of nanomaterials leads to biodegradable, recyclable, and even compostable paper packaging products that outperform conventional plastic-based packaging in key areas such as resistance to moisture, shelf life, and recyclability [3, 4]. While promising, the bulk use of nanotechnology on pulp and paper is hindered by cost, scalability, health concerns, and regulatory uncertainty [5].

This chapter is dedicated to providing an in-depth review of the current usage, future trends, and prospects of nanotechnology in the pulp and paper industry. It highlights trends like nanocoatings, intelligent papers, nanocellulose composites, and paper structures created using 3D printing. This chapter also comments on current limitations in a critical context and proposes avenues for future industrial collaboration and research that can drive the sustainable adoption and commercialization of these technologies.

2 Current State of Nanotechnology in Pulp and Paper

The practice of nanotechnology is the examination, design, and characterization of devices and systems at the nanoscale level in the pulp and paper Industry. It is capable of making energy more efficient, aiding in environmental cleanup, and addressing critical health problems. Supporters of nanotechnology claim that its products will be smaller in size, cost less, be more reliable, and work more effectively. Nanotechnology has applications in various industries, including health and sanitation, drug delivery, agricultural nanosensors, nanoparticle fertilizers, and the cultivation of crops in adverse conditions. Nanoscale filters, nanocomposites, and nanoparticles can be utilized for environmental remediation and sustainable energy solutions [6]. Research into nano-based treatment technologies has been a long-standing and extensive research topic. With the widespread application of nanotechnology, new techniques are being implemented in the pulp and paper industry, also resulting in several improvements and challenges, some of which are mentioned below:

- **Improved paper resistance and longevity:** Nanocellulose (such as CNC and CNF) enhances tensile strength, tear resistance, and flexibility, therefore reducing the need for artificial reinforcing additives [7, 8].
- **Higher protective properties:** Nanocoatings (such as nanoclay and nanocellulose) enhance resistance to moisture, oxygen, and oil. They are used in food and pharmaceutical packaging as a substitute for plastics [9, 10].
- **Enhanced printability and surface quality:** Nanoparticles produce smooth, bright surfaces for quality printing since they promote ink adhesion and minimize ink consumption [11, 12].
- **Antimicrobial activity and intelligent packaging:** Silver, copper, and zinc nanoparticles exhibit antibacterial properties and enable specialized paper for intelligent packaging, sensors, and radio frequency identification (RFID) uses [13, 14].
- **Environmentally friendly and sustainable methods:** Nanocellulose is renewable and biodegradable, eliminating the need for petroleum-based materials. This lowers energy and chemical usage in papermaking and bleaching [15].

3 Future Trends in Nanotechnology for Pulp and Paper

3.1 Nanocellulose Development

Nanocellulose has been identified as one of the key bio-based materials in the paper and pulp sectors. Nanocellulose is characterized by numerous properties such as high stiffness, surface reactivity, and, most importantly, it is predominantly biodegradable since it is derived from plants. Nanocellulose, in particular, nanofibrillated cellulose (NFC) and cellulose nanocrystals (CNC) have observed significant progress in manufacturing technology as well as functional uses (Fig. 1).

NFC is typically produced through the mechanical fibrillation of pulp fibers, which is an energy-consuming process. Furthermore, enzyme-based pretreatment, oxidative treatment (for example, 2,2,6,6-tetramethylpiperidine-1-oxyl [TEMPO]-mediated oxidation), and ultrasound-assisted treatments are also highly preferred, as these treatments focus on conserving energy and enhancing the efficacy of fiber fibrillation [16, 17]. In TEMPO-mediated oxidation, a carboxyl group is attached to the cellulose chain, which finally breaks the fiber bond and facilitates delamination into the nanofibrils more easily. This provides a stable NFC dispersion with enhanced colloidal stability, even at high concentrations (Table 1).

Alternatively, CNCs are typically produced through acid hydrolysis, which targets the amorphous part of the cellulosic chain and removes it, resulting in highly crystalline nanorods. The alteration of hydrolysis procedures using sulfuric acid, phosphoric acid, or ionic liquids has ensured that the procedure is both controllable

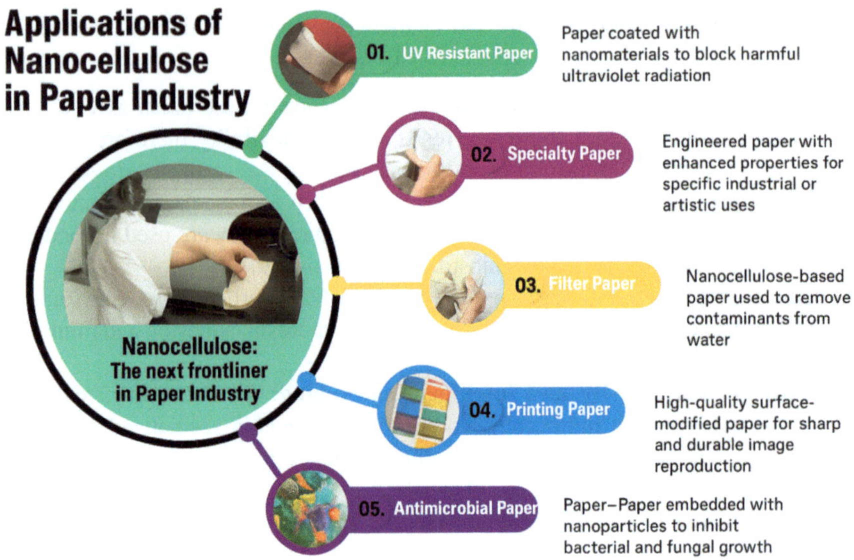

Fig. 1 Nanotechnology applications in the pulp and paper industry

and versatile [5]. Additionally, agricultural residues and nonwood products, such as sugarcane bagasse and wheat straw, are utilized to make the procedure cost-effective and economical [22]. Nanocellulose—particularly its nanofibrillated state (NFC)—has garnered increasing attention within the pulp and paper industry due to its unique physicochemical properties. Its nanoscale dimensions, high surface area, and high mechanical strength make nanocellulose a major driver for enhancing the performance of traditional paper products. Its most exciting advantage is that it can form dense hydrogen-bonding networks, significantly increasing interfiber bonding within paper matrices. Even at low concentrations (typically 1–5% by weight), NFC has been reported to significantly improve several key mechanical properties, including tensile strength, stiffness, and internal cohesion of paper sheets [3]. It does this without the use of synthetic or petroleum-based additives, thus improving the process from an environmental perspective.

At a microscopic level, nanocellulose is an excellent filler agent that takes up spaces between pulp fibers. It reduces porosity and enhances the paper's bulk density as much as possible, leading to smoother surfaces and denser textures. These aspects enable the creation of lighter, yet mechanically upgraded, packaging materials. Such properties are particularly beneficial in applications such as food packaging, folding cartons, and high-quality print media, where material usage can be maintained at low levels without compromising on strength [1]. In addition, nanocellulose enhances printability and surface gloss, properties highly desirable in graphic solutions and branding.

Besides mechanical reinforcement, nanocellulose has become a functional material design platform, particularly for smart and active packaging systems. By applying

Table 1 Emerging trends in nanotechnology in the pulp and paper industry

Future prospects/trends	Description	Benefits	Challenges	Potential applications	References
Nanocellulose-based paper products	Paper products are modified with cellulose nanocrystals and cellulose nanofibrils to enhance strength, flexibility, or barrier properties	High strength, highly flexible, improved moisture or gas barrier properties, and biodegradability	Production cost is higher, cannot hold water, and processing techniques are complex	Healthcare or medical product, food wraps, specialty paper like a currency note, or paper used for packaging	Moon et al. [18] (https://doi.org/https://doi.org/10.1039/c0cs00108b)
Nano-enhanced coatings for paper	Nanoparticle-based coatings, such as silica, titanium dioxide, and clay-based materials, to improve paper properties	Enhanced water and grease resistance, improved print quality, and increased durability	Coating uniformity and compatibility with existing coating systems	Food packaging, grease-proof papers, and high-quality printing papers	Sharma et al. [19] (https://doi.org/https://doi.org/10.1016/j.porgcoat.2021.106317)
Functionalized nanoparticles for paper strength enhancement	Nanoparticles like nanoclays and carbon nanotubes incorporated into paper fibers for strength improvement	Increased tensile strength, tear resistance, and dimensional stability	Nanoparticle dispersion, fiber compatibility, and production cost	Industrial packaging, construction materials, and security papers	Jonoobi et al. [65] (https://doi.org/https://doi.org/10.1007/s10570-015-0551-0)
Nanotechnology for sustainable paper manufacturing	Use of nanocellulose and nanoadditives to reduce chemical use and improve process efficiency	Reduced environmental footprint, lower energy consumption, and improved recycling rates	High initial investment, scaling challenges	Eco-friendly paper production, recycling enhancement	Dufresne [17] (https://doi.org/https://doi.org/10.1016/j.mattod.2013.06.004)

(continued)

Table 1 (continued)

Future prospects/trends	Description	Benefits	Challenges	Potential applications	References
Antimicrobial nanocoatings for packaging	Incorporation of silver, copper, and zinc oxide nanoparticles in paper coatings	Enhanced shelf life, reduced microbial contamination	Potential toxicity, regulatory challenges	Food packaging, medical packaging	Hubbe et al. [20] (https://bioresources.cnr.ncsu.edu/resources/nanocellulose-in-thin-films-coatings-and-plies-for-packaging-applications-a-review/)
Smart and intelligent paper products	Integration of nanotechnology for sensing and response properties in paper	Real-time tracking, environmental monitoring, and authentication	High production costs, technological integration issues	Smart labels, biosensors, and anti-counterfeiting measures	Singh et al. [21] (https://doi.org/https://doi.org/10.1016/j.porgcoat.2021.106317)
Biodegradable and compostable nano paper	Nanocellulose-based paper that degrades naturally in the environment	Eco-friendly, reduced plastic use, compostable	Production scalability, moisture resistance	Food packaging, disposable utensils, and sustainable materials	Moon et al. [18] (https://doi.org/https://doi.org/10.1039/c0cs00108b)
Graphene and carbon nanotubes for conductive papers	Incorporation of graphene and CNTs to provide electrical conductivity to paper products	Flexible electronics, lightweight, and high conductivity	Cost of graphene, dispersion challenges	Flexible displays, RFID tags, smart packaging	Jonoobi et al. (2015) (https://doi.org/https://doi.org/10.1007/s10570-015-0551-0)
Nanocellulose-based printing technologies	Use of nanocellulose to improve ink adhesion, drying rate, and print quality	Improved print resolution, faster drying, and eco-friendly materials	Compatibility with printing processes, production costs	Security printing, high-quality commercial printing	Sharma et al. [4] (https://doi.org/https://doi.org/10.1016/j.porgcoat.2021.106317)

surface modification methods such as polymer grafting, metal nanoparticle loading, or pH-sensitive dye conjugation, nanocellulose can be made responsive to external stimulants. These modifications enable the material to sense and respond to environmental stimuli, such as moisture, temperature, oxygen concentration, or the presence of spoilage gases. For instance, anthocyanin-functionalized nanocellulose films can detect the spoilage of perishables by visibly changing color in response to amine vapors that arise as foods spoil [23]. Such systems are especially appealing for use in the food and pharmaceutical industries, where product freshness and safety are monitored in real-time.

Furthermore, the blending of nanocellulose with conductive nanomaterials, such as carbon nanotubes (CNTs), silver nanowires, or graphene, has created new avenues for making electrically conductive paper-based composites. These composites are being developed for future technologies, including RFID antennas, electronic skins, flexible circuits, and smart labels [24]. Their biodegradable nature, lightweight, and electronic properties make these composites ideal for incorporation into green, intelligent packages. Not only do they enhance consumer interaction with interactive functions, but they also facilitate tracing in the supply chain, enabling industries to comply with regulations and optimize inventory management. Overall, the application of nanocellulose in pulp and paper technology represents a significant leap forward toward high-performance, environmentally friendly, and multifunctional packaging materials. As consumers demand innovative and green packaging, the future potential versatility of nanocellulose makes it a frontrunner material for next-generation bio-based innovations.

3.2 Nanocoatings for Enhanced Paper Properties

The need for high-performance, multifunctional, and sustainable paper-based products has spurred the adoption of nanotechnology in surface coatings, most notably nanocoatings. These exceptionally thin films of nanoengineered materials significantly enhance the barrier, antimicrobial, self-cleaning, and electrical conductivity properties of paper without compromising recyclability. This section explores the revolutionary potential of nanocoatings in the development of next-generation paper products. Conventional paper is hydrophilic and extremely porous, making it not ideal for uses in packaging that require water resistance, grease resistance, and gas permeability. Nanocoatings provide an effective solution as they create a dense, consistent film on the paper surface, sealing gaps among fibers and minimizing porosity. Nanoclay, silica nanoparticles, and cellulose nanocrystals (CNCs) have been utilized to develop gas barrier coatings, particularly for oxygen and carbon dioxide, which are crucial in food packaging [2]. CNCs are particularly effective because they are crystalline, biodegradable, and film-forming [3]. For moisture and oil resistance, hydrophobic compounds such as zinc oxide (ZnO), TiO_2 nanoparticles, or chitosan-based nano-emulsions have been used as coatings on paper. These lower water vapor transmission rates (WVTR) and enhanced grease resistance render paper suitable for use as food wrappers, takeaway boxes, and beverage cartons [25]. For further improving moisture resistance, hybrid coatings involving nanocellulose and biodegradable waxes or bioplastics have proved promising. These technologies are gaining popularity as environmentally friendly alternatives to traditional polyethylene-laminated paper [4].

Due to increasing concern about food safety and hygiene, particularly in the post-COVID-19 era, interest has grown in antimicrobial nanocoatings. Silver nanoparticles (AgNPs), zinc oxide (ZnO), and copper nanoparticles (CuNPs) exhibit wide-range antimicrobial efficacy and have been effectively integrated into paper coatings to prevent bacterial growth on packaging surfaces [23]. For instance, AgNPs incorporated into a nanocellulose matrix have exhibited sustained bactericidal activity against E. coli and S. aureus, making them suitable for food-contact packaging and medical wraps. The coatings also contribute to extending shelf life and reducing contamination risk. Furthermore, self-cleaning nanocoatings based on the lotus leaf effect are increasingly popular. TiO_2 nanoparticle coatings are not only water- and dirt-repellent but also photocatalytic, with the capability to degrade organic contaminants when exposed to UV light [26]. Such self-cleaning surfaces have applications in outdoor signage, disposable trays, and printed medical records.

Conductive papers for flexible electronics, sensors, and RFID use have been made possible through nanotechnology. By embedding conductive nanomaterials, such as carbon nanotubes (CNTs), graphene, or silver nanowires (AgNWs), into paper substrates, scientists have designed printable inks that are conductive without compromising the lightweight and biodegradable properties of paper [27]. These functional coatings are deposited via inkjet printing, screen printing, or roll-to-roll coating techniques, facilitating integration of circuits, antennas, or biosensors onto paper surfaces directly. These intelligent paper devices are particularly worthwhile in logistics, retail packaging, and wearable technology, where conventional plastic-based sensors are environmentally costly. In a recent development, graphene oxide and reduced graphene oxide inks were utilized to produce paper-based strain and humidity sensors, which remained flexible, inexpensive, and compostable [28]. These innovations usher in opportunities for fully recyclable smart packaging and environmentally friendly eco-electronics.

3.3 *Nanomaterials for Sustainability and Green Processing*

Biodegradable nanomaterials have played a crucial role in sustainable practices within the pulp and paper industry. As an alternative to petroleum-based plastics, biodegradable starch- or chitosan-based nanocomposites, polylactic acid (PLA), and other biopolymers offer hygienic, sustainable, less contaminated, and moisture-resistant packaging [29]. For instance, starch-based nanocomposites (composites with organized nanostructures) incorporating nanofillers (e.g., silicon dioxide [SiO_2] and ZnO) reduce greenhouse gas emissions (pp) by 80% and nonrenewable energy usage compared to traditional polymers such as low-density polyethylene (LDPE).

The aggregate environmental benefits of plant-based proteins are evident in life-cycle assessments, with environmental concerns, such as eutrophication potential, still requiring attention [30].

Nanotechnology plays a vital role in reducing energy consumption and chemical usage in pulp and bleaching processes. In sulfite pulp effluents, ceramic membrane filtration has been successfully implemented, replacing conventional wastewater treatments, and achieving more than a 35% reduction in chemical oxygen demand (COD) and over 70% lignin removal [31]. Kraft pulping can facilitate both efficient chemical recycling and biomass power generation by co-firing lignin-rich effluents, thus promoting sustainable energy recycling [32]. For instance, with the use of recombinant *Bacillus subtilis* expressing ligninolytic enzymes (laccase and lignin peroxidase), enzymatic bleaching has shown improved International Organization for Standardization (ISO) brightness values (as much as 66.45%) with a significant reduction in chlorinated waste. The delignification rates obtained using these enzymatic methods are higher than 60–79% as compared with conventional chemical pretreatment technologies [33].

Nanocatalysts improve the sustainability of paper production. The production of nanoparticles using microbes and plant extracts does not require hazardous chemical species, making the process more environmentally friendly [34]. Since metallic nanoparticles (e.g., gold and silver) catalyze reactions at lower temperatures (e.g., carbon monoxide [CO] oxidation at room temperature), this leads to lower energy demand (process efficiency) [35].

By using titanium dioxide nanocatalysts, the photocatalytic degradation of pollutants was efficiently countered, while by customizing nanocatalysts, selectivity in hydrogenation reactions was promoted, and less harmful byproducts were formed [36].

These innovations are reducing dependence on fossil fuels, decreasing the industrial level of carbon footprints, and implementing tenets of a circular economy. For widespread adoption, these systems must be optimized for nanocatalyst stability and scalability.

Microbial nanotechnology offers environmental advantages over conventional synthetic methods, as it eliminates dangerous reducing agents and stabilizers, thereby reducing environmental pollution and risks to human health [6]. Additionally, reactions can be carried out under mild conditions (room temperature, room pressure, and neutral pH), consuming significantly less energy than high-temperature chemical processes [37]. The production of toxic byproducts is less compared to physical and chemical methods, while the waste from microbes is biodegradable, which is compatible with zero-waste. These materials can sequester environmental pollutants (e.g., heavy metals, dyes, and organic compounds) and thus can be applied in pollution remediation. This approach utilizes naturally occurring microorganisms or agro-waste as raw materials, thereby reducing dependence on nonrenewable resources [38]. Microbe-based nanotechnology embodies principles of green chemistry, offering eco-friendly and biocompatible alternatives in a range of applications and lowering the overall cost of production by as much as 90% compared to chemical synthesis, allowing huge-scale applications to be economically viable [34].

While offering significant advantages, microbial nanotechnology faces certain limitations that affect its scalability and broader applicability: Nanoparticles produced via microbial synthesis usually have heterogeneous size, shape, and physicochemical properties, which can hinder uniformity and reproducibility [34]. The absence of fine-tuning of nanoparticle morphology inhibits their functionality and directly restricts their applications in biomedicine and electronics [39].

Nanoparticles produced by microbes frequently feature bio-capping layers that act as shields yet hinder their purification and functionalization. Strategies for isolating or altering these bio-capping layers while preserving the integrity of the NP remain poorly established. Microbial systems also exhibit high variability, resulting in heterogeneous products, which necessitate standardized protocols for microbial cultivation and production strain cultivation, as well as biosynthesis and extraction methods in microbial systems [34].

Microbial synthesis is tedious, and significantly lower quantities of nanoparticles are produced compared to chemical methods, which raises economic issues over large-scale production. Although the specific molecular architectures of nanostructures can be achieved through genetic engineering, some of the biosynthetic pathways are complicated, and genetic engineering raises ethical issues. Although it is environmentally friendly, the disposal of microbial cultures and nanoparticles afterward may harm the environment if not monitored. This is where regulatory structures are needed to govern these issues [39]. Overcoming these challenges benefits from the development of genetic engineering, optimization of microbial culture techniques, purification methods, and a deeper knowledge of biosynthetic mechanisms to enable industrial implementation.

3.4 Smart and Functionalized Papers

Modern innovative regimes, such as nanotechnology, thermal responsiveness, and anti-counterfeit, are revolutionizing all current paper-based products and making them smarter and more functional. By enhancing safety, sustainability, and consumerengagement in areas such as packaging, security, and branding, these innovations make it easier for consumers to stay informed about the brand or channel.

Intelligent packaging is the latest in the evolution of nanotechnology, as it can facilitate monitoring in real-time for improved safety and quality of food items. For monitoring food quality, embedded nanomaterials serve as pH-sensors, freshness indicators, time–temperature indicators (TTIs), allergens, toxins, and contaminants detectors to monitor food quality control throughout the food supply chain [13, 40]. New types of sensors, called "electronic tongue" sensors, can detect foodborne pathogens at very low concentrations (<0.01 colony-forming units [CFU]/g), and their presence and concentration are evident via visible indications, such as a color change on the packaging [41]. Gas sensors for spoilage detection are specialized nanomaterial-based sensors that can identify food spoilage by monitoring gases such as oxygen, carbon dioxide, or ethylene, which helps increase shelf life and minimizefood waste [42].

Inkless paper and thermochromic papers possessing dynamic visual indicators are created for consumers and industries by smart papersthat respond to environmental changes. Thermochromics are temperature-sensitivepapers that change from one color to another depending on the range of heat to which they are exposed, indicating whether perishable items, such as beverages and pharmaceuticals, are being stored or consumed at their optimal temperature [42].On the other hand, photochromic papers contain photochromic inks that react to light exposure, functioningas interactive design elements or freshness indicators. pH-sensitive papers are also used for signaling expiration or chemical contamination (e.g., in food and medical applications) via color changes [42]. Artificial intelligence (AI)-enhanced smart paper packaging can improve quality control across manufacturing and introduce personalized consumer engagement via augmented reality (AR) experiences [42].

Nanomaterials are also being highly looked upon for security, branding, and anti-counterfeiting applications. The majority of paper-based products are under threat from counterfeit goods, and thus arises a push for brand authenticity, where smart papers can help offer both. Anti-counterfeiting measures, such as physically unclonable identifiers (PUIs), possess unique surface irregularities that work as tamper-proof authentication markers to ensure product authenticity [43]. Connected packaging using QR codes and augmented reality (AR) elements enables consumersto instantaneously authenticate products, both of which provide interactive brand experiences. For branding innovations, interactive and smart packaging powered by AR offers virtual demonstrations, hidden messages, orpersonalized advertisements to build consumer confidence and commitment [44]. The market for smart paper will continue to innovate as industries push for increased sustainability anddigitization. From real-time safety monitoring in food packaging solutionsto anti-counterfeiting applications in luxury goods, the combination of nanotechnology, AI, and interactive design holds the potential to bring next-gen intelligent paper-based solutions.

3.5 Nano-enhanced Recycling and Waste Management

Nanotechnology is a key to new solutions in recycling and waste management in the pulp and paper industry. Nanoparticles enable better deinking, enhanced wastewater treatment, and effective recycling of coated and functionalized papers. For paper recycling, deinking—the removal of ink from paper fibers—is essential to making quality recycled paper. Conventional deinking processes often use chemicals that are unkind to the environment. An alternative presented by nanoparticles, which enhances the efficiency and sustainability of the deinking process. For instance, these are carbon-based or mineral-based functionalized nanoparticles that can chemically bond with the cellulosic fibers, leading to better mechanical strength and helping to separate ink residues during recycling. Incorporating nanoparticles into the deinking process has demonstrated benefits in terms of improved fiber recovery and paper quality [11, 12].

The pulp and paper industry produces a sizable amount of wastewater, which is loaded with multiple pollutants, including heavy metals, natural dyes, and pesticides. Due to their large surface area and high reactivity, nanoadsorbents have emerged as highly efficient candidates for the removal of these pollutants. Nanocellulose-based materials can absorb heavy metals and organic compounds and be chemically altered to attract specific pollutants. Nanomaterials, including metal oxides and zeolites, have been used in adsorption technologies and nanomembrane systems to effectively remove toxins, with the inhibition of bacterial growth [9, 10, 45].

Decoratively coated and tailored functionalized papers have recycling challenges in separating coatings and fibers. Nanotechnologies emerge as a potential engineering solution by modifying the surface properties of such papers to facilitate easier separation during recycling. For example, by the addition of nanoparticles (e.g., silver and titanium dioxide) into coatings, the hydrophilicity or hydrophobicity properties of the coating can be manipulated to facilitate the detachment of coatings from fibers. These improvements lead to higher fiber recovery and lower waste, which helps create a more sustainable recycling process [12, 46].

Some nanoparticles are remarkably effective in deinking and fiber recovery processes. Silver nanoparticles (AgNPs) are particularly effective due to their known antibacterial nature, which prevents microbial growth during deinking and helps reduce paper quality degradation. They also help separate the ink from fibers better, increasing the overall efficiency of the recycling process [10, 47]. Carbon nanotubes (CNTs) are used as adsorbents due to their high surface area and unique structural properties. They enable the removal of inks and coatings for better quality recycled fibers [48, 49]. TiO_2 nanoparticles have photocatalytic properties that contribute to the breakdown of organic pollutants during deinking. Under UV light, dyes and inks are broken down by these particles, resulting in cleaner fibers and improved fiber recovery [10, 50]. As a green alternative, lignin nanoparticles have been shown to selectively adsorb the ink for its removal without causing damage to cellulose fibers. This enables the recovery of fiber more efficiently, closing the loop in sustainable recycling [51].

Although nanoparticles of titanium dioxide and carbon nanotubes are highly suitable for deinking applications, the unique combination of antimicrobial properties culminated by silver nanoparticles and their physicochemical attributes makes them more effective. The various inconsistencies in particle size and shape, concentration of the ink, and the deinking process can lead to variabilityin their effectiveness. AgNPs can be identified assuspensions of very fine particles with distinctive characteristics when compared to other nanoparticles in deinking applications. AgNPs have broad-spectrum antimicrobial activity against bacteria, fungi, and viruses, which inhibits microbial degradation of paper quality during deinking [7, 64]. The large surface area of AgNPs increases their reactivity and enables the efficient separation of ink particles from paper fibers [52]. Moreover, AgNPs produce ROS that help decompose ink particles, allowing them to be easily removed from fibers. When used and combined with certain chemicals used during the deinking process, AgNPs can exert a synergistic effect, contributing to anincrease in overall deinking efficiency [7]. Due to the ability to synthesize AgNPs with a variety of coatings and stabilizers, these nanoparticles are more easily dispersed and stable in deinking solutions than some other nanomaterials [53].

As evident from the above examples, nanotechnology canbe an innovative approach to streamlining the recycling process and waste management in the pulp and paper industry. Nanoparticles play a significant role in enhancing deinking processes, wastewater treatment, and the recyclability of coated papers, thereby contributing to the overall sustainability andefficiency of paper recycling.

3.6 3D Printing and Additive Manufacturing with Nanocellulose

3D Printing, also referred to as additive printing, has been altering and greatly influencing our lives. It is among the greatest assets in sectors such as construction, aerospace, or healthcare. Within the pulp and paper industry, nanocellulose makes an important contribution to 3D printing, contributing to the construction of sustainable, functional materials with low weight. Nanocellulose, being a significant constituent of plants, i.e., derived from plant-based fibers, possesses different physical properties—such as high strength, high surface area, and film-forming properties—rendering it a promising material for bio-inks and composite filaments in additive manufacturing.

Nanocellulose predominantly exists in two forms: cellulose nanocrystals (CNCs) and cellulose nanofibrils (CNFs). Both structures have different mechanical properties, high aspect ratios, and shear-thinning behavior, which are all required criteria for 3D printing, especially in extrusion-based systems [54]. Nanocellulose's rheological behavior makes it possible to flow under pressure and retain structural integrity after deposition—a key requirement for layer-by-layer fabrication methods. The structural strength capability of nanocellulose makes it possible to create 3D-printed materials with enhanced structural performance Guan et al. [20]. The hydrogels that are derived from nanocellulose provide excellent shape accuracy and stability during and after printing, enabling complex design and use in load-bearing structures. Based on the recent studies and developments, the nanocellulose-based composite filaments are created by combining cellulose nanofibrils with high-density polyethylene (HDPE) and other biodegradable polymers [55]. These combined products gave rise to numerous challenges in 3D printing, including warping, shrinkage, and poor interlayer bonding. The altered products produced from nanocellulose are an excellent substitute for petroleum-based materials since they possess fine printability, thermal stability, and mechanical strength.

Additionally, advancements in nanocellulose ink formulation have significantly contributed to improvements in biomedical and packaging applications. For example, bio-inks containing nanocellulose and native polymers such as alginate or gelatin have been used in printing biocompatible tissue engineering scaffolds [56]. In packaging, 3D-printed nanocellulose designs offer tailored solutions that are lightweight and completely biodegradable, thereby eliminating plastic waste while enhancing the aesthetic appeal of the product and its appeal on shelves [19].

The application of nanocellulose plays a crucial role in widening its relevance in additive manufacturing. Different surface modification methods, such as oxidation, carboxylation, and polymer grafting, have altered the nanocellulose to improve its hydrophilicity, conductivity, and binding efficiency [57]. The different modified nanocellulose products find themselves as a base material in conductive inks, drug delivery systems, and electronic paper. For example, nanocellulose functionalized with silver nanoparticles has been successfully printed into antimicrobial coatings, while carbon-nanotube-reinforced nanocellulose architectures hold potential in flexible electronics [58]. The application of nanocellulose in 3D printing aligns robustly with the tenets of green chemistry and sustainable development. As a renewable and biodegradable resource, nanocellulose finds solutions to issues surrounding plastic pollution and nonrecyclable packaging on an international scale. The eco-friendliness, coupled with outstanding performance metrics, makes nanocellulose a core element in the next wave of environmentally friendly manufacturing. Even with these advantages, challenges remain—scaling up production, enhancing inter-layer adhesion, and establishing standardized procedures for print processes. The growing collaboration between universities and industry partners is, however, speeding up innovations, bringing nanocellulose-based 3D printing closer to commercial viability. In summary, the use of nanocellulose in additive manufacturing and 3D printing is a revolutionary movement in the pulp and paper industry. It allows for the production of biodegradable, high-performance materials for use in tailored packaging, biomedical equipment, and structural elements. As science continues to overcome existing limitations and refine processes, nanocellulose will be at the forefront of the revolutionization of eco-friendly manufacturing practices across many sectors.

4 Challenges and Barriers to the Adoption of Nanotechnology in the Pulp and Paper Industry

While promising, nanotechnology is transforming the pulp and paper sector; its large-scale implementation into commercial processes has several challenges. These challenges encompass a range of issues, including technical scalability and affordability, as well as environmental concerns and regulatory complexities. Knowledge of these hurdles is crucial for crafting realistic pathways to widespread adoption.

4.1 Technical and Scalability Challenges in Nanomaterial Production

One of the key challenges in implementing nanotechnology is the intricate and energy-intensive process of producing nanomaterials, such as cellulose nanofibrils (CNFs) and cellulose nanocrystals (CNCs). The majority of production methods, including mechanical fibrillation, acid hydrolysis, and TEMPO-mediated oxidation, consume high energy and have low throughput, particularly on industrial levels [1, 16]. Moreover, it is difficult to preserve uniformity, distribution, and integrity of these nanomaterials throughout large-scale production processes [3].

4.2 Cost Factor and Economic Viable Option

Nanomaterials' high production costs restrict their commercial applicability. The costs of raw material pretreatment, energy usage, and postprocessing typically outweigh the benefits for conventional paper manufacturers, who often have thin profit margins [5]. Although pilot plants and academic tests have demonstrated the viability of nanotechnology-treated paper products, their scalability into cost-saving solutions remains to be achieved [59].

4.3 Environmental and Health Issues of Engineered Nanoparticles

The ecological impact of engineered nanoparticles, particularly metallic nanoparticles such as AgNPs and ZnO NPs, is another concern. Although they help impart antimicrobial activity, these particles can get deposited in ecosystems and pose unknown toxicity threats to aquatic life and humans [26]. The life cycle analysis of nanotechnology-enhanced products is in its nascent stage, making it challenging to assess their overall sustainability [4].

4.4 Standardization and Regulation Barriers

Currently, there are no globally harmonized standards or regulatory systems in place for the use of nanomaterials in paper and packaging products. Inconsistent definitions, safety limits, and testing methodologies hinder the industry from commercializing nanotechnology-enabled products [22]. Additionally, labeling and consumer information requirements concerning nano-enabled products are either unclear or absent in the majority of jurisdictions, further undermining regulatory compliance [60].

5 Future Prospects and Research Directions

Although there are current challenges, the future of nanotechnology in the pulp and paper sector is promising, led by technological convergence, sustainability agendas, and innovation systems. As industries focus more on zero-waste and circular production, nanotechnology presents a range of transformative opportunities.

5.1 Integration of Artificial Intelligence and Machine Learning for Nano-optimized Paper Production

Artificial intelligence (AI) and machine learning (ML) will reshape the incorporation of nanomaterials in papermaking. AI can refine fiber dispersion, material blending, and drying conditions to yield better performance with reduced energy input [61]. ML can also track real-time production metrics to identify variability and regulate process conditions, allowing for better consistency in nano-enriched products [24].

5.2 Synergy of Nanotechnology with Biotechnology for Next-Gen Materials

Nanotechnology, when combined with biotechnology, offers the potential to develop responsive and intelligent materials. For instance, enzyme-functionalized nanocellulose can facilitate biodegradable, intelligent packaging that turns color when spoiled or responds to environmental stimuli [23]. Biofunctionalized nanocellulose composites also find uses in tissue engineering, wound dressings, and drug delivery systems, expanding the industry's product range [62].

5.3 Potential for Circular Economy and Closed-Loop Manufacturing

Nanotechnology fits with circular economy principles. Recyclable packaging, biodegradable nanocoatings, and compostable inks decrease environmental footprint and enable closed-loop manufacturing systems [63]. Agricultural residues and paper mill sludge can also be utilized to produce nanocellulose, providing a sustainable feedstock for next-generation paper materials [22].

5.4 *Industry–Academia Collaborations for Accelerated Innovation Adoption*

Academic-industry collaborations are critical for translating research into commercially viable products. Universities may specialize in basic research, whereas industrial collaborators would test at a pilot scale and offer market reach. Government and international research organization-supported programs have already been shown to succeed in speeding up nanotech advances [1, 17].

6 Conclusion

Nanotechnology has the ability to revolutionize the pulp and paper sector by offering tailored, multifunctional applications that address current demands for sustainability, biodegradability, and performance. As this chapter highlights, the addition of nanomaterials—most notably nanocellulose—has enabled significant enhancement in areas of paper strength, barrier protection, antimicrobial functionality, smart packaging, and additive manufacturing. From increased recyclability to in-time monitoring of the freshness of food, paper products created with nanotechnology are breaking the boundaries of what can be achieved in the industry.

Yet, there are still some crucial issues that must be addressed before nanotechnology becomes fully integrated at the industrial level. Excessive production costs of nanomaterials, low standardization, potential environmental hazards associated with metallic nanoparticles, and regulatory compliance issues continue to discourage commercialization processes. These limitations require multi-stakeholder strategies with industry–academia collaboration, government assistance, and good research and development to increase scalability and reduce costs. Emerging trends, nevertheless, will couple nanotechnology with artificial intelligence, biotechnology, and green chemistry such that the forthcoming generation of intelligent, sustainable, and high-performance paper products shall be engineered. As companies make the transition toward zero-waste practices and circular economy networks, nanotechnology stands to shape a new phase for the pulp and paper manufacturing sector, heralding a new world of innovation, stewardship, and resiliency.

Competing Interests The authors declare that they have no competing interests.

Declaration of Generative AI and AI-Assisted Technologies in the Writing Process During the preparation of this work, the book chapter author(s) have not taken any assistance from AI, and the book chapter author(s) reviewed and edited the content as needed and take(s) full responsibility for the content of the publication.

References

1. Klemm D, Kramer F, Moritz S, Lindström T, Ankerfors M, Gray D, Dorris A (2011) Nanocelluloses: a new family of nature-based materials. Angew Chem Int Ed 50(24):5438–5466
2. Hubbe MA, Ferrer A, Tyagi P, Yin Y, Salas C, Pal L, Rojas OJ (2017) Nanocellulose in thin films, coatings, and plies for packaging applications: a review. BioResources 12(1):2143–2233
3. Lavoine N, Desloges I, Dufresne A, Bras J (2012) Microfibrillated cellulose–Its barrier properties and applications in cellulosic materials: a review. Carbohyd Polym 90(2):735–764
4. Sharma M, Saini A, Nandanwar D (2020) Biodegradable coatings for sustainable packaging applications. Trends Food Sci Technol 100:140–152
5. Kargarzadeh H, Mariano M, Gopakumar D, Ahmad I, Thomas S, Dufresne A, Huang J (2018) Advances in cellulose nanomaterials. Cellulose 25(4):2151–2189
6. Kapoor RT, Salvadori MR, Rafatullah M, Siddiqui MR, Khan MA, Alshareef SA (2021) Exploration of microbial factories for synthesis of nanoparticles–a sustainable approach for bioremediation of environmental contaminants. Front Microbiol 12:658294
7. Rodrigues AS, Batista JG, Rodrigues MÁ, Thipe VC, Minarini LA, Lopes PS, Lugão AB (2024) Advances in silver nanoparticles: a comprehensive review on their potential as antimicrobial agents and their mechanisms of action elucidated by proteomics. Front Microbiol 15:1440065
8. Nguyen NP, Dang NT, Doan L, Nguyen TT (2023) Synthesis of silver nanoparticles: from conventional to 'modern'methods—a review. Processes 11:2617
9. Salama A, Abouzeid R, Leong WS, Jeevanandam J, Samyn P, Dufresne A, Bechelany M, Barhoum A (2021) Nanocellulose-based materials for water treatment: adsorption, photocatalytic degradation, disinfection, antifouling, and nanofiltration. Nanomaterials 11:3008
10. Tripathy J, Mishra A, Pandey M, Thakur RR, Chand S, Rout PR, Shahid MK (2024) Advances in nanoparticles and nanocomposites for water and wastewater treatment: a review. Water 16:1481
11. De Cicco D, Asaee Z, Taheri F (2017) Use of nanoparticles for enhancing the interlaminar properties of fiber-reinforced composites and adhesively bonded joints—a review. Nanomaterials 7:360
12. Rivero PJ, Urrutia A, Goicoechea J, Arregui FJ (2015) Nanomaterials for functional textiles and fibers. Nanoscale Res Lett 10:1–22
13. Mustafa F, Andreescu S (2020) Nanotechnology-based approaches for food sensing and packaging applications. RSC Adv 10:19309–19336
14. Garcia CV (2022) Nanomaterials in food packaging. Nanomater Food Packag 9:1083185. (Frontiers Media SA)
15. Huang Y, Mei L, Chen X, Wang Q (2018) Recent developments in food packaging based on nanomaterials. Nanomaterials 8:830
16. Isogai A, Saito T, Fukuzumi H (2011) TEMPO-oxidized cellulose nanofibers. Nanoscale 3(1):71–85. https://doi.org/10.1039/C0NR00583E
17. Dufresne A (2013) Nanocellulose: from nature to high performance tailored materials. De Gruyter
18. Klemm D, Kramer F, Moritz S, Lindström T, Ankerfors M, Gray D, Dorris A (2011) Nanocelluloses: a new family of nature-based materials. Angew Chem Int Ed 50(24):5438–5466. https://doi.org/10.1002/anie.201001273
19. Nogi M, Yano H (2008) Transparent nanocomposites based on cellulose produced by bacteria offer potential innovation in the electronics device industry. Adv Mater 20(10):1849–1852. https://doi.org/10.1002/adma.200800120
20. Guan Q, Yu Y, Lin X, Zhang Q, Liu J (2021) Advances in nanocellulose-based bioinks for 3D bioprinting. Biomater Sci 9(14):4806–4820. https://doi.org/10.1039/D1BM00440G
21. Anvar AA, Ahari H, Ataee M (2021) Antimicrobial properties of food nanopackaging: a new focus on foodborne pathogens. Front Microbiol 1–13
22. Abitbol T, Rivkin A, Cao Y, Nevo Y, Abraham E, Ben-Shalom T, Lapidot S, Shoseyov O, Dufresne A (2016) Nanocellulose, a tiny fiber with huge applications. Curr Opin Biotechnol 39:76–88. https://doi.org/10.1016/j.copbio.2016.01.002

23. Missoum K, Belgacem MN, Bras J (2013) Nanofibrillated cellulose surface modification: a review. Materials 6(5):1745–1766. https://doi.org/10.3390/ma6051745
24. Zhou C, Zhang L, Liu H (2021) Recent advances in flexible and wearable electronics based on nanocellulose. Carbohyd Polym 255:117377. https://doi.org/10.1016/j.carbpol.2020.117377
25. Nechita P, Roman M (2020) Review on food packaging polymeric materials and alternatives to improve food safety and eco-friendliness. Polym Bull 77:2229–2249. https://doi.org/10.1007/s00289-019-02830-7
26. Ma J, Wang H, He Y, Xu L (2016) Superhydrophobic and photocatalytic self-cleaning coating for wood protection based on TiO_2 and PDMS. Prog Org Coat 90:149–156. https://doi.org/10.1016/j.porgcoat.2015.09.025
27. Jiang Q, Huang Y, Wang D, Yang C (2020) Recent development of flexible and stretchable sensors for monitoring human health. Adv Mater Technol 5(3):1900692. https://doi.org/10.1002/admt.201900692
28. Wang Z, Yu H, Yang H, Wu H, Zhang Y (2021) Green and flexible paper-based electronics: from conductive inks to devices. ACS Sustain Chem & Eng 9(1):9–23. https://doi.org/10.1021/acssuschemeng.0c06713
29. El-Sayed SM, Youssef AM (2023) Eco-friendly biodegradable nanocomposite materials and their recent use in food packaging applications: a review. Sustain Food Technol 1:215–227
30. Gamage A, Thiviya P, Mani S, Ponnusamy PG, Manamperi A, Evon P, Merah O, Madhujith T (2022) Environmental properties and applications of biodegradable starch-based nanocomposites. Polymers 14:4578
31. Ebrahimi M, Busse N, Kerker S, Schmitz O, Hilpert M, Czermak P (2015) Treatment of the bleaching effluent from sulfite pulp production by ceramic membrane filtration. Membranes 6:7
32. Isogai A (2018) Development of completely dispersed cellulose nanofibers. Proc Jpn Acad Ser B 94:161–179
33. Ozer A, Uzuner U, Guler HI, Ay Sal F, Belduz AO, Deniz I, Canakci S (2018) Improved pulp bleaching potential of Bacillus subtilis WB800 through overexpression of three lignolytic enzymes from various bacteria. Biotechnol Appl Biochem 65:560–571
34. Grasso G, Zane D, Dragone R (2019) Microbial nanotechnology: challenges and prospects for green biocatalytic synthesis of nanoscale materials for sensoristic and biomedical applications. Nanomaterials 10:11
35. Sun O (2024) Nanocatalysts: driving chemical reactions toward sustainability. J Mater Sci Nanomater 8(4):144
36. Nasrollahzadeh M, Sajjadi M, Iravani S, Varma RS (2021) Green-synthesized nanocatalysts and nanomaterials for water treatment: current challenges and future perspectives. J Hazard Mater 401:123401
37. Koul B, Poonia AK, Yadav D, Jin J-O (2021) Microbe-mediated biosynthesis of nanoparticles: applications and future prospects. Biomolecules 11:886
38. Bhardwaj B, Singh P, Kumar A, Kumar S, Budhwar V (2020) Eco-friendly greener synthesis of nanoparticles. Adv Pharm Bull 10:566
39. Dhanker R, Hussain T, Tyagi P, Singh KJ, Kamble SS (2021) The emerging trend of bioengineering approaches for microbial nanomaterial synthesis and its applications. Front Microbiol 12:638003
40. Siddiqui J, Taheri M, Alam AU, Deen MJ (2022) Nanomaterials in smart packaging applications: a review. Small 18:2101171
41. Feature AE (2019) Food packaging using nanotechnology: 'Smart Packaging' and 'Active Packaging'. Technical report
42. Pandey D (2025) Color-changing smart packaging enhanced by AI. Technical report
43. Nakra A, Wu M, Wong C-W (2024) SoK: fighting counterfeits with cyber-physical synergy based on physically-unclonable identifiers of paper surface. arXiv:2408.02221
44. Creative A (2024) How connected packaging fights counterfeiting. Technical report
45. Mpongwana N, Rathilal S (2022) A review of the techno-economic feasibility of nanoparticle application for wastewater treatment. Water 14:1550

46. Chen X, Cao H, He Y, Zhou Q, Li Z, Wang W, He Y, Tao G, Hou C (2022) Advanced functional nanofibers: strategies to improve performance and expand functions. Front Optoelectron 15:50
47. Montes-Hernandez G, Di Girolamo M, Sarret G, Bureau S, Fernandez-Martinez A, Lelong C, Eymard Vernain E (2021) In situ formation of silver nanoparticles (Ag-NPs) onto textile fibers. ACS omega 6:1316–1327
48. Elhenawy S, Khraisheh M, AlMomani F, Al-Ghouti M, Selvaraj R, Al-Muhtaseb A (2024) Emerging nanomaterials for drinking water purification: A new era of water treatment technology. Nanomaterials 14:1707
49. Gehrke I, Geiser A, Somborn-Schulz A (2015) Innovations in nanotechnology for water treatment. Nanotechnol Sci Appl 1–17
50. Chalew TE, Ajmani GS, Huang H, Schwab KJ (2013) Evaluating nanoparticle breakthrough during drinking water treatment. Environ Health Perspect 121:1161–1166
51. Andeme Ela RC, Heiden PA (2024) Lignin smart nanoparticles for effective and eco-friendly resource recovery from wastewater: status, challenges, and opportunities. ACS Sustain Resour Manag 1:1302–1325
52. Almatroudi A (2020) Silver nanoparticles: synthesis, characterisation and biomedical applications. Open Life Sci 15:819–839
53. Marassi V, Di Cristo L, Smith SG, Ortelli S, Blosi M, Costa AL, Reschiglian P, Volkov Y, Prina-Mello A (2018) Silver nanoparticles as a medical device in healthcare settings: a five-step approach for candidate screening of coating agents. R Soc Open Sci 5:171113
54. Niemark LD, Tang J, Chen C, Wang Z (2021) Nanocellulose-based materials for 3D printing: current progress and future prospects. Carbohyd Polym 252:117148. https://doi.org/10.1016/j.carbpol.2020.117148
55. Moustafa H, El-Hakim AE, Abou-Kandil AI (2021) Eco-friendly polymer composites for 3D printing: a review. Environ Chem Lett 19:2135–2160. https://doi.org/10.1007/s10311-020-01145
56. Honarvar Z, Hadian Z, Mashayekh M (2016) Nanocomposites in food packaging applications and their risk assessment for health. Electron Physician 8:2531
57. Shanmugam K, Sathishkumar E, Venkatesh R (2021) Surface modifications of nanocellulose for diverse applications. ACS Omega 6(7):4920–4936. https://doi.org/10.1021/acsomega.0c05047
58. Saito T, Isogai A (2006) Introduction of aldehyde groups on surfaces of native cellulose fibers by TEMPO-mediated oxidation. Colloids Surf A 289(1–3):219–225. https://doi.org/10.1016/j.colsurfa.2006.04.027
59. Hubbe MA, Ferrer A, Tyagi P, Yin Y, Salas C, Pal L, Rojas OJ (2017) Nanocellulose in thin films, coatings, and plies for packaging applications: a review. BioResources 12(1):2143–2233. https://doi.org/10.15376/biores.12.1.2143-2233
60. Rosen D (2013) Woodrow Wilson International Center for Scholars, & Project on Emerging Nanotechnologies. Thinking big about things small: a policy agenda for the new federal nanotechnology initiative. Woodrow Wilson Center, Washington, D.C.
61. Singh N, Chand N, Yadav A (2021) Role of artificial intelligence and machine learning in pulp and paper industry. J Clean Prod 305:127117. https://doi.org/10.1016/j.jclepro.2021.127117
62. Mariano M, El Kissi N, Dufresne A (2016) Cellulose nanocrystals and related nanocomposites: review of some properties and challenges. J Polym Sci Part B Polym Phys 54(7):830–866. https://doi.org/10.1002/polb.23935
63. Sharma S, Barkauskas J, Qian X, Hubbe MA (2020) Recent advances in barrier coatings for paper-based packaging: a review. BioResources 15(4):8575–8625. https://doi.org/10.15376/biores.15.4.8575-8625
64. Deshmukh SP, Patil SM, Mullani SB, Delekar SD (2019) Silver nanoparticles as an effective disinfectant: a review. Mater Sci Eng C 97:954–965
65. Jonoobi M, Oladi R, Davoudpour Y, Oksman K, Dufresne A, Hamzeh Y, Davoodi R (2015) Different preparation methods and properties of nanostructured cellulose from various natural resources and residues: a review. Cellulose 22:935–969